Acta Numerica 2004

Managing editor

A. Iserles
DAMTP, University of Cambridge,
Centre for Mathematical Sciences, Wilberforce Road,
Cambridge CB3 0WA, England

Editorial Board

C. de Boor, *University of Wisconsin, Madison, USA*
F. Brezzi, *Instituto di Analisi Numerica del CNR, Italy*
J. C. Butcher, *University of Auckland, New Zealand*
P. G. Ciarlet, *City University of Hong Kong, China*
G. H. Golub, *Stanford University, USA*
H. B. Keller, *California Institute of Technology, USA*
H.-O. Kreiss, *University of California, Los Angeles, USA*
K. W. Morton, *University of Oxford, England*
M. J. D. Powell, *University of Cambridge, England*
R. Temam, *Université Paris-Sud, France*

Acta Numerica

Volume 13   2004

CAMBRIDGE
UNIVERSITY PRESS

CAMBRIDGE
UNIVERSITY PRESS

University Printing House, Cambridge CB2 8BS, United Kingdom

Cambridge University Press is part of the University of Cambridge.

It furthers the University's mission by disseminating knowledge in the pursuit of education, learning and research at the highest international levels of excellence.

www.cambridge.org
Information on this title: www.cambridge.org/9780521838115

© Cambridge University Press 2004

This publication is in copyright. Subject to statutory exception and to the provisions of relevant collective licensing agreements, no reproduction of any part may take place without the written permission of Cambridge University Press.

First published 2004
First paperback edition 2011

A catalogue record for this publication is available from the British Library

ISBN 978-0-521-83811-5 Hardback

Cambridge University Press has no responsibility for the persistence or accuracy of URLs for external or third-party internet websites referred to in this publication, and does not guarantee that any content on such websites is, or will remain, accurate or appropriate.

Contents

The calculation of linear least squares problems 1
Åke Björck

**The numerical analysis of functional integral and
integro-differential equations of Volterra type** 55
Hermann Brunner

Sparse grids 147
Hans-Joachim Bungartz and Michael Griebel

**Complete search in continuous global optimization
and constraint satisfaction** 271
Arnold Neumaier

Multiscale computational modelling of the heart 371
N. P. Smith, D. P. Nickerson, E. J. Crampin and P. J. Hunter

The calculation of linear least squares problems

Åke Björck
Department of Mathematics,
Linköping University,
SE-581 83 Linköping, Sweden
E-mail: akbjo@math.liu.se

Dedicated to Michael A. Saunders on the occasion of his sixtieth birthday.

We first survey componentwise and normwise perturbation bounds for the standard least squares (LS) and minimum norm problems. Then some recent estimates of the optimal backward error for an alleged solution to an LS problem are presented. These results are particularly interesting when the algorithm used is not backward stable.

The QR factorization and the singular value decomposition (SVD), developed in the 1960s and early 1970s, remain the basic tools for solving both the LS and the total least squares (TLS) problems. Current algorithms based on Householder or Gram–Schmidt QR factorizations are reviewed. The use of the SVD to determine the numerical rank of a matrix, as well as for computing a sequence of regularized solutions, is then discussed. The solution of the TLS problem in terms of the SVD of the compound matrix $(b \ A)$ is described.

Some recent algorithmic developments are motivated by the need for the efficient implementation of the QR factorization on modern computer architectures. This includes blocked algorithms as well as newer recursive implementations. Other developments come from needs in different application areas. For example, in signal processing rank-revealing orthogonal decompositions need to be frequently updated. We review several classes of such decompositions, which can be more efficiently updated than the SVD.

Two algorithms for the orthogonal bidiagonalization of an arbitrary matrix were given by Golub and Kahan in 1965, one using Householder transformations and the other a Lanczos process. If used to transform the matrix $(b \ A)$ to upper bidiagonal form, this becomes a powerful tool for solving various LS and TLS problems. This bidiagonal decomposition gives a core regular subproblem for the TLS problem. When implemented by the Lanczos process it forms the kernel in the iterative method LSQR. It is also the basis of the partial least squares (PLS) method, which has become a standard tool in statistics.

We present some generalized QR factorizations which can be used to solve different generalized least squares problems. Many applications lead to LS problems where the solution is subject to constraints. This includes linear equality and inequality constraints. Quadratic constraints are used to regularize solutions to discrete ill-posed LS problems. We survey these classes of problems and discuss their solution.

As in all scientific computing, there is a trend that the size and complexity of the problems being solved is steadily growing. Large problems are often sparse or structured. Algorithms for the efficient solution of banded and block-angular LS problems are given, followed by a brief discussion of the general sparse case. Iterative methods are attractive, in particular when matrix-vector multiplication is cheap.

CONTENTS

1	Introduction	2
2	Perturbation analysis and stability	4
3	Orthogonal decompositions	8
4	Generalized least squares problems	13
5	Blocked algorithms	16
6	Rank-revealing decompositions	20
7	Bidiagonal reduction	26
8	Constrained and regularized problems	32
9	Direct methods for sparse problems	38
10	Iterative methods	43
	References	47

1. Introduction

The method of least squares has been the standard procedure for the analysis of data from the beginning of 1800s. A famous example of its use is when Gauss successfully predicted the orbit of the asteroid Ceres in 1801. Two hundred years later, least squares remains a widely used computational principle in science and engineering.

In the simplest case the problem is, given $A \in \mathbb{R}^{m \times n}$ and $b \in \mathbb{R}^m$, to find a vector $x \in \mathbb{R}^n$ such that

$$\min_x \|b - Ax\|_2, \qquad (1.1)$$

where $\| \cdot \|_2$ denotes the Euclidean norm. A least squares solution x is characterized by $r \perp \mathcal{R}(A)$, where $r = b - Ax$ is the residual and $\mathcal{R}(A)$ the range space of A. The residual r is uniquely determined and the solution x is unique if and only if $\text{rank}(A) = n$. If $\text{rank}(A) < n$, we seek the unique least squares solution $x \perp \mathcal{N}(A)$, which is called the pseudo-inverse solution.

Under-determined systems arise from problems where there are more variables than needed to match the observed data. The model problem for this case is to find $y \in \mathbb{R}^m$ such that

$$\min \|y\|_2, \qquad A^T y = c, \tag{1.2}$$

where $c \in \mathbb{R}^n$. Here $y \in \mathbb{R}^m$, the minimum norm solution of the consistent under-determined system $A^T y = c$, is characterized by $y \perp \mathcal{N}(A^T)$. If the system $A^T y = c$ is not consistent we compute the pseudo-inverse solution.

When uncertainties are present also in the matrix A, the total least squares (TLS) model is more appropriate. The TLS problem is

$$\min \| (E \ r) \|_F, \qquad (A + E)x = b + r, \tag{1.3}$$

where $\| \cdot \|_F$ denotes the Frobenius matrix norm.

Models where the parameters x occur nonlinearly are common, but in this survey we will take the simplistic view that nonlinear problems can be solved by linearization.

From the time of Gauss until the computer age the basic computational tool for solving (1.1) was to form the normal equations $A^T A x = A^T b$ and solve these by symmetric Gaussian elimination (which Gauss did), or later by the Cholesky factorization (Benoit 1924). This approach has the drawback that forming the matrix $A^T A$ will square the condition number of the original problem. This can lead to difficulties since least squares problems are frequently ill-conditioned.

In the 1950s algorithms based on Gram–Schmidt orthogonalization were widely used, although their numerical properties were not well understood at the time. Björck (1967b) analysed the modified Gram–Schmidt algorithm and showed its stability for solving linear least squares problems.

A breakthrough came with the seminal paper by Golub (1965), where it was shown how to compute a QR factorization of A using Householder transformations. A backward stable algorithm for the linear least squares problems was given. Another important development, which took place around the same time, was that of a stable algorithm for computing the singular value decomposition (SVD); see Golub and Kahan (1965) and Golub (1968), and the Algol program for computing the SVD in Golub and Reinsch (1970).

Modern numerical methods for solving least squares problems are surveyed in the two comprehensive monographs by Lawson and Hanson (1995) and Björck (1996). The latter contains a bibliography of 860 references, indicating the considerable research interest in these problems. Hansen (1998) gives an excellent survey of numerical methods for the treatment of numerically rank-deficient linear systems arising, for example, from discrete ill-posed problems. A comprehensive discussion of theory and methods for solving TLS problems is found in Van Huffel and Vandewalle (1991).

Although methods continue to evolve, variations of the QR factorization and SVD remain the basic tools for solving least squares problems. Much of the algorithmic development taking place has been motivated by needs in different application areas, e.g., statistics, signal processing and control theory. For example, in signal processing data is often analysed in real time and estimates need to be updated at each time step. Other applications lead to generalized least squares problems, where the solution is subject to linear or quadratic constraints. A common trend, as in all scientific computing, is that the size and complexity of the problems being solved are steadily growing. There is also an increased need to take advantage of any structure that may exist in the model. Geodetic networks lead to huge sparse structured least squares problems, that have to be treated by sparse factorization methods. Other large-scale problems are better handled by a combination of direct and iterative methods.

The following survey of some areas of recent progress represents a highly subjective selection. Hopefully it will show that many interesting developments still take place in this field.

2. Perturbation analysis and stability

2.1. Perturbation analysis

Consider the least squares problem (1.1) with $\text{rank}(A) = n$ and solution x and residual vector $r = b - Ax$. Let the data A, b be perturbed to $A + \delta A$, $b + \delta b$ where $\text{rank}(A + \delta A) = n$. The perturbed solution by $x + \delta x$ and $r + \delta r$ satisfies the normal equations

$$(A + \delta A)^T (A + \delta A)(x + \delta x) = (A + \delta A)^T (b + \delta b).$$

Subtracting $A^T A x = A^T b$ and solving for δx gives

$$\delta x \approx A^\dagger (\delta b - \delta A \, x) + (A^T A)^{-1} \delta A^T r, \quad A^\dagger = (A^T A)^{-1} A^T,$$

where $r = b - Ax$ is the residual and second-order terms have been neglected. For $r = 0$ this reduces to the well-known first-order perturbation bound for a square nonsingular linear system. For the residual we have $\delta r \approx (\delta b - \delta A x) - A \delta x$ and hence

$$\delta r \approx P_{\mathcal{N}(A^T)}(\delta b - \delta A \, x) + (A^\dagger)^T \delta A^T \, r, \quad P_{\mathcal{N}(A^T)} = I - A A^\dagger.$$

Here $P_{\mathcal{N}(A^T)}$ is the orthogonal projection onto $\mathcal{N}(A^T)$. These equations yield the componentwise estimates (see Björck (1991))

$$|\delta x| \lesssim |A^\dagger|(|\delta b| + |\delta A| \, |x|) + |(A^T A)^{-1}| \, |\delta A|^T \, |r|, \qquad (2.1)$$

$$|\delta r| \lesssim |P_{\mathcal{N}(A^T)}|(|\delta b| + |\delta A| \, |x|) + |(A^\dagger)^T| \, |\delta A|^T \, |r|, \qquad (2.2)$$

where the inequalities are to be interpreted componentwise. Taking norms in (2.1) and using

$$\|A^\dagger\|_2 = 1/\sigma_n, \quad \|(A^TA)^{-1}\|_2 = 1/\sigma_n^2,$$

where σ_n is the smallest singular value of A, we obtain the approximate upper bound

$$\|\delta x\|_2 \lesssim \frac{1}{\sigma_n}(\|\delta b\|_2 + \|\delta A\|_2 \|x\|_2) + \frac{1}{\sigma_n^2}\|\delta A\|_2 \|r\|_2. \quad (2.3)$$

It can be shown that for an arbitrary matrix A and vector b there are perturbations δA and δb such that this upper bound is almost attained. Note that when the residual $r \neq 0$ there is an additional term not present for consistent linear systems. The presence of this term, which will dominate if $\|r\|_2 > \sigma_n \|x\|_2$, was first pointed out by Golub and Wilkinson (1966).

Setting $\delta b = 0$ and assuming $x \neq 0$, we get for the normwise relative perturbation in x

$$\frac{\|\delta x\|_2}{\|x\|_2} \lesssim \kappa(A) \frac{\|\delta A\|_2}{\|A\|_2}\left(1 + \frac{\|r\|_2}{\sigma_n \|x\|_2}\right), \quad (2.4)$$

where $\kappa(A) = \sigma_1/\sigma_n$ is the condition number of A.

For the minimum norm problem (1.2) with A^T of full row rank, the solution can be expressed in terms of the normal equation as $y = Az$, where $A^TAz = c$. Proceeding as before and neglecting second-order terms in the perturbation we obtain

$$\delta y \approx P_{\mathcal{N}(A^T)} \delta A\, A^\dagger y + (A^\dagger)^T(\delta c - \delta A^T y),$$

giving the componentwise approximate bound

$$|\delta y| \lesssim |P_{\mathcal{N}(A^T)}| |\delta A|\, |A^\dagger|\, |y| + |(A^\dagger)^T|(|\delta c| + |\delta A|^T |y|). \quad (2.5)$$

Taking norms we get

$$\|\delta y\|_2 \lesssim \frac{1}{\sigma_n}(\|\delta c\|_2 + 2\|\delta A\|_2 \|y\|_2). \quad (2.6)$$

The statistical model leading to the least squares problem (1.1) is that the vector b of observations is related to the solution x by a linear relation $Ax = b + \epsilon$, where ϵ is a random error vector with zero mean and whose components are uncorrelated and have equal variance. More generally, if the covariance matrix of ϵ equals a symmetric positive definite matrix W, then the best linear unbiased estimate of x is the solution to the least squares problem $\min_x (Ax - b)^T W^{-1}(Ax - b)$, or equivalently

$$\min_x \|W^{-1/2}(b - Ax)\|_2. \quad (2.7)$$

If the errors are uncorrelated then W is a diagonal matrix and we set $D = \mathrm{diag}(d_1,\ldots,d_m) = W^{-1/2}$. Then (2.7) is a weighted least squares problem. If some components of the error vector have much smaller variance than the rest, $\kappa(DA) \gg \kappa(A) \geq 1$. The perturbation bound (2.4) then seems to indicate that the problem is ill-conditioned. This is not necessarily so and for such problems it is preferable to use the componentwise bounds (2.1)–(2.2). Special methods for weighted problems are discussed in Björck (1996, Section 4.4).

2.2. Backward error and stability

Consider an algorithm for solving the linear least squares problem (1.1). The algorithm is said to be numerically stable if, for any data A and b, there exist small perturbation matrices and vectors δA and δb, such that the computed solution $\bar{x}$ is the *exact* solution to

$$\min_x \|(A + \delta A)x - (b + \delta b)\|_2, \tag{2.8}$$

where $\|\delta A\| \leq \tau$, $\|\delta b\| \leq \tau$, with τ being a small multiple of the unit round-off u. Any computed solution $\bar{x}$ is called a stable solution if it satisfies (2.8). This does not mean that $\bar{x}$ is close to the exact solution x. If the least squares problem is ill-conditioned then a stable solution can be very different from x. For a stable solution the error $\|x - \bar{x}\|$ can be estimated using the perturbation results given in Section 2.1.

The method by Golub (1965) based on Householder QR factorization is known to be numerically stable with $\delta b = 0$ (Higham 2002, Theorem 20.3). Methods which explicitly form the normal equations are not backward stable. This is because *round-off errors that occur in forming $A^T A$ and $A^T b$ are not in general equivalent to small perturbations in A and b*. Although the method of normal equations gives results of sufficient accuracy for many applications, its use can result in errors in the computed solution, which are of much larger size than for a stable method.

Many fast methods exist for solving structured least squares problems, e.g., when A is a Toeplitz or Cauchy matrix. These are not in general backward stable (see Gu (1998b)), which is one reason why the following results are of interest.

Given an alleged solution $\tilde{x}$, a backward error is a perturbation δA, such that $\tilde{x}$ is the exact solution to the perturbed problem

$$\min_x \|(b + \delta b) - (A + \delta A)x\|_2. \tag{2.9}$$

If we could find the backward error of smallest norm, this could be used to verify numerically the stability properties of an algorithm. There is not much loss in assuming that $\delta b = 0$ in (2.10). Then the optimal backward error in the Frobenius norm is

$$\eta_F(\tilde{x}) = \min\{\|\delta A\|_F \mid \tilde{x} \text{ solves } \min_x \|b - (A + \delta A)x\|_2\}. \tag{2.10}$$

How to find the optimal backward error for the linear least squares problem was an open problem for many years, until it was elegantly answered by Waldén, Karlsson and Sun (1995). They solved the problem by characterizing the set of all backward perturbations and by giving an optimal bound, which minimizes the Frobenius norm $\|\delta A\|_F$; see also Higham (2002, pp. 392–393). Their main result can be stated as follows.

Theorem 1. Let $\tilde{x}$ be an alleged solution and $\tilde{r} = b - A\tilde{x} \neq 0$. The optimal backward error in the Frobenius norm is

$$\eta_F(\tilde{x}) = \begin{cases} \|A^T \tilde{r}\|_2 / \|\tilde{r}\|_2, & \text{if } \tilde{x} = 0, \\ \min\{\eta, \sigma_{\min}([A \ \ C])\}, & \text{otherwise,} \end{cases} \tag{2.11}$$

where

$$\eta = \|\tilde{r}\|_2 / \|\tilde{x}\|_2, \qquad C = I - (\tilde{r}\tilde{r}^T)/\|\tilde{r}\|_2^2,$$

and $\sigma_{\min}([A \ \ C])$ denotes the smallest (nonzero) singular value of the matrix $[A \ \ C] \in \mathbb{R}^{m \times (n+m)}$.

The task of computing $\eta_F(\tilde{x})$ is thus reduced to that of computing $\sigma_{\min}(\mathcal{A})$. Since this is expensive, approximations that are accurate and less costly have been derived. Karlsson and Waldén (1997) assume that a QR factorization of A is available and give lower and upper bounds for $\eta_F(\tilde{x})$ that only require $O(mn)$ operations. Gu (1998a) gives several approximations to $\eta_F(\tilde{x})$ that are optimal up to a factor less than 2. His bounds are formulated in terms of the singular value decomposition of A but his Corollary 2.2 can also be stated as follows.

Let $r_1 = P_{\mathcal{R}(A)} \tilde{r}$ be the orthogonal projection of $\tilde{r}$ onto the range of A. If $\|r_1\|_2 \leq \alpha \|r\|_2$ it holds that

$$\frac{\sqrt{5} - 1}{2} \tilde{\sigma}_1 \leq \eta_F(\tilde{x}) \leq \sqrt{1 + \alpha^2} \, \tilde{\sigma}_1, \tag{2.12}$$

where

$$\tilde{\sigma}_1 = \left\| (A^T A + \eta I)^{-1/2} A^T \tilde{r} \right\|_2 / \|\tilde{x}\|_2. \tag{2.13}$$

Since $\alpha \to 0$ for small perturbations $\tilde{\sigma}_1$ is an asymptotic upper bound.

Optimal backward perturbation bounds for under-determined systems are derived in Sun and Sun (1997). The extension of backward error bounds to the case of constrained least squares problems is discussed by Cox and Higham (1999b).

3. Orthogonal decompositions

3.1. Algorithms using Householder reflections

The QR factorization of a matrix $A \in \mathbb{R}^{m \times n}$ is

$$A = Q \begin{pmatrix} R \\ 0 \end{pmatrix}, \qquad (3.1)$$

where $R \in \mathbb{R}^{n \times n}$ is upper triangular and $Q \in \mathbb{R}^{m \times m}$ is orthogonal. If A has linearly independent columns, i.e., rank$(A) = n$, then R is nonsingular. If we partition

$$Q = (Q_1 \ Q_2), \quad Q_1 \in \mathbb{R}^{m \times n}, \quad Q_2 \in \mathbb{R}^{m \times (m-n)},$$

we obtain the compact form $A = Q_1 R$ of the QR factorization. In the full rank case Q_1 and R are uniquely determined, provided R is normalized to have positive diagonal elements. Q_1 gives an orthogonal basis for $\mathcal{R}(A)$. Q_2, which is not uniquely determined, gives an orthogonal basis for $\mathcal{N}(A^T)$.

The standard method to compute the QR factorization (3.1) is to premultiply A with a product of Householder reflections $Q^T = P_n \cdots P_2 P_1$, where

$$P_j = I - 2 v_j v_j^T / \|v_j\|_2^2, \quad j = 1 : n,$$

is constructed to zero out the elements below the main diagonal in the jth column of A. Since a Householder reflection is symmetric and orthogonal,

$$Q = P_1 P_2 \cdots P_n. \qquad (3.2)$$

There is usually no need to form Q explicitly, since the matrix–vector products Qy and $Q^T z$ can be efficiently formed using only the Householder vectors $v_1, v_2, \ldots, v_n$. Since v_j only has nonzero elements in positions $j : m$, these can be stored in an $m \times n$ lower trapezoidal matrix. In the dense case this is the most compact representation possible of Q and Q^T.

Given the QR factorization (3.1), the solution x to the linear least squares problem (1.1) and the corresponding residual $r = b - Ax$ is computed:

$$\begin{pmatrix} d_1 \\ d_2 \end{pmatrix} = Q^T b, \quad x = R^{-1} d_1, \quad r = Q \begin{pmatrix} 0 \\ d_2 \end{pmatrix} = Q_2 d_2. \qquad (3.3)$$

This algorithm is backward stable (with $\delta b = 0$) both for computing the solution x and the residual $r = b - Ax$; see Higham (2002, Theorem 20.3).

Note that the residual r solves the problem of computing the orthogonal projection of b onto $\mathcal{N}(A^T)$:

$$\min_r \|b - r\|_2 \quad \text{subject to} \quad A^T r = 0.$$

In some applications we are more interested in the residual r than in the solution x. From the stability (see also the error analysis in Björck (1967a))

it follows that the computed residual $\bar{r}$ using (3.3) satisfies a relation
$$(A+E)^T \bar{r} = 0, \quad \|E\|_2 \leq cu\|A\|_2. \tag{3.4}$$
Here and in the following c is a generic constant that grows slowly with n. This implies
$$\|A^T \bar{r}\|_2 \leq cu\|\bar{r}\|_2\|A\|_2, \tag{3.5}$$
that is, the computed residual is accurately orthogonal to $\mathcal{R}(A)$. On the other hand, if $\bar{r} = \text{fl}(b - Ax)$, then the best bound we can guarantee is of the form $\|A^T \bar{r}\|_2 \leq cu\|b\|_2\|A\|_2$, even if x is the *exact* least squares solution, When $\|\bar{r}\|_2 \ll \|b\|_2$ this is a much weaker bound than (3.5).

The solution to the minimum norm problem (1.2) can be computed from the QR factorization (3.1) using
$$z = R^{-T}c, \quad y = Q\begin{pmatrix} z \\ 0 \end{pmatrix} = Q_1 z. \tag{3.6}$$

The fact that this algorithm is backward stable is a relatively new result and the first proof was published in Higham (1995, Theorem 20.3).

An implementation of Householder QR factorization is given in Businger and Golub (1965) (see Wilkinson and Reinsch (1971, Contribution I/8)). A more general implementation, that also solves least squares problems with linear constraints and performs a stable form of iterative refinement of the solution, is given in Björck and Golub (1967).

3.2. Algorithms using modified Gram–Schmidt

In Gram–Schmidt orthogonalization the kth column of Q in the QR factorization is computed as a linear combination of the first k columns in A. This is equivalent to computing the compact QR factorization[1]

$$A = (a_1, a_2, \ldots, a_n) = (q_1, q_2, \ldots, q_n) \begin{pmatrix} r_{11} & r_{12} & \cdots & r_{1n} \\ & r_{22} & \cdots & r_{2n} \\ & & \ddots & \vdots \\ & & & r_{2n} \end{pmatrix}.$$

Gram–Schmidt QR factorization can also be described as employing a sequence of elementary orthogonal projections to orthogonalize a given sequence of vectors For any nonzero vector $a \in \mathbb{R}^m$ the orthogonal projector P onto the orthogonal complement of a is given by
$$P = I_m - qq^T, \quad q = a/\|a\|_2. \tag{3.7}$$

---

[1] Trefethen and Bau, III (1997) aptly calls Householder QR orthogonal triangularization and Gram–Schmidt QR triangular orthogonalization.

Two versions of the Gram–Schmidt algorithm exist, usually called the Classical Gram–Schmidt (CGS) and the Modified Gram–Schmidt (MGS) algorithms. Although these only differ in the order in which the operations are performed, MGS has much better numerical stability properties.

Setting $a_j = a_j^{(1)}$, $j = 1:n$, in MGS at the beginning of step k, $k = 1:n$, we have computed

$$(q_1, \ldots, q_{k-1}, a_k^{(k)}, \ldots, a_n^{(k)}), \tag{3.8}$$

where $a_k^{(k)}, \ldots, a_n^{(k)}$ are orthogonal to $q_1, \ldots, q_{k-1}$. First the vector q_k is obtained by normalizing $a_k^{(k)}$. The remaining columns are then made orthogonal[2] to q_k, using orthogonal projections

$$a_j^{(k+1)} = (I - q_k q_k^T) a_j^{(k)} = a_j^{(k)} - q_k (q_k^T a_j^{(k)}), \quad j = k+1:n.$$

Owing to rounding errors the computed $Q_1 = (q_1, q_2, \ldots, q_n)$ will not be orthogonal to working accuracy. For MGS the loss of orthogonality can be bounded in terms of the condition number of A, namely,

$$\|I - Q_1^T Q_1\|_2 \leq c_1 u \kappa(A),$$

where u is the unit round-off; see Björck (1967b), Björck and Paige (1992).

Because of the loss of orthogonality care is needed in using the MGS factorization. Using a remarkable connection between MGS and Householder QR factorization, Björck and Paige (1992) were able to analyse MGS and rigorously prove the stability of several algorithm based on the MGS factorization. If these algorithms are used with MGS there is *no need for reorthogonalization of the q vectors* for computing least squares solutions, orthogonal projections or solving minimum norm problems. Since few textbooks describe these stable algorithms we present them again here.

Linear least squares solution by MGS
Carry out MGS on $A \in R^{m \times n}$, rank$(A) = n$, to give $Q_1 = (q_1, \ldots, q_n)$ and R, and put $b^{(1)} = b$. Compute the vector $z = (z_1, \ldots, z_n)^T$ by

$$\text{for } k = 1:n$$
$$z_k = q_k^T b^{(k)}; \quad b^{(k+1)} = b^{(k)} - z_k q_k;$$
$$\text{end}$$
$$r = b^{(n+1)};$$
$$\text{solve } Rx = z;$$

---

[2] MGS can also be organized so that all previous projections to a_k are applied in the kth step, but this version is not suitable for column pivoting.

This algorithm for solving linear least squares problems by MGS was quite widely used even in the 1960s. A common mistake, still to be found in some textbooks, is to compute x from $Rx = d$, where $d = Q_1^T b$, which may ruin the accuracy in the solution.

Orthogonal projection by MGS
To make MGS backward stable for r it suffices to add a loop where the vector $b^{(n+1)}$ is orthogonalized against $q_n, q_{n-1}, \ldots, q_1$:

$$\text{for } k = n, n-1, \ldots, 1$$
$$z_k = q_k^T b^{(k+1)}; \quad b^{(k)} = b^{(k+1)} - z_k q_k;$$
$$\text{end}$$
$$r = b^{(1)};$$

Note that the reorthogonalization has to be done *in reverse order* to prove that $\bar{r}$ is stable; see Björck and Paige (1992).

Minimum norm solution by MGS
Carry out MGS on $A^T \in R^{m \times n}$, with $\text{rank}(A) = n$ to give $Q_1 = (q_1, \ldots, q_n)$ and R. Then the minimum norm solution $y = y^{(0)}$ is obtained from

$$R^T (\zeta_1, \ldots, \zeta_n)^T = c;$$
$$y^{(n)} = 0;$$
$$\text{for } k = n, \ldots, 2, 1$$
$$\omega_k = q_k^T y^{(k)}; \quad y^{(k-1)} = y^{(k)} - (\omega_k - \zeta_k) q_k;$$
$$\text{end}$$

If the columns of Q_1 were orthogonal to working accuracy, then $\omega_k = 0$, $k = m, \ldots, 1$. Here ω compensates for the lack of orthogonality to make this algorithm backward stable!

There are a few applications where it is advantageous to compute a matrix Q_1 that is orthogonal to working precision; see Giraud, Langou, and Rozložnık (2002). The classical schemes for reorthogonalization described in Hoffman (1989) will approximately double the cost of the factorization. A new, more efficient reorthogonalization scheme, based on a low-rank update of Q_1, is described in Giraud, Gratton and Langou (2003).

3.3. Algorithms using SVD

The singular value decomposition is in general the most versatile decomposition for treating rank-deficient and severely ill-conditioned least squares problem. We write the SVD in the partitioned form

$$A = (U_1 \ U_2) \begin{pmatrix} \Sigma_1 & 0 \\ 0 & \Sigma_2 \end{pmatrix} \begin{pmatrix} V_1^T \\ V_2^T \end{pmatrix}, \tag{3.9}$$

$$\Sigma_1 = \text{diag}(\sigma_1, \ldots, \sigma_r), \quad \Sigma_2 = \text{diag}(\sigma_{r+1}, \ldots, \sigma_n), \tag{3.10}$$

where $\sigma_1 \geq \sigma_2 \geq \cdots \geq \sigma_n$.

If $\sigma_{r+1} = 0$, then A has rank r and

$$x = A^\dagger b = V_1 \Sigma_1^{-1} U_1^T b = \sum_{i=1}^{r} \frac{u_i^T b}{\sigma_i} v_i \tag{3.11}$$

is the pseudo-inverse solution. This is also the unique solution to the least squares problem

$$\min_{x \in S} \|x\|_2, \quad S = \{x \in \mathbb{R}^n \mid \|b - Ax\|_2 = \min\}. \tag{3.12}$$

The mathematical concept of rank is not, in general, computationally useful. The *numerical rank* of a matrix $A \in \mathbb{R}^{n \times m}$ is defined in terms of its singular values. A is said to have the numerical ϵ-rank equal to $r < n$ if its singular values satisfy

$$\sigma_r > \epsilon \geq \sigma_{r+1},$$

where ϵ is a problem-dependent parameter.

If A is ill-conditioned, but there is a gap between σ_r and σ_{r+1}, then the numerical rank r is well defined. The SVD can be used to extract the linearly independent information in A, to arrive at a more well-conditioned problem. We have

$$A = U_1 \Sigma_1 V_1^T + E, \quad E = U_2 \Sigma_2 V_2^T. \tag{3.13}$$

Among the perturbation that makes $A + E$ have exact rank r, (3.13) minimizes both $\|E\|_2 = \sigma_{r+1}$, and $\|E\|_F$. The approximate least squares solution (3.11) is called the truncated SVD (TSVD) solution. It is the least squares solution restricted to the subspace $\mathcal{R}(V_1)$. The matrices U_1 and V_2 give orthogonal bases for the numerical range space and null space, respectively, of A.

The total least squares problem is best analysed in terms of the SVD

$$(b \ A) = \hat{U} \hat{\Sigma} \hat{V}^T, \quad \hat{\Sigma} = \text{diag}(\hat{\sigma}_1, \ldots, \hat{\sigma}_{n+1}).$$

Assume for simplicity that $\text{rank}(A) = n$ and $\hat{\sigma}_{n+1} < \hat{\sigma}_n$. Then the unique perturbation of minimum norm $\|(r \ E)\|_F$ that makes $(A + E)x = b + r$ consistent is the rank one perturbation

$$(r \ E) = -\hat{\sigma}_{n+1} \hat{u}_{n+1} \hat{v}_{n+1}^T = -(b \ A) \hat{v}_{n+1} \hat{v}_{n+1}^T. \tag{3.14}$$

Multiplying (3.14) from the right with $\hat{v}_{n+1}$ gives
$$\begin{pmatrix} b & A \end{pmatrix} \hat{v}_{n+1} = - \begin{pmatrix} r & E \end{pmatrix} \hat{v}_{n+1}. \qquad (3.15)$$

Writing the relation $(A + E)x = b + r$ in the form
$$\begin{pmatrix} b & A \end{pmatrix} \begin{pmatrix} 1 \\ -x \end{pmatrix} = - \begin{pmatrix} r & E \end{pmatrix} \begin{pmatrix} 1 \\ -x \end{pmatrix}$$

and comparing with (3.15), it is easily seen that the TLS solution
$$x = \gamma(\hat{v}_{2,n+1}, \ldots, \hat{v}_{n+1,n+1})^T, \quad \gamma = -1/\hat{v}_{1,n+1}$$

is obtained from the right singular vector $\hat{v}_{n+1}$.

In the classical algorithm (Golub and Reinsch 1970) the SVD is computed in three steps. In the first step A is transformed into upper bidiagonal form $U_1^T A V_1 = B$ using orthogonal transformations (see Section 7). In the second step the shifted QR algorithm is applied *implicitly* to the matrix $B^T B$ giving the SVD $B = U_2 \Sigma V_2^T$. Finally, with $U = U_1 U_2$ and $V = V_1 V_2$ we obtain the SVD $A = U\Sigma V^T$.

Subroutines for computing the SVD for dense rectangular matrices are available in most mathematical software libraries; see Table 2.1 of Hansen (1998) for a list. In MATLAB the command [U,S,V] = svd(A) computes the SVD of a matrix of dimension 500 in only about 12 seconds on a modest SUN-server (Eldén 2004).

A survey of direct methods for computing the SVD is given in Bai, Demmel, Dongarra, Ruhe and van der Vorst (2000, Section 6.2). A divide-and-conquer method for finding the the SVD of a bidiagonal matrix is implemented in the LAPACK subroutine xGESDD. This is faster than the QR algorithm for bidiagonal matrices larger than about 25×25. Bisection and inverse iteration can be used when we only want to compute the singular values in an interval and the corresponding left and right singular vectors. (This option is suitable for the TLS problem.) Bisection methods, analysed in Fernando (1998), rely on a very accurate algorithm for counting singular values of a bidiagonal matrix.

4. Generalized least squares problems

4.1. Generalized orthogonal decompositions

The motivation for introducing different generalizations of orthogonal decompositions is basically to avoid the explicit computation of matrix products and quotients of matrices. For example, let A and B be square and nonsingular matrices and assume we need the SVD of AB^{-1} (or AB). Then the explicit calculation of AB^{-1} (or AB) may result in a loss of precision and should be avoided.

An early application of generalized QR decomposition (GQR) is described in Hammarling (1976). The systematic use of GQR as a basic conceptual and computational tool are explored by Paige (1990), who shows that these decompositions allow the solution of very general formulations of several least squares problems. Further generalizations are discussed in De Moor and Van Dooren (1992), where the QR, URV and SVD decompositions are generalized to any number of matrices.

Routines for computing a GQR decomposition of a pair of matrices $A \in \mathbb{R}^{m \times n}$ and $B \in \mathbb{R}^{m \times p}$ are included in LAPACK; see Anderson, Bai, Bischof, Demmel, Dongarra, Du Croz, Greenbaum, Hammarling, McKenney, Ostrouchov and Sorensen (1995, Section 2.3.3). The GQR decomposition is given by

$$A = QR, \qquad B = QTZ, \qquad (4.1)$$

where $Q \in \mathbb{R}^{m \times m}$ and $Z \in \mathbb{R}^{p \times p}$ are orthogonal matrices and R and T have one of the forms

$$R = \begin{pmatrix} R_{11} \\ 0 \end{pmatrix} \quad (m \geq n), \qquad R = \begin{pmatrix} R_{11} & R_{12} \end{pmatrix} \quad (m < n), \qquad (4.2)$$

and

$$T = \begin{pmatrix} 0 & T_{12} \end{pmatrix} \quad (m \leq p), \qquad T = \begin{pmatrix} T_{11} \\ T_{21} \end{pmatrix} \quad (m > p). \qquad (4.3)$$

If B is square and nonsingular GQR implicitly gives the QR factorization of $B^{-1}A$. There is also a similar generalized RQ factorization related to the QR factorization of AB^{-1}. These generalized decompositions and their applications are discussed in Anderssen, Bai and Dongarra (1992).

Similar generalizations for the SVD were first discussed in Van Loan (1976) and Paige and Saunders (1981). Paige (1986) gave an algorithm for computing the the quotient SVD (QSVD) of two matrices $A \in \mathbb{R}^{m \times n}$ and $B \in \mathbb{R}^{p \times n}$; when B is square and nonsingular this is equivalent to the SVD of AB^{-1}. The computation of the SVD of AB, the product SVD (PSVD), is discussed in Heath, Laub, Paige and Ward (1986). These generalized SVDs are special cases of a more general theory developed by De Moor and Zha (1991). The GSVD algorithms in LAPACK are based on Bai and Demmel (1993), where several improvements of Paige's algorithm are given. An important role in these algorithms is played by a new accurate algorithm for the 2×2 triangular GSVD.

In Golub, Solna and Van Dooren (1995), an algorithm is developed for computing the SVD of an expression of the form

$$A = A_p^{s_p} \cdots A_2^{s_2} A_1^{s_1}, \qquad s_i = \pm 1,$$

that is, a sequence of products or quotients of matrices A_i of compatible dimensions. To illustrate the idea, consider for simplicity the case when $s_i = 1$. Then it is possible to construct orthogonal matrices Q_i, $i = 0 : p$, such that the product

$$B = Q_p^T A_p Q_{p-1} \cdots Q_2^T A_2 Q_1 Q_1^T A_1 Q_0$$

is a bidiagonal matrix. The SVD of B can then be found by standard methods.

4.2. Generalized least squares problems

An important class of generalized least squares problems is related to symmetric linear systems of the form

$$\begin{pmatrix} V & A \\ A^T & 0 \end{pmatrix} \begin{pmatrix} y \\ x \end{pmatrix} = \begin{pmatrix} b \\ c \end{pmatrix}, \qquad (4.4)$$

where $A \in \mathbb{R}^{m \times n}$ ($m \geq n$), and $V \in \mathbb{R}^{m \times m}$, is symmetric and positive semi-definite. The system matrix in (4.4) is symmetric but indefinite; it is nonsingular if and only if A has full column rank and $(V \; A)$ full row rank. Linear systems of this form (4.4) represent the condition for equilibrium of a physical system and therefore occur in numerous applications. It is also called a saddle point system and in optimization it is known as a KKT (Karush–Kuhn–Tucker) system. In applications V and A are often large and sparse matrices.

If V is positive definite and A has full column rank, then the system (4.4) is nonsingular and gives the first-order conditions for the solution of the following two optimization problems.

1. *Generalized linear least squares problem* (GLLS)

$$\min_x (Ax - b)^T V^{-1}(Ax - b) + 2c^T x. \qquad (4.5)$$

If $c = 0$, the solution x gives the best linear unbiased estimate for the linear model $Ax + \epsilon = b$, where $V = \sigma^2 V$ is the covariance matrix of the error vector ϵ.

2. *Equality-constrained quadratic optimization* (ECQO)

$$\min_y \frac{1}{2} y^T V y - b^T y, \qquad A^T y = c. \qquad (4.6)$$

This problem occurs as a subproblem in linearly constrained optimization. Another application, for which $c = 0$, is structural optimization (Heath, Plemmons and Ward 1984). Here A^T is called the equilibrium matrix, V the element flexibility matrix, y is the force, and x a Lagrange multiplier vector.

There are two different approaches to the solution of systems of the form (4.4). In the *range space method* the y variables are eliminated to obtain

$$A^T V^{-1} A x = A^T V^{-1} b - c, \qquad y = V^{-1}(b - Ax). \tag{4.7}$$

For $V = I$ the first equation in (4.7) is the normal equations for the least squares problem. If V is positive definite then one way to solve these equations is to compute the Cholesky factorization $V = BB^T$ and then solve

$$\min_x \|B^{-1}(Ax - b)\|_2 \tag{4.8}$$

using the QR factorization of $B^{-1}A$. However, a more stable approach is to use a GQR factorization of the matrix pair A, B.

In the *null space method* the solution y to (4.7) is split as

$$y = y_1 + y_2, \qquad y_1 \in \mathcal{R}(A), \qquad y_2 \in \mathcal{N}(A^T). \tag{4.9}$$

Let y_1 be the minimum norm solution of $A^T y = c$. This can be computed using the QR factorization of A. If we set $Q = (Q_1 \ Q_2)$ then

$$y_1 = Q_1 z_1, \qquad z_1 = R^{-T} c.$$

Next y_2 is obtained by solving the reduced system

$$Q_2^T V Q_2 z_2 = Q_2^T (b - V y_1), \qquad y_2 = Q_2 z_2. \tag{4.10}$$

Finally, form

$$y = Q_1 z_1 + Q_2 z_2 \quad \text{and} \quad x = R^{-1} Q_1^T (b - V y).$$

In the special case that $V = I$, the numerical stability of methods which use Q and R, or only R, in the QR factorization of A are studied in Björck and Paige (1994). Backward stability was proved for several methods.

Recently perturbation analyses and condition numbers for problem (4.4) have been given in Arioli (2000) and Gulliksson and Wedin (2000). Arioli (2000) also gives a round-off error analysis of a null space method for solving (4.4) and applies this to developing methods for solving a problem arising from a mixed finite element discretization of a magnetostatic problem.

5. Blocked algorithms

5.1. Partitioned algorithms

The impact of the architecture of modern computers on algorithms of numerical linear algebra is surveyed in depth in Dongarra, Duff, Sorensen and van der Vorst (1998). One conclusion is that to obtain near-peak performance for large dense matrix computations on current computing architectures requires code that is dominated by level 3 Basic Linear Algebra Subroutines (BLAS 3). These kernels perform various types of matrix–matrix multiplication and involve less data movement per floating point computation; see Dongarra, Du Croz, Duff and Hammarling (1990). The subroutines in LAPACK, including those for QR factorization, are therefore organized in partitioned or blocked form, in which the operations have been reordered and grouped into matrix operations. These partitioned algorithms are as stable as their point counterparts. This is not the case for all block algorithms; see Higham (1997) for the distinction between block and partitioned algorithms.

For the QR factorization $A \in \mathbb{R}^{m \times n}$ ($m \geq n$) is partitioned as

$$A = (A_1, A_2), \qquad A_1 \in \mathbb{R}^{m \times nb}, \tag{5.1}$$

where nb is a suitable block size and the QR factorization

$$Q_1^T A_1 = \begin{pmatrix} R_1 \\ 0 \end{pmatrix}, \qquad Q_1 = P_1 P_2 \cdots P_{nb}, \tag{5.2}$$

is computed, where $P_i = I - u_i u_i^T$ are Householder reflections. Then the remaining columns A_2 are are updated

$$Q_1^T A_2 = Q_1^T \begin{pmatrix} A_{12} \\ A_{22} \end{pmatrix} = \begin{pmatrix} R_{12} \\ \tilde{A}_{22} \end{pmatrix}. \tag{5.3}$$

In the next step we partition $\tilde{A}_{22} = (B_1, B_2)$, and compute the QR factorization of $B_1 \in \mathbb{R}^{(m-r) \times r}$. Then B_2 is updated as above, and we continue in this way until the columns in A are exhausted.

A major part of the computation is spent in the updating step (5.3). As written this step cannot use BLAS-3, which slows down the execution. To achieve better performance it is essential to speed this part up. The solution is to aggregate the Householder transformations so that their application can be expressed as matrix operations; see Schreiber and Van Loan (1989). For use in the next subsection, we show a slightly more general result due to Elmroth and Gustavson (2000).

Assume that $r = r_1 + r_2$, and

$$Q_1 = P_1 \cdots P_{r_1} = I - Y_1 T_1 Y_1^T, \qquad Q_2 = P_{r_1+1} \cdots P_r = I - Y_2 T_2 Y_2^T,$$

where $T_1, T_2 \in \mathbb{R}^{r \times r}$ are upper triangular. Then

$$Q = Q_1 Q_2 = (I - Y_1 T_1 Y_1^T)(I - Y_2 T_2 Y_2^T) = (I - Y T Y^T), \tag{5.4}$$

where
$$\hat{Y} = (Y_1, Y_2), \quad \hat{T} = \begin{pmatrix} T_1 & -(T_1 Y_1^T)(Y_2 T_2) \\ 0 & T_2 \end{pmatrix}. \quad (5.5)$$

Note that Y is formed by concatenation, but computing the off-diagonal block in T requires extra operations.

For the partitioned algorithm we use the special case when $r_2 = 1$ to aggregate the Householder transformations for each processed block. Starting with $Q_1 = I - \tau_1 u_1 u_1^T$, we set $Y = u_1$, $T = \tau_1$ and update

$$Y := (Y, u_{k+1}), \quad T := \begin{pmatrix} T & -\tau_k T Y^T u_k \\ 0 & \tau_k \end{pmatrix}, \quad k = 2 : nb. \quad (5.6)$$

Note that Y will have a trapezoidal form and thus the matrices Y and R can overwrite the matrix A. With the representation $Q = (I - YTY^T)$ the updating of A_2 becomes

$$B = Q_1^T A = (I - YT^T Y^T) A_2 = A_2 - YT^T Y^T A_2,$$

which now involves only matrix operations. An analogous partitioned version of MGS is discussed in Björck (1994).

This partitioned algorithm requires more storage and operations than the point algorithm, namely those needed to produce and store the T matrices. However, for large matrices this is more than offset by the increased rate of execution.

5.2. Recursive algorithms

As shown by Elmroth, Gustavson, Jonsson and Kågström (2004) recursive algorithms can be developed into highly efficient algorithms for high performance computers and are an alternative to the partitioned algorithms currently used by LAPACK. The reason for this is that recursion leads to automatic variable blocking that dynamically adjusts to an arbitrary number of levels of memory hierarchy.

The recursive QR factorization

$$A = (A_1 \quad A_2) = Q \begin{pmatrix} R_{11} & R_{12} \\ 0 & R_{22} \end{pmatrix}$$

starts with a QR factorization of the first $\lfloor n/2 \rfloor$ columns of A and updating of the remaining part of the matrix

$$Q_1^T A_1 = \begin{pmatrix} R_{11} \\ 0 \end{pmatrix}, \quad Q_1^T A_2 = Q_1^T \begin{pmatrix} A_{12} \\ A_{22} \end{pmatrix} = \begin{pmatrix} R_{12} \\ \tilde{A}_{22} \end{pmatrix}.$$

Next $\tilde{A}_{22}$ is recursively QR-decomposed, giving Q_2, R_{22}, and $Q = Q_1 Q_2$.

As an illustration we give a simple implementation in MATLAB, which is convenient to use since it allows for the definition of recursive functions.

```
function [Y,T,R] = recqr(A);
%
% RECQR computes the QR factorization of the m by n matrix A,
% (m >= n). Output is the n by n triangular factor R, and
% Q = (I - YTY')  represented in aggregated form, where Y is
% m by n and unit lower trapezoidal, and T is n by n upper
% triangular It uses [u,tau,sigma] = house(a) to compute
% a Householder transformation P = I - tau uu', such that
% Pa = sigma e1, sigma = -sign(a_1)norm(a).
    [m,n] = size(A);
    if n == 1
    [Y,T,R] = house(A);
    else
        n1 = floor(n/2);
        n2 = n - n1; j = n1+1;
        [Y1,T1,R1]= recqr(A(1:m,1:n1));
        B = A(1:m,j:n) - (Y1*T1')*(Y1'*A(1:m,j:n));
        [Y2,T2,R2] = recqr(B(j:m,1:n2));
        R = [R1, B(1:n1,1:n2); zeros(n-n1,n1), R2];
        Y2 = [zeros(n1,n2); Y2];
        Y = [Y1, Y2];
        T = [T1, -T1*(Y1'*Y2)*T2; zeros(n2,n1), T2];
    end
%
```

The above algorithm is just a prototype and needs to be improved and tuned in several ways. A serious defect is the overhead in storage and operations caused by the T matrices. In the partitioned algorithm n/nb T-matrices of size $nb \times nb$ are formed and stored, giving a storage overhead of $\frac{1}{2}n \cdot nb$. In the recursive QR algorithm in the end a T-matrix of size $n \times n$ is formed and stored, leading to a much too large storage and operation overhead.

Elmroth and Gustavson (2000) develop and analyse recursive algorithms for the QR factorization. They find that the best option is a hybrid between the partitioned and the recursive algorithm, where the recursive QR algorithm is used to factorize the blocks in the partitioned algorithm. In Elmroth and Gustavson (2001) these hybrid QR algorithms are used to implement recursive algorithms for computing least squares solutions to overdetermined linear systems and minimum norm solutions to under-determined linear systems. These implementations are shown to be significantly faster – usually 50–100% and sometimes much more – than the corresponding current LAPACK algorithms based on the partitioned approach.

6. Rank-revealing decompositions

6.1. Column pivoting

Golub (1965) remarks that the accuracy in the QR factorization is slightly improved if the following column pivoting strategy is used. Assume that after k steps we have computed the partial QR factorization

$$A^{(k)} = (P_k \cdots P_1) A (\Pi_1 \cdots \Pi_k) = \begin{pmatrix} R_{11}^{(k)} & R_{12}^{(k)} \\ 0 & \tilde{A}^{(k)} \end{pmatrix}, \qquad (6.1)$$

where $\Pi_1, \ldots, \Pi_k$ are permutation matrices performing the column interchanges. The remaining steps will only affect the submatrix $\tilde{A}^{(k)}$. The pivot column in the step $k+1$ is chosen as a column of largest norm in the submatrix

$$\tilde{A}^{(k)} = (\tilde{a}_{k+1}^{(k)}, \ldots, \tilde{a}_n^{(k)}) \in \mathbb{R}^{(m-k) \times (n-k)},$$

i.e., Π_{k+1} interchanges columns p and $k+1$, where p is the smallest index such that

$$s_p^{(k)} \geq s_j^{(k)}, \qquad s_j^{(k)} = \|\tilde{a}_j^{(k)}\|_2, \quad j = k+1:n. \qquad (6.2)$$

If $s_p^{(k)} = 0$ then the algorithm terminates with $\tilde{A}^{(k)} = 0$ in (6.1), which implies that $\mathrm{rank}(A) = k$. This pivoting strategy can be viewed as choosing a remaining column of largest distance to the subspace spanned by the previously chosen columns and is equivalent to maximizing the diagonal element $r_{k+1,k+1}$.

Golub (1965) also notes that 'the strategy above is most appropriate when one has a sequence of vectors $b_1, b_2, \ldots, b_p$ for which one desires a least squares estimate. In many problems there is one vector b and one wishes to express it in as few columns of A as possible.' For this case one should at each stage choose the column of $A^{(k)}$ that will maximally reduce the sum of squares of the residuals after the kth stage. If

$$(P_k \cdots P_1) b = \begin{pmatrix} c_1^{(k)} \\ c_2^{(k)} \end{pmatrix},$$

this is equivalent to choosing a pivot which maximizes

$$t_j^{(k)} = |(c_2^{(k)})^T \tilde{a}_j^{(k)}| / \|\tilde{a}_j^{(k)}\|_2, \quad k < j \leq n.$$

The partitioned algorithm as reviewed in Section 5 cannot easily be implemented for the pivoted QR factorization. This is because in order to choose a pivot column all remaining columns need first to be updated. Therefore it is not possible to accumulate several Householder transformations and perform the update simultaneously. Quintana-Ortí, Sun and Bischof (1998) show how the pivoted QR algorithm can be implemented so that half of the work is performed in BLAS-3 kernels.

6.2. Rank-revealing QR decompositions

Rank-deficient problems are common, e.g., in statistics, where the term collinearity is used. Although the SVD is generally the most reliable method for computing the numerical rank of a matrix it has the disadvantage of a high computational cost. Alternative decompositions based on QR factorization with column pivoting were first proposed in Faddeev, Kublanovskaya and Faddeeva (1968) and Hanson and Lawson (1969).

Let $A \in \mathbb{R}^{m \times n}$ be a matrix with singular values $\sigma_1 \geq \sigma_1 \geq \cdots \geq \sigma_n \geq 0$. By a rank-revealing QR (RRQR) decomposition of A we mean a decomposition of the form

$$A\Pi = (Q_1 \ Q_2) \begin{pmatrix} R_{11} & R_{12} \\ 0 & R_{22} \end{pmatrix}, \tag{6.3}$$

where Π is a permutation matrix, $R_{11} \in \mathbb{R}^{k \times k}$ upper triangular, and

$$\sigma_k(R_{11}) \geq \frac{1}{c_1}\sigma_k, \qquad \|R_{22}\|_2 \leq c_2\sigma_{k+1}. \tag{6.4}$$

for some not too large constants c_1 and c_2. If σ_{k+1} is small and $\sigma_k \gg \sigma_{k+1}$, then the factorization (6.4) will reveal that the numerical rank of A is k.

In (6.3) the matrix Q_1 gives an orthogonal basis for the numerical range of A and

$$W = \begin{pmatrix} R_{11}^{-1} R_{12} \\ -I \end{pmatrix} \in \mathbb{R}^{n \times (n-k)} \tag{6.5}$$

gives a basis for the numerical null space of AP. If a more accurate basis than (6.5) is desired this can be computed using a few inverse iterations, as suggested in Chan and Hansen (1990).

Hong and Pan (1992) prove the existence of a decomposition (6.3)–(6.4) with

$$c = \sqrt{k(n-k) + \min(k, n-k)}, \quad \forall k, \quad 0 < k < n.$$

Thus, whenever there is a well-determined gap in the singular value spectrum, $\sigma_r \gg \sigma_{r+1}$, there exists for $k = r$ an RRQR decomposition that reveals the numerical rank of A.

To find a permutation Π such that the rank of A is revealed is not always simple. Note that an exhaustive search has combinatorial complexity! It is still an open question if an algorithm of polynomial complexity exists for finding an optimal permutation. Fortunately there are algorithms that in practice work almost always.

From the interlacing properties of singular values (Golub and Van Loan 1996, Corollary 8.6.3) it follows by induction that, for any decomposition of the form (6.3), we have the inequalities

$$\sigma_{\min}(R_{11}) \leq \sigma_k(A), \quad \sigma_{\max}(R_{22}) \geq \sigma_{k+1}(A).$$

Hence, to achieve a rank-revealing QR decomposition we want to find a permutation P that aims to solve the two problems

(i) $\max_\Pi \sigma_{\min}(R_{11})$;   (ii) $\min_\Pi \sigma_{\max}(R_{22})$.

These two problems are dual in a certain sense.

Problem (i) is equivalent to the subset selection problem of determining the $k < n$ most linearly independent columns of A. An SVD-based algorithm for solving this problem was given in Golub, Klema and Stewart (1976) and an RRQR algorithm in Chan and Hansen (1992). Although the methods will not in general compute equivalent solutions, the subspaces spanned by the two sets of selected columns can be shown to be almost identical whenever the ratio σ_{k+1}/σ_k is small.

The column pivoting strategy used in Golub (1965) chooses as the next pivot column the one having maximum distance from the subspace spanned by the already chosen columns. Hence this is a greedy algorithm that addresses problem (i). While this strategy can fail on certain matrices (see Golub and Van Loan (1996, Section 5.5.7)), it is widely used due to its simplicity and practical reliability. In several other RRQR algorithms this pivoting strategy is used in a preprocessing stage. In the second stage a new permutation matrix is determined so that the rank-revealing property is improved.

Several algorithms have been suggested, which address problem (ii). These are based on the following property.

Lemma 1. (Chan and Hansen (1990)) Given any column permutation P and $V \in \mathbb{R}^{n \times p}$, the QR factorization of AP yields an R_{22} such that

$$\|R_{22}\|_2 \leq \|AV\|_2 \|W_2^{-1}\|_2, \quad W = P^T V = \begin{pmatrix} W_1 \\ W_2 \end{pmatrix}. \quad (6.6)$$

This means that if the columns of V lie approximately in the numerical null space of A and we can find a permutation such that $\|W_2^{-1}\|_2$ is not large, then the bottom $p \times p$ block of R in the QR factorization will be small. Ideally, for $p = 1$, we take $v \approx v_n$, the right singular vector corresponding to σ_n. The permutation P is chosen so that the component of largest absolute value is moved to the end.

Early algorithms based on the above lemma include Golub et al. (1976) and Chan (1987). Algorithms that satisfy (6.3)–(6.4) with c_1, c_2 equal to low-order polynomials in n and m are developed in Pan and Tang (1999) and Chandrasekaran and Ipsen (1994). These consist of two main stages: an initial pivoted QR factorization of A followed by a rank-revealing stage in which the triangular factor is modified. Chandrasekaran and Ipsen (1994) provide a common framework for these algorithms.

In Bischof and Quintana-Ortí (1998b) more efficiently implementable variants of RRQR algorithm for triangular matrices are developed. In the first stage a pivoted QR factorization is computed where the pivoting is restricted to a pivot window. In the post-processing stage either Algorithm 3 in Pan and Tang (1999) or Hybrid III in Chandrasekaran and Ipsen (1994) is used. These hybrid algorithms are nearly as fast as current partitioned QR algorithms without pivoting; see Bischof and Quintana-Ortí (1998a)

6.3. The URV and ULV decompositions

In signal processing problems the data analysed often arrives in real time and it is necessary to update matrix decompositions at each time step. For such applications the SVD has the disadvantage that it cannot in general be updated in less than $O(n^3)$ operations, when rows and columns are added or deleted to A. In special cases simplified updating schemes may be viable, such as the fast SVD updating algorithm in Moonen, Van Dooren and Vandewalle (1992), which is a combination of a QR updating followed by a Jacobi-type SVD method.

Although the RRQR decomposition can be updated, it is less suitable in applications where a basis for the approximate null space of A is needed, since the matrix W in (6.5) cannot easily be updated. For this reason Stewart (1991) introduced the URV rank-revealing decomposition

$$A = URV^T = \begin{pmatrix} U_1 & U_2 \end{pmatrix} \begin{pmatrix} R_{11} & R_{12} \\ 0 & R_{22} \end{pmatrix} \begin{pmatrix} V_1^T \\ V_2^T \end{pmatrix}, \qquad (6.7)$$

where U and V are orthogonal matrices, $R_{11} \in \mathbb{R}^{k \times k}$, and

$$\sigma_k(R_{11}) \geq \frac{1}{c}\sigma_k, \qquad (\|R_{12}\|_F^2 + \|R_{22}\|_F^2)^{1/2} \leq c\sigma_{k+1}. \qquad (6.8)$$

Note that here both submatrices R_{12} and R_{22} have small elements.

From (6.7) we have

$$\|AV_2\|_2 = \left\|\begin{pmatrix} R_{12} \\ R_{22} \end{pmatrix}\right\|_F \leq c\sigma_{k+1},$$

and hence the orthogonal matrix V_2 can be taken as an approximation to the numerical null space $\mathcal{N}_k$.

Algorithms for computing a URV decomposition start with an initial QR decomposition, followed by a rank-revealing stage in which singular vectors corresponding to the smallest singular values of R are estimated. Assume that w is a unit vector such that $\|Rw\| = \sigma_n$. Let P and Q be orthogonal matrices such that $Q^T w = e_n$ and $P^T RQ = \hat{R}$, where $\hat{R}$ is upper triangular. Then

$$\|\hat{R}e_n\| = \|P^T RQQ^T w\| = \|P^T Rw\| = \sigma_n,$$

which shows that the entire last column in $\hat{R}$ is small. Given w the matrices P and Q can be constructed as a sequence of Givens rotations (see Stewart (1992), where algorithms are also given for updating a URV decomposition when a new row is appended).

As for the RRQR decompositions, the URV decomposition yields approximations to the singular values. Mathias and Stewart (1993) derive the following bounds:

$$f\sigma_i \leq \sigma_i(R_{11}) \leq \sigma_i, \quad i = 1:r,$$

and

$$\sigma_i \leq \sigma_{i-k}(R_{22}) \leq \sigma_i/f, \quad i = r+1:n,$$

where

$$f = \left(1 - \frac{\|R_{12}\|_2^2}{\sigma_{\min}(R_{11}^2 - \|R_{22}\|_2^2)}\right)^{1/2}.$$

Hence the smaller the norm of the off-diagonal block R_{12}, the better the bounds will be. Similar bounds can be given for the angle between the range of V_2 and the right singular subspace corresponding to the smallest $n-r$ singular values of A.

Stewart (1993) gives an alternative decomposition that is more satisfactory for applications where an accurate approximate null space is needed, as in subspace tracking. This is the rank-revealing ULV decomposition

$$A = U \begin{pmatrix} L_{11} & 0 \\ L_{21} & L_{22} \end{pmatrix} V^T, \tag{6.9}$$

where the middle matrix has lower triangular form. For this decomposition

$$\|AV_2\|_2 = \|L_{22}\|_F, \quad V = (V_1, V_2),$$

and hence the size of $\|L_{21}\|$ does not adversely affect the null space approximation. On the other hand the URV decomposition usually gives a superior approximation for the numerical range space and the updating algorithm for URV is much simpler.

We finally mention that rank-revealing QR decompositions can be effectively computed only if the numerical rank r is either high, $r \approx n$ or low, $r \ll n$. The low rank case is discussed in Chan and Hansen (1994). MATLAB templates for rank-revealing UTV decompositions are described in Fierro, Hansen and Hansen (1999).

6.4. Stewart's QLP decomposition

The ULV and URV decompositions are rank-revealing, but do not attempt to give good approximations to the singular values. The pivoted QLP decomposition, also introduced by Stewart (1999), yields for the extra cost of one more QR decomposition quite accurate approximations to the singular values of A. The QLP decomposition can be considered as the first step in a rapidly converging iterative algorithm for computing the SVD of A, which is analysed in Huckaby and Chan (2003).

The QLP algorithm starts by computing the pivoted QR factorization

$$Q^T A\Pi = \begin{pmatrix} R \\ 0 \end{pmatrix}, \quad R \in \mathbb{R}^{n \times n}. \tag{6.10}$$

In the second step the *upper* triangular matrix R is transformed into a *lower* triangular matrix L, using postmultiplication by a product P of Householder transformations,

$$RP = L, \quad L \in \mathbb{R}^{n \times n}. \tag{6.11}$$

No pivoting is used in this step. (Transposing (6.11) shows that this LQ factorization of R is equivalent to a QR factorization of the lower triangular matrix R^T.) Combining these two factorizations (6.10) and (6.11) we obtain

$$A\Pi = Q \begin{pmatrix} L \\ 0 \end{pmatrix} P^T. \tag{6.12}$$

To compute the QLP decomposition requires roughly $mn^2 - n^3/3$ flops for the decomposition (6.10) and $2n^3/3$ flops for the decomposition (6.11).

Suppose that after k steps of the pivoted Householder QR algorithm (6.10) we have computed the partial QR factorization

$$Q_k A \Pi_k = A^{(k+1)} = \begin{pmatrix} R_{11} & R_{12} \\ 0 & \tilde{A}_{22} \end{pmatrix},$$

where $(R_{11} \quad R_{12})$ are the first k rows of R in the QR factorization of A. By postmultiplying with k Householder transformations we obtain

$$(R_{11} \quad R_{12}) P_k = (L_{11} \quad 0),$$

where L_{11} is the first k rows of L in the QLP decomposition. This observation shows that the two factorization can be interleaved, *i.e.*, in the kth step we first compute the kth row of R and then the kth row of L. To determine the first k diagonal elements of L, which give the QLP approximations to the first k singular values of A, it is only necessary to perform k steps in each of the two factorization. This is advantageous when the numerical rank is much less than n.

Despite the simplicity of the QLP decomposition the diagonal elements of L usually give remarkably good approximations to all the singular values

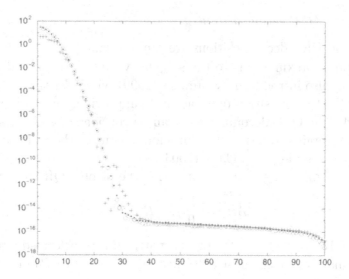

Figure 6.1. Diagonal elements of L (circles) in the QLP and in R (plus) in pivoted QR compared with singular values (points) of the matrix K.

of A. In particular a good estimate of $\sigma_1 = \|A\|_2$ can be obtained in $O(n^2)$ operations from the first row $(r_{11} \ r_{12})$ of R setting

$$\sigma_1 \approx l_{11} = (r_{11}^2 + \|r_{12}\|_2^2)^{1/2}.$$

As an illustration we consider the integral equation of the first kind

$$\int_{-1}^{1} k(s,t) f(s) \, ds = g(t), \quad k(s,t) = e^{-(s-t)^2},$$

on $-1 \leq t \leq 1$. If this equation is discretized using a uniform mesh on $[-1, 1]$ and the trapezoidal rule, the resulting linear system $Kf = g$, $K \in \mathbb{R}^{n \times n}$, is very ill-conditioned. In Figure 6.1 the singular values σ_k of the matrix K_n, $n = 100$, are displayed together with (absolute values of) the diagonal elements of R and L in the QLP decomposition. The diagonal elements of L are seen to track both large and small singular values much more accurately than those of R.

7. Bidiagonal reduction

Any matrix $A \in \mathbb{R}^{m \times n}$ can be decomposed as

$$A = UBV^T, \qquad (7.1)$$

where B is an upper (or lower) bidiagonal matrix and U and V are orthogonal matrices. This important decomposition first appeared in Golub and Kahan (1965). It is usually the first step in computing the SVD of A, but is also a powerful tool in itself for solving various least squares problems. It is the core decomposition used in the iterative method LSQR (Paige and Saunders 1982b) for solving least squares problems. Recently Paige and Strakoš (2002) (see also Paige (2002)) have shown that the reduction to upper bidiagonal form of $(b \ A)$ provides an elegant way to extract a core problem, both for the linear least squares problem (1.1) and the total least squares problem (1.3). Because of its importance we review this decomposition in some detail below.

Golub and Kahan (1965) gave two quite different algorithms for computing the decomposition (7.1). In the first algorithm U and V are formed as products of two sequences of Householder transformations. In the second algorithm the successive columns in U and V are generated by a Lanczos process. Once the first column $u_1 = Ue_1$ (or $v_1 = Ve_1$) is fixed, the decomposition (7.1) is uniquely determined in the nondegenerate case. Therefore the two algorithms will, using exact arithmetic, produce the same bidiagonal decomposition. However, the Lanczos method is less stable numerically and is mainly of interest when A is a large and sparse matrix.

7.1. Bidiagonalization using Householder transformations

With no loss of generality we assume that $m \geq n$. Following the first algorithm in Golub and Kahan (1965), for $k = 1, 2, \ldots$, we alternately multiply $(b, \ A)$ from the left and right with Householder transformations Q_k and P_k, respectively. Here Q_k is chosen to zero the last $m - k$ elements in the kth column of $(b, \ A)$ and P_k is chosen to zero the last $n - k$ elements in the kth row of A. The final result is the decomposition

$$U^T (b \ AV) = \begin{pmatrix} \beta_1 e_1 & B \\ 0 & 0 \end{pmatrix}, \quad B = \begin{pmatrix} \alpha_1 & & & \\ \beta_2 & \alpha_2 & & \\ & \ddots & \ddots & \\ & & \beta_n & \alpha_n \\ & & & \beta_{n+1} \end{pmatrix}, \quad (7.2)$$

where e_1 is the first unit vector, and

$$U = Q_1 Q_2 \cdots Q_{n+1} \in \mathbb{R}^{m \times (n+1)}, \quad V = P_1 P_2 \cdots P_{n-1} \in \mathbb{R}^{n \times n}. \quad (7.3)$$

Note that the first column in U equals $u_1 = Ue_1 = b/\beta_1$ and that $U^T AV = B \in \mathbb{R}^{(n+1) \times n}$ is a *lower* bidiagonal matrix.

Setting $y = V^T x$ and using the invariance of the Euclidean norm it follows that

$$\|b - Ax\|_2 = \left\| (b \ A) \begin{pmatrix} -1 \\ x \end{pmatrix} \right\|_2 = \left\| U^T (b \ AV) \begin{pmatrix} -1 \\ V^T x \end{pmatrix} \right\|_2$$

$$= \|\beta_1 e_1 - By\|_2. \quad (7.4)$$

Hence, if y solves the bidiagonal least squares problem

$$\min_y \|By - \beta_1 e_1\|_2, \qquad (7.5)$$

then $x = Vy$ solves $\min_x \|Ax - b\|_2$.

After $k < n$ steps of the bidiagonal reduction we have computed an upper bidiagonal matrix B_k and orthogonal matrices $U_k = Q_1 \cdots Q_k$, and $V_k = P_1 \cdots P_k$, such that $U_k^T A V_k = B_k$. From the construction of the Householder matrices P_j it follows that

$$U \begin{pmatrix} I_k \\ 0 \end{pmatrix} = Q_1 \cdots Q_k Q_{k+1} \cdots Q_{n+1} \begin{pmatrix} I_k \\ 0 \end{pmatrix} = Q_1 \cdots Q_k \begin{pmatrix} I_k \\ 0 \end{pmatrix} = U_k. \qquad (7.6)$$

This shows that the first k columns in U_k and the final matrix U are equal. Similarly the first k columns in V_k equals those in V.

7.2. The core subproblem

Assume first that $\alpha_j, \beta_{j+1} \neq 0$, $j = 1 : k - 1$, for some $1 \leq k \leq n$, but $\alpha_k = 0$. Then after k steps we have obtained a decomposition of the form

$$U_k^T A V_k = \begin{pmatrix} B_k & 0 \\ 0 & A_k \end{pmatrix},$$

where $B_k \in \mathbb{R}^{k \times (k-1)}$ is a leading submatrix of B and $A_k \in \mathbb{R}^{(m-k) \times (n-k+1)}$. The resulting transformed least squares problem

$$\min_y \left\| \begin{pmatrix} B_k & 0 \\ 0 & A_k \end{pmatrix} \begin{pmatrix} y_1 \\ y_2 \end{pmatrix} - \begin{pmatrix} \beta_1 e_1 \\ 0 \end{pmatrix} \right\|_2, \quad y = \begin{pmatrix} y_1 \\ y_2 \end{pmatrix}, \qquad (7.7)$$

is separable and decomposes into the two independent subproblems

$$\min_{y_1} \|B_k y_1 - \beta_1 e_1\|_2 \quad \text{and} \quad \min_{y_2} \|A_k y_2\|_2. \qquad (7.8)$$

The first subproblem is similar to (7.5) and since B_k has full column rank the solution y_1 is unique. We call this the *core subproblem*. Clearly the minimum norm solution $x = Vy$ is obtained by taking $y_2 = 0$.

Assume next that also $\alpha_k \neq 0$, but $\beta_{k+1} = 0$. Then the reduced matrix again has the separable form (7.7), where now $B_k \in \mathbb{R}^{k \times k}$ and $A_k \in \mathbb{R}^{(m-k) \times (n \times k)}$. The core subproblem then simplifies to $B_k y_1 = \beta_1 e_1$, with B_k square, nonsingular and lower triangular. The unique solution y_1 is obtained simply by forward substitution. Taking $y_2 = 0$, the corresponding residual $b - AV_k y$ is zero and hence in this case the original system $Ax = b$ is consistent.

We give two simple examples of termination. Assume first that $b \perp \mathcal{R}(A)$. Then the reduction will terminate with $\alpha_1 = 0$, and $x = 0$ is the minimal norm least squares solution. As a second example, assume that the bidiagonalization terminates with $\beta_2 = 0$. Then the system $Ax = b$ is consistent and the minimum norm solution equals

$$x = (\beta_1/\alpha_1)v_1, \quad v_1 = V_1 e_1 = P_1 e_1.$$

The minimally dimensioned core subproblem, obtained by terminating the bidiagonalization of $(b\ A)$ when the first zero element is encountered, has several important properties. The matrix B_k has full column rank and its singular values are simple. Further, the right-hand side βe_1 has non-zero components along each left singular vector of B_k. These properties considerably simplify the solution of the LS or TLS subproblem; see Paige and Strakoš (2002).

7.3. Bidiagonalization using a Lanczos process

In the second approach to bidiagonalization in Golub and Kahan (1965), the columns of U and V are generated sequentially by a Lanczos process. The following algorithm is identical to the procedure Bidiag 1 in Paige and Saunders (1982b). From (7.2) we get the equations

$$AV = UB \quad \text{and} \quad A^T U = V B^T. \tag{7.9}$$

Setting

$$U = (u_1\ u_2 \cdots u_{n+1}), \quad V = (v_1\ v_2 \cdots v_n),$$

and equating columns in the two equations (7.9), we obtain the relations

$$A^T u_1 = \alpha_1 v_1, \quad A^T u_j = \beta_j v_{j-1} + \alpha_j v_j, \quad j = 2, \ldots, n,$$
$$A v_j = \alpha_j u_j + \beta_{j+1} u_{j+1}, \quad j = 1, \ldots, n.$$

Given the unit starting vector $u_1 = b/\beta_1$, $\beta_1 = \|b\|_2$, these relations can be used to recursively compute the column vectors $v_1, u_2, v_2, \ldots, u_{n+1}$. We get

$$v_j = r_j/\alpha_j, \quad u_{j+1} = s_j/\beta_{j+1}, \quad j = 1, \ldots, n, \tag{7.10}$$

where

$$r_j = A^T u_j - \beta_j v_{j-1}, \quad \alpha_j = \|r_j\|_2, \tag{7.11}$$
$$s_j = A v_j - \alpha_j u_j, \quad \beta_{j+1} = \|s_j\|_2. \tag{7.12}$$

The advantage of using this process to generate B_k is that we only need to be able to compute matrix–vector products with A and A^T.

The recurrence relations can also be written in matrix form as

$$U_{k+1}(\beta_1 e_1) = b, \tag{7.13}$$
$$AV_k = U_{k+1} B_k, \tag{7.14}$$
$$A^T U_{k+1} = V_k B_k^T + \alpha_{k+1} v_{k+1} e_{k+1}^T. \tag{7.15}$$

In exact arithmetic it holds that $V_k^T V_k = U_k^T U_k = I_k$. From the uniqueness of the bidiagonal decomposition it follows that B_k, U_k and V_k, are the same as generated by the Householder algorithm.

If $\|r_j\|_2$ and $\|s_j\|_2 = 0$ for $j \leq n$, this process will terminate prematurely with $\alpha_j = 0$ and $\beta_{j+1} = 0$, respectively. However, as shown previously, in these cases the least squares problem is separable and the minimum norm solution can be obtained from the partial decomposition computed so far.

7.4. A Krylov subspace method for least squares

For a square matrix $C \in \mathbb{R}^{n \times n}$ and a vector $z \in \mathbb{R}^n$ we define the Krylov subspace

$$\mathcal{K}_k(C, z) = \mathrm{span}\left\{z, Cz, \ldots, C^{k-1}z\right\}. \tag{7.16}$$

From the Lanczos recurrence relations (7.10)–(7.12) it follows by induction that $u_{j+1} \in \mathcal{K}_j(AA^T, u_1)$ and $v_j \in \mathcal{K}_j(A^T A, A^T u_1)$, $j = 1 : n$. Hence the columns of U_{k+1} and V_k form orthonormal bases for the Krylov subspaces

$$\mathcal{R}(U_k) = \mathcal{K}_k(AA^T, u_1), \qquad \mathcal{R}(V_k) = \mathcal{K}_k(A^T A, A^T u_1). \tag{7.17}$$

Suppose that after performing the first k steps of bidiagonalization we seek an approximate least squares solution of the form

$$x_k = V_k y_k. \tag{7.18}$$

By (7.17) this is equivalent to restricting x_k to lie in the Krylov subspace $\mathcal{K}_k(A^T A, A^T b)$. Using (7.14) we obtain

$$b - Ax_k = b - AV_k y_k = U_{k+1}(\beta_1 e_1 - B_k y_k)$$

and from the orthogonality of the columns of U_{k+1} it follows that

$$\|b - Ax_k\|_2 = \|\beta_1 e_1 - B_k y_k\|_2.$$

Hence $\|b - Ax_k\|_2$ is minimized over all $x_k \in \mathcal{R}(V_k)$ by taking y_k as a solution to the least squares problem

$$\min_{y_k} \|\beta_1 e_1 - B_k y_k\|_2. \tag{7.19}$$

This problem is of exactly the same form as that obtained when the reduction is completed. Since we are minimizing $\|b - Ax\|_2$ over an increasing nested set of subspaces it follows that the sequence $\|b - Ax_k\|_2$, $k = 1 : n$, will be non-increasing. Hence we can solve (7.19) for $k = 1, 2, \ldots$ and stop when the residual norm of the solution is small enough. We then accept $x = V_k y_k$ as an approximate solution of the original least squares problem. Except for some details this is essentially the LSQR algorithm by Paige and Saunders (1982b). The convergence properties of this algorithm will be discussed in Section 10.2.

7.5. Solving the sequence of bidiagonal problems

The bidiagonal least squares problem can easily be solved by reducing B_k to upper bidiagonal form. The QR decomposition of B_k is computed by premultiplication with a sequence of Givens rotations. If these are also applied to the right-hand side $\beta_1 e_1$ we obtain

$$Q_k^T \begin{pmatrix} B_k & \beta_1 e_1 \end{pmatrix} = \begin{pmatrix} R_k & d_k \\ 0 & \bar{\phi}_{k+1} \end{pmatrix}, \qquad (7.20)$$

where

$$Q_k^T = G_{k,k+1} \cdots G_{23} G_{12}, \quad G_{j,j+1} = \begin{pmatrix} c_j & s_j \\ -s_j & c_j \end{pmatrix}, \quad j = 1:k, \qquad (7.21)$$

is a product of Givens rotations and

$$R_k = \begin{pmatrix} \rho_1 & \theta_2 & & & \\ & \rho_2 & \theta_3 & & \\ & & \rho_3 & \ddots & \\ & & & \ddots & \theta_k \\ & & & & \rho_k \end{pmatrix}, \quad f_k = \begin{pmatrix} \phi_1 \\ \phi_2 \\ \phi_3 \\ \vdots \\ \phi_k \end{pmatrix}. \qquad (7.22)$$

The solution and residual to (7.19) can then be computed from upper bidiagonal linear system

$$R_k y_k = d_k, \qquad s_k = \bar{\phi}_{k+1} Q_k e_{k+1}. \qquad (7.23)$$

The residual $A^T r_k$ to the normal equations will be zero for the exact solution x and this quantity can therefore be used as a stopping criterion. Using (7.15) we obtain

$$A^T r_k = \bar{\phi}_{k+1} A^T U_{k+1} Q_k e_{k+1} = \bar{\phi}_{k+1} \alpha_{k+1} c_k v_{k+1},$$

where $c_k = e_{k+1}^T Q_k e_{k+1}$, the $(k+1)$st diagonal element of Q_k, equals the element c_k in $G_{k,k+1}$. Hence the norm

$$\|A^T r_k\|_2 = \bar{\phi}_{k+1} \alpha_{k+1} |c_k| \qquad (7.24)$$

is cheaply computable.

Paige and Saunders (1982b) showed an ingenious way to interleave the solution of (7.19) with the reduction to bidiagonal form. The QR decomposition (7.20) can be efficiently updated. Assume that we have computed R_{k-1}, f_{k-1} and $\bar{\phi}_k$ in (7.21)–(7.22). To update these quantities when a column is added to B_k we first apply the Givens rotation to rows $k-1, k$ in the last column in B_k:

$$G_{k-1,k} \begin{pmatrix} 0 \\ \alpha_k \end{pmatrix} = \begin{pmatrix} \theta_k \\ \bar{\rho}_k \end{pmatrix}.$$

Next we construct and apply a Givens rotation $G_{k,k+1}$ to zero out the element β_{k+1}

$$G_{k,k+1}\begin{pmatrix} \bar{\rho}_k & \bar{\phi}_k \\ \beta_{k+1} & 0 \end{pmatrix} = \begin{pmatrix} \rho_k & \phi_k \\ 0 & \bar{\phi}_{k+1} \end{pmatrix}.$$

(Here only elements affected by the rotation are shown.)

From (7.23) we note that y_k will normally differ from y_{k-1} in all its elements. However, since R_k and f_k differ from R_{k-1} and f_{k-1} only in the last row and column, we can write

$$x_k = V_k R_k^{-1} f_k = D_k f_k = x_{k-1} + \phi_k d_k,$$

where $D_k = (d_1 \; d_2 \; \cdots \; d_k)$ is obtained from $R_k^T D_k = V_k^T$ by forward substitution, giving

$$d_k = \rho_k^{-1}(v_k - \theta_k d_{k-1}). \tag{7.25}$$

Only the last iterates d_k and x_k have to be saved. Although the residual $r_k = U_{k+1} s_k$ is not cheaply computable, by (7.23) its norm equals

$$\|r_k\|_2 = |\bar{\phi}_{k+1}|.$$

The Householder algorithm gives a backward stable algorithm for computing the sequence of Krylov subspace approximations x_k, $k = 1, 2, \ldots$. For problems where A is dense the cost is comparable to that for the Lanczos approach.

In many least squares problems the 'effective rank' of the problem is much smaller than n, i.e., a good approximate solution can be found in a subspace of much smaller dimension than n. For example, this is the case in multiple linear regression problems, where many columns of A are nearly linearly dependent. The Krylov subspace method described above is a standard tool for regression in chemometrics. In this context it is known as the partial least squares (PLS) method; see Wold, Ruhe, Wold and Dunn (1984). It is known that PLS often gives a faster reduction of the residual than TSVD; see Eldén (2004). PLS is often implemented by a deflation method called NIPALS (Nonlinear Iterative Partial Least Squares), or using the LSQR. Since many of these problems are neither sparse nor particularly large an implementation based on Householder bidiagonalization should be preferred.

8. Constrained and regularized problems

8.1. Constrained least squares problems

In various applications the solution to a least squares problem is required to satisfy a subsystem of linear equations exactly. This is the least squares problem with equality constraints (LSE)

$$\min \|b - Ax\|_2 \quad \text{subject to} \quad Bx = d, \tag{8.1}$$

where $A \in \mathbb{R}^{m \times n}$ and $B \in \mathbb{R}^{p \times n}$, with $m + p \geq n \geq p$. If we assume that

$$\text{rank}(B) = p, \qquad \mathcal{N}(A) \cap \mathcal{N}(B) = 0,$$

then this problem has a unique solution. A survey of solution methods is given in Björck (1996, Section 5.1). Perturbation bounds for problem LSE are derived in Eldén (1980). An analysis of the accuracy and stability of three different implementations of the null space method is given in Cox and Higham (1999a), where a forward error bound suitable for practical use is also derived.

Problem LSE arises, e.g., in the solution of inequality-constrained least squares problem (LSI):

$$\min \|b - Ax\|_2 \quad \text{subject to} \quad l \leq Bx \leq u, \qquad (8.2)$$

where the inequalities are to be interpreted componentwise, $l_i \leq (Bx)_i \leq u_i$, $i = 1:p$. We assume that linear equality constraints, if present, have been eliminated and that the set $\mathcal{M} = \{x \mid l \leq Bx \leq u\}$ is not empty.

A special case is the least distance problem (LDP)

$$\min \|x_1\|_2 \quad \text{subject to} \quad l \leq Bx \leq u, \qquad (8.3)$$

where $x^T = (x_1^T \ x_2^T)$.

Questions of existence, uniqueness and boundedness of solutions to problem LSI are given by Lötstedt (1983). It is convenient to split the solution into two mutually orthogonal components

$$x = x_R + x_N, \qquad x_R \in \mathcal{R}(A^T), \qquad x_N \in \mathcal{N}(A). \qquad (8.4)$$

The existence of a bounded solution x to (8.2) follows from the facts that the objective function $\|Ax - b\|_2$ is bounded below by 0 and that the constraint set $l \leq Cx \leq u$ is convex and polyhedral. It can further be shown that x_R and Ax are uniquely determined; see Lötstedt (1983, Theorem 1). In particular, if $\text{rank}(A) = n$ then $\mathcal{N}(A)$ is empty and the solution is unique. The sensitivity of the solution of problem LSI to perturbations in the data A, B, b is also studied in Lötstedt (1983).

An important special case of problem LSI is when the inequalities are simple bounds, problem BLS:

$$\min_{l \leq x \leq u} \|Ax - b\|_2. \qquad (8.5)$$

(Some lower and upper bounds may not be present.) For reasons of computational efficiency it is essential that such constraints be considered separately from more general constraints in (8.2). If $\text{rank}(A) = n$ the BLS problem is a strictly convex optimization problem and there exists a unique solution for any vector b.

Sometimes only one-sided bounds apply. After a shift these can then be transformed into $x \geq 0$ and we have a least squares problems with nonnegativity constraints (NNLS):

$$\min_{x \geq 0} \|Ax - b\|_2. \tag{8.6}$$

Problems BLS and NNLS arise naturally in many applications, e.g., reconstruction problems in geodesy and tomography, contact problems for mechanical systems, control problems, etc. It can often be argued that a linear model is only realistic when the variables are constrained within meaningful intervals.

To determine a unique solution for the BLS problem when $\text{rank}(A) < n$, we may look for a solution to the problem

$$\min_{x \in \mathcal{M}} \|x\|_2, \quad \mathcal{M} = \left\{ x \mid \min_{l \leq x \leq u} \|Ax - b\|_2 \right\}. \tag{8.7}$$

Lötstedt (1984) developed a two-stage algorithm to solve problem (8.7). In the first stage a particular solution x to (8.5) is determined. If x is decomposed according to (8.4) then x_R is uniquely determined, but any x_N such that x remains feasible is admissible. Since $\|x\|_2^2 = \|x_R\|_2^2 + \|x_N\|_2^2$ in the second stage we need to solve

$$\min_{l \leq x \leq u} \|x_N\|_2, \quad x_N \in \mathcal{N}(A). \tag{8.8}$$

Let

$$A^T = (U_1 \; U_2) \begin{pmatrix} S & 0 \\ 0 & 0 \end{pmatrix} \begin{pmatrix} V_1^T \\ V_2^T \end{pmatrix} \tag{8.9}$$

be a full orthogonal decomposition of A^T with S nonsingular (this could be the SVD of A^T). Then $\mathcal{R}(A^T)$ is spanned by U_1 and $\mathcal{N}(A)$ by U_2 and we have

$$x_R = U_1(U_1^T x), \quad x_N = U_2(U_2^T x) = U_2 z.$$

Since U_2 has orthonormal columns, problem (8.8) is equivalent to

$$\min \|z\|_2, \quad l - x_R \leq U_2 z \leq u - x_R, \tag{8.10}$$

which is a least distance problem for z.

In general, methods for problems with inequality constraints are iterative in nature. At the solution only a certain subset of the inequalities will be active, i.e., satisfied with equality. If this set was known the solution to the LSI problem could be found from a problem with equality constraints, for which efficient solution techniques exist. In active set methods a sequence of equality-constrained problems are solved corresponding to predictions of the correct active set.

An implementation of an active set method for NNLS is given in Lawson and Hanson (1995). This can also be used to solve problem LDP using a dual approach; see Cline (1975). Consider the least distance problem with lower bounds

$$\min_x \|x\|_2, \quad \text{subject to} \quad c \le Bx. \tag{8.11}$$

Let $u \in \mathbb{R}^{m+1}$ be the solution to the NNLS problem

$$\min_u \|Au - b\|_2, \quad \text{subject to} \quad u \ge 0, \tag{8.12}$$

where

$$A = \begin{pmatrix} B^T \\ c^T \end{pmatrix}, \quad b = \begin{pmatrix} 0 \\ 1 \end{pmatrix} \begin{matrix} \}n \\ \}1 \end{matrix}. \tag{8.13}$$

Let the residual corresponding to the solution be

$$r \equiv b - Au = \begin{pmatrix} r_1 \\ \gamma \end{pmatrix} \begin{matrix} \}n \\ \}1 \end{matrix}, \quad \sigma = \|r\|_2.$$

If $\sigma \ne 0$, then the vector $x = -r_1/\gamma$, is the unique solution to (8.11). If $\sigma = 0$, then the constraints $g \le Bx$ are inconsistent and (8.11) has no solution. Hence this relation also gives a method to determine if a set of linear inequalities has a feasible solution by solving an NNLS problem.

Problem LSI can be transformed into an LDP problem by using the orthogonal decomposition (8.9). An algorithm based on this transformation and the dual approach for solving the LDP problem is given by Lawson and Hanson (1995, Chapter 23) (see also Haskell and Hanson (1981)). The method proposed by Schittkowski (1983) for solving the LDP problem is a primal method.

8.2. Regularization of discrete ill-posed problems

Inverse problems are problems where we want to determine the structure of a physical system from its measured behaviour. Such problems are often ill-posed in the sense that their solution does not depend continuously on the data. Inverse problems arise in many application such as astronomy, computerized tomography, geophysics, signal processing, *etc.*

The discretization of ill-posed problems gives rise to a class of least squares problems

$$\min_x \|Ax - b\|_2, \quad A \in \mathbb{R}^{m \times n}, \tag{8.14}$$

that share a number of properties. The singular values of A decay gradually and cluster at zero, resulting in a huge condition number; *cf.* Figure 6.1. However, the components of the right-hand side b along singular vectors corresponding to small singular values decay rapidly, so that the systems are effectively well-conditioned in the sense of Chan and Foulser (1988).

Owing to the huge condition number and the presence of noise in the right-hand side b, additional information, e.g., in the form of constraints, must be imposed on the solution in order to get a regularized problem with a well-determined solution. Neglecting this can be catastrophic, since it may lead to a meaningless solution of huge norm, or even to failure of the algorithm.

One common regularization method is to project the linear system onto a smaller dimensional problem by solving

$$\min_{x \in \mathcal{V}_k} \|b - Ax\|_2,$$

where $\mathcal{V}_k$ is a suitably chosen subspace of dimension $k < n$. One possible choice is the subspace spanned by the first k right-singular vectors of A, which leads to a TSVD solution (3.11). Another choice is to use the Krylov subspaces $\mathcal{V}_k = \mathcal{K}_k(A^T A, A^T u_1)$ (7.17), which corresponds to using LSQR or the PLS method. There is some evidence that for ill-posed problems the Krylov subspaces often have better approximation properties than the singular subspaces used in TSVD; see Hanke (2001).

Another widely used regularization method is Tikhonov regularization (Tikhonov 1963). In this method an approximate solution is obtained by solving a least squares problem with a quadratic constraint

$$\min_x \|Ax - b\|_2 \quad \text{subject to} \quad \|Lx\|_2 \leq \gamma, \quad L \in \mathbb{R}^{p \times n}. \tag{8.15}$$

Here $\gamma > 0$ is the regularization parameter, which is used to find a balance between the size of the residual $\|Ax - b\|_2$ and size of the solution as measured by the norm (or seminorm) $\|Lx\|_2$. This makes it possible to include *a priori* information about the size or smoothness of the solution. In statistics Tikhonov's method is known as ridge regression. A survey of the properties of least squares problems with a quadratic constraint is given by Gander (1981). Problem (8.15) is related to, but less general than, the trust-region subproblem in optimization; see Rojas and Sorensen (2002).

In the following we discuss the simple but important case when $L = I$. Methods to transform (8.15) into this standard form are described in Hansen (1998, Section 2.3). In (8.15) the constraint is binding if $\|A^\dagger b\|_2 > \gamma$, which is invariably the case in regularization of ill-posed problems. Then $x = x(\lambda)$ solves the least squares problem

$$\min_x \left\| \begin{pmatrix} A \\ \sqrt{\lambda} L \end{pmatrix} x - \begin{pmatrix} b \\ 0 \end{pmatrix} \right\|_2, \tag{8.16}$$

where the Lagrange multiplier λ is determined by the secular equation $g(\lambda) = \|x(\lambda)\|_2 - \gamma = 0$.

Using the singular value decomposition $A = U\Sigma V^T$ and setting $c = U^T b$, the secular equation becomes

$$g(\lambda) = \left(\sum_{i=1}^{n} \frac{\sigma_i^2 c_i^2}{(\sigma_i^2 + \lambda)^2}\right)^{1/2} - \gamma = 0. \qquad (8.17)$$

Here each term is a convex and strictly decreasing function of λ. Since the Euclidean norm is monotonic, the same then also holds for $g(\lambda)$. This shows that if $\gamma < g(0)$ then the secular equation has a unique root $\lambda > 0$.

To determine λ by an iterative method, faster convergence is obtained by writing the secular equation in the form

$$f(\lambda) = \frac{1}{\|x(\lambda)\|_2} - \frac{1}{\gamma} = 0. \qquad (8.18)$$

Reinsch (1971) showed that $f(\lambda)$ is a concave and strictly increasing function for $\lambda > 0$. It follows that Newton's method is monotonically convergent to the solution λ^* from any starting value $\lambda_0 \in [0, \lambda^*]$. Usually about 4–6 iterations suffice, even when $\lambda_0 \ll \lambda^*$.

Writing $x(\lambda) = (A^T A + \lambda I)^{-1} A^T b$ and taking the derivative with respect to λ, we find that

$$f'(\lambda) = -\frac{x^T(\lambda) x'(\lambda)}{\|x(\lambda)\|_2^3}, \qquad x'(\lambda) = -(A^T A + \lambda I)^{-1} x(\lambda). \qquad (8.19)$$

Here

$$x(\lambda)^T x(\lambda)' = -x(\lambda)^T (A^T A + \lambda I)^{-1} x(\lambda) = -\|z(\lambda)\|_2^2, \qquad (8.20)$$

and Newton's method for equation (8.18) becomes

$$\lambda_{k+1} = \lambda_k + \left(\frac{\|x(\lambda_k)\|_2}{\gamma} - 1\right) \frac{\|x(\lambda_k)\|_2^2}{\|z(\lambda_k)\|_2^2}. \qquad (8.21)$$

Given the QR decomposition

$$Q(\lambda_k)^T \begin{pmatrix} A \\ \sqrt{\lambda_k} I \end{pmatrix} = \begin{pmatrix} R(\lambda_k) \\ 0 \end{pmatrix}, \qquad Q(\lambda_k)^T \begin{pmatrix} b \\ 0 \end{pmatrix} = \begin{pmatrix} c_1(\lambda_k) \\ c_2(\lambda_k) \end{pmatrix}, \qquad (8.22)$$

and using (8.20) we obtain

$$x(\lambda_k) = R(\lambda_k)^{-1} c_1(\lambda_k), \qquad z(\lambda_k) = R(\lambda_k)^{-T} x(\lambda_k).$$

The main cost per iteration step in Newton's method is the QR decomposition (8.22). Computing the derivative costs only one triangular solve. Hence Newton's method is to be preferred to the secant method and other methods based on interpolation.

When $A \in \mathbb{R}^{m \times n}$ is a full matrix computing the Householder QR decomposition in each iteration requires about mn^2 multiplications. As pointed out by Moré (1978), if $m > n$ it is more efficient to initially compute the

QR of A at a cost of $mn^2 - n^3/3$ multiplications. For each value λ_k we can then compute $R(\lambda_k)$ from the QR decomposition of

$$\begin{pmatrix} R \\ \sqrt{\lambda_k} I \end{pmatrix}$$

in $n^3/3$ multiplications. However, if A is sparse R may have many more nonzero elements than A and then this modification will instead *increase* the amount of work.

The repeated QR decomposition can be avoided by computing the SVD or a bidiagonal decomposition of A. The bidiagonal form can be updated with $O(n)$ multiplications when λ changes, as shown by Eldén (1977). Since the initial reduction requires $4n^3/3$ flops ($2n^3/3$ if $m = n$) the reduction to bidiagonal form pays off in the dense case if more than four Newton iterations are needed.

9. Direct methods for sparse problems

A matrix A is called sparse if many of its entries are zero. Clearly a square and banded matrix is sparse, but sparse matrices that have much more irregular sparsity pattern occur in many applications in science and engineering. Often these problems are huge and it is essential that advantage is taken of sparsity for storage and operations. Matrix operations on general sparse matrices are supported in MATLAB: see Gilbert, Moler and Schreiber (1992) for a discussion of design and implementation of these algorithms.

9.1. QR factorization of banded matrices

A banded matrix $A \in \mathbb{R}^{m \times n}$, $m \geq n$, is a matrix for which in each row the nonzero elements lie in a narrow band. By the bandwidth of A we mean smallest number w such that

$$|j - k| \geq w \quad \Rightarrow \quad a_{ij} a_{ik} = 0 \quad i = 1 : n. \tag{9.1}$$

It is easy to deduce that if A has bandwidth w then the upper triangular part of $A^T A$ and its Cholesky factor R also have bandwidth w. Forming $A^T A$ and computing its Cholesky factorization therefore requires only about $\frac{1}{2}(m+n)w^2$ multiplications. The QR decomposition of a banded matrix can also be computed efficiently, but the implementation is not quite trivial! The standard Householder QR factorization algorithm can be very inefficient and cause unnecessary intermediate fill-in. Similarly, if a row-wise reduction with Givens rotations is used the operation count and intermediate storage requirement can differ strongly for different row orderings of A.

Reid (1967) showed that for banded rectangular matrices the QR factorization can be obtained very efficiently by sorting the rows of A and suitably subdividing the Householder transformations. The rows of A are first sorted by leading entry order (i.e., increasing minimum column subscript order) so that the matrix is represented as q blocks

$$A = \begin{pmatrix} A_1 \\ A_2 \\ \vdots \\ A_q \end{pmatrix}, \quad q \leq n,$$

where in block A_i the first nonzero element of each row is in column i. The Householder QR process is then applied to the matrix in q major steps. In the first step a QR decomposition of the first block A_1 is computed, yielding R_1. Next, at step k, $k = 2 : q$, R_{k-1} will be merged with A_k, yielding

$$Q_k^T \begin{pmatrix} R_{k-1} \\ A_k \end{pmatrix} = R_k.$$

Since the rows of block A_k have their first nonzero elements in column k, the first $k-1$ rows of R_{k-1} will not be affected. The matrix Q can be implicitly represented in terms of the Householder vectors of the factorization of the subblocks. This sequential Householder algorithm, which is also described in Lawson and Hanson (1995, Chapter 27), requires $(m + 3n/2)w(w + 1)$ multiplications or about twice the work of the less stable Cholesky approach.

A banded upper triangular matrix can be reduced to bidiagonal form using an algorithm similar to the one used by Schwarz (1968) for reducing a symmetric banded matrix to tridiagonal form. However, because each zero element introduced generates a new nonzero element that has to be 'chased' across the border of the matrix, the reduction is much more expensive than the QR decomposition The reduction of a banded upper triangular matrix to bidiagonal form requires $\approx 4n^2(w-2)$ multiplications. A computational routine for this called xGBBRD in LAPACK uses a vectorized version due to Kaufman (1984), in which several elements are chased in parallel.

A special case occurs in regularization, where we need the repeated QR decomposition of

$$\begin{pmatrix} R_1 \\ \sqrt{\lambda} R_2 \end{pmatrix},$$

for banded matrices R_1 and R_2. Note that if the above row ordering algorithm is applied this will interleave the rows of R_1 and R_2. Eldén (1984) gives a row-wise Givens algorithm which requires approximately $2n(w_1^2 + w_2^2)$ multiplications and is optimal in that no unnecessary fill-in is created.

9.2. Block angular least squares problems

There is often a substantial similarity in the structure of many large-scale sparse least squares problems. In particular, the problem can often be put in the following bordered block diagonal or block angular form:

$$A = \begin{pmatrix} A_1 & & & & B_1 \\ & A_2 & & & B_2 \\ & & \ddots & & \vdots \\ & & & A_M & B_M \end{pmatrix}, \quad x = \begin{pmatrix} x_1 \\ x_2 \\ \vdots \\ x_M \\ \hline z \end{pmatrix}, \quad b = \begin{pmatrix} b_1 \\ b_2 \\ \vdots \\ b_M \end{pmatrix}, \quad (9.2)$$

where $A \in \mathbb{R}^{m \times n}$,

$$A_i \in \mathbb{R}^{m_i \times n_i}, \quad B_i \in \mathbb{R}^{m_i \times p}, \quad i = 1, 2, \ldots, M.$$

We assume in the following, for simplicity, that $\text{rank}(A) = n$. Note that the variables $x_1, \ldots, x_M$ are coupled only to the variables z, which reflects a 'local connection' structure in the underlying physical problem. There is usually further structure in the individual blocks A_i and B_i that should be taken advantage of.

Applications where the form (9.2) arises naturally in many applications including photogrammetry (Golub, Luk and Pagano 1979), Doppler radar positioning (Manneback, Murigande and Toint 1985), geodetic survey problems (Golub and Plemmons 1980) and GPS positioning (Chang and Paige 2003),

It is easily seen that then the factor R in the QR decomposition of A will have the block structure

$$R = \begin{pmatrix} R_1 & & & & S_1 \\ & R_2 & & & S_2 \\ & & \ddots & & \vdots \\ & & & R_M & S_M \\ \hline & & & & R_{M+1} \end{pmatrix}, \quad (9.3)$$

where by assumption $R_i \in \mathbb{R}^{n_i \times \tilde{n}_i}$, $i = 1, \ldots, M+1$ is nonsingular.

The following algorithm for solving least squares problems of block angular form by QR decomposition is given in Golub et al. (1979).

(1) For $i = 1, 2, \ldots, M$ reduce the diagonal block A_i to upper triangular form by a sequence of orthogonal transformations applied to (A_i, B_i) and the right-hand side b_i, yielding

$$Q_i^T (A_i, B_i) = \begin{pmatrix} R_i & S_i \\ 0 & T_i \end{pmatrix}, \quad Q_i^T b_i = \begin{pmatrix} c_i \\ d_i \end{pmatrix}.$$

It is usually advantageous to continue the reduction in step (1) so that the matrices T_i, $i = 1, \ldots, M$, are brought into upper trapezoidal form.

(2) Compute the QR decomposition

$$\tilde{Q}_{M+1}^T \begin{pmatrix} T_1 \\ \vdots \\ T_M \end{pmatrix} = \begin{pmatrix} R_{M+1} \\ 0 \end{pmatrix}, \quad \tilde{Q}_{M+1}^T \begin{pmatrix} d_1 \\ \vdots \\ d_M \end{pmatrix} = \begin{pmatrix} c_{M+1} \\ d_{M+1} \end{pmatrix}.$$

Then the linking variables z are obtained from the triangular system

$$R_{M+1} z = c_{M+1}$$

and the residual norm equals $\rho = \|d_{M+1}\|_2$.

(3) For $i = M, \ldots, 1$ compute $x_M, \ldots, x_1$ by back-substitution in the triangular systems

$$R_i x_i = c_i - S_i z.$$

Note that in steps (1) and (3) the computations can be performed in parallel on the M independent subsystems.

There are many alternative ways to organize this algorithm. Cox (1990) considers the following modifications to reduce the storage requirement. By merging steps (1) and (2) it is not necessary to hold all blocks T_i simultaneously in memory. Even more storage can be saved by discarding R_i and S_i after T_i has been computed in step (1). These matrices are then recomputed when needed for step (3). Indeed, only R_i needs to be recomputed, since when z has been computed in step (2), we can determine x_i by solving the least squares problems

$$\min_{x_i} \|A_i x_i - g_i\|_2, \quad g_i = b_i - B_i z, \quad i = 1, \ldots, M.$$

Hence, to determine x_i we only need to (re-)compute the QR factorization of (A_i, g_i). In some practical problems this modification can reduce the storage requirement by an order of magnitude, while the recomputation of R_i only increases the operation count by a few per cent.

9.3. QR decomposition for general sparse matrices

The factor R in the QR decomposition of A is mathematically equivalent to the Cholesky factor of the cross product matrix $A^T A$. Although this is true in exact arithmetic the difficulty in recognizing numerical cancellation means that the computed structure of the Cholesky factor can overestimate the structure of R. However, in many cases it accurately predicts the structure of R.

A reordering of the rows and columns of A can be written $\hat{A} = P_r A P_c$, where P_r and P_c are permutation matrices. Since

$$\hat{A}^T \hat{A} = P_c^T A^T P_r^T P_r A P_c = P_c^T A^T A P_c,$$

this corresponds to a symmetric reordering of $A^T A$. Hence the column ordering of A will affect the structure and number of nonzeros nnz(R) in R. A reordering of the rows in A has no effect on the final R, but influences the sparsity of Q and the number of operations needed to perform the decomposition.

Before computing R numerically, it is important to find a to find a column ordering that approximately minimizes the number of nonzero elements in R. (Finding the optimal ordering is known to be an NP-complete problem.) The simplest ordering methods use *a priori* information, such as ordering the columns in order of increasing column count. Such orderings are usually inferior to ordering methods obtained from a symmetric ordering on the structure of the normal matrix, using minimum degree or nested dissection. The graph $G(A^T A)$ representing the structure of $A^T A$ can be constructed directly from the structure of the matrix A as being the direct sum of all the subgraphs $G(a_i a_i^T)$, $i = 1, \ldots, m$. Note that the nonzeros in any row a_i^T will generate a subgraph where all pairs of nodes are connected. Good surveys of the state of the art in symmetric ordering algorithms and direct methods for sparse linear systems are given in Duff (1997) and Dongarra *et al.* (1998, Chapter 6).

After a column ordering P_c has been determined, the rows in the permuted matrix AP_c should be reordered to minimize intermediate fill-in. The following heuristic row ordering method usually works well. Denote the column index for the first and last nonzero elements in the ith row by f_i and l_i, respectively. The rows are ordered by increasing values of f_i, and in each group of rows with equal f_i after increasing l_i. (This is the row ordering used for rectangular banded matrices.)

For the numerical QR decomposition several implementations of multifrontal methods for QR have been developed; see Matstoms (1995), Amestoy, Duff and Puglisi (1996). Multifrontal methods have the advantage of allowing dense matrix kernels to be used in the sparse matrix code and can be considered as a generalization of methods for banded matrices.

A problem with the QR decomposition is that for a large class of sparse matrices the matrix Q will be much less sparse than R. If $A \in \mathbb{R}^{m \times n}$ is a matrix of full column rank, such that its column intersection graph is a member of a $\sqrt{n}$-separable class of graphs, then it is shown in Gilbert, Ng and Peyton (1997) that there exists a column permutation P such that

$$\text{nnz}(R) = O(n \log n), \quad \text{nnz}(Q_1) = O(n\sqrt{n}),$$

where $Q = (Q_1 \; Q_2)$. These bounds are best possible within a constant factor for a large class of matrices. It is therefore not advisable to store Q_1 (or Q) explicitly. In MATLAB sparse QR the matrix Q *is* provided if wanted, but this option should be used with care!

Because of the lack of sparsity in Q a common practice when solving sparse least squares problems is to compute $Q^T b$ 'on the fly' for any right-hand side available at the time of decomposition and discard the orthogonal transformations after they have been used. This approach, advocated in George and Heath (1980), is also taken in several later multifrontal QR codes; see Matstoms (1994), Sun (1996) and Pierce and Lewis (1997).

To discard Q creates a problem if additional right-hand sides b are to be treated later. George and Heath (1980) suggested that if the original matrix A is saved one can use R from the QR decomposition, rewriting the normal equations as

$$R^T R x = A^T b, \qquad (9.4)$$

known as the seminormal equations (SNE). Even with R from the QR decomposition in (9.4) this is not an acceptable-error stable algorithm for the least squares problem. However, if improved by one step of iterative refinement in fixed precision it is, in most cases, numerically acceptable. A detailed error analysis of this algorithm, called CSNE, is given in Björck (1987).

Lu and Barlow (1996) use the multifrontal QR factorization method and represent Q implicitly by storing the frontal Householder vectors. Let A_i be the matrix consisting of those rows of A that have their leading nonzero element in column i. Provided that the number of rows of each A_i is bounded by a constant, Lu and Barlow (1996) show that their method requires only $O(n \log n)$ storage if $A \in \mathbb{R}^{m \times n}$ is a member of a $\sqrt{n}$-separable class of graphs. Adlers (2000) has developed a similar implementation for MATLAB, which provides Q by storing the frontal Householder vectors. Operator overloading is used to implement the matrix–vector products Qy and $Q^T y$.

10. Iterative methods

10.1. Introduction

For several classes of large sparse least squares problems iterative methods are useful alternatives to direct methods. Iterative methods for the linear least squares problem (1.1) can be derived by applying an iterative method for symmetric positive definite linear systems to the normal equations $A^T A x = A^T b$. However, it is important to avoid the explicit formation of $A^T A$ since this leads to a loss of stability. Also, $A^T A$ can be much less sparse than A, leading to higher cost of storage and operations. Instead the factored form of the normal equations $A^T(Ax - b) = 0$ should be employed. This is well illustrated by the non-stationary Richardson's method for least squares problems, which should be written

$$x_{k+1} = x_k + \omega_k A^T r_k, \quad r_k = b - Ax_k. \tag{10.1}$$

This method, also known as Landweber's method, can be shown to converge, provided that, for some $\epsilon > 0$,

$$0 < \epsilon < \omega_k < (2 - \epsilon)/\sigma_{\max}^2(A), \quad \forall k.$$

As is typical for iterative methods for least squares problems the matrix A need only accessed through its action in the matrix–vector operations Ax_k and $A^T r_k$. This can also be an advantage for problems where A is dense but structured. Consider, e.g., a rectangular Toeplitz matrix

$$T = \begin{pmatrix} t_0 & t_1 & \cdots & t_n \\ t_{-1} & t_0 & \cdots & t_{n-1} \\ \vdots & \vdots & \ddots & \vdots \\ t_{-m} & t_{-m+1} & \cdots & t_0 \end{pmatrix} \in \mathbb{R}^{(m+1) \times (n+1)}, \quad \text{for } m \geq n,$$

defined by the $(n+m+1)$ values of $t_{-m}, \ldots, t_0, \ldots, t_n$. Toeplitz linear least squares problems of large dimension arise in many applications, e.g., in signal restoration, acoustics, and seismic exploration. Matrix–vector products Tx_k and $T^T r_k$ can be computed at a cost of $O(m \log m)$ by embedding T in a circulant matrix and using FFT; see Chan, Nagy and Plemmons (1994). Fast multiplication algorithms also exist for several other classes of structured matrices, e.g., Vandermonde and Cauchy matrices; see Gohberg and Olshevsky (1994).

10.2. Implementation of Krylov subspace methods

In Sections 7.4–7.5 we outlined the LSQR Krylov subspace algorithm for the least squares problem. A different way to compute the same sequence of approximations x_k is to use the conjugate gradient method (CG), developed in the early 1950s. This has become a basic tool for solving large sparse symmetric positive definite linear systems. If applied to the normal equations it can also be used to solve linear least squares problems. The conjugate gradient algorithm for the normal equations generates approximations x_k in the Krylov subspace

$$x_k \in \mathcal{K}_k(A^T A, s_0), \quad s_0 = A^T b. \tag{10.2}$$

The iterates x_k generated are optimal in the sense that, for each k, $y = x_k$ minimizes the error functional

$$E(y) = (x - y)^T (A^T A)(x - y), \quad y \in \mathcal{K}_k(A^T A, s_0). \tag{10.3}$$

Using $A(x - x_k) = b - r - Ax_k = r_k - r$, we obtain

$$E(x_k) = \|r - r_k\|^2 = \|r_k\|^2 - \|r\|^2, \qquad (10.4)$$

where the second expression follows from the fact that $r \perp r - r_k$.

The following version of CGLS was originally given by Hestenes and Stiefel (1952, p. 424) and Stiefel (1952/53).

Initialize

$$r_0 = b, \quad s_0 = p_1 = A^T r_0, \quad \gamma_0 = \|s_0\|_2^2, \qquad (10.5)$$

and for $k = 1, 2, \ldots$ compute

$$q_k = Ap_k,$$
$$\alpha_k = \gamma_{k-1}/\|q_k\|_2^2,$$
$$x_k = x_{k-1} + \alpha_k p_k,$$
$$r_k = r_{k-1} - \alpha_k q_k, \quad s_k = A^T r_k,$$
$$\gamma_k = \|s_k\|_2^2,$$
$$\beta_k = \gamma_k/\gamma_{k-1},$$
$$p_{k+1} = s_k + \beta_k p_k.$$

Each iteration requires two matrix vector products and $2m + 3n$ multiplications. Storage is required for the n-vectors x, p and m-vectors r, q.

The variational property of the conjugate gradient method implies that, in exact arithmetic, the error functional $\|r - r_k\|_2$ (as well as $\|r_k\|_2$) decreases monotonically as a function of k. In Hestenes and Stiefel (1952, p. 416) it is proved that the error functional $\|x - x_k\|_2$ also decreases monotonically. However, when $\kappa(A)$ is large, $\|A^T(r - r_k)\|_2$ will often exhibit large oscillations. We stress that this behaviour is *not* a result of rounding errors.

If A has $t \leq n$ distinct singular values, then (in exact arithmetic) the solution is obtained in at most t steps. It is well known that an upper bound on the rate of convergence is given by

$$\|r - r_k\|_2^2 \leq 2\left(\frac{\kappa - 1}{\kappa + 1}\right)^k \|r_0\|_2^2, \qquad (10.6)$$

where $\kappa = \kappa(A)$; see Björck (1996, Section 7.4). The convergence also depends on the distribution of the singular values of A and often superlinear convergence is observed.

The LSQR algorithm (Paige and Saunders 1982a) generates, in exact arithmetic, the same sequence of approximations x_k as CGLS. In LSQR the Lanczos process described in Section 7.3 is used to compute u_k, v_k and B_k, for $k = 1, 2, \ldots$. This is interleaved with the solution of the bidiagonal systems as described in Section 7.5.

In addition to two matrix vector products LSQR requires $3m + 5n$ multiplications and storage of two m-vectors u, Av, and three n-vectors x, v, w. Although this is slightly more storage and operations than CGLS, this is partly offset by the fact that viable rules for stopping the iterations are more costly for CGLS than for LSQR. For LSQR Paige and Saunders (1982b) consider several stopping rules which use (estimates of) $\|r_k\|, \|x_k\|, \|A^T r_k\|, \|A\|$, and $\|A^\dagger\|$. All these quantities can be obtained at minimal cost in LSQR. For CGLS $\|s_k\|$ is available but $\|r_k\|$ has to be separately computed, if needed, at an extra cost of m multiplications.

In finite precision orthogonality will be lost for U_k and V_k. This causes a slowdown of convergence, but does not affect the final accuracy. A comparison of the stability of LSQR and CGLS is given in Björck, Elfving and Strakoš (1998). It may be believed that LSQR, since its derivation avoids any references to the normal equations, should have better stability properties than CGLS. To some extent that is true, but the difference is rather small. Björck et al. (1998) compares the achievable accuracy in finite precision of different implementations of Krylov subspace methods for solving (1.1). The conclusion from both theoretical analysis and experimental evidence is that LSQR and CGLS are both well behaved and achieve a final accuracy consistent with a backward stable method.

A slightly different version of CGLS is obtained if, instead of r_k, the residual of the normal equations $s_k = A^T r_k$ is recurred. For this, line 4 in algorithm CGLS becomes

$$s_k = s_{k-1} - \alpha_k(A^T q_k).$$

As shown in Björck et al. (1998), this small change can substantially lower the achievable final accuracy in CGLS. This can be explained by noting that in this version the right-hand side b is used only in the initialization $p_0 = s_0 = A^T b$ and no reference to b is made in the iterative phase. A componentwise round-off analysis shows that round-off occurring in computing $A^T b$ perturbs the solution by δx, where

$$|\delta x| \leq \gamma_m |(A^T A)^{-1}| |A^T| |b|. \tag{10.7}$$

Using norms we obtain the bound

$$\|\delta x\|_2 \leq \gamma_m \|(A^T A)^{-1}\|_2 \|A^T\|_2 \|b\|_2 = \gamma_m \kappa^2(A) \|b\|_2 / \|A\|_2, \tag{10.8}$$

where $\gamma_m = mu/(1 - mu)$. Comparing with the perturbation bounds in Section 2 we see that this error term is a factor $\|b\|_2/\|r\|_2$ larger than is allowed for a backward stable algorithm. Note that for nearly consistent systems we obtain $\|b\|_2 \gg \|r\|_2$.

REFERENCES

M. Adlers (2000), Topics in sparse least squares problems, PhD thesis, Linköping Studies in Science and Technology, Linköping University, Sweden.

P. R. Amestoy, I. S. Duff and C. Puglisi (1996), 'Multifrontal QR factorization in a multiprocessor environment', *Numer. Linear Algebra Appl.* **3**, 275–300.

E. Anderson, Z. Bai, C. Bischof, J. Demmel, J. Dongarra, J. Du Croz, A. Greenbaum, S. Hammarling, A. McKenney, S. Ostrouchov and D. Sorensen, eds (1995), *LAPACK Users' Guide*, 2nd edn, SIAM, Philadelphia.

E. Anderssen, Z. Bai and J. Dongarra (1992), 'Generalized QR factorization and its applications', *Linear Algebra Appl.* **162–164**, 243–271.

M. Arioli (2000), 'The use of QR factorization in sparse quadratic programming and backward error issues', *SIAM J. Matrix Anal. Appl.* **21**, 825–839.

Z. Bai and J. W. Demmel (1993), 'Computing the generalized singular value decomposition', *SIAM J. Sci. Comput.* **14**, 1464–1486.

Z. Bai, J. Demmel, J. Dongarra, A. Ruhe and H. van der Vorst, eds (2000), *Templates for the Solution of Algebraic Eigenvalue Problems: A Practical Guide*, SIAM, Philadelphia, PA.

C. Benoit (1924), 'Sur la méthode de résolution des équations normales, etc. (procédés du commandant Cholesky)', *Bull. Géodésique* **2**, 67–77.

C. H. Bischof and G. Quintana-Ortí (1998a), 'Algorithm 782: Codes for rank-revealing QR factorizations of dense matrices', *ACM Trans. Math. Software* **24**, 254–257.

C. H. Bischof and G. Quintana-Ortí (1998b), 'Computing rank-revealing QR factorizations of dense matrices', *ACM Trans. Math. Software* **24**, 226–253.

Å. Björck (1967a), 'Iterative refinement of linear least squares solutions, I', *BIT* **7**, 257–278.

Å. Björck (1967b), 'Solving linear least squares problems by Gram–Schmidt orthogonalization', *BIT* **7**, 1–21.

Å. Björck (1987), 'Stability analysis of the method of semi-normal equations for least squares problems', *Linear Algebra Appl.* **88/89**, 31–48.

Å. Björck (1991), Error analysis of least squares algorithms, in *Numerical Linear Algebra, Digital Signal Processing and Parallel Algorithms* (G. H. Golub and P. Van Dooren, eds), Vol. 70 of *NATO ASI Series*, Springer, Berlin, pp. 41–73.

Å. Björck (1994), 'Numerics of Gram–Schmidt orthogonalization', *Linear Algebra Appl.* **197–198**, 297–316.

Å. Björck (1996), *Numerical Methods for Least Squares Problems*, SIAM, Philadelphia.

Å. Björck and G. H. Golub (1967), 'Iterative refinement of linear least squares solution by Householder transformation', *BIT* **7**, 322–337.

Å. Björck and C. C. Paige (1992), 'Loss and recapture of orthogonality in the modified Gram–Schmidt algorithm', *SIAM J. Matrix Anal. Appl.* **13**, 176–190.

Å. Björck and C. C. Paige (1994), 'Solution of augmented linear systems using orthogonal factorizations', *BIT* **34**, 1–26.

Å. Björck, T. Elfving and Z. Strakoš (1998), 'Stability of conjugate gradient and Lanczos methods for linear least squares problems', *SIAM J. Matrix Anal. Appl.* **19**, 720–736.

P. Businger and G. H. Golub (1965), 'Linear least squares solutions by Householder transformations', *Numer. Math.* **7**, 269–276.

R. H. Chan, J. G. Nagy and R. J. Plemmons (1994), 'Circulant preconditioned Toeplitz least squares iterations', *SIAM J. Matrix Anal. Appl.* **15**, 80–97.

T. F. Chan (1987), 'Rank revealing QR-factorizations', *Linear Algebra Appl.* **88/89**, 67–82.

T. F. Chan and D. E. Foulser (1988), 'Effectively well-conditioned linear systems', *SIAM J. Sci. Statist. Comput.* **9**, 963–969.

T. F. Chan and P. C. Hansen (1990), 'Computing truncated SVD least squares solutions by rank revealing QR factorizations', *SIAM J. Sci. Statist. Comput.* **11**, 519–530.

T. F. Chan and P. C. Hansen (1992), 'Some applications of the rank revealing QR factorization', *SIAM J. Sci. Statist. Comput.* **13**, 727–741.

T. F. Chan and P. C. Hansen (1994), 'Low-rank revealing QR factorizations', *Numer. Linear Algebra Appl.* **1**, 33–44.

S. Chandrasekaran and I. C. F. Ipsen (1994), 'On rank-revealing factorizations', *SIAM J. Matrix Anal. Appl.* **15**, 592–622.

X.-W. Chang and C. C. Paige (2003), 'An orthogonal transformation algorithm for GPS positioning', *SIAM J. Sci. Comput.* **24**, 1710–1732.

A. K. Cline (1975), The transformation of a quadratic programming problem into solvable form, Technical Report ICASE 75-14, NASA, Langley Research Center, Hampton, VA.

A. J. Cox and N. J. Higham (1999a), 'Accuracy and stability of the null space method for solving equality constrained least squares problems', *BIT* **39**, 34–50.

A. J. Cox and N. J. Higham (1999b), 'Backward error bounds for constrained least squares problems', *BIT* **39**, 210–227.

M. G. Cox (1990), The least-squares solution of linear equations with block-angular observation matrix, in *Reliable Numerical Computation* (M. G. Cox and S. J. Hammarling, eds), Oxford University Press, UK, pp. 227–240.

B. De Moor and P. Van Dooren (1992), 'Generalizations of the singular value and QR decompositions', *SIAM J. Matrix. Anal. Appl.* **13**, 993–1014.

B. De Moor and H. Zha (1991), 'A tree of generalizations of the ordinary singular value decomposition', *Linear Algebra Appl.* **147**, 469–500.

J. Dongarra, I. S. Duff, D. Sorensen and H. van der Vorst (1998), *Numerical Linear Algebra for High-Performance Computers*, SIAM, Philadelphia, PA.

J. J. Dongarra, J. Du Croz, I. S. Duff and S. Hammarling (1990), 'A set of level 3 basic linear algebra subprograms', *ACM Trans. Math. Software* **16**, 1–17.

I. S. Duff (1997), Sparse numerical linear algebra: Direct methods and preconditioning, in *The State of the Art in Numerical Analysis* (I. S. Duff and G. A. Watson, eds), Oxford University Press, London, UK, pp. 27–62.

L. Eldén (1977), 'Algorithms for the regularization of ill-conditioned least squares problems', *BIT* **17**, 134–145.

L. Eldén (1980), 'Perturbation theory for the least squares problem with linear equality constraints', *SIAM J. Numer. Anal.* **17**, 338–350.

L. Eldén (1984), 'An efficient algorithm for the regularization of ill-conditioned least squares problems with a triangular Toeplitz matrix', *SIAM J. Sci. Statist. Comput.* **5**, 229–236.

L. Eldén (2004), 'Partial least squares vs Lanczos bidiagonalization, I: Analysis of a projection method for multiple projection', *Comp. Stat. Data Anal.* **46**, 11–31.

E. Elmroth and F. G. Gustavson (2000), 'Applying recursion to serial and parallel QR factorization leads to better performance', *IBM J. Res. Develop.* **44**, 605–624.

E. Elmroth and F. G. Gustavson (2001), 'A faster and simpler recursive algorithm for the LAPACK routine DGELS', *BIT* **41**, 936–949.

E. Elmroth, F. G. Gustavson, I. Jonsson and B. Kågström (2004), 'Recursive blocked algorithms and hybrid data structures for dense matrix library software', *SIAM Review* **46**, 3–45.

D. K. Faddeev, V. N. Kublanovskaya and V. N. Faddeeva (1968), 'Solution of linear algebraic systems with rectangular matrices', *Proc. Steklov Inst. Math.* **96**, 93–111.

K. V. Fernando (1998), 'Accurately counting singular values of bidiagonal matrices and eigenvalues of skew-symmetric tridiagonal matrices', *SIAM J. Matrix Anal. Appl.* **20**, 373–399.

R. D. Fierro, P. Hansen, and P. S. K. Hansen (1999), 'UTV tools: Matlab templates for rank revealing UTV decompositions', *Numer. Algorithms* **20**, 165–194.

W. Gander (1981), 'Least squares with a quadratic constraint', *Numer. Math.* **36**, 291–307.

J. A. George and M. T. Heath (1980), 'Solution of sparse linear least squares problems using Givens rotations', *Linear Algebra Appl.* **34**, 69–83.

J. R. Gilbert, C. Moler and R. Schreiber (1992), 'Sparse matrices in MATLAB: Design and implementation', *SIAM J. Matrix. Anal. Appl.* **13**, 333–356.

J. R. Gilbert, E. G. Ng and B. W. Peyton (1997), 'Separators and structure prediction in sparse orthogonal factorization', *Linear Algebra Appl.* **262**, 83–97.

L. Giraud, J. Langou, and M. Rozložnık (2002), On the loss of orthogonality in the Gram–Schmidt orthogonalization process, Technical report TR/PA/02/33, CERFACS, Toulouse, France.

L. Giraud, S. Gratton and J. Langou (2003), A reorthogonalization procedure for modified Gram–Schmidt algorithm based on a rank-k update, Technical Report TR/PA/03/11, CERFACS, Toulouse, France.

I. Gohberg and V. Olshevsky (1994), 'Complexity of multiplication with vectors for structured matrices', *Linear Algebra Appl.* **202**, 163–192.

G. H. Golub (1965), 'Numerical methods for solving least squares problems', *Numer. Math.* **7**, 206–216.

G. H. Golub (1968), 'Least squares, singular values and matrix approximations', *Aplikace Matematiky* **13**, 44–51.

G. H. Golub and W. Kahan (1965), 'Calculating the singular values and pseudo-inverse of a matrix', *SIAM J. Numer. Anal. Ser. B* **2**, 205–224.

G. H. Golub and R. J. Plemmons (1980), 'Large-scale geodetic least-squares adjustment by dissection and orthogonal decomposition', *Linear Algebra Appl.* **34**, 3–28.

G. H. Golub and C. Reinsch (1970), 'Singular value decomposition and least squares solution', *Numer. Math.* **14**, 403–420.

G. H. Golub and C. F. Van Loan (1996), *Matrix Computations*, 3rd edn, Johns Hopkins University Press, Baltimore.

G. H. Golub and J. H. Wilkinson (1966), 'Note on the iterative refinement of least squares solution', *Numer. Math.* **9**, 139–148.

G. H. Golub, V. Klema and G. W. Stewart (1976), Rank degeneracy and least squares problems, Technical Report STAN-CS-76-559, August 1976, Computer Science Department, Stanford University, CA.

G. H. Golub, F. T. Luk and M. Pagano (1979), A sparse least squares problem in photogrammetry, in *Proceedings of the Computer Science and Statistics 12th Annual Symposium on the Interface* (J. F. Gentleman, ed.), University of Waterloo, Canada, pp. 26–30.

G. H. Golub, K. Solna and P. Van Dooren (1995), A QR-like SVD algorithm for a product/quotient of several matrices, in *SVD and Signal Processing, III: Algorithms, Architectures and Applications* (M. Moonen and B. De Moor, eds), Elsevier Science BV, Amsterdam, pp. 139–147.

M. Gu (1998a), 'Backward perturbation bounds for linear least squares problems', *SIAM J. Matrix. Anal. Appl.* **20**, 363–372.

M. Gu (1998b), 'New fast algorithms for structured linear least squares problems', *SIAM J. Matrix. Anal. Appl.* **20**, 244–269.

M. Gulliksson and P.-Å. Wedin (2000), 'Perturbation theory for generalized and constrained linear least squares', *Numer. Linear Algebra Appl.* **7**, 181–196.

S. Hammarling (1976), The numerical solution of the general Gauss–Markoff linear model, in *Mathematics in Signal Processing* (T. S. Durrani et al., eds), Clarendon Press, Oxford.

M. Hanke (2001), 'On Lanczos methods for the regularization of discrete ill-posed problems', *BIT* **41**, 1008–1018.

P. C. Hansen (1998), *Rank-Deficient and Discrete Ill-Posed Problems: Numerical Aspects of Linear Inversion*, SIAM, Philadelphia.

R. J. Hanson and C. L. Lawson (1969), 'Extensions and applications of the Householder algorithm for solving linear least squares problems', *Math. Comput.* **23**, 787–812.

K. H. Haskell and R. J. Hanson (1981), 'An algorithm for linear least squares problems with equality and nonnegativity constraints', *Math. Program.* **21**, 98–118.

M. T. Heath, A. J. Laub, C. C. Paige and R. C. Ward (1986), 'Computing the SVD of a product of two matrices', *SIAM J. Sci. Statist. Comput.* **7**, 1147–1149.

M. T. Heath, R. J. Plemmons and R. C. Ward (1984), 'Sparse orthogonal schemes for structural optimization using the force method', *SIAM J. Sci. Statist. Comput.* **5**, 514–532.

M. R. Hestenes and E. Stiefel (1952), 'Methods of conjugate gradients for solving linear systems', *J. Res. Nat. Bur. Standards* **B49**, 409–436.

N. J. Higham (1995), *Accuracy and Stability of Numerical Algorithms*, 1st edn, SIAM, Philadelphia, PA.

N. J. Higham (1997), Recent developments in dense numerical linear algebra, in *The State of the Art in Numerical Analysis* (I. S. Duff and G. A. Watson, eds), Oxford University Press, pp. 1–26.

N. J. Higham (2002), *Accuracy and Stability of Numerical Algorithms*, 2nd edn, SIAM, Philadelphia, PA.

W. Hoffman (1989), 'Iterative algorithms for Gram–Schmidt orthogonalization', *Computing* **41**, 335–348.

H. P. Hong and C. T. Pan (1992), 'Rank-revealing QR factorization and the singular value decomposition', *Math. Comput.* **58**, 213–232.

D. A. Huckaby and T. F. Chan (2003), 'On the convergence of Stewart's QLP algorithm for approximating the SVD', *Numer. Algorithms* **32**, 287–316.

R. Karlsson and B. Waldén (1997), 'Estimation of optimal backward perturbation bounds for the linear least squares problem', *BIT* **37**, 862–869.

L. Kaufman (1984), 'Banded eigenvalue solvers on vector machines', *ACM Trans. Math. Software* **10**, 73–86.

C. L. Lawson and R. J. Hanson (1995), *Solving Least Squares Problems*, Classics in Applied Mathematics, SIAM, Philadelphia. Revised republication of work first published by Prentice-Hall, Englewood Cliffs, NJ (1974).

P. Lötstedt (1983), 'Perturbation bounds for the linear least squares problem subject to linear inequality constraints', *BIT* **23**, 500–519.

P. Lötstedt (1984), 'Solving the minimal least squares problem subject to bounds on the variables', *BIT* **24**, 206–224.

S.-M. Lu and J. L. Barlow (1996), 'Multifrontal computation with the orthogonal factors of sparse matrices', *SIAM J. Matrix Anal. Appl.* **17**, 658–679.

P. Manneback, C. Murigande and P. L. Toint (1985), 'A modification of an algorithm by Golub and Plemmons for large linear least squares in the context of Doppler positioning', *IMA J. Numer. Anal.* **5**, 221–234.

R. Mathias and G. W. Stewart (1993), 'A block QR algorithm and the singular value decompositions', *Linear Algebra Appl.* **182**, 91–100.

P. Matstoms (1994), 'Sparse QR factorization in MATLAB', *ACM Trans. Math. Software* **20**, 136–159.

P. Matstoms (1995), 'Parallel sparse QR factorization on shared memory architectures', *Parallel Comput.* **21**, 473–486.

M. Moonen, P. Van Dooren and J. Vandewalle (1992), 'An SVD updating algorithm for subspace tracking', *SIAM J. Matrix Anal. Appl.* **13**, 1015–1038.

J. J. Moré (1978), The Levenberg–Marquardt algorithm: Implementation and theory, in *Numerical Analysis: Proceedings Biennial Conference Dundee 1977* (G. A. Watson, ed.), Vol. 630 of *Lecture Notes in Mathematics*, Springer, Berlin, pp. 105–116.

C. C. Paige (1986), 'Computing the generalized singular value decomposition', *SIAM J. Sci. Statist. Comput.* **7**, 1126–1146.

C. C. Paige (1990), Some aspects of generalized QR factorizations, in *Reliable Numerical Computation* (M. G. Cox and S. J. Hammarling, eds), Clarendon Press, Oxford, pp. 71–91.

C. C. Paige (2002), Unifying least squares, total least squares and data least squares, in *Total Least Squares and Errors-in-Variables Modeling* (S. V. Huffel and P. Lemmerling, eds), Kluwer Academic Publishers, Dordrecht, pp. 25–34.

C. C. Paige and M. A. Saunders (1981), 'Toward a generalized singular value decomposition', *SIAM J. Numer. Anal.* **18**, 398–405.

C. C. Paige and M. A. Saunders (1982a), 'Algorithm 583 LSQR: Sparse linear equations and sparse least squares', *ACM Trans. Math. Software* **8**, 195–209.

C. C. Paige and M. A. Saunders (1982b), 'LSQR: An algorithm for sparse linear equations and sparse least squares', *ACM Trans. Math. Software* **8**, 43–71.

C. C. Paige and Z. Strakoš (2002), 'Scaled total least squares fundamentals', *Numer. Math.* **91**, 117–146.

C. T. Pan and P. T. P. Tang (1999), 'Bounds on singular values revealed by QR factorizations', *BIT* **39**, 740–756.

D. J. Pierce and J. G. Lewis (1997), 'Sparse multifrontal rank revealing QR factorization', *SIAM J. Matrix Anal. Appl.* **18**, 159–181.

G. Quintana-Ortí, X. Sun and C. H. Bischof (1998), 'A BLAS-3 version of the QR factorizations with column pivoting', *SIAM J. Sci. Comput.* **19**, 1486–1494.

J. K. Reid (1967), 'A note on the least squares solution of a band system of linear equations by Householder reductions', *Comput. J.* **10**, 188–189.

C. H. Reinsch (1971), 'Smoothing by spline functions', *Numer. Math.* **16**, 451–454.

M. Rojas and D. C. Sorensen (2002), 'A trust region approach to the regularization of large-scale discrete forms of ill-posed problems', *SIAM J. Sci. Comput.* **23**, 1843–1861.

K. Schittkowski (1983), 'The numerical solution of constrained linear least-squares problems', *IMA J. Numer. Anal.* **3**, 11–36.

R. Schreiber and C. F. Van Loan (1989), 'A storage efficient WY representation for products of Householder transformations', *SIAM J. Sci. Statist. Comput.* **10**, 53–57.

H. R. Schwarz (1968), 'Tridiagonalization of a symmetric bandmatrix', *Numer. Math.* **12**, 231–241.

G. W. Stewart (1991), On an algorithm for refining a rank-revealing URV factorization and a perturbation theorem for singular values, Technical Report UMIACS-TR-91-38, CS-TR-2626, Department of Computer Science, University of Maryland, College Park, MD.

G. W. Stewart (1992), 'An updating algorithm for subspace tracking', *IEEE Trans. Signal Process.* **40**, 1535–1541.

G. W. Stewart (1993), 'Updating a rank-revealing ULV decomposition', *SIAM J. Matrix Anal. Appl.* **14**, 494–499.

G. W. Stewart (1999), 'The QLP approximation to the singular value decomposition', *SIAM J. Sci. Comput.* **20**, 1336–1348.

E. Stiefel (1952/53), 'Ausgleichung ohne Aufstellung der Gausschen Normalgleichungen', *Wiss. Z. Tech. Hochsch. Dresden* **2**, 441–442.

C. Sun (1996), 'Parallel sparse orthogonal factorization on distributed-memory multiprocessors', *SIAM J. Sci. Comput.* **17**, 666–685.

J.-G. Sun and Z. Sun (1997), 'Optimal backward perturbation bounds for underdetermined systems', *SIAM J. Matrix Anal. Appl.* **18**, 393–402.

A. N. Tikhonov (1963), 'Regularization of incorrectly posed problems', *Soviet Math.* **4**, 1624–1627.

L. N. Trefethen and D. Bau, III (1997), *Numerical Linear Algebra*, SIAM, Philadelphia.

S. Van Huffel and J. Vandewalle (1991), *The Total Least Squares Problem: Computational Aspects and Analysis*, Vol. 9 of *Frontiers in Applied Mathematics*, SIAM, Philadelphia.

C. F. Van Loan (1976), 'Generalizing the singular value decomposition', *SIAM J. Numer. Anal.* **13**, 76–83.

B. Waldén, R. Karlsson and J.-G. Sun (1995), 'Optimal backward perturbation bounds for the linear least squares problem', *Numer. Linear Algebra Appl.* **2**, 271–286.

J. H. Wilkinson and C. Reinsch (1971), *Handbook for Automatic Computation, Vol. II: Linear Algebra*, Springer, New York.

S. Wold, A. Ruhe, H. Wold and W. J. Dunn (1984), 'The collinearity problem in linear regression: The partial least squares (PLS) approach to generalized inverses', *SIAM J. Sci. Statist. Comput.* **5**, 735–743.

The numerical analysis of functional integral and integro-differential equations of Volterra type

Hermann Brunner
Department of Mathematics and Statistics,
Memorial University of Newfoundland,
St. John's, NL, Canada A1C 5S7
E-mail: hermann@math.mun.ca

The qualitative and quantitative analysis of numerical methods for delay differential equations is now quite well understood, as reflected in the recent monograph by Bellen and Zennaro (2003). This is in remarkable contrast to the situation in the numerical analysis of functional equations, in which delays occur in connection with memory terms described by Volterra integral operators. The complexity of the convergence and asymptotic stability analyses has its roots in new 'dimensions' not present in DDEs: the problems have distributed delays; kernels in the Volterra operators may be weakly singular; a second discretization step (approximation of the memory term by feasible quadrature processes) will in general be necessary before solution approximations can be computed.

The purpose of this review is to introduce the reader to functional integral and integro-differential equations of Volterra type and their discretization, focusing on collocation techniques; to describe the 'state of the art' in the numerical analysis of such problems; and to show that – especially for many 'classical' equations whose analysis dates back more than 100 years – we still have a long way to go before we reach a level of insight into their discretized versions to compare with that achieved for DDEs.

CONTENTS

1. Introduction   56
2. Basic theory of Volterra functional integral equations I: non-vanishing delays   60
3. Collocation methods for VFIEs with non-vanishing delays   72
4. Basic theory of Volterra functional integral equations II: (vanishing) proportional delays   96
5. Collocation methods for pantograph-type VFIEs   107
6. Summary and outlook   128
References   132

1. Introduction

1.1. Early Volterra functional integral equations

1.1.1. Volterra integral equations with proportional delays

In his paper of 1897 (a sequel to his four fundamental papers that appeared in 1896), Vito Volterra studied the 'invertibility' of the 'definite integral' (using his notation)

$$f(y) - f(0) = \int_{\alpha y}^{y} \theta(x) H(x, y)\, dx, \quad 0 < y < a, \tag{1.1}$$

where $0 < \alpha < 1$; the functions f, f', H, H_y are assumed to be continuous on their respective domains. The integral operator describing this first-kind integral equation has two variables of integration, and the lower limit represents a *proportional delay* vanishing at $t = 0$.

Volterra preceded the analysis of the existence and uniqueness of the solution $\theta \in C[0, a]$ by the following observation (Volterra 1897, pp. 156–157). Suppose that the given (real-valued) functions λ and φ are continuous on $[0, a]$, with $|\lambda(0)| \leq 1$, and consider the infinite series

$$\theta(x) := \varphi(x) + \sum_{j=1}^{\infty} \alpha^j \left(\prod_{l=0}^{j-1} \lambda(\alpha^l x) \right) \varphi(\alpha^j x), \quad x \in [0, a]. \tag{1.2}$$

This series converges uniformly, and hence its limit θ lies in $C[0, a]$. On the other hand, if $\theta \in C[0, a]$ is given, replacing x in (1.2) by αx and then multiplying by $\alpha \lambda(x)$ readily leads to an expression for the unknown φ,

$$\theta(x) - \alpha \lambda(x) \theta(\alpha x) = \varphi(x), \quad x \in [0, a]. \tag{1.3}$$

In other words, the pair of equations (1.2) and (1.3) are reciprocal to each other. This observation was then used by Volterra to establish the desired result for the delay integral equation (1.1) in a rather elegant way. We shall encounter (1.3) again later, as a special case of (2.12); see also Liu (1995b).

Volterra's analysis – which relies on Picard iteration techniques – was extended by Lalesco (1908, 1911) (see also Volterra (1913, pp. 92–101) and Fenyö and Stolle (1984, pp. 324–327)) to first-kind integral equations with more general vanishing delays, and by Andreoli (1913, 1914) to closely related integral equations of the second kind,

$$\varphi(x) + \lambda \int_0^{g(x)} N(x,y)\varphi(x)\,\mathrm{d}x = f(x), \quad x \in [0,a]. \tag{1.4}$$

Andreoli observed that '*la $g(x)$ avrà un'enorme influenza sulle formole di soluzione ...*' (the truth of this visionary remark regarding the analysis of discretized versions of such equations – especially when $g(x) = \alpha x$ ($0 < \alpha < 1$) – will become apparent in Section 4.2!), and he illustrated it by means of two examples: $g(x) = \alpha x$ ($0 < \alpha < 1$) and $g(x) = x^m$ ($m > 0$; $x \in [0,1]$).

1.1.2. The Volterra delay VIDEs of population dynamics

In Part IV ('Studio delle azioni ereditarie') of his 1927 paper Volterra refined his earlier celebrated (ODE) 'predator–prey' model to include situations where 'historical actions cease after a certain interval of time' (see also Volterra (1939, p. 8)). This leads to a system of nonlinear Volterra integro-differential equations with constant delay $T_0 > 0$ (again using Volterra's notation),

$$\frac{\mathrm{d}N_1}{\mathrm{d}t} = N_1(t)\left(\varepsilon_1 - \gamma_1 N_2(t) - \int_{t-T_0}^{t} F_1(t-\tau)N_1(\tau)\,\mathrm{d}\tau\right), \tag{1.5}$$

$$\frac{\mathrm{d}N_2}{\mathrm{d}t} = N_2(t)\left(-\varepsilon_2 + \gamma_2 N_1(t) + \int_{t-T_0}^{t} F_2(t-\tau)N_2(\tau)\,\mathrm{d}\tau\right),$$

with $\varepsilon_i > 0$, $\gamma_i \geq 0$, and continuous $F_i(t) \geq 0$. Volterra later extended this model and its analysis to n interacting populations (see also his survey paper of 1939). Cushing (1977) is an excellent source on the further development of such population models based on VIDEs with delays; see also Bocharov and Rihan (2000) and its bibliography.

1.2. Volterra functional equations as mathematical models

Many basic mathematical models in epidemiology and population growth (Cooke and Yorke 1973, Waltman 1974, Cooke 1976, Smith 1977, Busenberg and Cooke 1980, Metz and Diekmann 1986 (especially Chapter IV), Hethcote and van den Driessche 2000, Brauer and van den Driessche 2003 (see also the extensive bibliographies in the last two papers)) are described by nonlinear Volterra integral equations of the second kind with (constant) delay $\tau > 0$:

$$y(t) = \int_{t-\tau}^{t} P(t-s)G(s,y(s))\,ds + g(t), \quad t > t_0, \tag{1.6}$$

or

$$y(t) = \int_{t-\tau}^{t} P(t-s)G(y(s) + g(s))\,ds, \quad t > t_0. \tag{1.7}$$

Here, g is usually assumed to be such that $\lim_{t\to\infty} g(t) =: g(\infty)$ exists. These delay integral equations model the deterministic growth of a population $y = y(t)$ (e.g., of animals, or cells) or the spread of an epidemic with *immigration* into the population; it also has applications in economics.

A generalization of the above model is discussed in Bélair (1991): here, the delay τ in the delay (or lag) function $\theta(t) := t - \tau(y(t))$ (life span) is no longer constant but depends on the size $y(t)$ of the population at time t (reflecting, e.g., crowding effects). Bélair's model corresponds to the delay VIE with *state-dependent delay*,

$$y(t) = \int_{t-\tau(y(t))}^{t} P(t-s)G(y(s))\,ds, \quad t > 0, \tag{1.8}$$

with $P(t) \equiv 1$. Here it is assumed that the number of births is a function of the population size only (that is, the birth rate is density-dependent but not age-dependent). For this choice of the kernel P it is tempting to 'simplify' the delay VIE, by differentiating it with respect to t, to obtain the state-dependent (but 'local') DDE

$$y'(t) = \frac{G(y(t)) - G(y(t-\tau(y(t))))}{1 - \tau'(y(t))G(y(t-\tau(y(t))))}. \tag{1.9}$$

While *any* constant $y(t) = y_c$ solves the above DDE, this is *not* true in the original DVIE (1.8): it is easily verified that $y(t) = y_c$ is a solution if and only if $y_c = G(y_c)\tau(y_c)$. This simple example also contains a warning: the use of the the DDE (1.9) as the basis for the ('indirect') numerical solution of the delay VIE (1.8) may lead to approximations for $y(t)$ that do not correctly reflect the dynamics of the original (highly nonlinear) delay integral equation.

The elastic motions of a three-degree-of-freedom airfoil section with a flap in a two-dimensional incompressible flow can be described by a system of neutral functional integro-differential equations of the form

$$\frac{d}{dt}\left(A_0 x(t) - \int_{-\tau}^{0} A_1(s)x(t+s)\,ds\right) \tag{1.10}$$

$$= B_0 x(t) + B_1 x(t-\tau) + \int_{-\tau}^{0} K(s)x(t+s)\,ds + F(t), \quad t > 0,$$

with $x(t) = \phi(t)$ ($-\tau \leq t \leq 0$) and $\tau > 0$. Here, the matrices A_0, $A_1(\cdot)$, B_0, B_1 and $K(\cdot)$ in $L(\mathbb{R}^d)$ (with $d = 8$) are given. (Here, $L(\mathbb{R}^d)$ denotes the linear space of all real square matrices of order d.) The matrix A_0 is singular: typically, its last row consists of zeros, and some of the elements of $A_1(s) = (a_{ij}^{(1)}(s))$ are weakly singular, e.g.,

$$a_{88}^{(1)}(s) = C(s)(-s)^{-\alpha} + p(s), \quad 0 < \alpha < 1,$$

with smooth c and p. (See Burns, Cliff and Herdmann (1983a, 1983b), Burns, Herdman and Stech (1983c), Burns, Herdman and Turi (1987, 1990), and Herdman and Turi (1991) for details on the derivation and the mathematical framework of (1.10)).

The NFIDE (1.10) contains two new ingredients that make its analysis and the analysis of collocation methods significantly more difficult. The first complication is related to the occurrence of weakly singular kernel functions: they lead to solutions with unbounded derivatives at $t = 0^+$) and hence, on uniform meshes, to low order of convergence in collocation methods, regardless of the degree of the underlying piecewise polynomials. While there are ways to deal with this problem (compare Section 6.2 and, e.g., Chapters 6 and 7 in Brunner (2004b)), it is not yet known how to overcome it when it occurs in conjunction with the (special) singular matrix A_0, since we are now facing a so-called *integro-differential algebraic system* (see März (2002a) for examples and a possible framework for their numerical analysis). For such problems (even when the kernel K is smooth) the analysis of numerical methods (based on a generalization of the notion of a *numerically properly formulated* DAE; see März (2002a, 2002b) and references) is very much in its infancy, but the subject of current joint work by R. Lamour, R. März, C. Tischendorf (Humboldt University, Berlin) and the author.

We conclude this section with a brief survey of the literature on applications of functional integral and integro-differential equations of Volterra type. Although this selection is necessarily subjective, taken together with the information contained in these books and papers (and their bibliographies) it will serve as a guide to the history and the present state of affairs of Volterra functional equations.

Starting with population dynamics (one of the major sources of Volterra integral and integro-differential equations with delay arguments) we mention the monographs by Volterra (1931), Volterra and d'Ancona (1935), Cushing (1977), Webb (1985), Brauer and Castillo-Chávez (2001), and Zhao (2003); the proceedings edited by Schmitt (1972), Metz and Diekmann (1986), and Ruan, Wolkowicz and Wu (2003); and the survey papers by Cooke and Yorke (1973), Busenberg and Cooke (1980), Ruan and Wu (1994), and Brauer and van den Driessche (2003). Among the milestone papers on this subject are the papers by Volterra (1927, 1928, 1934, 1939), Cooke (1976), Cooke and

Kaplan (1976), Smith (1977), Hethcote and Tudor (1980), Hethcote, Lewis and van den Driessche (1989), Cañada and Zertiti (1994), Hethcote and van den Driessche (1995, 2000). In addition, the reader may find it worthwhile to look at Tychonoff (1938) (for early applications of Volterra functional equations), Corduneanu and Lakshmikantham (1980) (on functional equations with unbounded delays), Ruan and Wu (1994) (on non-standard Volterra integro-differential equations), and Thieme and Zhao (2003), not least because of the numerous additional references contained in these papers.

Detailed treatments (and numerous additional applications) of nonlinear delay VIEs and VIDEs can be found in Marshall (1979), Lakshmikantham (1987), Györi and Ladas (1991), Yoshizawa and Kato (1991), Kolmanovskii and Myshkis (1992), Yatsenko (1995), Hritonenko and Yatsenko (1996), Piila (1996), Ruan and Wolkowicz (1996), and Corduneanu and Sandberg (2000). Compare also the papers by Tavernini (1978), and Cahlon and Nachman (1985), and their lists of references, on Volterra equations with state-dependent delays. The second chapter in Vogel (1965) contains an illuminating survey of the historical development of Volterra equations with delays and corresponding detailed references. Finally, the recent monograph by Ito and Kappel (2002) is the authoritative source for information on the mathematical framework for, and applications of, neutral functional integro-differential equations of the type (1.10).

2. Basic theory of Volterra functional integral equations I: non-vanishing delays

It goes without saying that a thorough understanding of the quantitative and qualitative properties of solutions to Volterra functional equations is essential for the design and the analysis of numerical methods for such problems. We therefore precede the sections dealing with the analysis of collocation methods (Sections 3 and 5) by brief sections giving an introduction to relevant theory of VFIEs, complemented by suggestions for additional readings. In this section we consider Volterra functional integral and integro-differential equations with *non-vanishing* delays.

The reader who – in order to see the subsequent analysis in a wider perspective – wishes to acquire a broader knowledge of the theory of delay differential equations is referred to, *e.g.*, the monographs by Myshkis (1972) (in Russian, with German translation of 1955), Bellman and Cooke (1963), El'sgol'ts and Norkin (1973), Kolmanovskii and Myshkis (1992), and – especially – Hale (1977), Hale and Verduyn Lunel (1993), Diekmann, van Gils, Verduyn Lunel and Walther (1995), and Wu (1996). Regularity results for solutions of DDEs may be found in Neves and Feldstein (1976), de Gee (1985), and Willé and Baker (1992).

2.1. Second-kind Volterra integral equations with non-vanishing delays

The general linear Volterra integral equation with delay $\theta(t)$ has the form

$$y(t) = g(t) + (\mathcal{V}y)(t) + (\mathcal{V}_\theta y)(t), \quad t \in (t_0, T]. \tag{2.1}$$

Here, $\mathcal{V}: C(I) \to C(I)$ denotes the classical (linear) Volterra integral operator,

$$(\mathcal{V}y)(t) := \int_{t_0}^{t} K_1(t,s) y(s) \, ds, \tag{2.2}$$

with kernel $K_1 \in C(D)$, $D := \{(t,s) : t_0 \leq s \leq t \leq T\}$. The kernel K_2 of the delay integral operator

$$(\mathcal{V}_\theta y)(t) := \int_{t_0}^{\theta(t)} K_2(t,s) y(s) \, ds \tag{2.3}$$

is assumed to be continuous in $D_\theta := \{(t,s) : \theta(t_0) \leq s \leq \theta(t), t \in I\}$, with $I := [t_0, T]$. Throughout this and the next section the delay (or lag) function θ will be subject to the following conditions:

(D1) $\theta(t) = t - \tau(t)$, with $\tau \in C^d(I)$ for some $d \geq 0$;
(D2) $\tau(t) \geq \tau_0 > 0$ for $t \in I$;
(D3) θ is strictly increasing on I.

We will refer to the function $\tau = \tau(t)$ as the delay.

Remark. The subsequent discussion will reveal that condition (D3) has been introduced mainly to simplify the description and the analysis of the collocation methods; the recent monograph by (Bellen and Zennaro 2003) on the numerical solution of DDEs deals with many of the complications that can arise if (D3) does not hold, providing illuminating examples and remarks.

We have seen in Section 1.2 that in applications (for example, in mathematical models for population growth or the spreading of an epidemic one encounters delay integral equations of the type

$$y(t) = g(t) + (\mathcal{W}_\theta y)(t), \quad t \in (t_0, T], \tag{2.4}$$

corresponding to the delay integral operator

$$(\mathcal{W}_\theta y)(t) := \int_{\theta(t)}^{t} K(t,s) y(s) \, ds, \tag{2.5}$$

or its nonlinear (Hammerstein) version,

$$(\mathcal{W}_\theta y)(t) := \int_{\theta(t)}^{t} K(t,s) G(s, y(s)) \, ds. \tag{2.6}$$

The (linear) delay equation (2.4) may be viewed as a particular case of (2.1), obtained formally by setting $K_2 = -K_1 =: -K$.

These delay integral equation are complemented, in analogy to DDEs, by an appropriate initial condition,

$$y(t) = \phi(t), \quad t \in [\theta(t_0), t_0].$$

We observe that, in contrast to initial-value problems for DDEs and DVIDEs with non-vanishing delays, the interval in which (2.1) and (2.4) are considered is the left-open interval $(t_0, T]$: we shall see below (Theorem 2.1) that solutions to Volterra integral equations with non-vanishing delays typically possess a finite (jump) discontinuity at $t = t_0$, while for first-order DDEs (and DVIDEs) the solution y is continuous at this point, with the discontinuity occurring in y'.

However, in complete analogy to DDEs the non-vanishing delay $\tau(\cdot)$ gives rise to the *primary discontinuity points* $\{\xi_\mu\}$ for the solution y: they are determined by the recursion

$$\theta(\xi_\mu) = \xi_\mu - \tau(\xi_\mu) = \xi_{\mu-1}, \quad \mu \geq 1, \quad \xi_\mu = t_0$$

(see, for example, Section 2.2 in Bellen and Zennaro (2003)). Condition (D2) ensures that these discontinuity points have the (uniform) separation property

$$\xi_\mu - \xi_{\mu-1} = \tau(\xi_\mu) \geq \tau_0 > 0, \quad \text{for all } \mu \geq 1.$$

This implies that the number of primary discontinuity points in any bounded interval I remains finite: there is no clustering of the $\{\xi_\mu\}$.

Theorem 2.1. Assume that the given functions in (2.1)–(2.3) are continuous on their respective domains and that the delay function θ satisfies the above conditions (D1)–(D3). Then, for any initial function $\phi \in C[\theta(t_0), t_0]$, there exists a unique (bounded) $y \in C(t_0, T]$ solving the delay integral equation (2.1) on $(t_0, T]$ and coinciding with ϕ on $[\theta(t_0), t_0]$. In general, this solution has a finite (jump) discontinuity at $t = t_0$:

$$\lim_{t \to t_0^+} y(t) \neq \lim_{t \to t_0^-} y(t) = \phi(t_0).$$

The solution is continuous at $t = t_0$ if and only if the initial function is such that

$$g(t_0) - \int_{\theta(t_0)}^{t_0} K_2(t_0, s)\phi(s)\, ds = \phi(t_0).$$

Proof. For $t \in I^{(\mu)} := [\xi_\mu, \xi_{\mu+1}]$ ($\mu \geq 1$) the initial-value problem for (2.1) may be written as a Volterra integral equation of the second kind,

$$y(t) = g_\mu(t) + \int_{\xi_\mu}^{t} K_1(t, s) y(s)\, ds, \tag{2.7}$$

with $g_\mu(t) := g(t) + \Phi_\mu(t)$ and

$$\Phi_\mu(t) := \int_{t_0}^{\xi_\mu} K_1(t,s)y(s)\,ds + \int_{t_0}^{\theta(t)} K_2(t,s)y(s)\,ds.$$

For $\mu = 0$ this function is known and given by

$$\Phi_0(t) = -\int_{\theta(t)}^{t_0} K_2(t,s)\phi(s)\,ds, \quad t \in I_0 := (t_0, \xi_1];$$

by our assumptions we have $\Phi_0 \in C(I^{(0)})$. It follows from the classical Volterra theory (Volterra (1896, 1913, 1959) and Miller (1971); see also Brunner and van der Houwen (1986), or Brunner (2004b)) that the integral equation (2.7) possesses a unique continuous (bounded) solution on each interval $I^{(\mu)}$ ($\mu \geq 0$).

As for its regularity, we first observe that for $\mu = 0$ (with $\xi_0 = t_0$),

$$\lim_{t \to t_0^+} y(t) = g(t_0) + \Phi_0(t_0) = g(t_0) - \int_{\theta(t_0)}^{t_0} K_2(t_0, s)\phi(s)\,ds$$

which, for arbitrary (continuous) data g, K_2, ϕ, will not coincide with the value $\phi(t_0)$. For $\mu \geq 1$ we derive

$$y(\xi_\mu^-) = g(\xi_\mu) + \int_{t_0}^{\xi_\mu} K_1(\xi_\mu, s)y(s)\,ds + \int_{t_0}^{\theta(\xi_\mu)} K_2(\xi_\mu, s)y(s)\,ds$$

and

$$y(\xi_\mu^+) = g(\xi_\mu) + \int_{t_0}^{\xi_\mu} K_1(\xi_\mu, s)y(s)\,ds + \int_{t_0}^{\theta(\xi_\mu)} K_2(\xi_\mu, s)y(s)\,ds.$$

Hence,

$$y(\xi_\mu^+) - y(\xi_\mu^-) = 0,$$

whenever g, K_1, K_2 and θ are continuous functions. This completes the proof of Theorem 2.1. □

The solution of a linear Volterra integral equation of the second kind,

$$y(t) = g(t) + \int_{t_0}^{t} K(t,s)y(s)\,ds, \quad t \in I,$$

with continuous g and K, can be expressed in term of the resolvent kernel $R = R(t,s)$ and the nonhomogeneous term g, namely,

$$y(t) = g(t) + \int_{t_0}^{t} R(t,s)g(s)\,ds, \quad t \in I.$$

This 'variation-of-constants' formula is the key to establishing (global and local) superconvergence results for collocation solutions to such equations.

As the above proof implicitly shows, an analogous representation can be derived for the solution of the delay Volterra integral equation (2.1), since by (D2) the delay $\tau = \tau(t)$ in $\theta(t) = t - \tau(t)$ does not vanish in I. Suppose, for ease of notation and without loss of generality, that T in $I = [t_0, T]$ is such that $\xi_{M+1} = T$ (or, alternatively, $T \in (\xi_M, \xi_{M+1}])$ for some $M \geq 1$.

Theorem 2.2. Let (D1)–(D3) and the assumptions of Theorem 2.1 hold, and set

$$g_0(t) := g(t) - \int_{\theta(t)}^{t_0} K_2(t,s)\phi(s)\,ds \quad \text{for } t \in [t_0, \xi_1].$$

Then, for $t \in I^{(\mu)} := [\xi_\mu, \xi_{\mu+1}]$ $(\mu \geq 1)$, the unique (bounded) solution y of (2.1) corresponding to the initial function ϕ can be expressed in the form

$$y(t) = g(t) + \int_{\xi_\mu}^{t} R_1(t,s)g(s)\,ds + F_\mu(t) + \Phi_\mu(t), \tag{2.8}$$

with

$$F_\mu(t) := \int_{t_0}^{\xi_1} R_{\mu,0}(t,s)g_0(s)\,ds + \sum_{\nu=1}^{\mu-1} \int_{\xi_\nu}^{\xi_{\nu+1}} R_{\mu,\nu}(t,s)g(s)\,ds,$$

$$\Phi_\mu(t) := \int_{t_0}^{\theta^\mu(t)} Q_{\mu,0}(t,s)g_0(s)\,ds + \sum_{\nu=1}^{\mu-1} \int_{\xi_\nu}^{\theta^{\mu-\nu}(t)} Q_{\mu,\nu}(t,s)g(s)\,ds.$$

On the initial interval $I^{(0)} := (\xi_0, \xi_1]$ (with $\xi_0 = t_0$) the solution y is given by

$$y(t) = g_0(t) + \int_{t_0}^{t} R_1(t,s)g_0(s)\,ds. \tag{2.9}$$

Here, R_1 is the resolvent kernel associated with the given kernel K_1 of the Volterra integral operator (2.2), $R_{\mu,\nu}$ and $Q_{\mu,\nu}$ denote functions which are continuous on their respective domains and depend on K_1, K_2, R_1 and θ, and $\theta^k := \underbrace{\theta \circ \cdots \circ \theta}_{k}$.

Remark. The structure of the above 'variation-of-constants' formula (2.8) and (2.9) clearly reveals the interaction between the classical lag term $F_\mu(t)$ (governed by the classical Volterra operator $\mathcal{V}$) and the delay term $\Phi_\mu(t)$ (which reflects the action of the non-vanishing lag function θ in $\mathcal{V}_\theta$). The structure of the latter will play a crucial role in the selection of appropriate ('θ-invariant') meshes for which local superconvergence results are possible (Sections 3.4.2 and 3.4.3).

Proof. The solution of the 'local' integral equation

$$y(t) = g_\mu(t) + \int_{\xi_\mu}^t K_1(t,s) y(s) \, ds, \quad t \in I^{(\mu)},$$

is given by

$$y(t) = g_\mu(t) + \int_{\xi_\mu}^t R_1(t,s) g_\mu(s) \, ds, \quad t \in I^{(\mu)}, \tag{2.10}$$

with $R_1(t,s)$ defined by the resolvent equation

$$R_1(t,s) = K_1(t,s) + \int_s^t R_1(t,v) K_1(v,s) \, dv, \quad (t,s) \in D^{(\mu)},$$

where $D^{(\mu)} := \{(t,s) : \xi_\mu \leq s \leq t \leq \xi_{\mu+1}\}$. The expression (2.9) for the solution on the interval $I^{(0)}$ thus follows immediately.

On $I^{(1)}$ ($\mu = 1$) we thus have, using Dirichlet's formula,

$$g_1(t) = g(t) + \int_{t_0}^{\xi_1} \left(K_1(t,s) + \int_s^{\xi_1} K_1(t,v) R_1(v,s) \, dv \right) g_0(s) \, ds$$

$$+ \int_{t_0}^{\theta(t)} \left(K_2(t,s) + \int_s^{\theta(t)} K_2(t,v) R_1(v,s) \, dv \right) g_0(s) \, ds$$

$$=: g(t) + \int_{t_0}^{\xi_1} Q_{1,1}^{(1)}(t,s) g_0(s) \, ds + \int_{t_0}^{\theta(t)} Q_{1,0}^{(1)}(t,s) g_0(s) \, ds,$$

with obvious meaning of the (continuous) functions $Q_{1,0}^{(1)}$ and $Q_{1,1}^{(1)}$.

Recall now the representation (2.10) with $\mu = 1$ of the solution y on $I^{(1)}$: after trivial algebraic manipulation it can be written as

$$y(t) = g(t) + \int_{t_0}^t R_1(t,s) g(s) \, ds + \int_{t_0}^{\xi_1} (Q_{1,1}^{(1)}(t,s) + \hat{Q}_{1,1}^{(1)}(t,s)) g_0(s) \, ds$$

$$+ \int_{t_0}^{\theta(t)} (Q_{1,0}^{(1)}(t,s) + \hat{Q}_{1,0}^{(1)}(t,s)) g_0(s) \, ds.$$

This yields (2.8) with $\mu = 1$, by setting

$$R_{1,0}(t,s) := Q_{1,1}^{(1)}(t,s) + \hat{Q}_{1,1}^{(1)}(t,s), \qquad Q_{1,0}(t,s) := Q_{1,0}^{(1)}(t,s) + \hat{Q}_{1,0}^{(1)}(t,s).$$

Clearly, the functions describing this expression for y are continuous in the region where they are defined.

The proof is now concluded by a simple but notationally tedious induction argument. This argument reveals that, in the variation-of-constants formula (2.8), the integrals over $I^{(\mu)} = [\xi_\mu, \xi_{\mu+1}]$ with $\mu \geq 1$ will contribute terms involving only $g(t)$, while the integrals over $[\xi_0, \xi_1]$ and $[\xi_0, \theta^\mu(t)]$ contain the 'entire' initial function $g_0(t)$ which includes the contribution of the initial function ϕ. □

The result of Theorem 2.2 and its proof lead to the following result on the regularity of solutions of (2.1).

Theorem 2.3. Assume that (D1)–(D3) are satisfied, with $d \geq m \geq 1$ in (D1), and that the functions describing the delay Volterra integral equation (2.1) all possess continuous derivatives of at least order $m \geq 1$ on their respective domains. Then the following properties hold.

(a) The (unique) solution y of the initial-value problem for (2.1) is in $C^m(\xi_\mu, \xi_{\mu+1}]$ for each $\mu = 0, 1, \ldots, M$ and is bounded on $Z_M := \{\xi_\mu : \mu = 0, 1, \ldots, M\}$.

(b) At $t = \xi_\mu$ ($\mu = 1, \ldots, \min\{m, M\}$) we have

$$\lim_{t \to \xi_\mu^-} y^{(\mu-1)}(t) = \lim_{t \to \xi_\mu^+} y^{(\mu-1)}(t),$$

while the μth derivative of y is in general not continuous at ξ_μ. However, for $t \in [\xi_{m+1}, T]$ the solution lies in $C^m[\xi_{m+1}, T]$.

Remark. Differentiation of the Volterra delay integral equation of the first kind,

$$\int_{\theta(t)}^{t} H(t, s) y(s) \, ds = f(t), \quad t \in I, \quad f(0) = 0, \tag{2.11}$$

leads – under appropriate regularity assumptions for H and f (see Section 2.2) – to a second-kind delay VIE that is somewhat more general than (2.1), namely

$$y(t) = g(t) + b(t) y(\theta(t)) + (\mathcal{W}_\theta y)(t), \quad t \in (\theta(t_0), t_0], \tag{2.12}$$

where we have set

$$g(t) := \frac{f'(t)}{K(t,t)}, \quad b(t) := \frac{H(t, \theta(t)) \theta'(t)}{H(t,t)},$$

and

$$K(t, s) := -\frac{\partial H(t,s)/\partial t}{H(t,t)}$$

in (2.12) and in the Volterra operator $\mathcal{W}_\theta$ (cf. (2.5)). Since the delay τ in $\theta(t) = t - \tau(t)$ does not vanish on I, the above result on the existence and uniqueness of a solution of the corresponding initial-value problem (Theorem 2.1), the variation-of-constants formula (Theorem 2.2), and the regularity properties (Theorem 2.3) can be generalized to encompass (2.11). We leave the proofs of these generalizations as an exercise. Note that for kernels with $\partial H(t,s)/\partial t \equiv 0$, equation (2.11) is closely related to (1.3).

2.2. First-kind Volterra integral equations with non-vanishing delays

For the sake of comparison we briefly consider the linear first-kind Volterra integral equation with delay function θ satisfying (D1)–(D3),

$$(\mathcal{V}y)(t) + (\mathcal{V}_\theta y)(t) = g(t), \quad t \in (t_0, T], \tag{2.13}$$

subject to the initial condition $y(t) = \phi(t)$, $t \in [\theta(t_0), t_0]$. The (linear) Volterra integral operators are those in (2.2) and (2.3).

Using the notation of the previous section we can write (2.13) in the local form

$$\int_{\xi_\mu}^t K_1(t,s) y(s) \, ds = g_\mu(t), \quad t \in (\xi_\mu, \xi_{\mu+1}], \tag{2.14}$$

with

$$g_\mu(t) := g(t) - \int_{t_0}^{\xi_\mu} K_1(t,s) y(s) \, ds - \int_{t_0}^{\theta(t)} K_2(t,s) y(s) \, ds, \tag{2.15}$$

for $\mu \geq 1$. For $t \in (\xi_0, \xi_1]$ this becomes

$$g_0(t) := g(t) + \int_{\theta(t)}^{t_0} K_2(t,s) \phi(s) \, ds. \tag{2.16}$$

This reveals that for arbitrary continuous K_2, g, ϕ, θ, we have

$$g_0(t_0) = g(t_0) + \int_{\theta(t_0)}^{t_0} K_2(t_0, s) \phi(s) \, ds \neq 0.$$

Hence, according to the classical Volterra theory of 1896, it follows that typically the solution of (2.13) (with $\mu = 0$) will be unbounded at $t = \xi_0^+ = t_0^+$:

$$\lim_{t \to t_0^-} y(t) = \phi(t_0) \neq \lim_{t \to t_0^+} y(t) = \pm\infty.$$

For the solution to be bounded at $t = t_0^+$ the initial function must be such that

$$\int_{\theta(t_0)}^{t_0} K_2(t_0, s) \phi(s) \, ds = -g(t_0). \tag{2.17}$$

We summarize these observations in the following theorem.

Theorem 2.4. Assume:

(a) $K_1 \in C^1(D)$, with $|K_1(t,t)| \geq \kappa_0 > 0$, $t \in I := [t_0, T]$;
(b) $K_2 \in C^1(D_\theta)$;
(c) $g \in C^1(I)$;
(d) $\theta \in C^1$ is subject to (D1)–(D3) of Section 2.1, with $d = 1$ in (D1).

Then, for any $\phi \in C[\theta(t_0), t_0]$, there exists a unique y with $y \in C(\xi_\mu, \xi_{\mu+1}]$ ($\mu = 0, 1, \ldots, M$) which solves (2.13) on $(t_0, T]$ and coincides with ϕ on $[\theta(t_0), t_0]$. This solution y remains bounded at $t = t_0 = \xi_0$ if and only if (2.17) holds.

Is the smoothing property we encountered in solutions of delay Volterra integral equations of the second kind (Theorem 2.3) also present in solutions of the first-kind delay equation (2.13)? The simple but representative example

$$\int_{t_0}^{t} y(s)\,ds + \int_{t_0}^{\theta(t)} \lambda_2 y(s)\,ds = g(t), \quad t \in (t_0, T], \tag{2.18}$$

with $y(t) = \phi(t) = \phi_0$ for $t \in [\theta(t_0), t_0]$, whose solution can easily be found explicitly, shows that this is not so. The following theorem describes the general situation.

Theorem 2.5. Let the assumptions of Theorem 2.4 for the given functions in (2.13) hold, and assume that the initial function $\phi \in C[\theta(t_0), t_0]$ is such that the solution y of the initial-value problem for (2.13) is bounded at $t = t_0^+$ (cf. (2.17)). If y possesses a finite discontinuity at $t = t_0$, then it also has finite jumps at the other points of Z_M.

The extension of this regularity result to first-kind Volterra integral equations and to a class of related neutral functional integro-differential equations with weakly singular kernels will play an important role in the (not yet fully understood) analysis of convergence of collocation methods for such equations. See Brunner (1999a, 1999b) and the remarks in Section 6.3 below.

2.3. VIDEs with non-vanishing delays

We now turn to the (regularity) properties of solutions to the linear first-order delay VIDE

$$y'(t) = a(t)y(t) + b(t)y(\theta(t)) + g(t) + (\mathcal{V}y)(t) + (\mathcal{V}_\theta y)(t), \quad t \in I := [t_0, T], \tag{2.19}$$

corresponding to the Volterra integral operators $\mathcal{V}$ and $\mathcal{V}_\theta$ introduced in (2.2) and (2.3). It includes the analogue of the particular delay VIE (2.4),

$$y'(t) = a(t)y(t) + b(t)y(\theta(t)) + g(t) + (\mathcal{W}_\theta y)(t), \quad t \in I, \tag{2.20}$$

with $\mathcal{W}_\theta$ given by (2.5) or (2.6).

The solutions y of the delay VIDE (2.19) (and hence those of (2.20)) will in general again have lower regularity at the *primary discontinuity points* $\{\xi_\mu\}$ defined by the recursion

$$\theta(\xi_\mu) = \xi_{\mu-1}, \quad \mu = 1, \ldots, \quad (\xi_0 = t_0).$$

We start with a basic result on the existence and uniqueness of solutions of the initial-value problem for (2.19).

Theorem 2.6. Assume:

(a) $a, b, g, \theta \in C(I)$, $K_1 \in C(D)$, $K_2 \in C(D_\theta)$;
(b) $\theta(t) = t - \tau(t)$ satisfies the conditions (D1)–(D3) of Section 2.1.

Then, for any initial function $\phi \in C[\theta(t_0), t_0]$, there exists a unique function $y \in C(I) \cap C^1(t_0, T]$ which satisfies the delay VIDE (2.19) on I and coincides with ϕ on $[\theta(t_0), t_0]$. At $t = t_0$ its derivative is, in general, discontinuous (but bounded):

$$\lim_{t \to t_0^+} y'(t) \neq \lim_{t \to t_0^-} y'(t) = \phi'(t_0)$$

(assuming that $\theta'(t_0)$ exists).

The (unique) solution y of the initial-value problem for (2.19) can be expressed by a variation-of-constants formula, analogous to the one in Theorem 2.2 for the delay VIE (2.1). This result is based on the 'local' form of the above delay VIDE, that is, on the initial-value problem with respect to the interval $I^{(\mu)} := [\xi_\mu, \xi_{\mu+1}]$ ($\mu = 1, \ldots, M$):

$$y'(t) = a(t)y(t) + g_\mu(t) + \int_{\xi_\mu}^{t} K_1(t,s)y(s)\,ds, \quad t \in I^{(\mu)}, \tag{2.21}$$

where $y(\xi_\mu)$ is known and g_μ is defined by

$$g_\mu(t) := g(t) + b(t)y(\theta(t)) + \int_{t_0}^{\xi_\mu} K_1(t,s)y(s)\,ds$$

$$+ \int_{t_0}^{\theta(t)} K_2(t,s)y(s)\,ds. \tag{2.22}$$

For $\mu = 0$ the above lag term reduces to

$$g_0(t) := g(t) + b(t)\phi(\theta(t)) - \int_{\theta(t)}^{t_0} K_2(t,s)\phi(s)\,ds, \quad t \in I^{(0)}. \tag{2.23}$$

The solution of the (local) VIDE (2.21) has the form

$$y(t) = r_1(t, \xi_\mu)y(\xi_\mu) + \int_{\xi_\mu}^{t} r_1(t, s)g_\mu(s)\,ds, \quad t \in I^{(\mu)}, \tag{2.24}$$

with the resolvent kernel r_1 given by the solution of the resolvent equation

$$\frac{\partial r_1(t,s)}{\partial s} = -r_1(t,s)a(s) - \int_{s}^{t} r_1(t,v)K_1(v,s)\,dv, \quad (t,s) \in D^{(\mu)}, \tag{2.25}$$

subject to the initial condition $r_1(t,t) = 1$ for $t \in I^{(\mu)}$. (Compare also Grossman and Miller (1970), Brunner and van der Houwen (1986), or Brunner (2004b) for the theory of classical linear VIDEs.)

The following variation-of-constants formula is the analogue of the one presented in Theorem 2.2. Observe, however, that we now have additional terms involving the values of y at the primary discontinuity points $\{\xi_\mu\}$.

Theorem 2.7. Let the given functions a, b, g, K_1, K_2, ϕ be continuous, and assume that θ is subject to (D1)–(D3). Then on the interval $I^{(\mu)} := [\xi_\mu, \xi_{\mu+1}]$ ($\mu \geq 1$) the solution of the initial-value problem for (2.19) can be written as

$$y(t) = r_1(t, \xi_\mu) y(\xi_\mu) + \int_{\xi_\mu}^{t} r_1(t,s) g(s)\, ds + F_\mu(t) + \Phi_\mu(t), \qquad (2.26)$$

with

$$F_\mu(t) := \sum_{\nu=1}^{\mu-1} \rho_{\mu,\nu}(t) y(\xi_\nu) + \int_{\xi_0}^{\xi_1} r_{\mu,0}(t,s) g_0(s)\, ds + \sum_{\nu=1}^{\mu-1} \int_{\xi_\nu}^{\xi_{\nu+1}} r_{\mu,\nu}(t,s) g(s)\, ds,$$

$$\Phi_\mu(t) := \int_{\xi_0}^{\theta^\mu(t)} q_{\mu,0}(t,s) g_0(s)\, ds + \sum_{\nu=1}^{\mu-1} \int_{\xi_\nu}^{\theta^{\mu-\nu}(t)} q_{\mu,\nu}(t,s) g(s)\, ds.$$

On the first interval $I^{(0)}$ this representation reduces to

$$y(t) = r_1(t, t_0) y(t_0) + \int_{t_0}^{t} r_1(t,s) g_0(s)\, ds, \qquad (2.27)$$

where $y(t_0) = \phi(t_0)$. The functions $\rho_{\mu,\nu}$, $r_{\mu,\nu}$, and $q_{\mu,\nu}$ depend on a, b, K_1, K_2, r_1 and θ and are continuous on their respective domains; $r_1 = r_1(t,s)$ denotes the resolvent kernel for $K_1 = K_1(t,s)$ defined by the resolvent equation (2.25).

Proof. The basic idea governing the proof of the above result is essentially the one used to establish Theorem 2.2, except that now the variation-of-constants formula is based on the resolvent representation of the solution of the 'local' VIDE (2.21) and will thus reflect the initial values $y(\xi_\mu)$. We leave the details of this simple proof to the reader. □

Remarks. **(1)** As in Theorem 2.2 we see again how the presence of the delay term $(\mathcal{V}_\theta y)(t)$ in (2.19) influences the resolvent representation of the classical (non-delay) VIDE on the macro-interval $I^{(\mu)}$. In addition, we now have terms reflecting the initial values $y(\xi_\nu)$ ($0 \leq \nu \leq \mu$).

(2) There is a close connection between the representation of the solution of certain classes of Volterra functional (integro-)differential equations and the *semigroup framework* into which such equations can be embedded. Among the many papers dealing with this framework and corresponding solution representations we mention Burns et al. (1983c), Staffans (1985a, 1985b), Kappel and Zhang (1986), Burns, Herdman and Turi (1987), Clément, Desch and Homan (2002), and Ito and Kappel (2002).

2.4. Volterra functional equations with weakly singular kernels

As we have briefly seen in the remarks following equation (1.10), Volterra functional integral and integro-differential equations with weakly singular (*i.e.*, unbounded but integrable) kernels occur in many applications. Owing to limitations of space we shall not be able to say much about them in this paper, except to comment on open problems in their collocation analysis (Section 6.3). Here, we introduce relevant notation and point to papers in which the reader will find additional information.

Assume that $\alpha \in (0,1)$ is given, and define the delay integral operators

$$(\mathcal{V}_{\theta,\alpha} y)(t) := \int_0^{\theta(t)} (t-s)^{-\alpha} K(t,s) y(s) \, ds, \qquad (2.28)$$

and

$$(\mathcal{W}_{\theta,\alpha} y)(t) := \int_{\theta(t)}^t (t-s)^{-\alpha} K(t,s) y(s) \, ds, \qquad (2.29)$$

corresponding to continuous kernel functions K satisfying $|K(t,t)| \geq k_0 > 0$ when $s = t$. In Section 6.2 we will comment on some of the open problems arising for the corresponding functional equations

$$y(t) = g(t) + (\mathcal{T}_{\theta,\alpha} y)(t) \qquad (2.30)$$

and

$$y'(t) = f(t, y(t), y(\theta(t))) + (\mathcal{T}_{\theta,\alpha} y)(t), \qquad (2.31)$$

with $\mathcal{T}_{\theta,\alpha}$ representing one of the Volterra integral operators $\mathcal{V}_{\theta,\alpha}$ or $\mathcal{W}_{\theta,\alpha}$.

In Section 1.2 (equation (1.10)) we have encountered a (system of a) functional integro-differential equations of neutral type whose scalar counterpart may be written as (now using our standard notation),

$$\frac{d}{dt}(a_0 y(t) - (\mathcal{W}_{\theta,\alpha} y)(t)) = F(t, y(t), y(\theta(t)), y'(\theta(t))), \quad t \geq 0,$$

with $0 < \alpha < 1$ and

$$(\mathcal{W}_{\theta,\alpha} y)(t) := \int_{\theta(t)}^t (t-s)^{-\alpha} K(t,s) y(s) \, ds, \quad \theta(t) = t - \tau.$$

A related (but more complex) Volterra functional integro-differential equation is

$$\frac{d}{dt}((\mathcal{W}_{\theta,\alpha} y)(t)) = f(t). \qquad (2.32)$$

The mathematical analysis of functional equations of this type may be found in, *e.g.*, Kappel and Zhang (1986), Ito and Kappel (1991), Clément et al. (2002), and Ito and Kappel (2002). We will briefly return to these two classes of functional equations in Section 6.3. The mathematical (semigroup) framework for such equations has been developed in, *e.g.*, the papers and the monograph mentioned at the end of Section 2.3 (Remark (2)); results on the regularity of their solutions may be found in Brunner and Ma (2004).

3. Collocation methods for VFIEs with non-vanishing delays

3.1. Numerical analysis of VFIEs: an overview

We will use this section to sketch briefly the development of numerical methods for solving delay differential equations and more general functional integral and integro-differential equations of Volterra type. In the subsequent sections we shall then focus on collocation methods for such problems.

Most of the early discretization schemes for delay problems are based on 'classical' linear multistep and Runge–Kutta methods for ODEs. These methods have to be complemented by a suitable interpolation procedure (*e.g.*, by a *natural continuous extension* (NCE)), to generate approximations at certain non-mesh points $\theta(t)$. One of the principal merits of a collocation method is that the NCE is part of the method itself.

3.1.1. DDEs

The monograph by Myshkis (1972) (first published in Russian in 1955; see also the German translation of 1955) stands at the beginning of a sequence of distinguished monographs on the theory and applications of delay differential equations. Of these we mention Bellman and Cooke (1963), El'sgol'ts and Norkin (1973), Hale (1977), Kolmanovskii and Myshkis (1992), Hale and Verduyn Lunel (1993), Diekmann et al. (1995), Wu (1996) (on partial DDEs), and Ito and Kappel (2002) (on more general functional equations). The early survey papers by Halanay and Yorke (1971), Cryer (1972) and Bellen (1985), when read in 'hand-in-hand' with the recent ones by Zennaro (1995), Baker (1997, 2000), Baker and Paul (1997) and Bocharov and Rihan (2000), give a good idea of how the interest in theory, numerical analysis, and applications of DDE has grown since the early 1970s.

The monograph by Bellen and Zennaro (2003) provides not only a good introduction, by means of numerous illuminating examples, to the theory of DDEs but gives a state-of-the-art treatment of numerical methods for DDEs. Focusing on Runge–Kutta-type methods, we see that, beginning in the early 1980s, one can discern two main trends in the analysis of such methods. The first is concerned with the adaptation of (explicit and implicit) RK methods to DDEs and the construction of various interpolants, including NCEs.

Typical contributions are those by Bellen and Zennaro (1985), Zennaro (1986), in 't Hout (1992), and Vermiglio and Zennaro (1993) (Chapters 5 and 6 in Bellen and Zennaro (2003) contain a description of these quantitative aspects). Computational aspects are discussed in detail in Bellen and Zennaro (2003); compare also Neves and Thompson (1992) and Guglielmi and Hairer (2001b).

The second aspect is the study of asymptotic stability and contractivity properties of RK methods. Early milestones in the qualitative analysis of such methods are the papers by Reverdy (1981, 1990) and Torelli (1989). Of the many later contributions extending these results, the reader may also wish to consult those by Zennaro (1993, 1997), Torelli and Vermiglio (2003) Spijker (1997), Vermiglio and Torelli (1998), Guglielmi (1998) (dealing with delay-dependent stability), Guglielmi and Hairer (2001a) and Maset (2003). Finally, we mention the papers by Ascher and Petzold (1995) and Hauber (1997) on related numerical aspects for differential-algebraic equations with delays.

We shall see in Section 5.8 that the analogous analysis of the qualitative behaviour of collocation solutions for Volterra-type functional integral and integro-differential equations remains largely open.

Most of these papers consider only DDEs with constant delay $\tau > 0$. For general lag functions $\theta(t) = t - \tau(t)$ with *nonlinear* delay $\tau(t)$, the analysis and the implementation of IRK methods become much more complex (compare Lemma 3.1 below). This problem is not present in piecewise polynomial collocation methods since, as we have indicated before, they are global methods and thus automatically include an NCE. Bellen (1984) gave the first complete (super-) convergence analysis for such methods when applied to nonlinear DDEs with general nonlinear (non-vanishing) delays; his analysis is complemented in Vermiglio (1985). These collocation methods employ distinct collocation points; Hermite-type collocation for DDEs (and the attainable order of convergence) was studied by Oberle and Pesch (1981). More recent work on various aspects of collocation methods are studied in Enright and Hayashi (1998), Liu (1999a, 1999b), Engelborghs, Luzyanina, in 't Hout and Roose (2000), Engelborghs and Doedel (2002), and in Guglielmi and Hairer (2001a) (collocation at Radau II points).

We will not mention any of the superconvergence results here, since they can be obtained as particular cases of those for Volterra integro-differential equations with delays (see Section 3.1.3).

3.1.2. VFIEs of the second kind

As we saw in Section 1, the first papers on the theory of delay integral equations (Volterra (1897), Lalesco (1908, 1911), and Andreoli (1913, 1914)) considered the case of vanishing (proportional) delays. The development of the early theory for Volterra equations with non-vanishing delays is well sketched in Vogel (1965). Also of interest is the paper by Lin (1963), which contains a comparison result for solutions of systems of second-kind Volterra integral equations with constant delay. Additional results on the existence, uniqueness, and representation of solutions to such functional equations can be found in Levin and Nohel (1964), Bownds, Cushing and Schutte (1976), Cerha (1976) and Mureşan (1984, 1999), as well as in Cooke (1976), Esser (1976, 1978), Meis (1976), Busenberg and Cooke (1980), Cahlon, Nachman and Schmidt (1984) and Cañada and Zertiti (1994) (also for additional references). Chapter 4 in Brunner (2004b) contains an introduction to the theory of VIEs and VIDEs with non-vanishing delays.

The numerical analysis of Volterra integral equations with delays can be traced back to Esser (1976, 1978), Vâţă (1978), Wolff (1982), Cahlon *et al.* (1984) and Cahlon and Nachman (1985). More recent contributions on Runge–Kutta methods are those by Arndt and Baker (1988), Baker and Derakhshan (1990), Vermiglio (1992), as well as those by Cahlon (1990, 1992, 1995), Cahlon and Schmidt (1997), and Tian and Kuang (1995) (on the stability of numerical approximations). The reader may also wish to look at the survey papers by Cryer (1972) and Baker (1997, 2000).

Collocation methods in piecewise polynomial spaces occur in Vermiglio (1992), and their superconvergence properties are studied in detail in Brunner (1994a), Baddour and Brunner (1993), Hu (1997, 1999), and Brunner (2004b, Chapter 4).

3.1.3. VFIDEs

The literature on the theory and the numerical solution of VIDEs with delays is more extensive. It starts of course with Volterra's work (Volterra (1909, 1912) and, especially, (1927), (1931)). Of the numerous books we list the ones by Cushing (1977), Györi and Ladas (1991), Lakshmikantham, Wen and Zhang (1994), Ito and Kappel (2002), and Zhao (2003); see also the surveys by Corduneanu and Lakshmikantham (1980) and Jackiewicz and Kwapisz (1991), and their bibliographies. The regularity of solutions is analysed in, *e.g.*, Willé and Baker (1992) and in Brunner and Zhang (1999).

Important early contributions to the numerical solution of VFIDEs are due to Thompson (1968) and Tavernini (1971, 1973, 1978) (linear multistep and general one-step methods). We also mention the papers by Jackiewicz (1984), Arndt and Baker (1988), Jackiewicz and Kwapisz (1991), Makroglou (1983) (block methods for VIDEs with constant delay), Kazakova and Bainov (1990), Enright and Hu (1997) (continuous Runge–Kutta methods), and Baker and Tang (1997, 2000). Most of these methods are based on ODE schemes and hence they require an appropriate interpolation scheme to produce 'dense' data. The construction of NCEs for RK methods applied to classical VIDEs is the subject in Vermiglio (1988) (see also Bellen, Jackiewicz, Vermiglio and Zennaro (1989) for the case of delay VIEs); it can be extended to VIDEs with non-vanishing delays. Bellen (1985) and Baker (1997, 2000) contain comprehensive surveys and extensive lists of references on the numerical treatment of functional differential equations.

The numerical treatment of *partial* VIDEs with delay arguments have received increased attention in recent years. This topic is beyond the scope of this article (and the expertise of its author); the interested reader may consult Zubik-Kowal (1999) and Zubik-Kowal and Vandewalle (1999) for results and additional references.

Cryer and Tavernini (1972) study Euler's method for very general Volterra functional equations. This method may of course be interpreted as a simple collocation method. The (super-) convergence properties of piecewise polynomial collocation methods for delay VIDEs are described in Brunner (1994b), Burgstaller (1993, 2000), and Hu and Peng (1999); see also Chapter 4 in Brunner (2004b). The papers by Koto (2002) and by Brunner and Vermiglio (2003) investigate stability and contractivity properties of solutions to VIDEs with constant delays and neutral VFIDEs of 'Hale's form'. However, much work remains to be done before a good understanding of the qualitative (asymptotic) properties of collocation solutions to general (nonlinear) VFIDEs is obtained.

Finally, we mention another, important approach to the numerical solution of VFIDEs: it is based on a semigroup framework generated by the given functional equation (*cf.* also Clément *et al.* (2002) and references) and is able to deal with a rather general class of (linear) neutral VFIDEs. This approach originated in the work of Banks and Kappel (1979); see also Ito and Kappel (1989, 1991), Ito and Turi (1991), Clément *et al.* (2002), and, especially, the recent monograph by Ito and Kappel (2002).

3.2. *Collocation methods for VFIEs with non-vanishing delays*

In order to lead the reader not familiar with collocation methods for classical Volterra integral and integro-differential equations to their application to Volterra-type functional equations, we briefly summarize the principal ideas and mathematical tools underlying these global discretization methods.

3.2.1. *Collocation spaces for classical Volterra equations*
Let $I_h := \{t_n : 0 = t_0 < t_1 < \cdots < t_N = T\}$ be a mesh on the interval $I := [0, T]$, and set

$$\sigma_n := (t_n, t_{n+1}], \quad \bar{\sigma}_n := [t_n, t_{n+1}], \quad h_n := t_{n+1} - t_n \quad (0 \leq n \leq N-1);$$

the diameter of the mesh I_h is $h := \max_{(n)} h_n$. For given integers $m \geq 1$ and $d \geq -1$ we let

$$S_{m+d}^{(d)}(I_h) := \{u_h \in C^d(I) : u_h|_{\sigma_n} \in \pi_{m+d} \ (0 \leq n \leq N-1)\} \qquad (3.1)$$

denote the linear space of (real) piecewise polynomials with respect to the mesh I_h whose degree does not exceed $m+d$. If $d = -1$ then $u_h \in S_{m-1}^{(-1)}(I_h)$ will in general have finite (jump) discontinuities at the interior points of I_h; the space of step functions, $S_0^{(-1)}(I_h)$, is the most obvious example of such a discontinuous piecewise polynomial space.

The dimension of the linear space defined by (3.1) is given by

$$\dim S_{m+d}^{(d)}(I_h) = Nm + (d+1).$$

The choice of d, the degree of regularity, will be governed by the type of functional equation whose solution will be approximated by collocation in the linear space $S_{m+d}^{(d)}(I_h)$: for the functional integral equations not containing derivatives of the unknown solution the 'natural' piecewise polynomial space is $S_{m-1}^{(-1)}(I_h)$ ($d = -1$), while for functional integro-differential equations in which the highest derivative of the unknown solution is $y^{(k)}$ ($k \geq 1$) we choose $d = k - 1$.

The desired collocation solution $u_h \in S_{m+d}^{(d)}(I_h)$ will be determined by requiring that it satisfy the given functional equation on the set of collocation points

$$X_h := \{t_{n,i} := t_n + c_i h_n : 0 < c_1 < \cdots < c_m \leq 1 \ (0 \leq n \leq N-1)\}, \quad (3.2)$$

described by given *collocation parameters* $\{c_i\}$. Clearly,

$$\dim S_{m+d}^{(d)}(I_h) = Nm + (d+1) = |X_h| + (d+1).$$

If $d \geq 0$ the collocation solution will also be required to coincide, at $t = 0$, with the prescribed initial value(s); e.g., in the case of the DVIDEs (2.19) and (2.20) ($k = 1$) we have $u_h(0) = y_0$.

3.2.2. Constrained and θ-invariant meshes

Assume that the given lag function $\theta(t) = t - \tau(t)$ satisfies the assumptions (D1)–(D3) of Section 2.1, which we will recall for the convenience of the reader:

(D1) $\theta \in C^d(I)$ for some $d \geq 0$, with $I := [t_0, T]$;
(D2) $\tau(t) \geq \tau_0 > 0$ for $t \in I$;
(D3) θ is strictly increasing on I.

We have seen, in the comments preceding Theorem 2.1, that the primary discontinuity points $\{\xi_\mu\}$, induced by θ and given by $\theta(\xi_\mu) = \xi_{\mu-1}$ ($\mu = 1, \cdots$; $\xi_0 := t_0$), possess the (uniform) separation property $\xi_\mu - \xi_{\mu-1} \geq \tau_0 > 0$ for all $\mu \geq 1$. For ease of notation we will again assume that T defining $I = [t_0, T]$ is such that

$$T = \xi_{M+1} \quad \text{for some} \ M \geq 1,$$

FUNCTIONAL INTEGRAL AND INTEGRO-DIFFERENTIAL VOLTERRA EQUATIONS

and we set $Z_M := \{\xi_\mu : \mu = 0, 1, \ldots, M\}$.

Since solutions of delay problems with non-vanishing delays generally suffer from a loss of reguarity at the primary discontinuity points $\{\xi_\mu\}$, the mesh I_h underlying the collocation space will have to include these points if the collocation solution is to attain its optimal global (or local) order (of superconvergence. Thus, we shall employ meshes of the form

$$I_h := \bigcup_{\mu=0}^{M} I_h^{(\mu)}, \quad I_h^{(\mu)} := \{t_n^{(\mu)} : \xi_\mu = t_0^{(\mu)} < t_1^{(\mu)} < \cdots < t_{N_\mu}^{(\mu)} = \xi_{\mu+1}\}. \quad (3.3)$$

Such a mesh is called a *constrained mesh* (with respect to θ) for I. We will refer to I_h as the *macro-mesh* and call the $I_h^{(\mu)}$ the underlying *local meshes*.

Definition. A mesh I_h for $I := [t_0, T]$ is said to be θ-*invariant* if it is constrained (that is, given by (3.3)) and if

$$\theta(I_h^{(\mu)}) = I_h^{(\mu-1)}, \quad \mu = 1, \ldots, M. \quad (3.4)$$

We then have $N_\mu = N$ for all $\mu \geq 0$.

Observe that if I_h is θ-invariant then

$$t \in I_h^{(\mu)} \implies \theta^{\mu-\nu}(t) \in I_h^{(\nu)}, \quad \nu = 0, 1, \ldots, \mu. \quad (3.5)$$

In analogy to Section 3.2.1 we will use the following notation:

$$\sigma_n^{(\mu)} := (t_n^{(\mu)}, t_{n+1}^{(\mu)}], \quad h_n^{(\mu)} := t_{n+1}^{(\mu)} - t_n^{(\mu)}, \quad h^{(\mu)} := \max_{(n)} h_n^{(\mu)}, \quad h := \max_{(\mu)} h^{(\mu)},$$

and $\bar{\sigma}_n^{(\mu)} := [t_n^{(\mu)}, t_{n+1}^{(\mu)}]$.

For a given θ-invariant mesh I_h the collocation solution u_h will be an element of a piecewise polynomial space

$$S_{m+d}^{(d)}(I_h) := \{v \in C^d(I_h) : v|_{\sigma_n^{(\mu)}} \in \pi_{m+d} \ (0 \leq n < N; \ 0 \leq \mu \leq M)\}. \quad (3.6)$$

It follows from Section 3.2.1 that this linear space has the dimension

$$\dim S_{m+d}^{(d)}(I_h) = (M+1)Nm + d + 1.$$

Hence the collocation points will now be chosen as

$$X_h := \bigcup_{\mu=0}^{M} X_h^{(\mu)}; \quad (3.7)$$

they are based on the $M+1$ local sets

$$X_h^{(\mu)} := \{t_n^{(\mu)} + c_i h_n^{(\mu)} : 0 < c_1 < \cdots < c_m \leq 1 \ (0 \leq n \leq N-1)\}$$

of cardinality Nm. In the collocation equation for a given delay equation with *non-vanishing* delay $\tau(t)$, we shall encounter the mapping $\theta(X_h^{(\mu)})$

(see, for example, (3.9) below). It is clear that for *linear lag functions* θ and given θ-invariant mesh I_h the set X_h defined by (3.7) is also θ-invariant. However, for *nonlinear* delays this will no longer be true. We record this important fact – which will affect the computational form of the collocation equation – in the following lemma. Its proof is straightforward and is left as an exercise.

Lemma 3.1. Assume that the delay function θ satisfies (D1)–(D3), and let I_h be a θ-invariant mesh on $I = [t_0, T]$.

(a) If θ is linear, then

$$\theta(X_h^{(\mu)}) = X_h^{(\mu-1)}, \quad \mu = 1, \ldots, M,$$

and the set X_h of collocation points is also θ-invariant.

(b) For nonlinear θ this is no longer true: setting

$$\theta(t_n^{(\mu)} + c_i h_n^{(\mu)}) = t_n^{(\mu-1)} + \tilde{c}_i h_n^{(\mu-1)} =: \tilde{t}_{n,i}^{(\mu-1)}, \quad i = 1, \ldots, m,$$

the images $\{\tilde{c}_i\}$ of the $\{c_i\}$ satisfy

$$0 \leq \tilde{c}_1 < \cdots < \tilde{c}_m \leq 1 \quad \text{(with } \tilde{c}_i \neq c_i \text{ in general)},$$

and they depend on the micro-interval $\sigma_n^{(\mu)}$ and the macro-interval $I^{(\mu)}$:

$$\tilde{c}_i = \tilde{c}_i(n; \mu), \quad i = 1, \ldots, m.$$

3.3. Delay integral equations of the second kind

3.3.1. The collocation equations

The collocation solution $u_h \in S_{m-1}^{(-1)}(I_h)$ for the delay integral equation

$$y(t) = g(t) + (\mathcal{V}y)(t) + (\mathcal{V}_\theta y)(t), \quad t \in (t_0, T], \tag{3.8}$$

with

$$(\mathcal{V}y)(t) := \int_{t_0}^{t} K_1(t, s) y(s)\, ds, \quad (\mathcal{V}_\theta y)(t) := \int_{t_0}^{\theta(t)} K_2(t, s) y(s)\, ds,$$

and with initial condition $y(t) = \phi(t)$, $t \leq t_0$, is defined by the collocation equation

$$u_h(t) = g(t) + (\mathcal{V}u_h)(t) + (\mathcal{V}_\theta u_h)(t), \quad t \in X_h. \tag{3.9}$$

The values of u_h at $t \in [\theta(t_0), t_0]$ are determined by the given initial function for (3.8), $u_h(t) = \phi(t)$. As for classical second-kind Volterra integral equations we will also consider the *iterated collocation solution* corresponding to u_h:

$$u_h^{\text{it}}(t) := g(t) + (\mathcal{V}u_h)(t) + (\mathcal{V}_\theta u_h)(t), \quad t \in (t_0, T]. \tag{3.10}$$

The lag function $\theta = \theta(t) = t - \tau(t)$ will be assumed to satisfy the conditions (D1)–(D3) of Section 3.2.2, and the mesh I_h on $I := [t_0, T]$ will be assumed to be the θ-invariant mesh defined by (3.3) and (3.4).

On $\sigma_n^{(\mu)} := (t_n^{(\mu)}, t_{n+1}^{(\mu)}]$ the collocation solution will have the usual local Lagrange representation,

$$u_h(t_n^{(\mu)} + vh_n^{(\mu)}) = \sum_{j=1}^{m} L_j(v) U_{n,j}^{(\mu)}, \quad v \in (0,1], \quad \text{with } U_{n,j}^{(\mu)} := u_h(t_{n,j}^{(\mu)}). \tag{3.11}$$

Since the contribution of the classical Volterra term $\mathcal{V}u_h$ to the computational form of the collocation equation is obvious, we will focus here on the terms induced by the delay part $(\mathcal{V}_\theta u_h)(t)$ with $t = t_{n,i}^{(\mu)}$.

Assume first that the delay θ is *linear*. As we have seen in Lemma 3.1, the θ-invariance of the mesh I_h implies the θ-invariance of the set X_h of collocation points; thus we may write, using the fact that $\theta(t_{n,i}^{(\mu)}) = t_{n,i}^{(\mu-1)}$,

$$(\mathcal{V}_\theta u_h)(t_{n,i}^{(\mu)}) = \int_{t_0}^{\theta(t_{n,i}^{(\mu)})} K_2(t_{n,i}^{(\mu)}, s) u_h(s) \, ds \tag{3.12}$$

$$= \int_{t_0}^{t_{n,i}^{(\mu-1)}} K_2(t_{n,i}^{(\mu)}, s) u_h(s) \, ds,$$

and hence, recalling the local representation (3.11) of u_h,

$$(\mathcal{V}_\theta u_h)(t_{n,i}^{(\mu)}) = \Psi_n^{(\mu-1)}(t_{n,i}^{(\mu)}) \tag{3.13}$$

$$+ h_n^{(\mu-1)} \sum_{j=1}^{m} \left(\int_0^{c_i} K_2(t_{n,i}^{(\mu)}, t_n^{(\mu-1)} + s h_n^{(\mu-1)}) L_j(s) \, ds \right) U_{n,j}^{(\mu-1)},$$

with lag term

$$\Psi_n^{(\mu-1)}(t) := \int_{t_0}^{\xi_{\mu-1}} K_2(t, s) u_h(s) \, ds \tag{3.14}$$

$$+ \int_{\xi_{\mu-1}}^{t_n^{(\mu-1)}} K_2(t, s) u_h(s) \, ds, \quad t \in \sigma_n^{(\mu)}.$$

If the delay θ is *nonlinear*, then the above terms have to be modified: by the (strict) monotonicity assumption (D3) for θ the image of $t_{n,i}^{(\mu)} \in \sigma_n^{(\mu)}$ under θ lies in $\sigma_n^{(\mu-1)}$ (but will be different from the collocation point $t_{n,i}^{(\mu-1)} = t_n^{(\mu-1)} + c_i h_n^{(\mu-1)}$); that is,

$$\theta(t_{n,i}^{(\mu)}) = t_n^{(\mu-1)} + \tilde{c}_i h_n^{(\mu-1)} =: \tilde{t}_{n,i}^{(\mu-1)}, \quad i = 1, \ldots, m, \tag{3.15}$$

with
$$0 < \tilde{c}_1 < \cdots < \tilde{c}_m \le 1 \quad \text{and} \quad \tilde{c}_i = \tilde{c}_i(n;\mu)$$

(cf. Lemma 3.1). Accordingly, the expression (3.13) for $(\mathcal{V}_\theta u_h)(t_{n,i}^{(\mu)})$ now reads

$$(\mathcal{V}_\theta u_h)(t_{n,i}^{(\mu)}) = \Psi_n^{(\mu-1)}(t_{n,i}^{(\mu)}) \tag{3.16}$$
$$+ h_n^{(\mu-1)} \sum_{j=1}^m \left(\int_0^{\tilde{c}_i} K_2(t_{n,i}^{(\mu)}, t_n^{(\mu-1)} + sh_n^{(\mu-1)}) L_j(s)\, ds \right) U_{n,j}^{(\mu-1)}.$$

Hence, the collocation equation (3.9) at $t = t_{n,i}^{(\mu)}$ ($i = 1, \ldots, m$) can now be written as

$$U_{n,i}^{(\mu)} = h_n^{(\mu)} \sum_{j=1}^m \left(\int_0^{c_i} K_1(t_{n,i}^{(\mu)}, t_n^{(\mu)} + sh_n^{(\mu)}) L_j(s)\, ds \right) U_{n,j}^{(\mu)} \tag{3.17}$$
$$+ g(t_{n,i}^{(\mu)}) + F_n^{(\mu)}(t_{n,i}^{(\mu)}) + (\mathcal{V}_\theta u_h)(t_{n,i}^{(\mu)}).$$

The classical lag term (corresponding to the Volterra operator $\mathcal{V}$ in (3.9)) has, for $t \in \sigma_n^{(\mu)}$, the form

$$F_n^{(\mu)}(t) := \int_{t_0}^{\xi_\mu} K_1(t,s) u_h(s)\, ds + \int_{\xi_\mu}^{t_n^{(\mu)}} K_1(t,s) u_h(s)\, ds. \tag{3.18}$$

Let $\mathbf{U}_n^{(\mu)} := (U_{n,1}^{(\mu)}, \ldots, U_{n,m}^{(\mu)})^T \in \mathbb{R}^m$ and define the matrices in $L(\mathbb{R}^m)$,

$$B_n^{(\mu)} := \left(\int_0^{c_i} K_1(t_{n,i}^{(\mu)}, t_n^{(\mu)} + sh_n^{(\mu)}) L_j(s)\, ds \right)_{i,j=1}^m,$$

$$\tilde{B}_n^{(\mu-1)} := \left(\int_0^{\tilde{c}_i} K_2(t_{n,i}^{(\mu)}, t_n^{(\mu-1)} + sh_n^{(\mu-1)}) L_j(s)\, ds \right)_{i,j=1}^m.$$

Finally, set
$$\mathbf{g}_n^{(\mu)} := (g(t_{n,1}^{(\mu)}), \ldots, g(t_{n,m}^{(\mu)}))^T,$$
$$\mathbf{G}_n^{(\mu)} := (F(t_{n,1}^{(\mu)}), \ldots, F_n^{(\mu)}(t_{n,m}^{(\mu)}))^T,$$

and
$$\mathbf{Q}_n^{(\mu-1)} := (\Psi_n^{(\mu-1)}(t_{n,1}^{(\mu)}), \ldots, \Psi_n^{(\mu-1)}(t_{n,m}^{(\mu)}))^T.$$

Thus, the collocation solution $u_h \in S_{m-1}^{(-1)}(I_h)$ to (3.8) on $\sigma_n^{(\mu)}$ is described by (3.11), in which $\mathbf{U}_n^{(\mu)}$ is the solution of the linear algebraic system

$$[I_m - h_n^{(\mu)} B_n^{(\mu)}] \mathbf{U}_n^{(\mu)} = \mathbf{g}_n^{(\mu)} + \mathbf{G}_n^{(\mu)} + \mathbf{Q}_n^{(\mu-1)} + h_n^{(\mu-1)} \tilde{B}_n^{(\mu)} \mathbf{U}_n^{(\mu-1)}, \tag{3.19}$$

where $n = 0, 1, \ldots, m$ and $\mu = 0, 1, \ldots, M$. The matrix I_m denotes the identity operator in $L(\mathbb{R}^m)$.

The following theorem on the existence of a unique collocation solution is an obvious consequence of the uniform boundedness of the inverses of the matrices $\mathcal{B}_n^{(\mu)} := I_m - h_n^{(\mu)} B_n^{(\mu)}$ for sufficiently small mesh diameters h.

Theorem 3.2. Assume that g, θ, K_1 and K_2 are continuous on their respective domains I, D and D_θ, with the lag function θ satisfying (D1)–(D3). Then there exists an $\bar{h} > 0$ such that, for any θ-invariant mesh I_h with $h \in (0, \bar{h})$ and any initial function $\phi \in [\theta(t_0), t_0]$, each of the linear algebraic systems (3.19) possesses a unique solution $\mathbf{U}_n^{(\mu)} \in \mathbb{R}^m$. Hence, the collocation equation (3.9) defines a unique collocation solution $u_h \in S_{m-1}^{(-1)}(I_h)$ for (3.8) whose local representation on the subintervals $\sigma_n^{(\mu)}$ is given by (3.11).

The computational form of the *iterated collocation solution* (3.10) at $t = t_n^{(\mu)} + v h_n^{(\mu)} \in \bar{\sigma}_n^{(\mu)}$ can be written as

$$u_h^{it}(t) = g(t) + F_n^{(\mu)}(t) + \Psi_n^{(\mu-1)}(t) \tag{3.20}$$

$$+ h_n^{(\mu)} \sum_{j=1}^{m} \left(\int_0^v K_1(t, t_n^{(\mu)} + s h_n^{(\mu)}) L_j(s) \, ds \right) U_{n,j}^{(\mu)}$$

$$+ h_n^{(\mu-1)} \sum_{j=1}^{m} \left(\int_0^{\tilde{v}} K_2(t, t_n^{(\mu-1)} + s h_n^{(\mu-1)}) L_j(s) \, ds \right) U_{n,j}^{(\mu-1)}.$$

Recall that the lag term $\Psi_n^{(\mu-1)}(t)$ corresponding to the delay operator $\mathcal{V}_\theta$ is given above by (3.14). The image $\tilde{t} := t_n^{(\mu-1)} + \tilde{v} h_n^{(\mu-1)}$ of $t = t_n^{(\mu)} + v h_n^{(\mu)}$ under θ depends on the nature of the lag function θ: if θ is *linear* then we have $\tilde{v} = v$; for *nonlinear* θ the value of $\tilde{v} \in [0, 1]$ must be obtained from

$$\theta(t_n^{(\mu)} + v h_n^{(\mu)}) =: t_n^{(\mu-1)} + \tilde{v} h_n^{(\mu-1)}, \quad v \in (0, 1]. \tag{3.21}$$

We note in passing that $u_h^{it} \in C[t_0, T]$ whenever the given data defining the initial-value problem for (3.8) are continuous functions and we have

$$u_h^{it}(t_0) = g(t_0) - \int_{\theta(t_0)}^{t_0} K_2(t_0, s) \phi(s)) \, ds \; (= y(t_0^+)).$$

Moreover,

$$u_h^{it}(t) = u_h(t) \quad \text{for all } t \in X_h.$$

Since second-kind Volterra integral equations with non-vanishing delays often arise in the particular form

$$y(t) = g(t) + (\mathcal{W}_\theta y)(t), \quad t \in (t_0, T], \tag{3.22}$$

where
$$(\mathcal{W}_\theta y)(t) := \int_{\theta(t)}^{t} K(t,s) y(s)\, ds,$$

we present the corresponding computational form of the collocation equation defining $u_h \in S_{m-1}^{(-1)}(I_h)$ in some detail (although it could of course be formally obtained by setting $K_2 = -K_1 =: -K$ in (3.17). We first note that for $t = t_{n,i}^{(\mu)}$ we have

$$(\mathcal{W}_\theta u_h)(t) = \int_{\theta(t)}^{t_{n+1}^{(\mu-1)}} K(t,s) u_h(s)\, ds \qquad (3.23)$$
$$+ \int_{t_{n+1}^{(\mu-1)}}^{\xi_\mu} K(t,s) u_h(s)\, ds + \int_{\xi_\mu}^{t_n^{(\mu)}} K(t,s) u_h(s)\, ds$$
$$+ h_n^{(\mu)} \int_0^{c_i} K(t, t_n^{(\mu)} + s h_n^{(\mu)}) u_h(t_n^{(\mu)} + s h_n^{(\mu)})\, ds,$$

where
$$\theta(t) = \theta(t_{n,i}^{(\mu)}) = \begin{cases} t_{n,i}^{(\mu-1)} = t_n^{(\mu-1)} + c_i h_n^{(\mu-1)}, & \text{if } \theta \text{ is } \textit{linear}, \\ \tilde{t}_{n,i}^{(\mu-1)} := t_n^{(\mu-1)} + \tilde{c}_i h_n^{(\mu-1)}, & \text{if } \theta \text{ is } \textit{nonlinear}. \end{cases}$$

Define, for $t = t_n^{(\mu)} + c_i h_n^{(\mu)}$,
$$\bar{\Psi}_n^{(\mu-1)}(t) := h_n^{(\mu-1)} \int_{\tilde{c}_i}^{1} K(t, t_n^{(\mu-1)} + s h_n^{(\mu-1)}) u_h(t_n^{(\mu-1)} + s h_n^{(\mu-1)})\, ds$$
$$+ \int_{t_{n+1}^{(\mu-1)}}^{\xi_\mu} K(t,s) u_h(s)\, ds + \int_{\xi_\mu}^{t_n^{(\mu)}} K(t,s) u_h(s)\, ds. \qquad (3.24)$$

The collocation equation for (3.22) on $\sigma_n^{(\mu)}$ then becomes
$$U_{n,i}^{(\mu)} = g(t_{n,i}^{(\mu)}) + \bar{\Psi}_n^{(\mu-1)}(t_{n,i}^{(\mu)}) \qquad (3.25)$$
$$+ h_n^{(\mu)} \sum_{j=1}^{m} \left(\int_0^{c_i} K(t_{n,i}^{(\mu)}, t_n^{(\mu)} + s h_n^{(\mu)}) L_j(s)\, ds \right) U_{n,j}^{(\mu)}, \quad i = 1, \ldots, m.$$

Hence, the resulting linear algebraic system for $U_n^{(\mu)} \in \mathbb{R}^m$ defining the local representation of u_h on $\sigma_n^{(\mu)}$ (cf. (3.11)) has the form
$$[I_m - h_n^{(\mu)} B_n^{(\mu)}] U_n^{(\mu)} = \mathbf{g}_n^{(\mu)} + \bar{\mathbf{G}}_n^{(\mu-1)}, \qquad (3.26)$$

with
$$\mathbf{g}_n^{(\mu)} := (g(t_{n,1}^{(\mu)}), \ldots, g(t_{n,m}^{(\mu)}))^T$$

and
$$\bar{\mathbf{G}}_n^{(\mu-1)} := (\bar{\Psi}_n^{(\mu-1)}(t_{n,1}^{(\mu)}), \ldots, \bar{\Psi}_n^{(\mu-1)}(t_{n,m}^{(\mu)}))^T.$$

The corresponding *iterated collocation solution* at $t = t_n^{(\mu)} + vh_n^{(\mu)} \in \bar{\sigma}_n^{(\mu)}$ can then be computed via

$$u_h^{it}(t) = g(t) + \bar{\Psi}_n^{(\mu-1)}(t) \qquad (3.27)$$
$$+ h_n^{(\mu)} \sum_{j=1}^m \left(\int_0^{c_i} K(t, t_n^{(\mu)} + sh_n^{(\mu)}) L_j(s)\, ds \right) U_{n,j}^{(\mu)}.$$

3.3.2. Global convergence results

The collocation error $e_h := y - u_h$ associated with the collocation solution $u_h \in S_{m-1}^{(-1)}(I_h)$ for the delay integral equation (3.8) solves the initial-value problem

$$e_h(t) = \delta_h(t) + (\mathcal{V}e_h)(t) + (\mathcal{V}_\theta e_h)(t), \quad t \in (t_0, T], \qquad (3.28)$$

with initial condition $e_h(t) = 0$ for $t \in [\theta(t_0), t_0]$. The defect δ_h, defined by

$$\delta_h(t) := -u_h(t) + g(t) + (\mathcal{V}u_h)(t) + (\mathcal{V}_\theta u_h)(t), \quad t \in I,$$

vanishes on the set X_h. For $t \in \sigma_n^{(\mu)}$ ($\mu \geq 1$) the above error equation can be written as

$$e_h(t) = E_\mu(t) + \delta_h(t) + \int_{\xi_\mu}^t K_1(t,s) e_h(s)\, ds, \qquad (3.29)$$

where

$$E_\mu(t) := \sum_{\nu=0}^{\mu-1} \int_{\xi_\nu}^{\xi_{\nu+1}} K_1(t,s) e_h(s)\, ds + (\mathcal{V}_\theta e_h)(t). \qquad (3.30)$$

On the first macro-interval $(t_0, \xi_1]$ we have

$$E_0(t) := (\mathcal{V}_\theta e_h)(t) = -\int_{\theta(t)}^{t_0} K_2(t,s) e_h(s)\, ds = 0.$$

If the given functions in (3.8) have continuous derivatives of at least order m on their respective domains, the global convergence and order analysis can be based on the (local) representation of the collocation error based on the Peano Kernel Theorem for polynomial interpolation. This representation has the form

$$e_h(t_n^{(\mu)} + vh_n^{(\mu)}) = \sum_{j=1}^m L_j(v) \mathcal{E}_{n,j}^{(\mu)} + (h_n^{(\mu)})^m R_{m,n}^{(\mu)}(v), \quad v \in (0,1], \qquad (3.31)$$

with $\mathcal{E}_{n,j}^{(\mu)} := e_h(t_{n,j}^{(\mu)})$ and Peano remainder term $R_{m,n}^{(\mu)}(v)$ (see Brunner (2004b, Chapters 1 and 2) for details). On the first macro-interval $[\xi_0, \xi_1]$

the estimate for e_h is the one for classical Volterra integral equations of the second kind (Brunner and van der Houwen 1986, Chapter 5):

$$\|e_h\|_{0,\infty} := \sup_{t \in I^{(0)}} |e_h(t)| \leq C_0(h^{(0)})^m, \quad n = 0, 1, \ldots, N-1;$$

it is a consequence of the estimate $\|\mathcal{E}_n^{(0)}\|_1 = \mathcal{O}((h^{(0)})^m)$ (where $\mathcal{E}_n^{(\mu)} := (\mathcal{E}_{n,1}^{(\mu)}, \ldots, \mathcal{E}_{n,m}^{(\mu)})^T$). A simple induction argument, employing the estimates for the terms $E_\mu(t)$ ($t \in I^{(\mu)}$) in (3.29) and (3.30), together with the observation that by the conditions (D1)–(D3) for the delay θ the number $(M+1)$ of macro-intervals $I^{(\mu)} := [\xi_\mu, \xi_{\mu+1}]$ is finite, yields the results summarized in the following theorem.

Theorem 3.3. Let the following conditions be satisfied.

(a) The given functions g, K_1, K_2 and ϕ in (3.8) all possess continuous derivatives of order m on their respective domains.

(b) The delay function $\theta(t) = t - \tau(t)$ is subject to the conditions (D1)–(D3) of Section 3.2.2, with $d \geq m$ in (D1).

(c) $u_h \in S_{m-1}^{(-1)}(I_h)$ is the collocation solution to (4.8) corresponding to a θ-invariant mesh I_h with $h \in (0, \bar{h}$, where $\bar{h}$ is defined in Theorem 3.2.

Then, for any set of collocation parameters $\{c_i : 0 \leq c_1 < \cdots < c_m \leq 1\}$, the collocation error admits the estimate

$$\|y - u_h\|_\infty := \sup_{t \in (t_0, T]} |e_h(t)| \leq Ch^m. \tag{3.32}$$

The constant C depends on the $\{c_i\}$ but not on $h := \max_{(n,\mu)} h_n^{(\mu)}$.

Although it follows from (3.28) and Theorem 3.3 that, in general, $\|\delta_h\|_\infty = \mathcal{O}(h^m)$ only, a judicious choice of the collocation parameters $\{c_i\}$ leads (not too surprisingly, if we look at the close connection between the degree of precision of interpolatory m-point quadrature formulas based on these abscissas and the variation-of-constants formula of Theorem 2.2 adapted to the error equation!) to *global superconvergence* on I for the *iterated collocation solution* u_h^{it}.

Theorem 3.4. Suppose that the assumptions (a)–(c) of Theorem 3.3 hold, but with $m+1$ replacing m in (a) and (b). If the collocation parameters $\{c_i\}$ are chosen so that the orthogonality condition

$$J_0 := \int_0^1 \prod_{i=1}^m (s - c_i) \, ds = 0 \tag{3.33}$$

is satisfied, then the iterated collocation solution corresponding to the collocation solution $u_h \in S_{m-1}^{(-1)}(I_h)$ for (3.8) is globally superconvergent on I_h:

$$\|y - u_h^{\text{it}}\|_\infty \leq Ch^{m+1},$$

with C depending on the $\{c_i\}$ but not on h.

Proof. The key to the proof of Theorem 3.4 (and Theorem 3.6 below) on global superconvergence is the variation-of-constants formula (or 'resolvent representation') for e_h, together with the general global convergence result of Theorem 3.3 and the observation that

$$e_h^{\text{it}}(t) := y(t) - u_h^{\text{it}}(t) = e_h(t) - \delta_h(t), \quad t \in I.$$

For $t = t_n^{(\mu)} + vh_n^{(\mu)} \in \bar{\sigma}_n^{(\mu)}$ Theorem 2.2 yields, with e_h and δ_h replacing y, g and $g_0 = g$, respectively,

$$e_h^{\text{it}}(t) = \int_{\xi_\mu}^t R_1(t,s)\delta_h(s)\,ds + \sum_{\nu=0}^{\mu-1} \int_{\xi_\nu}^{\xi_{\nu+1}} R_{\mu,\nu}(t,s)\delta_h(s)\,ds \qquad (3.34)$$

$$+ \sum_{\nu=0}^{\mu-1} \int_{\xi_\nu}^{\theta^{\mu-\nu}(t)} Q_{\mu,\nu}(t,s)\delta_h(s)\,ds.$$

The integrals, having as lower and upper limits points of the (θ-invariant) mesh I_h, can be written as sums of integrals over individual micro-intervals $\bar{\sigma}_n$, and each of these integrals can then be replaced by the sum of an interpolatory m-point quadrature formula with respect to the collocation points in that interval and the corresponding quadrature error. The expression given by the quadrature formula has value zero, since $\delta_h(t) = 0$ for $t \in X_h$. Owing to the assumed regularity of the data (which is inherited on D by the resolvent R_1 and, piecewise on D, by the functions $R_{\mu,\nu}$, $Q_{\mu,\nu}$), the orthogonality condition (3.33) implies that all quadrature errors are $\mathcal{O}(h^{m+1})$. Here, we have used the result that, by definition, the defect δ_h and its derivatives $\delta_h^{(\nu)}$ ($\nu \leq m+1$), are uniformly bounded on each interval $I^{(\mu)}$.

It remains to deal with the integrals

$$\int_{t_n^{(\mu)}}^t R_1(t,s)\delta_h(s)\,ds \quad \text{and} \quad \int_{t_n^{(\nu)}}^{\theta^{\mu-\nu}(t)} Q_{\mu,\nu}(t,s)\delta_h(s)\,ds$$

(recall from (3.5) that $\theta^{\mu-\nu}(t) \in \sigma_n^{(\nu)}$ if $t \in \sigma_n^{(\mu)}$). As we have observed before, the defect δ_h induced by the collocation solution satisfies $\|\delta_h\|_\infty = \mathcal{O}(h^m)$. Thus, in the estimation of the above integrals (via the usual scaling) the uniform estimate for δ_h is multiplied by h, leading to the required $\mathcal{O}(h^{m+1})$-term in Theorem 3.4. $\square$

Corollary 3.5. In the particular delay integral equation (3.22) assume that $g \in C^{m+1}(I)$ and $K \in C^{m+1}(\bar{D}_\theta)$, with $\bar{D}_\theta := \{(t,s) : \theta(t) \leq s \leq t, \ t \in I\}$. Then the iterated collocation solution based on $u_h \in S_{m-1}^{(-1)}(I_h)$ and defined by (3.27) has the global superconvergence property

$$\|y - u_h^{it}\|_\infty \leq Ch^{m+1}$$

provided the mesh I_h is θ-invariant, the $\{c_i\}$ underlying the set X_h of collocation points satisfy $J_0 = 0$ (cf. (3.33)), and $\phi \in C^{m+1}[\theta(t_0), t_0]$.

3.3.3. Local superconvergence results

The proof of the global superconvergence result in Theorem 3.4 indicates that we can readily refine it so as to establish stronger *local* superconvergence properties for u_h and u_h^{it} at the *mesh points* $t = t_n^{(\mu)}$.

Theorem 3.6. Let the given functions g, K_1, K_2 and ϕ in the delay integral equation (3.8) have continuous derivatives of order $m + \kappa$ in their respective domains I, D, D_θ and $[\theta(t_0), t_0]$, and assume that the delay function θ is subject to the conditions (D1)–(D3) of Section 3.2.2, with $d \geq m+\kappa$ in (D1). If $u_h \in S_{m-1}^{(-1)}(I_h)$ denotes the collocation solution for a θ-invariant mesh I_h, with corresponding iterated collocation solution u_h^{it}, and if the collocation parameters satisfy the orthogonality conditions

$$J_\nu := \int_0^1 s^\nu \prod_{i=1}^m (s - c_i) \, ds = 0, \quad 0 \leq \nu \leq \kappa - 1,$$

with $J_\kappa \neq 0$, then

$$\max_{t \in I_h \setminus \{t_0\}} |y(t) - u_h^{it}| \leq Ch^{m+\kappa}, \quad \text{for } h \in (0, \hat{h}).$$

If, in addition, we have $c_m = 1$ (implying $\kappa < m$), then u_h itself exhibits local superconvergence at the mesh points, that is,

$$\max_{t \in I_h \setminus \{t_0\}} |y(t) - u_h(t)| \leq Ch^{m+\kappa}.$$

Proof. Our starting point is (3.34) in the proof of Theorem 3.4 where we now set $t = t_n^{(\mu)}$. Hence,

$$e_h^{it}(t_n^{(\mu)}) = \int_{\xi_\mu}^{t_n^{(\mu)}} R_1(t_n^{(\mu)}, s)\delta_h(s)\,ds + \sum_{\nu=0}^{\mu-1} \int_{\xi_\nu}^{\xi_{\nu+1}} R_{\mu,\nu}(t_n^{(\mu)}, s)\delta_h(s)\,ds$$

$$+ \sum_{\nu=0}^{\mu-1} \int_{\xi_\nu}^{\theta^{\mu-\nu}(t_n^{(\mu)})} Q_{\mu,\nu}(t_n^{(\mu)}, s)\delta_h(s)\,ds,$$

where $0 \leq n < N$, $0 \leq \mu \leq M$, with $\theta^{\mu-\nu}(t_n^{(\mu)}) = t_n^{(\nu)}$ (cf. (3.5)). Hence, the familiar quadrature argument is applicable: since the defect δ_h vanishes on X_h (and possesses uniformly bounded derivatives of order $m = \kappa$ on each $I^{(\mu)}$), and since the orthogonality and regularity conditions imply that the quadrature errors induced by the interpolatory m-point quadrature formulas based on the $\{c_i\}$ are all of order $\mathcal{O}(h^{m+\kappa})$, with the number $M+1$ of macro-intervals $I^{(\mu)}$ being finite, the first assertion in Theorem 4.6 follows immediately.

The second assertion is based on the fact that when $c_m = 1$, each mesh point $t_n^{(\mu)}$ ($1 \le n \le N$) is a collocation point and thus $u_h^{it}(t_n^{(\mu)}) = u_h(t_n^{(\mu)})$, since $\delta_h(t_n^{(\mu)}) = 0$. Note also that $e_h^{it}(t_0) = 0$ because $u_h^{it}(t_0) = y(t_0^+)$. □

Corollary 3.7. Assume $\kappa = m$ in Theorem 3.6. Then collocation in $S_{m-1}^{(-1)}(I_h)$ at the Gauss points leads to an iterated collocation solution with the property that

$$\max_{t \in I_h \setminus \{t_0\}} |y(t) - u_h^{it}(t)| \le Ch^{2m},$$

while

$$\max_{t \in I_h \setminus \{t_0\}} |y(t) - u_h(t)| \le Ch^m \quad \text{only.}$$

Corollary 3.8. Suppose that $\kappa = m-1$ and $c_m = 1$. The optimal order of convergence of the collocation solution $u_h \in S_{m-1}^{(-1)}(I_h)$ corresponding to the Radau II points is then given by

$$\max_{t \in I_h \setminus \{t_0\}} |y(t) - u_h(t)| \le Ch^{2m-1}.$$

Recall that we have $u_h^{it}(t) = u_h(t)$ for $t \in I_h \setminus \{t_0\}$ whenever $c_m = 1$ (i.e., when $t_n \in X_h$, $n = 1, \ldots, N$).

We illustrate these results by an example; it is also introduced, in view of Sections 4 and 5, to remind the reader that the nature of a given delay $\tau(t)$ (non-vanishing versus vanishing) is often governed by location of the initial point t_0 in $I = [t_0, T]$.

Example 3.1. (Non-vanishing proportional delay) On $I = [t_0, T]$ with $t_0 > 0$, the delay function $\theta(t) = qt (0 < q < 1)$ corresponds to a non-vanishing delay $\tau(t)$ since

$$\theta(t) = qt = t - (1-q)t =: t - \tau(t),$$

with $\tau(t) \ge (1-q)t_0 > 0$ for $t \in I$. Hence, the primary discontinuity points $\{\xi_\mu\}$ are given by

$$\xi_\mu = q^{-\mu} t_0, \quad \mu \ge 0.$$

We will assume, for ease of exposition and without loss of generality, that T is such that $\xi_{M+1} = T$ for some $M > 1$. Hence, we may write

$$\xi_\mu = q^{M+1-\mu}T, \quad \mu = 0, 1, \ldots, M+1.$$

Suppose that the mesh I_h is constrained, and let each local mesh $I_h^{(\mu)}$ be *uniform*:

$$I_h^{(\mu)} := \{t_n^{(\mu)} := \xi_\mu + nh^{(\mu)} : n = 0, 1, \ldots, N \; (h^{(\mu)} = q^{-(\mu+1)}(1-q)t_0/N)\}.$$

A mesh of this type is often called a *quasi-geometric mesh* (see also Liu (1995a), Bellen, Guglielmi and Torelli (1997), Bellen (2001), Bellen, Brunner, Maset and Torelli (2002), and Guglielmi and Zennaro (2003). The linearity of θ then implies that I_h is θ-invariant, and the same is true for the set X_h of collocation points.

This choice of the local meshes defining I_h implies that

$$h = h^{(M)} = \frac{1}{N}(\xi_{M+1} - \xi_M) = (1-q)\frac{T}{N},$$

and

$$h^{(\mu)} = \frac{1}{N}(\xi_{\mu+1} - \xi_\mu) = q^{M+1-\mu-1}(1-q)\frac{T}{N}, \quad \mu = 0, 1, \ldots, M.$$

The result of, *e.g.*, Theorem 3.6 then becomes

$$\max_{t \in I_h \setminus \{t_0\}} |y(t) - u_h^{\mathrm{it}}(t)| \leq C(q)N^{-(m+\kappa)}.$$

Note that this result also holds for the delay VIE (3.22) on intervals $I = [t_0, T]$ with $t_0 > 0$.

3.3.4. Nonlinear delay VIEs

Since the extension of the convergence analysis presented in the previous sections to the general nonlinear version of (3.8),

$$y(t) = g(t) + (\mathcal{V}y)(t) + (\mathcal{V}_\theta y)(t), \quad t \in (t_0, T], \tag{3.35}$$

is rather straightforward, we will omit it and instead focus on a class of nonlinear delay VIEs occurring frequently in applications. These functional equations have the form

$$y(t) = g(t) + (\mathcal{W}_\theta y)(t), \quad t \in (t_0, T], \tag{3.36}$$

where the (nonlinear) Volterra operator $\mathcal{W}_\theta$ is of *Hammerstein type*:

$$(\mathcal{W}_\theta y)(t) := \int_{\theta(t)}^{t} k(t-s)G(s, y(s))\,ds. \tag{3.37}$$

There are two ways to generate collocation approximations to solutions of Volterra–Hammerstein integral equations of the second kind. In the 'direct' approach discussed above we approximate y by $u_h \in S_{m-1}^{(-1)}(I_h)$, followed by the iterated collocation solution u_h^{it} based on u_h. Alternatively, we can resort to what is called *implicitly linear collocation*. Setting $z(t) := (\mathcal{N}y)(t) := G(t, y(t))$, where $\mathcal{N}$ is the *Niemytzki operator* (or substitution operator), the nonlinear delay VIE (3.36) becomes an *implicitly linear delay VIE* for z,

$$z(t) = G\left(t, g(t) + \int_{\theta(t)}^{t} k(t-s)z(s)\,ds\right), \quad t \in (t_0, T], \tag{3.38}$$

with initial condition $z(t) = G(t, \phi(t))$, $t \in [\theta(t_0), t_0]$. The solution of the original DVIE is then obtained via the recursion

$$y(t) = g(t) + (\mathcal{L}_\theta z)(t), \quad t \in (t_0, T], \tag{3.39}$$

where $\mathcal{L}_\theta$ denotes the linear delay Volterra operator

$$(\mathcal{L}_\theta y)(t) := \int_{\theta(t)}^{t} k(t-s)z(s)\,ds.$$

We note in passing that the survey paper by Brezis and Browder (1975) and the monograph by Krasnosel'skii and Zabreiko (1984) contain many results relevant in the analysis of solvability of Hammerstein integral equations and operator equations (*e.g.*, (3.38)) involving the Niemytzki operator. Compare also Kumar and Sloan (1987) and Brunner (1992) for details and additional references.

The solution z of (3.38) will be approximated by $z_h \in S_{m-1}^{(-1)}(I_h)$, using the same collocation points X_h as in the direct approach: it is defined by the *implicit linear collocation equation*

$$z_h(t) = G\left(t, g(t) + \int_{\theta(t)}^{t} k(t-s)z_h(s))\,ds\right), \quad t \in X_h, \tag{3.40}$$

with initial values $z_h(t) = G(t, \phi(t))$, $t \in [\theta(t_0), t_0]$. This leads to the approximation y_h for the solution y of the orginal DVIE,

$$y_h(t) := g(t) + (\mathcal{L}_\theta z_h)(t), \quad t \in [t_0, T]. \tag{3.41}$$

Setting

$$z_h(t_n^{(\mu)} + vh_n^{(\mu)}) = \sum_{j=1}^{m} L_j(v) Z_{n,j}^{(\mu)}, \quad v \in (0,1], \quad \text{with } Z_{n,i}^{(\mu)} := z_h(t_{n,j}^{(\mu)}), \tag{3.42}$$

the computational forms of these equations at $t = t_{n,i}^{(\mu)}$ and at $t = t_n^{(\mu)} + vh_n^{(\mu)}$, respectively, are

$$Z_{n,i}^{(\mu)} = G\left(t_{n,i}^{(\mu)}, g(t_{n,i}^{(\mu)}) + \bar{\Psi}_n^{(\mu-1)}(t_{n,i}^{(\mu)})\right) \tag{3.43}$$

$$+ h_n^{(\mu)} \sum_{j=1}^m \left(\int_0^{c_i} k((c_i - s)h_n^{(\mu)}) L_j(s)\,ds\right) Z_{n,j}^{(\mu)}$$

$(i = 1, \ldots, m)$, where for $t = t_n^{(\mu)} + v h_n^{(\mu)} \in \bar{\sigma}_n^{(\mu)}$ we have

$$\bar{\Psi}_n^{(\mu-1)}(t) := h_n^{(\mu-1)} \int_{\tilde{v}}^1 k(t - t_n^{(\mu-1)} - s h_n^{(\mu-1)}) z_h(t_n^{(\mu-1)} + s h_n^{(\mu-1)})\,ds$$

$$[2pt] \quad + \int_{t_{n+1}^{(\mu-1)}}^{\xi_\mu} k(t-s) z_h(s)\,ds + \int_{\xi_\mu}^{t_n^{(\mu)}} k(t-s) z_h(s)\,ds,$$

and

$$y_h(t) = g(t) + \bar{\Psi}_n^{(\mu-1)}(t) \tag{3.44}$$

$$+ h_n^{(\mu)} \sum_{j=1}^m \left(\int_0^v k((v-s)h_n^{(\mu)}) L_j(s)\,ds\right) Z_{n,j}^{(\mu)}, \quad v \in [0,1].$$

Recall that the number $\tilde{v} \in [0,1]$ is obtained from

$$\theta(t_n^{(\mu)} + v h_n^{(\mu)}) =: t_n^{(\mu-1)} + \tilde{v} h_n^{(\mu-1)}, \quad v \in [0,1],$$

with $\tilde{v} = v$ if the delay function θ is linear.

The merits of this indirect collocation approach are twofold. Since in many applications the convolution kernel $k(t-s)$ is given by an elementary function like $\exp(\gamma(t-s))$ or $(t-s)^{-\alpha}$ ($\alpha < 1$), the integrals in (3.43) and (3.44) can be found analytically. Perhaps more importantly, the 'decoupling' of the nonlinear G and z in (3.38) implies that during the iteration process for solving the nonlinear algebraic systems (3.43) we do not have to re-compute the integrals in each iteration step, in contrast to direct collocation for (3.36).

3.4. Collocation for VIDEs with delay arguments

3.4.1. The collocation equations

The description and analysis of piecewise polynomial collocation solutions for delay integral equations have provided all the essential ideas required to deal with collocation solutions for the initial-value problem for delay VIDEs,

$$y'(t) = f(t, y(t), y(\theta(t))) + (\mathcal{V}y)(t) + (\mathcal{V}_\theta y)(t), \quad t \in I := [t_0, T],$$
$$y(t) = \phi(t), \quad t \in [\theta(t_0), t_0], \tag{3.45}$$

with Volterra integral operators $\mathcal{V}$ and $\mathcal{V}_\theta$ as in (3.8). The lag function θ

will again be assumed to satisfy conditions (D1)–(D3) of Section 3.2.2. We will usually employ its linear counterpart, corresponding to

$$f(t, y, z) = a(t)y + b(t)z + g(t), \qquad (3.46)$$

to illustrate the essential ideas of the analysis.

The natural collocation space is now $S_m^{(0)}(I_h)$, and hence the collocation equation defining u_h with respect of the θ-invariant mesh I_h is

$$u_h(t) = f(t, u_h(t), u_h(\theta(t))) + (\mathcal{V}u_h)(t) + (\mathcal{V}_\theta u_h)(t), \quad t \in X_h, \qquad (3.47)$$

with $u_h(t) := \phi(t)$ if $t \leq t_0$. For $t \in \sigma_n^{(\mu)}$ we define the lag term approximations

$$F_n^{(\mu)}(t) := \int_{t_0}^{\xi_\mu} K_1(t, s) u_h(s) \, ds + \int_{\xi_\mu}^{t_n^{(\mu)}} K_1(t, s) u_h(s) \, ds, \qquad (3.48)$$

and

$$(\mathcal{V}_\theta u_h)(t) = \Psi_n^{(\mu-1)}(t) + \int_{t_n^{(\mu-1)}}^{\theta(t)} K_2(t, s) u_h(s) \, ds. \qquad (3.49)$$

In analogy to (3.14) we have

$$\Psi_n^{(\mu-1)}(t) = \int_{t_0}^{\xi_{\mu-1}} K_2(t, s) u_h(s) \, ds + \int_{\xi_{\mu-1}}^{t_n^{(\mu-1)}} K_2(t, s) u_h(s) \, ds.$$

With the local Lagrange representation of u_h on $\bar{\sigma}_n^{(\mu)}$,

$$u_h(t_n^{(\mu)} + s h_n^{(\mu)}) = y_n^{(\mu)} + h_n^{(\mu)} \sum_{j=1}^m \beta_j(v) Y_{n,j}^{(\mu)}, \qquad (3.50)$$

$$v \in [0, 1], \quad \text{with } Y_{n,j}^{(\mu)} := u_h'(t_n^{(\mu)}),$$

the computational form of (3.36) becomes

$$Y_{n,i}^{(\mu)} = f(t_{n,i}^{(\mu)}, y_n^{(\mu)} + h_n^{(\mu)} \sum_{j=1}^m a_{i,j} Y_{n,j}^{(\mu)}, u_h(\theta(t_{n,i}^{(\mu)}))) \qquad (3.51)$$

$$+ h_n^{(\mu)} \int_0^{c_i} K_1(t_{n,i}^{(\mu)}, t_n^{(\mu)} + s h_n^{(\mu)}) \left(y_n^{(\mu)} + h_n^{(\mu)} \sum_{j=1}^m \beta_j(s) Y_{n,j}^{(\mu)} \right) ds$$

$$+ F_n^{(\mu)}(t_{n,i}^{(\mu)}) + (\mathcal{V}_\theta u_h)(t_{n,i}^{(\mu)}), \quad i = 1, \ldots, m$$

(which is reminiscent of the natural interpolant for an m-stage continuous implicit Runge–Kutta method for a DDE on a constrained mesh; see Bellen and Zennaro (2003, Chapter 6)). Recall from Lemma 3.1 that

$$\theta(t_{n,i}^{(\mu)}) = t_n^{(\mu-1)} + \tilde{c}_i h_n^{(\mu-1)},$$

which coincides with the collocation point $t_{n,i}^{(\mu-1)}$ ($i = 1, \ldots, m$) only if θ is linear.

On the subinterval $\bar{\sigma}_n^{(\mu)}$ the collocation solution $u_h \in S_m^{(0)}(I_h)$ is defined by the local representation (3.50). Hence, the solution $\mathbf{Y}_n^{(\mu)} \in \mathbb{R}^m$ of the linear algebraic system

$$[I_m - h_n^{(\mu)}(A_n^{(\mu)} + h_n^{(\mu)}C_n^{(\mu)})]\mathbf{Y}_n^{(\mu)} = \mathbf{g}_n^{(\mu)} + \mathbf{G}_n^{(\mu)} + \kappa_n^{(\mu)}y_n^{(\mu)} \qquad (3.52)$$
$$+ \mathbf{Q}_n^{(\mu-1)} + \tilde{\kappa}_n^{(\mu-1)}y_n^{(\mu-1)} + (h_n^{(\mu-1)})^2 \tilde{C}_n^{(\mu-1)}\mathbf{Y}_n^{(\mu-1)},$$

for $n = 0, 1, \ldots, N - 1$, $\mu = 0, 1, \ldots, M$. The matrices in $L(\mathbb{R}^m)$ defining the left-hand side of (3.52) are

$$A_n^{(\mu)} := \mathrm{diag}(a(t_{n,i}^{(\mu)}))A, \quad \text{with} \ A := (a_{i,j});$$

$$\tilde{A}_n^{(\mu)} := \mathrm{diag}(b(t_{n,i}^{(\mu)}))\tilde{A}, \quad \text{with} \ \tilde{A} := (\beta_j(\tilde{c}_i);$$

$$C_n^{(\mu)} := \left(\int_0^{c_i} K_1(t_{n,i}^{(\mu)}, t_n^{(\mu)} + sh_n^{(\mu)})\beta_j(s)\,ds \right)_{i,j=1}^m$$

$$\tilde{C}_n^{(\mu-1)} := \left(\int_0^{\tilde{c}_i} K_2(t_{n,i}^{(\mu)}, t_n^{(\mu-1)} + sh_n^{(\mu-1)})\beta_j(s)\,ds \right)_{i,j=1}^m,$$

and we have set

$$\kappa_n^{(\mu)} := \mathbf{a}_n^{(\mu)} + h_n^{(\mu)}\left(\int_0^{c_i} K_1(t_{n,i}^{(\mu)}, t_n^{(\mu)} + sh_n^{(\mu)})\,ds \right)_{i=1}^m,$$

$$\tilde{\kappa}_n^{(\mu-1)} := \mathbf{b}_n^{(\mu)} + h_n^{(\mu-1)}\left(\int_0^{\tilde{c}_i} K_2(t_{n,i}^{(\mu)}, t_n^{(\mu-1)} + sh_n^{(\mu-1)})\,ds \right)_{i=1}^m,$$

with

$$\mathbf{a}_n^{(\mu)} := \left(a(t_{n,i}^{(\mu)})\right)_{i=1}^m, \qquad \mathbf{b}_n^{(\mu)} := \left(b(t_{n,i}^{(\mu)})\right)_{i=1}^m.$$

The vectors $\mathbf{G}_n^{(\mu)}$ and $\mathbf{Q}_n^{(\mu-1)}$ are defined by

$$\mathbf{G}_n^{(\mu)} := \left(F_n^{(\mu)}(t_{n,1}^{(\mu)}), \ldots, F_n^{(\mu)}(t_{n,m}^{(\mu)}) \right)^T,$$

$$\mathbf{Q}_n^{(\mu-1)} := \left(\Psi_n^{(\mu-1)}(t_{n,1}^{(\mu)}), \ldots, \Psi_n^{(\mu-1)}(t_{n,m}^{(\mu)}) \right)^T;$$

for $t \in \bar{\sigma}_n^{(\mu)}$ their components are given by

$$F_n^{(\mu)}(t) := \int_{t_0}^{\xi_\mu} K_1(t,s) u_h(s)\,ds + \int_{\xi_\mu}^{t_n^{(\mu)}} K_1(t,s) u_h(s)\,ds,$$

$$\Psi_n^{(\mu-1)}(t) := \int_{t_0}^{\xi_{\mu-1}} K_2(t,s) u_h(s)\,ds + \int_{\xi_{\mu-1}}^{t_n^{(\mu-1)}} K_2(t,s) u_h(s)\,ds,$$

respectively (cf. (3.18) and (3.14)).

Theorem 3.9. Assume that the given functions a, b, g, K_1, K_2 describing the linear delay VIDE (3.45), (3.46) are continuous on their respective domains, and let the delay functions θ be subject to the hypotheses (D1)–(D3) in Section 3.2.2. Then there exists a $\bar{h} > 0$ so that for any θ-invariant mesh I_h with $h \in (0, \bar{h})$ and any initial function $\phi \in C[\theta(t_0), t_0]$ each of the linear algebraic systems (3.52) possesses a unique solution $Y_n^{(\mu)} \in \mathbb{R}^m$. Therefore, the collocation equation (3.47) defines a unique collocation solution $u_h \in S_m^{(0)}(I_h)$ whose local representation on $\bar{\sigma}_n^{(\mu)}$ is given by (3.50).

3.4.2. Global convergence results

The collocation error $e_h := y - u_h$ associated with the collocation solution $u_h \in S_m^{(0)}(I_h)$ to the linear DVIDE (3.45), (3.46) solves the initial-value problem

$$e_h'(t) = a(t) e_h(t) + b(t) e_h(\theta(t)) + \delta_h(t) + (\mathcal{V} e_h)(t) + (\mathcal{V}_\theta e_h)(t), \quad t \in I,$$
$$e_h(t) = 0, \quad t \in [\theta(t_0), t_0], \tag{3.53}$$

where the defect δ_h vanishes on X_h, the set of collocation points. For $t \in I^{(\mu)} := [\xi_\mu, \xi_{\mu+1}]$ we write the above error equation in the form

$$e_h'(t) = a(t) e_h(t) + \delta_h(t) + G_\mu(t) + \int_{\xi_\mu}^{t} K_1(t,s) e_h(s)\,ds, \quad t \in I^{(\mu)}, \tag{3.54}$$

with given initial value $e_h(\xi_\mu)$ and lag term

$$G_\mu(t) := b(t) e_h(\theta(t)) + \int_{t_0}^{\xi_\mu} K_1(t,s) e_h(s)\,ds + (\mathcal{V}_\theta e_h)(t).$$

When $\mu = 0$ we have

$$e_h'(t) = a(t) e_h(t) + \delta_h(t) + \int_{t_0}^{t} K_1(t,s) e_h(s)\,ds, \quad t \in I^{(0)}, \tag{3.55}$$

since the initial condition $e_h(t) = 0$, $t \leq 0$ implies $G_0(t) = 0$ in $[\theta(t_0), t_0]$.

Thus, on the first macro-interval $I^{(0)}$ the gobal convergence result for classical VIDEs holds: under appropriate assumptions on the regularity of the solution (see Theorem 3.12 below) the collocation error can be estimated by

$$\|e_h^{(\nu)}\|_{0,\infty} := \sup_{t \in I^{(0)}} |e_h^{(\nu)}(t)| \leq C_\nu (h^{(0)})^m, \quad \nu = 0, 1.$$

It follows in particular that $e_h^{(\nu)}(\xi_1) = \mathcal{O}((h^{(0)})^m)$.

This result allows us to derive an similar global error estimate on each macro-interval $I^{(\mu)}$ ($1 \leq \mu \leq M$). We leave the detailed steps in this recursive argument to the reader and simply summarize the result in the following theorem.

Theorem 3.10. Let the following conditions hold.

(a) $a, b, g \in C^m(I)$, and $\phi \in C^{m+1}[\theta(t_0), t_0]$.

(b) $K_1 \in C^m(D)$, $K_2 \in C^m(D_\theta)$.

(c) θ satisfies conditions (D1)–(D3) of Section 3.2.2, with $d \geq m$ in (D1).

(d) $u_h \in S_m^{(0)}(I_h)$ is the collocation solution to the delay VIDE (3.45), (3.46) where I_h is θ-invariant and $h \in (0, \bar{h})$ so that the linear algebraic systems (3.52) all have unique solutions.

Then the estimates

$$\|y^{(\nu)} - u_h^{(\nu)}\|_\infty \leq C_\nu h^m, \quad \nu = 0, 1 \quad (3.56)$$

hold for any set $\{c_i\}$ of distinct collocation parameters in $[0, 1]$. The constants C_ν depend on these parameters but are independent of h.

As for second-kind VIEs with non-vanishing delays, a gain of one can be achieved in the global order of convergence of u_h by a judicious choice of the $\{c_i\}$, thus extending the global superconvergence result for classical VIDEs (Brunner and van der Houwen (1986) or Brunner (2004b, Chapter 3)).

Theorem 3.11. Let the assumed degree of regularity for the given functions in the initial-value problem for the linear delay VIDE (3.45), (3.46) be raised by one (to $m+1$ and $m+2$, respectively) in Theorem 3.10. If the collocation parameters satisfy the orthogonality condition

$$J_0 := \int_0^1 \prod_{i=1}^m (s - c_i) \, ds = 0$$

then, for all θ-invariant meshes I_h with $h \in (0, \bar{h})$, the collocation solution $u_h \in S_m^{(0)}(I_h)$ is globally superconvergent on I, that is,

$$\|y - u_h\|_\infty \leq C h^{m+1}, \quad (3.57)$$

with C depending on the $\{c_i\}$ but not on h.

FUNCTIONAL INTEGRAL AND INTEGRO-DIFFERENTIAL VOLTERRA EQUATIONS   95

Proof. The key to establishing this global superconvergence result (and the local superconvergence results in the next section) is the variation-of-constants result of Theorem 2.7, where y and g are replaced, respectively, by e_h and δ_h, and where the initial conditions are given by $e_h(t) = 0$ $(t \leq t_0)$ and $e_h(\xi_\mu) = \mathcal{O}((h^{(\mu)})^{m+1})$ $(1 \leq \mu \leq M;\ h^{(\mu)} \leq h)$. We note once more that the image of a point $t = t_n^{(\mu)} + vh_n^{(\mu)} \in \sigma_n^{(\mu)}$ under $\theta^{\mu-\nu}$ $(0 \leq \nu \leq \mu - 1)$ is given either by $t_n^{(\nu)} + vh_n^{(\nu)}$ $(v \in [0,1])$ if θ is linear, or by $t_n^{(\nu)} + \tilde{v}h_n^{(\nu)}$ (for some $\tilde{v} \in [0,1]$, with $\tilde{v} \neq v$) if θ is *nonlinear*.

Details of the proof are left as an exercise. □

Remark. The convergence results of Theorems 3.12 and 3.13 contain, as special cases, global convergence and superconvergence results for DDEs (for $K_i = 0$ on D and D_θ, respectively).

3.4.3. Local superconvergence results

In the previous section we described the foundation for proving optimal superconvergence results on I_h for the collocation solution $u_h \in S_m^{(0)}(I_h)$ to the linear delay VIDE (3.45), (3.46): it is given by the variation-of-constants formula (or 'resolvent representation') for the collocation error $e_h := y - u_h$. The essential ingredients of the proof of the local superconvergence result are thus all in place: the θ-invariance of the mesh I_h and the resulting mapping (3.4), (3.5) of mesh points $t_n^{(\mu)}$ into corresponding previous mesh points $t_n^{(\nu)}$ (which is of course true regardless of whether the delay function θ is linear or nonlinear) and the order of the quadrature errors corresponding to the interpolatory m-point quadrature formulas based on the collocation points and depending on the familiar orthogonality conditions for the collocation parameters $\{c_i\}$. Thus, without any more ado we state the following result.

Theorem 3.12. Let the following be satisfied.

(a) The given functions a, b, g and K_1, K_2 in the DVIDE (3.45), (3.46) are in $C^{m+\kappa}$ on their respective domains, for some κ with $1 \leq \kappa \leq m$, as specified in (d).

(b) The delay function θ is subject to (D1)–(D3), with $d \geq m+\kappa$ in (D1).

(c) $u_h \in S_m^{(0)}(I_h)$ is the collocation solution, with θ-invariant mesh I_h, for the given delay VIDE.

(d) The collocation parameters $\{c_i\}$ are such that the orthogonality conditions of Theorem 3.6,

$$J_\nu := \int_0^1 s^\nu \prod_{i=1}^m (s - c_i)\, ds = 0, \quad \nu = 0, 1, \ldots, \kappa - 1,$$

with $J_\kappa \neq 0$, hold.

Then, for all $h \in (0, \bar{h})$ (Theorem 3.9), the collocation error $e_h := y - u_h$ satisfies

$$\max_{t \in I_h} |e_h(t)| \leq Ch^{m+\kappa}, \tag{3.58}$$

for some constant C which depends on the $\{c_i\}$ but not on h.

If, in addition, $c_m = 1$ (implying $\kappa \leq m - 1$), then we also have

$$\max_{t \in I_h \setminus \{t_0\}} |e'_h(t)| \leq C_1 h^{m+\kappa}. \tag{3.59}$$

4. Basic theory of Volterra functional integral equations II: (vanishing) proportional delays

4.1. The pantograph equation: ca. 1971

The linear DDE with constant coefficients,

$$y'(t) = ay(t) + by(qt), \quad t \in I := [0, T], \quad 0 < q < 1, \tag{4.1}$$

arose in the mathematical modelling of the wave motion in the supply line to an overhead current collector (*pantograph*) of an electric locomotive (see Ockendon and Tayler (1971) and Fox, Mayers, Ockendon and Tayler (1971); also Tayler (1986, pp. 40–45, 50–53)). The resulting *pantograph equation* is a (seemingly!) very simple example of a DDE with *vanishing* variable delay: here, we have $\theta(t) = t - \tau(t)$, with $\tau(t) = (1 - q)t \geq 0$.

A special case of (4.1) is the 'pure delay' equation

$$y'(t) = by(qt), \quad t \geq 0, \quad y(0) = y_0, \quad b \neq 0. \tag{4.2}$$

Its (unique) solution is given by

$$y(t) = \sum_{j=0}^{\infty} \frac{q^{j(j-1)/2}}{j!} (bt)^j \cdot y_0, \quad t \geq 0. \tag{4.3}$$

The following result can be found in Kato and McLeod (1971); compare also Frederickson (1971), Morris, Feldstein and Bowen (1972), Carr and Dyson (1976), Derfel (1990), Iserles (1993), and Terjéki (1995).

Theorem 4.1. For any $q \in (0, 1)$ and any y_0 the delay differential equation (4.1) possesses a unique solution $y \in C^1(I)$ with $y(0) = y_0$, regardless of the choice of a, $b \neq 0$, and $T > 0$. It is given by

$$y(t) = \sum_{n=0}^{\infty} \gamma_n(q) t^n,$$

where

$$\gamma_n(q) := \frac{1}{n!} \prod_{j=1}^{n} (a + bq^{j-1}).$$

Proof. We apply Picard iteration to the equivalent Volterra integral equation,

$$y(t) = y_0 + \int_0^t (ay(s) + by(qs))\,ds, \quad t \in I.$$

It can be shown that the resulting sequence $\{y_n(t)\}$ ($n \geq 0$, $y_0(t) := y_0$) converges uniformly on any interval I. Moreover, setting

$$y(t) := \sum_{n=0}^{\infty} \gamma_n(q) t^n,$$

one verifies that the power series has infinite radius of convergence, since its coefficients satisfy

$$\frac{\gamma_n}{\gamma_{n-1}} = \frac{1}{n}[a + bq^{n-1}], \quad n \geq 1. \qquad \square$$

Remarks. **(1)** The survey paper by Iserles (1993) presents an illuminating introduction into the complex world of solutions to (4.1) and its generalizations; it also contains an extensive bibliography. The pantograph equation and its matrix version is also studied, within the framework of Volterra functional integral equations, in Chambers (1990, pp. 40–43).

(2) The above result on the existence and uniqueness of solutions remains true for (4.1) with variable coefficients a, $b \in C(I)$. More precisely, if a, $b \in C^m(I)$ then, for any $q \in (0,1)$ and any y_0, the solution y lies in $C^{m+1}(I)$. Properties of solutions of nonlinear versions of these equations (*e.g.*, Riccati-type equations) can be found in Iserles (1994a) and Iserles and Terjéki (1995).

(3) These results confirm a crucial difference between the regularity of solutions to DDEs with non-vanishing delays and those of pantograph-type DDEs: for the latter, smooth data lead to solutions that are smooth on the *entire* interval $[0, T]$. In particular, solutions to (4.2) are *entire functions* of *order zero*. It follows from classical complex function theory (Ahlfors' theorem) that an entire function of order zero cannot have finite asymptotes. This implies that, for $b < 0$, nontrivial solutions of (4.2) are not bounded on $\mathbb{R}^+$; also, the number of sign changes (zeros) is infinite. (See also Iserles (1993), Iserles (1997b), and Liu (1997).) To give the reader an idea of how these zeros depend on q, Table 4.1 exhibits a representative sample of zeros of y. Additional information (for $q = 1/4$, $q = 3/4$) can be found in Iserles (1993, p. 5).

The maximum of $|y(t)|$ in the interval given by the last listed zero and the following one exceed 10^{15}. We note in passing that the papers by Iserles (1997b) and Liu (1997) nicely describe and illustrate the various difficulties

Table 4.1. Zeros of $y(t)$ for $b = -1$.

$q = 0.05$	$q = 0.5$	$q = 0.95$
$z_1 = 1.02631$	$z_1 = 1.48808$	$z_1 = 8.96684$
$z_2 = 40.3651$	$z_2 = 4.88114$	$z_2 = 10.8942$
$\vdots$	$\vdots$	$\vdots$
$z_3 = 1205.57$	$z_{10} = 5223.38$	$z_{46} = 5258.99$

one encounters in the long-time approximation of solutions to the 'innocent' pantograph equation (4.1).

The reader interested in details on the asymptotic distribution of the zeros of such solutions may wish to consult the 1992 paper by Elbert (which includes a reference to the first study of this subject, a 1967 report by Feldstein and Kolb). The papers by Iserles (1994b), Iserles and Terjéki (1995), and Feldstein and Liu (1998) contain a wealth of results on *nonlinear* pantograph DDEs, including Riccati-type functional equations.

4.2. Linear Volterra integral equations with proportional delays

We now return to one of the particular delay VIEs considered by Andreoli (1913, 1914), and to his important remark regarding the effect the (vanishing) proportional delay has on the representation of its solution. In the notation employed in this paper this equation is

$$y(t) = g(t) + \int_0^{qt} K(t,s)y(s)\,ds, \quad t \in I := [0,T], \quad 0 < q < 1, \qquad (4.4)$$

where g and K are continuous functions.

Theorem 4.2. *Let g and K in (4.4) satisfy $g \in C(I)$ and $K \in C(D_\theta)$, where $D_\theta := \{(t,s) : 0 \le s \le \theta(t),\ t \in I\}$. Then, for any $\theta(t) := qt$ with $q \in (0,1)$, the delay integral equation (4.4) possesses a unique solution $y \in C(I)$. This solution is given by*

$$y(t) = g(t) + \sum_{n=1}^{\infty} \int_0^{q^n t} K_n(t,s)g(s)\,ds$$

$$= g(t) + \int_0^t \left(\sum_{n=1}^{\infty} q^n K_n(t, q^n s) g(q^n s) \right) ds, \quad t \in I. \qquad (4.5)$$

The iterated kernels $K_n(t,s) = K_n(t,s;q)$ $(n \geq 1)$ are obtained recursively by

$$K_{n+1}(t,s) := \int_{q^{-n}s}^{qt} K(t,v)K_n(v,s)\,dv, \quad (t,s) \in D_\theta^{(n+1)}, \quad n \geq 1,$$

with $K_1(t,s) := K(t,s)$ and

$$D_\theta^{(k)} := \{(t,s) : 0 \leq s \leq q^k t,\ t \in I\}.$$

Remark. For $q = 1$, the solution representation (4.5) reduces to the classical 'separable' expression involving the resolvent kernel $R(t,s)$ (as the limit of the Neumann series) of $K(t,s)$. Theorem 4.2 shows that for $0 < q < 1$ such a resolvent representation of the solution no longer exists: the values of the iterated kernels 'overlap' with those of g. However, the infinite series in (4.5) still converges uniformly on any compact interval I, as Lemma 4.3 below will make clear.

Proof. The Picard iteration process we applied to the integrated form of the pantograph DDE can of course be used for the delay VIE (4.4), with suitably adapted Dirichlet's formula when changing the order of integration in the double integrals: here, the resulting limits of integration now depend on the iteration number n. To see this in some more detail, we have, setting $y_0(t) := g(t)$,

$$y_1(t) := g(t) + \int_0^{qt} K_1(t,s)g(s)\,ds$$

and hence

$$y_2(t) := g(t) + \int_0^{qt} K_1(t,s)\left(g(s) + \int_0^{qs} K_1(s,v)g(v)\,dv\right)ds$$

$$= g(t) + \int_0^{qt} K_1(t,s)g(s)\,ds + \int_0^{qt}\left(\int_{q^{-1}v}^{qt} K_1(t,s)K_1(s,v)\,ds\right)y(v)\,dv.$$

It is now easily verified by induction that the iterated kernels $K_n(t,s)$ of the given kernel $K(t,s) =: K_1(t,s)$ are generated recursively by

$$K_{n+1}(t,s) = \int_{q^{-n}s}^{qt} K(t,v)K_n(v,s)\,dv, \quad (t,s) \in D_\theta^{(n+1)}, \quad n \geq 1$$

(see also Chambers (1990)). Hence the iterate $y_k(t)$ can be expressed in the form (4.5) where the index of summation ranges from 1 to k.

Lemma 4.3. *Uniform bounds on $I = [0,T]$ for the iterated kernels $K_n(t,s)$ defined in Theorem 4.2 are given by*

$$|K_n(t,s)| \leq \frac{q^{n(n-1)/2}}{(n-1)!}T^{n-1}\bar{K}_\theta^n, \quad (t,s) \in D_\theta^{(n)}, \quad n \geq 1,$$

where we have set $\bar{K}_\theta := \max_{(D_\theta)} |K(t,s)|$.

We leave the proof of this simple result, as well as that of the uniqueness of the solution (4.5), as an exercise. □

The existence, uniqueness and regularity properties hold also for the more general linear delay VIE with proportional delay,

$$y(t) = g(t) + (\mathcal{V}y)(t) + (\mathcal{V}_\theta y)(t), \quad t \in I, \tag{4.6}$$

corresponding to the Volterra integral operators

$$(\mathcal{V}y)(t) := \int_0^t K_1(t,s) y(s)\, ds, \quad (\mathcal{V}_\theta y)(t) := \int_0^{\theta(t)} K_2(t,s) y(s)\, ds,$$

with $\theta(t) := qt$ $(0 < q < 1)$, $K_1 \in C(D)$ and $K_2 \in C(D_\theta)$.

Theorem 4.4. Assume that $K_1 \in C^d(D)$ and $K_2 \in C^d(D_\theta)$, for some $d \geq 0$. Then the delay integral equation (4.6) with $\theta(t) = qt$ $(0 < q < 1)$ has a unique solution $y \in C^d(I)$ for any g with $g \in C^d(I)$.

Proof. Theorem 4.2 shows that the iterated kernels $K_n(t,s)$ associated with the kernel K of the special delay integral equation (4.4) inherit the regularity of K. Since the additional term $(\mathcal{V}y)(t)$ in the general linear delay VIE (4.6) will not lead to lower regularity in the Picard iteration process, the assertion of Theorem 4.4 follows from the uniform convergence of the Picard iterates on I, for any $q \in (0,1)$. □

We shall see in Section 4.5 that this regularity result can also be derived by means of embedding techniques.

Remark. The paper by Morris et al. (1972, pp. 518–523) contains an illuminating discussion of the connection between general pantograph DDEs and certain Volterra integral and integro-differential equations with (multiple) proportional delays. Compare also Iserles and Liu (1994).

4.3. First-kind Volterra integral equations with vanishing delays

In Section 1.1.1 we briefly alluded to the fact that in his *Nota I* of 1896 Volterra studied the problem of 'inverting' definite integrals of the form

$$(\mathcal{V}y)(t) := \int_0^t K(t,s) y(s)\, ds = g(t), \quad t \in I := [0,T], \quad g(0) = 0,$$

where $K \in C(D)$, and that he then went on (Volterra (1897), also Volterra (1913, pp. 92–101)) to analyse the more general functional equation

$$(\mathcal{W}_\theta y)(t) = g(t), \quad t \in I, \quad g(0) = 0, \tag{4.7}$$

where the Volterra integral operator $\mathcal{W}_\theta : C(I) \to C(I)$ is

$$(\mathcal{W}_\theta \phi)(t) := \int_{\theta(t)}^t K(t,s)\phi(s)\,ds, \quad \text{with } \theta(t) := qt \ (0 < q < 1). \quad (4.8)$$

Under suitable conditions on K and g this equation can be transformed into the equivalent second-kind equation

$$K(t,t)y(t) - qK(t,qt)y(qt) + \int_{qt}^t \frac{\partial K(t,s)}{\partial t} y(s)\,ds = g'(t), \quad t \in I \quad (4.9)$$

(see also Fenyö and Stolle (1984, pp. 324–327) and Brunner (1997b)), to which Picard iteration techniques can be applied. This reformulation was the basis for Volterra's 1897 result, which we now state. We set $\bar{D}_\theta := \{(t,s) : 0 \leq \theta(t) \leq s \leq t \leq T\}$.

Theorem 4.5. Assume:

(a) $g \in C^1(I)$, with $g(0) = 0$;
(b) $K \in C(\bar{D}_\theta)$, $\partial K/\partial t \in C(\bar{D}_\theta)$, with $|K(t,t)| \geq k_0 > 0$ $(t \in I)$.

Then, for each $\theta(t) = qt$ with $q \in (0,1)$, the first-kind delay integral equation (4.7) possesses a unique solution $y \in C(I)$.

The above result was generalized by Lalesco (1908, 1911) and – much later – by Denisov and Korovin (1992) and Denisov and Lorenzi (1995). From the latter paper we cite the following result.

Theorem 4.6. Assume the lag function θ satisfies

(a) $\theta \in C^3(I)$, with $\theta(0) = 0$, $\theta'(0) = 1$, $\theta''(0) < 0$, $\theta(t) < t$ $(t \in (0,T])$, $\theta'(t) > 0$ for $t \in I$,

and let

(b) $g \in C^2(I)$, with $g(0) = g'(0) = 0$;
(c) $K \in C^3(\bar{D}_\theta)$, with $|K(t,t)| \geq k_0 > 0$ $(t \in I)$.

Then the first-kind delay integral equation $(\mathcal{W}_\theta y)(t) = g(t)$ has a unique solution $y \in C(I)$.

Remark. A similar result was proved in Denisov and Korovin (1992), but under the hypothesis that $\theta'(0) < 1$. If, as in the above theorem, $\theta'(0) = 1$, the domain $\bar{D}_\theta$ has a cusp at the point $(t,s) = (0,0)$, and new techniques are needed to deal with this situation. We note that the case $\theta'(0) = 1$ was already treated, albeit in a somewhat sketchy way, by Lalesco in 1911.

4.4. Volterra integro-differential equations with proportional delays

In order to obtain some first insight into the properties of solutions of linear VIDEs with proportional delays we will first consider the 'pure delay' problem

$$y'(t) = g(t) + \int_0^{qt} K(t,s)y(s)\,ds, \quad t \in I := [0,T], \quad y(0) = y_0, \quad (4.10)$$

assuming that $g \in C(I)$, $K \in C(D_\theta)$, with $\theta(t) = qt$ and $0 < q < 1$. This initial-value problem is equivalent to the delay VIE

$$y(t) = g_0(t) + \int_0^{qt} H(t,s;q)y(s)\,ds, \quad t \in I, \quad (4.11)$$

where

$$g_0(t) := y_0 + \int_0^t g(s)\,ds, \quad H(t,s;q) := \int_{q^{-1}s}^t K(v,s)\,dv.$$

We now apply Theorem 4.2: setting $H_1(t,s) := H(t,s;q)$, and denoting by $H_n(t,s)$ the corresponding iterated kernels, the (unique) solution y of (4.11) (which, since g_0 and $H(\cdot,\cdot;q)$ are continuously differentiable functions, lies in $C^1(I)$) can be expressed in the form

$$y(t) = g_0(t) + \sum_{n=1}^\infty \int_0^{q^n t} H_n(t,s)g_0(s)\,ds, \quad t \in I,$$

where the infinite series converges absolutely and uniformly. If we now substitute the expressions for $g_0(t)$, an obvious rearrangement (using Dirichlet's formula) leads to the following result.

Theorem 4.7. *Under the above assumptions on g and K, the unique solution $y \in C^1(I)$ to the initial-value problem (4.10) has the representation*

$$y(t) = \left(1 + \sum_{n=1}^\infty \tilde{H}_n(t,0)\right) y_0 + \sum_{n=0}^\infty \int_0^{q^n t} \tilde{H}_n(t,s)g(s)\,ds, \quad t \in I.$$

Here, we have set $\tilde{H}_0(t,s) := 1$ $((t,s) \in D)$,

$$\tilde{H}_n(t,s) := \int_s^{q^n t} H_n(t,v)\,dv, \quad n \geq 1,$$

and we note that

$$\tilde{H}_n(t,0) = \int_0^{q^n t} H_n(t,v)\,dv, \quad n \geq 1.$$

The initial-value problem for the *general* linear VIDE with proportional delay,

$$y'(t) = a(t)y(t) + b(t)y(qt) + g(t) + (\mathcal{V}y)(t) + (\mathcal{V}_\theta y)(t), \quad t \in I, \quad (4.12)$$

with $\theta(t) = qt$ $(0 < q < 1)$, is equivalent to the delay VIE

$$y(t) = g_0(t) + \int_0^t \left(a(s) + \int_s^t K_1(v,s)\,dv \right) y(s)\,ds$$
$$+ \int_0^{qt} \left((1/q)b(s/q) + \int_{q^{-1}s}^t K_2(v,s)\,dv \right) y(s)\,ds$$
$$=: g_0(t) + \int_0^t G_1(t,s)y(s)\,ds + \int_0^{qt} G_2(t,s;q)y(s)\,ds,$$

where

$$g_0(t) := y_0 + \int_0^t g(s)\,ds.$$

The regularity of the kernels G_1 and $G_2(\cdot;\cdot;q)$ is determined by that of the original data a, b and K_1, K_2. Thus, Theorem 4.4 implies the following result.

Theorem 4.8. Assume:

(a) $a, b, g \in C^d(I)$ for some $d \geq 0$;
(b) $K_1 \in C^d(D)$ and $K_2 \in C^d(D_\theta)$.

Then, for each initial value y_0, the delay VIDE (4.12) possesses a unique solution $y \in C^{d+1}(I)$.

4.5. Embedding techniques

The embedding of a (proportional) delay differential equation into an infinite system of ordinary differential equations was studied in detail by Feldstein, Iserles and Levin (1995). The motivation behind their approach was to explore another way for obtaining results on the asymptotic stability (or the boundedness) of solutions of such DDEs, and for constructing feasible methods for their numerical solution. It also permits the derivation of existence and regularity results for the exact solutions.

Here, we extend these embedding techniques to the delay Volterra integral equation (4.6) and to the delay Volterra integro-differential equation (4.12). Note that these Volterra functional equations contain the important special cases characterized by $K_2(t,s) = -K_1(t,s) =: -K(t,s)$:

$$y(t) = g(t) + (\mathcal{W}_\theta y)(t), \quad t \in I, \tag{4.13}$$

and

$$y'(t) = a(t)y(t) + b(t)y(qt) + (\mathcal{W}_\theta y)(t), \quad t \in I, \quad y(0) = y_0, \tag{4.14}$$

corresponding to the delay Volterra operator $\mathcal{W}_\theta$ defined in (4.8). The following embedding results (which can be extended to the nonlinear counterparts of the above pantograph-type Volterra equations) contain the key not only to establishing results on the existence, uniqueness, and regularity of solutions but also to the analysis of the local superconvergence properties of collocation solutions to such functional equations.

4.5.1. Embedding results for the delay VIE (4.6)

Lemma 4.9. The delay VIE (4.6) can be embedded into an infinite-dimensional system of 'classical' VIEs of the second kind,

$$z_\nu(t) = g_\nu(t) + \int_0^t (K_{1,\nu}(t,s)z_\nu(s) + K_{2,\nu}(t,s)z_{\nu+1}(s))\,ds, \quad \nu \in \mathbb{N}_0, \quad (4.15)$$

where

$$z_\nu(t) := y(q^\nu t), \qquad g_\nu(t) := g(q^\nu t)$$

and

$$K_{1,\nu}(t,s) := q^\nu K_1(q^\nu t, q^\nu s), \qquad K_{2,\nu}(t,s) := q^{\nu+1} K_2(q^\nu t, q^{\nu+1} s).$$

The proof of this embedding result is straightforward and will thus be left as an exercise.

Consider now the *truncated* (finite) system corresponding to (4.15),

$$z_{M,\nu}(t) = g_\nu(t) + \int_0^t (K_{1,\nu}(t,s)z_{M,\nu}(s) + K_{2,\nu}(t,s)z_{M,\nu+1}(s))\,ds, \quad (4.16)$$

$$\nu = 0, 1, \ldots, M-1,$$

$$z_{M,M}(t) = g_M(t) + \int_0^t K_{1,M}(t,s)z_{M,M}(s)\,ds, \quad t \in I. \quad (4.17)$$

Lemma 4.10. Assume that $g \in C(I)$, $K_1 \in C(D)$, $K_2 \in C(D_\theta)$. Then, for $\nu = M, M-1, \ldots, 0$, the (unique) solution of (4.16) and (4.17) satisfies

$$\|z_\nu - z_{M,\nu}\|_\infty \leq Cq^{\tilde{M}}, \quad \text{with } \tilde{M} \geq M+1.$$

Proof. Setting $\varepsilon_{M,\nu} := z_\nu - z_{M,\nu}$, it follows from (4.16) and (4.17) that

$$\varepsilon_{M,\nu}(t) = \int_0^t K_{1,\nu}(t,s)\varepsilon_{M,\nu}(s)\,ds + \Phi_{M,\nu}(t), \quad t \in I, \quad (4.18)$$

for $\nu = 0, 1, \ldots, M$, with

$$\Phi_{m,\nu}(t) := \begin{cases} \int_0^t K_{2,M}(t,s)z_{M+1}(s)\,ds, & \text{if } \nu = M, \\ \int_0^t K_{2,\nu}(t,s)\varepsilon_{M,\nu+1}(s)\,ds, & \text{if } M-1 \geq \nu \geq 0. \end{cases}$$

Let $R_{1,\nu} = R_{1,\nu}(t,s)$ denote the resolvent kernel associated with the kernel $K_{1,\nu}$ in (4.15); we know from classical Volterra theory that $K_1 \in C(D)$

implies $R_{1,\nu} \in C(D)$ for all $\nu \geq 0$. The (unique) solution of the finite system (4.18) may thus be written as

$$\varepsilon_{M,\nu}(t) = \int_0^t R_{1,\nu}(t,s)\Phi_{M,\nu}(s)\,ds + \Phi_{M,\nu}(t), \quad t \in I, \tag{4.19}$$

for $\nu = M, M-1, \ldots, 0$. Since $|K_{2,\nu}(t,s)| \leq \bar{K}_2 q^{\nu+1}$, $(t,s) \in D_\theta$, where $\bar{K}_2 := \max_{D_\theta} |K_2(t,s)|$, setting $\nu = M$ in (4.18) leads to

$$|\varepsilon_{M,M}(t)| \leq Cq^{M+1}, \quad t \in I.$$

Thus, assuming that $\|\varepsilon\|_\infty \leq Cq^{\tilde{M}}$ ($\tilde{M} \geq M$) for $\nu = M, M-1, \ldots, M_0+1$, we find

$$|\Phi_{M_0,\nu}(t)| \leq \bar{K}_2 T q^{\nu+1} C_0 q^{\tilde{M}} =: Cq^{\tilde{M}+\nu+1}, \quad \nu \geq \tilde{M}+1,$$

and hence,

$$|\varepsilon_{M,M_0}(t)| \leq Cq^{\tilde{M}}, \quad t \in I, \quad \text{with } \tilde{M} \geq M+1.$$

This establishes the uniform bounds in Lemma 4.10. $\square$

4.5.2. Embedding results for the delay VIDE (4.12)
Lemma 4.11. The delay VIDE (4.12) can be embedded into an infinite-dimensional system of 'classical' VIDEs, namely,

$$z'_\nu(t) = \tilde{a}_\nu(t)z_\nu(t) + \tilde{b}_\nu(t)z_{\nu+1}(t) + \int_0^t \left(\tilde{K}_{1,\nu}(t,s)z_\nu(s) + \tilde{K}_{2,\nu}(t,s)z_{\nu+1}(s)\right) ds, \tag{4.20}$$

for $\nu \in \mathbb{N}_0$, with

$$\tilde{a}_\nu(t) := q^\nu a(q^\nu t), \quad \tilde{b}_\nu(t) := q^\nu b(q^\nu t),$$

and

$$\tilde{K}_{i,\nu}(t,s) := q^\nu K_{i,\nu}(t,s), \quad i = 1,2.$$

The kernels $K_{i,\nu}$ are those defined in Lemma 4.9.

This easily verified result leads to the VIDE analogue of Lemma 4.10.

Lemma 4.12. Assume that $a, b \in C(I)$, $K_1 \in C(D)$, and $K_2 \in C(D_\theta)$. Then the (unique) solution of the truncated (finite) system of VIDEs corresponding to (4.20),

$$z'_{M,\nu}(t) = \tilde{a}_\nu(t)z_{M,\nu}(t) + \tilde{b}_\nu(t)z_{M,\nu+1}(t)$$
$$+ \int_0^t \left(\tilde{K}_{1,\nu}(t,s)z_{M,\nu}(s) + \tilde{K}_{2,\nu}(t,s)z_{M,\nu+1}(s)\right) ds, \tag{4.21}$$
$$\nu = 0, 1, \ldots, M-1,$$

$$z'_{M,M}(t) = \tilde{a}_\nu(t)z_{M,M}(t) + \int_0^t \tilde{K}_{1,M}(t,s)z_{M,M}(s)\,ds, \quad t \in I, \tag{4.22}$$

with $z_{M,\nu}(0) = y_0$, satisfies

$$\|z_\nu(t) - z_{M,\nu}(t)\|_\infty \leq Cq^{\tilde{M}}, \quad \nu = 0, 1, \ldots, M, \quad \text{with } \tilde{M} \geq M.$$

Proof. Setting $\varepsilon_{M,\nu} := z_\nu - z_{M,\nu}$, we have

$$\varepsilon'_{M,\nu}(t) = \tilde{a}_\nu(t)\varepsilon_{M,\nu}(t) + \int_0^t \tilde{K}_{1,\nu}(t,s)\varepsilon_{M,\nu}(s)\,ds + \Psi_{M,\nu}(t), \quad t \in I, \quad (4.23)$$

with $\varepsilon_{M,\nu}(0) = 0$, for $\nu = M, M-1, \ldots, 0$. Here,

$$\Psi_{M,\nu}(t) := \begin{cases} \tilde{b}_M(t)z_{M+1}(t) + \int_0^t \tilde{K}_{2,M}(t,s)z_{M+1}(s)\,ds, & \text{if } \nu = M, \\ \tilde{b}_\nu(t)\varepsilon_{M,\nu+1}(t) + \int_0^t \tilde{K}_{2,\nu}(t,s)\varepsilon_{M,\nu+1}(s)\,ds, & \text{if } \nu < M. \end{cases}$$

Let $r_{1,\nu} = r_{1,\nu}(t,s)$ denote the (differential) resolvent kernel corresponding to the functions $\tilde{a}_\nu$ and $\tilde{K}_{1,\nu}$ in (4.20); that is, $r_{1,\nu}$ is defined by the (unique) solution of the (differential) resolvent equation

$$\frac{\partial r_{1,\nu}(t,s)}{\partial s} = -r_{1,\nu}(t,s)\tilde{a}_\nu(s) - \int_s^t r_{1,\nu}(t,z)\tilde{K}_{1,\nu}(z,s)\,dz, \quad (t,s) \in D,$$

with $r_{1,\nu}(t,t) = 1$, $t \in I$. The solution of the initial-value problem (4.21) can then be written in the form

$$\varepsilon_{M,\nu}(t) = r_{1,\nu}(t,0)\varepsilon_{M,\nu}(0) + \int_0^t r_{1,\nu}(t,s)\Psi_{M,\nu}(s)\,ds, \quad t \in I, \quad (4.24)$$

for $\nu = M, M-1, \ldots, 0$, where $\varepsilon_{M,\nu}(0) = 0$ for all ν.

Using the estimate

$$|\Psi_{M,M}(t)| \leq \gamma_0 q^M + \gamma_1 q^{2M+1}, \quad t \in I,$$

for finite constants γ_1 (recall that $\tilde{K}_{2,\nu}(t,s) = q^\nu K_{2,\nu}(t,s)$, with $|K_{2,\nu}(t,s)| \leq \bar{K}_2 q^{\nu+1}$), we derive for $\nu = M$ that

$$|\varepsilon_{M,M}(t)| \leq C_0 q^M + C_1 q^{2M+1} =: Cq^M, \quad t \in I,$$

where $C = C(q, M) < \infty$ for $M \in \mathbb{N}_0$ and $q \in (0,1)$.

For $\nu < M$ the argument for bringing the proof to its conclusion is analogous to the one in the proof of Lemma 4.10, except that now we employ the representation (4.24) and the estimate for $\|\varepsilon_{M,M}\|_\infty$. Details are left to the reader. □

Remark. The (uniform) convergence results in Lemmas 4.10 and 4.12 allow us not only to deduce the existence of unique solutions to the delay problems (4.6)) and (4.12) but also to establish the global regularity results already alluded to: C^m-data imply that the solutions of the DVIE and the DVIDE lie, respectively, in $C^m(I)$ and $C^{m+1}(I)$. We also encourage the reader to verify that the above embedding techniques can be extended to nonlinear Volterra equations with proportional delays, or with more general *nonlinear vanishing delays* described in Section 5.7.

5. Collocation methods for pantograph-type VFIEs

5.1. Numerical analysis of pantograph-type equations: an overview

The systematic study of the theory and the numerical analysis of the pantograph DDE and its various generalizations began with the papers by Ockendon and Tayler (1971), Fox et al. (1971), and Kato and McLeod (1971). While the theory of such functional equations almost immediately received much attention (see, for example, Frederickson (1971), Kato (1972), Bélair (1981), Derfel (1990, 1991), Kuang and Feldstein (1990), Derfel and Molchanov (1990), Iserles (1993, 1997a, 1994b), Terjéki (1995), Iserles and Terjéki (1995), Derfel and Vogl (1996), Liu (1996a), Iserles and Liu (1997), Feldstein and Liu (1998)), numerical analysts were singularly inattentive to the challenges of their numerical analysis: the fundamental paper on the numerical solution of the pantograph DDE (and its formulation as a Volterra functional equation) by Fox et al. (1971) stood alone until the early 1990s, when Buhmann and Iserles (1992, 1993), Iserles (1993), and Buhmann, Iserles and Nørsett (1993) understood that this class of functional differential equations represents a rich source of deep mathematical problems, both for the 'pure' and the numerical analyst.

5.1.1. Numerical analysis of the pantograph DDE
In the contributions just mentioned, the focus was on the asymptotic properties of numerical approximations, by linear multistep and simple collocation methods, for the pantograph equation (4.1). The survey by Iserles (1994a) and the papers of Iserles (1994c, 1997a, 1997b), Y. Liu (1995a, 1995b, 1996a, 1996b, 1997), Liang and Liu (1996), Liang, Qiu and Liu (1996), Bellen et al. (1997), Carvalho and Cooke (1998), Koto (1999), Liang and Liu (1996), Bellen (2001), Liu and Clements (2002), and Guglielmi and Zennaro (2003) describe various extensions of these early stability results, both on uniform and (quasi-) geometric meshes. Compare also the monograph by Bellen and Zennaro (2003) for a survey of many of these results, and Brunner (2004a) for additional references.

Collocation methods and their (super-) convergence properties are considered in Buhmann et al. (1993) (for $u_h \in S_1^{(0)}(I_h)$ and $q = 1/2$), Brunner (1997a), Zhang (1998), Zhang and Brunner (1998), Takama, Muroya and Ishiwata (2000), and Brunner (2004b, Chapter 5). While these properties are now reasonably well understood, this is not true for the qualitative aspects of piecewise polynomial (and continuous Runge–Kutta) methods: as shown in, *e.g.*, Buhmann *et al.* (1993), the present understanding is still at a very primitive level (except possibly when $q = 1/2$).

5.1.2. Volterra functional equations with proportional delays

Fox *et al.* (1971, pp. 292–295) used the integrated form of the pantograph equation, *i.e.*, a Volterra functional integral equation, to analyse the error induced by a variant of the classical Lanczos τ-method. Collocation methods for Volterra integral and integro-differential equations with proportional delays were studied in detail in Brunner (1997a), Zhang (1998), Zhang and Brunner (1998) (for second-order Volterra functional integro-differential equations), Takama *et al.* (2000), Ishiwata (2000), Muroya, Ishiwata and Brunner (2002), and Bellen *et al.* (2002). In these papers the focus is on the attainable orders of global and local (super-) convergence in collocation solutions. See also the survey by Brunner (2004a).

As we shall see in Section 5.8 the analysis of the asymptotic behaviour of collocation solutions to pantograph-type Volterra integral and integro-differential equations is completely open.

5.2. Piecewise polynomial collocation methods

5.2.1. Discretization on uniform meshes: overlap

The collocation equations corresponding to the pantograph-type delay Volterra equations to be discussed in Sections 5.2–5.3 will contain the delay integral terms $(\mathcal{V}_\theta u_h)(t)$ and $(\mathcal{W}_\theta u_h)(t)$, where $\theta(t) = qt$ ($0 < q < 1$) and $t = t_{n,i} := t_n + c_i h_n \in X_h \cap \sigma_n$. Thus, the structure of the difference equations corresponding to these collocation points will be governed by the location of the images of these collocation points,

$$\theta(t_{n,i}) = q(t_n + c_i h_n), \quad i = 1, \ldots, m.$$

For arbitrary non-uniform meshes these difference equations are obviously very complex. Therefore, in order to make the analysis tractable, we will for the present assume that the mesh I_h is *uniform*:

$$I_h := \{t_n := nh : n = 0, 1, \ldots, N; \ t_N = T\}.$$

For a uniform mesh I_h and $t = t_{n,i} := t_n + c_i h \in X_h$ we will write

$$\theta(t_{n,i}) = q(t_n + c_i h) =: h\{q_{n,i} + \gamma_{n,i}\} = t_{q_{n,i}} + \gamma_{n,i} h, \quad (5.1)$$

where

$$q_{n,i} := \lfloor q(n + c_i) \rfloor \in \mathbb{N}_0, \quad \gamma_{n,i} := q(n + c_i) - q_{n,i} \in [0, 1).$$

For given collocation parameters with $0 < c_1 < \cdots < c_m \leq 1$ and $q \in (0, 1)$ define

$$q^I := \lceil qc_1/(1 - q) \rceil, \quad q^{II} := \lceil qc_m/(1 - q) \rceil. \quad (5.2)$$

FUNCTIONAL INTEGRAL AND INTEGRO-DIFFERENTIAL VOLTERRA EQUATIONS 109

Here, $\lfloor x \rfloor$ is the greatest integer not exceeding $x \in \mathbb{R}$, and $\lceil x \rceil$ denotes the least upper integer bound for x.

The validity of the following lemma – which characterizes the 'overlap' of the images of the collocation points under the mapping θ – is easily verified.

Lemma 5.1. Let $q \in (0,1)$ and $0 < c_1 < \cdots < c_m \leq 1$, and assume that I_h is a uniform mesh on $I := [0,T]$ with diameter $h := T/N$. Then:

(i) for $n = 0$ we have $q(t_n + c_i h) \in (t_n, t_{n+1})$ for $i = 1, \ldots, m$,

(ii) if $n \geq 1$ we have $q(t_n + c_i h) \in (t_n, t_{n+1})$ if and only if $n < q^I$,

(iii) $q(t_n + c_i h) \leq t_n$ for $i = 1, \ldots, m$ if and only if $q^{II} \leq n \leq N - 1$.

This result reveals that the recursive computation of the collocation solution for functional equations with (vanishing) proportional delay consists in general of *three phases*.

Phase I. During this 'initial phase' we have *complete overlap* which is described by the values of n satisfying

$$0 \leq n < \left\lceil \frac{q}{1-q} c_1 \right\rceil =: q^I.$$

Phase II. The 'transition phase' corresponds to those n where *partial overlap* occurs; they are given by

$$q^I \leq n < \left\lceil \frac{q}{1-q} c_m \right\rceil =: q^{II}.$$

If this set of indices n is non-empty, there exists a $\nu_n \in \{1, \ldots, m-1\}$ so that

$$q(t_n + c_i h) \leq t_n \quad (i \leq \nu_n) \quad \text{and} \quad q(t_n + c_i h) > t_n \quad (i > \nu_n).$$

Phase III. In this 'pure delay phase', described by

$$q^{II} \leq n \leq N - 1,$$

there is no longer any *overlap* of $\theta(t_{n,i})$ and $t_{n,i}$: we have

$$q(t_n + c_i h) \leq t_n \quad \text{for} \quad i = 1, \ldots, m.$$

More precisely, for such a value of n there exist integers $\nu_n \in \{1, \ldots, m\}$ and $q_n < n - 1$ such that

$$q_{n,i} = q_n \quad (i \leq \nu_n) \quad \text{and} \quad q_{n,i} = q_{n+1} \quad (i > \nu_n).$$

Note that the integers q^I and q^{II} do not depend on the underlying mesh and are thus independent of N.

Illustration. We list a selection values of q^I and q^{II} for $m = 2$ and $m = 3$, corresponding to the

Gauss points:

$$m = 2: \quad c_1 = (3 - \sqrt{3})/6, \quad c_2 = (3 + \sqrt{3})/6,$$
$$m = 3: \quad c_1 = (5 - \sqrt{15})/10, \quad c_2 = 1/2, \quad c_3 = (5 + \sqrt{15})/10.$$

Radau II points:

$$m = 2: \quad c_1 = 1/3, \quad c_2 = 1,$$
$$m = 3: \quad c_1 = (4 - \sqrt{6})/10, \quad c_2 = (4 + \sqrt{6})/10, \quad c_3 = 1.$$

Table 5.1. $m = 2$.

	Gauss points				Radau II points			
q	1/2	2/3	0.9	0.99	1/2	2/3	0.9	0.99
q^I	1	1	2	21	1	1	3	33
q^{II}	1	2	8	79	1	2	9	99

Table 5.2. $m = 3$.

	Gauss points				Radau II points			
q	1/2	2/3	0.9	0.99	1/2	2/3	0.9	0.99
q^I	1	1	2	12	1	1	2	16
q^{II}	1	2	8	88	1	2	9	99

5.3. Second-kind VIEs with proportional delays

5.3.1. The structure of the collocation equations

The collocation solution $u_h \in S_{m-1}^{(-1)}(I_h)$ for the delay integral equation (4.6) is defined by

$$u_h(t) = g(t) + (\mathcal{V}u_h)(t) + (\mathcal{V}_\theta u_h)(t), \quad t \in X_h, \tag{5.3}$$

with $\theta(t) = qt$ ($0 < q < 1$) and X_h given by (3.2) ($h_n = h$). The contribution of the delay term $(\mathcal{V}_\theta u_h)(t_{n,i})$ to the collocation equation (5.3) will depend on n and the location of the collocation parameters $\{c_i\}$: it follows from the definition of $q_{n,i}$ and $\gamma_{n,i}$ in (5.1) that

$$(\mathcal{V}_\theta u_h)(t_{n,i}) = \int_0^{t_{q_{n,i}}} K(t_{n,i}, s) u_h(s)\, ds \qquad (5.4)$$

$$+ h \int_0^{\gamma_{n,i}} K(t_{n,i}, t_{q_{n,i}} + sh) u_h(t_{q_{n,i}} + sh)\, ds.$$

In order to obtain suitable computational forms of the collocation equation (5.3) (leading to systems of difference equations whose solution describes the collocation solution on I), we again express u_h on the subinterval σ_n by the local Lagrange representation,

$$u_h(t_n + v h_n) = \sum_{j=1}^m L_j(v) U_{n,j}, \quad v \in (0,1], \quad 0 \leq n \leq N-1, \qquad (5.5)$$

with

$$L_j(v) := \prod_{k \neq j}^m \frac{s - c_k}{c_j - c_k} \quad \text{and} \quad U_{n,j} := u_h(t_n + c_j h_n),$$

and where we have assumed that $h_n = h = T/N$ for $n = 0, 1, \ldots, N-1$. The above expression for $(\mathcal{V}_\theta u_h)(t_{n,i})$ now becomes

$$(\mathcal{V}_\theta u_h)(t_{n,i}) = h \sum_{\ell=0}^{q_{n,i}-1} \sum_{j=1}^m \left(\int_0^1 K(t_{n,i}, t_\ell + sh) L_j(s)\, ds \right) U_{\ell,j} \qquad (5.6)$$

$$+ h \sum_{j=1}^m \left(\int_0^{\gamma_{n,i}} K(t_{n,i}, t_{q_{n,i}} + sh) L_j(s)\, ds \right) U_{q_{n,i},j}$$

($i = 1, \ldots, m$). Thus, the collocation equation (5.3) leads to a system of linear difference equations for the vectors $\mathbf{U}_n \in \mathbb{R}^m$ ($0 \leq n \leq N-1$) whose structure will change as we pass from Phase I of *complete overlap* ($0 \leq n < q^I$) via Phase II of *partial overlap* ($q^I \leq n < q^{II}$) to the *pure delay* Phase III ($q^{III} \leq n \leq N-1$). To make this more precise, we set $\mathbf{g}_n := (g(t_{n,1}), \ldots, g(t_{n,m}))^T \in \mathbb{R}^m$ and introduce the matrices

$$B_n := \left(\int_0^{c_i} K_1(t_{n,i}, t_n + sh) L_j(s)\, ds \right)_{i,j=1}^m, \qquad (5.7)$$

$$[2pt] B_{n,\ell} := \left(\int_0^1 K_1(t_{n,i}, t_\ell + sh) L_j(s)\, ds \right)_{i,j=1}^m, \quad \ell < n, \qquad (5.8)$$

in $L(\mathbb{R}^m)$. These matrices correspond to the contribution to the difference equation due to the 'classical' (non-delay) Volterra operator $\mathcal{V}$ in (5.3). For the concise formulation of the terms in the difference equation describing

Phases I, II and III and corresponding to the delay operator $\mathcal{V}_\theta$, we introduce the following matrices in $L(\mathbb{R}^m)$.

Phase I.

$$B_n^I(q) := \left(\int_0^{\gamma_{n,i}} K_2(t_{n,i}, t_n + sh) L_j(s) \, ds \right)_{i,j=1}^m, \tag{5.9}$$

$$B_{n,\ell}^I(q) := \left(\int_0^1 K_2(t_{n,i}, t_\ell + sh) L_j(s) \, ds \right)_{i,j=1}^m, \quad \ell < n. \tag{5.10}$$

Here, $\mathbf{U}_n$ $(0 \le n < q^I)$ is given by the solution of

$$[I_m - h(B_n + B_n^I(q))]\mathbf{U}_n = \mathbf{g}_n + h \sum_{\ell=0}^{n-1} (B_{n,\ell} + B_{n,\ell}^I(q))\mathbf{U}_\ell. \tag{5.11}$$

Phase II.

$$B_n^{II}(q) := \text{diag}(\underbrace{0, \ldots, 0}_{\nu_n}, 1, \ldots, 1) B_n^I(q), \tag{5.12}$$

$$B_{n-1}^{II}(q) := \left(\int_0^{\gamma_{n,i}} K_2(t_{n,i}, t_{n-1} + sh) L_j(s) \, ds \right)_{i,j=1}^m, \tag{5.13}$$

$$S_{n-1}^{II}(q) := \text{diag}(\underbrace{0, \ldots, 0}_{\nu_n}, 1, \ldots, 1) B_{n,n-1}^I(q), \tag{5.14}$$

$$\hat{S}_{n-1}^{II}(q) := \text{diag}(\underbrace{1, \ldots, 1}_{\nu_n}, 0, \ldots, 0) B_{n-1}^{II}(q). \tag{5.15}$$

It is readily verified that now $\mathbf{U}_n$ $(q^I \le n < q^{II})$ is determined by the solution of

$$[I_m - h(B_n + B_n^{II}(q))]\mathbf{U}_n = \mathbf{g}_n + h \sum_{\ell=0}^{n-1} B_{n,\ell}\mathbf{U}_\ell + h \sum_{\ell=0}^{n-2} B_{n,\ell}^I(q)\mathbf{U}_\ell$$
$$+ h[\hat{S}_{n-1}^{II}(q) + S_{n-1}^{II}(q)]\mathbf{U}_{n-1}. \tag{5.16}$$

Phase III.

$$B_{q_n}^{III}(q) := \left(\int_0^{\gamma_{n,i}} K_2(t_{n,i}, t_{q_n} + sh) L_j(s) \, ds \right)_{i,j=1}^m, \tag{5.17}$$

$$S_{q_n+1}^{III}(q) := \text{diag}(\underbrace{0, \ldots, 0}_{\nu_n}, 1, \ldots, 1) B_{q_n+1}^{III}(q), \tag{5.18}$$

$$\hat{S}_{q_n}^{III}(q) := \text{diag}(\underbrace{1, \ldots, 1}_{\nu_n}, 0, \ldots, 0) B_{q_n}^{III}(q). \tag{5.19}$$

Once we have reached this pure delay phase the vector $\mathbf{U}_n$ ($q^{II} \leq n \leq N-1$) is given by the solution of

$$[I_m - hB_n]\mathbf{U}_n = \mathbf{g}_n + h\sum_{\ell=0}^{n-1} B_{n,\ell}\mathbf{U}_\ell + h\sum_{\ell=0}^{q_n-1} B_{n,\ell}^I(q)\mathbf{U}_\ell \qquad (5.20)$$
$$+ h[\hat{S}_{q_n}^{III}(q) + B_{n,q_n}^I(q)]\mathbf{U}_{q_n} + hS_{q_n+1}^{III}(q)\mathbf{U}_{q_n+1}.$$

The (different) integers ν_n occurring in Phases II and III, and the integer $q_n < n - 1$ in Phase III were defined following Lemma 5.1.

Example 5.1. Suppose that the second-kind delay VIE (4.4) is solved by collocation in $S_0^{(-1)}(I_h)$, with uniform mesh I_h and collocation points $X_h = \{t_n + c_1 h : 0 < c_1 \leq 1 \ (0 \leq n \leq N - 1)\}$. Since u_h is constant on each subinterval σ_n we write $y_{n+1} := u_h(t_n + vh)$ ($v \in (0, 1]$).

For $m = 1$ we have $q^I = q^{II} = \lceil qc_1/(1-q) \rceil$. Hence, the collocation equation (with $\mathcal{V} = 0$) becomes

$$\left(1 - h\int_0^{\gamma_{n,1}} K(t_{n,1}, t_n + sh)\,ds\right) y_{n+1}$$
$$= h\sum_{\ell=0}^{n-1}\left(\int_0^1 K(t_{n,1}, t_\ell + sh)\,ds\right) y_{\ell+1} + g(t_{n,1})$$

when $0 \leq n < q^I$. For $q^I \leq n \leq N - 1$ it reads

$$y_{n+1} = g(t_{n,1}) + h\sum_{\ell=0}^{q_{n,1}-1}\left(\int_0^1 K(t_{n,1}, t_\ell + sh)\,ds\right) y_{\ell+1}$$
$$+ h\left(\int_0^{\gamma_{n,1}} K(t_{n,1}, t_{q_{n,1}+1} + sh)\,ds\right) y_{q_{n,1}+1}.$$

Thus, if $K(t,s) \equiv b/q$ and $g(t) \equiv 1$ the collocation solution to the resulting DVIE

$$y(t) = 1 + \int_0^{qt} (b/q)y(s)\,ds, \quad t \in I$$

(which is equivalent to the initial-value problem $y'(t) = by(qt)$, $y(0) = 1$) is determined by the solution of the difference equation

$$y_{n+1} = 1 + \frac{hb}{q}\sum_{\ell=0}^{q_{n,1}-1} y_{\ell+1} + \frac{hb}{q}\gamma_{n,1}y_{q_{n,1}+1}, \qquad (5.21)$$

where $q_{n,1} := \lfloor qc_1/(1-q) \rfloor$ and $\gamma_{n,1} := q(n+c_1) - q_{n,1}$.

We list, also for use in Example 5.2, a sample of values of $q_{n,1}$ and $\gamma_{n,1}$ for $c_1 = 1/2$, to provide some insight into the structure of the above difference equations when collocation is at the Gauss point $t_n + h/2$.

Table 5.3. $q = 1/2$, $c_1 = 1/2$ ($q^I = q^{II} = 1$).

n	0	1	2	3	4	5	6
$q_{n,1}$	0	0	1	1	2	2	3
$\gamma_{n,1}$	1/4	3/4	1/4	3/4	1/4	3/4	1/4

Table 5.4. $q = 0.9$, $c_1 = 1/2$ ($q^I = q^{II} = 5$).

n	0	1	2	3	4	5	6
$q_{n,1}$	0	1	2	3	4	4	5
$\gamma_{n,1}$	0.45	0.35	0.25	0.15	0.05	0.95	0.85

For Volterra functional equations governed by the special delay operator $\mathcal{W}_\theta$ (cf. (4.8)) we find – in complete analogy to the above – that $(\mathcal{W}_\theta u_h)(t_{n,i})$ is given by

$$(\mathcal{W}_\theta u_h)(t_{n,i}) = h \int_{\gamma_{n,i}}^{1} K(t_{n,i}, t_{q_{n,i}} + sh) u_h(t_{q_{n,i}} + sh) \, ds \qquad (5.22)$$

$$+ \int_{t_{q_{n,i}+1}}^{t_n} K(t_{n,i}, s) u_h(s) \, ds$$

$$+ h \int_{0}^{c_i} K(t_{n,i}, t_n + sh) u_h(t_n + sh) \, ds.$$

This can be written as

$$(\mathcal{W}_\theta u_h)(t_{n,i}) = h \sum_{j=1}^{m} \left(\int_{\gamma_{n,i}}^{1} K(t_{n,i}, t_{q_{n,i}} + sh) L_j(s) \, ds \right) U_{q_{n,i},j} \qquad (5.23)$$

$$+ h \sum_{\ell=q_{n,i}+1}^{n-1} \left(\int_{0}^{1} K(t_{n,i}, t_\ell + sh) L_j(s) \, ds \right) U_{\ell,j}$$

$$+ h \sum_{j=1}^{m} \left(\int_{0}^{c_i} K(t_{n,i}, t_n + sh) K L_j(s) \, ds \right) U_{n,j}.$$

Hence, the collocation equation associated with $y = g + \mathcal{W}_\theta y$,
$$u_h(t) = g(t) + (\mathcal{W}_\theta u_h)(t), \quad t \in X_h,$$
leads to the following systems of linear algebraic equations for $\mathbf{U}_n \in \mathbb{R}^m$ describing the local representation (5.5) of $u_h \in S_{m-1}^{(-1)}(I_h$ (again with uniform I_h).

Phase I.
$$[I_m - h\bar{B}_n^I(q)]\mathbf{U}_n = \mathbf{g}_n, \quad 0 \le n < q^I, \tag{5.24}$$
with
$$\bar{B}_n^I(q) = \left(\int_{\gamma_{n,i}}^{c_i} K(t_{n,i}, t_n + sh) L_j(s)\,ds \right)_{i,j=1}^m,$$
which is formally equivalent to $B_n + B_n^I(q)$ in (5.11) when $K_2 = -K_1 =: -K$.

Phase II.
$$[I_m - h\bar{B}_n^{II}(q)]\mathbf{U}_n = \mathbf{g}_n + h\bar{S}_{n-1}^{II}(q)\mathbf{U}_{n-1}, \quad q^I \le n < q^{II}, \tag{5.25}$$
where
$$\bar{B}_n^{II}(q) := \mathrm{diag}(\underbrace{1,\ldots,1}_{\nu_n},0,\ldots,0)B_n + \mathrm{diag}(0,\ldots,0,\underbrace{1,\ldots,1}_{\nu_n})\bar{B}_n^I(q),$$
and
$$\bar{S}_{n-1}^{II}(q) := \mathrm{diag}(\underbrace{1,\ldots,1}_{\nu_n},0,\ldots,0)\left(\int_{\gamma_{n,i}}^1 K(t_{n,i}, t_{n-1} + sh) L_j(s)\,ds \right).$$

Phase III.
$$[I_m - hB_n]\mathbf{U}_n = \mathbf{g}_n + h[\bar{S}_{q_n}^{III}(q)\mathbf{U}_{q_n} + \sum_{\ell=q_n+1}^{n-1} B_{n,\ell}\mathbf{U}_\ell + S_{q_n+1}^{III}(q)\mathbf{U}_{q_n+1}], \tag{5.26}$$
with
$$\bar{S}_{q_n}^{III}(q) := \mathrm{diag}(\underbrace{1,\ldots,1}_{\nu_n},0,\ldots,0)\left(\int_{\gamma_{n,i}}^1 K(t_{n,i}, t_{q_n} + sh) L_j(s)\,ds \right),$$
$$S_{q_n+1}^{III}(q) := \mathrm{diag}(\underbrace{1,\ldots,1}_{\nu_n},0,\ldots,0)B_{n,q_n+1}$$
$$+ \mathrm{diag}(0,\ldots,0,\underbrace{1,\ldots,1}_{\nu_n})\left(\int_{\gamma_{n,i}}^1 K(t_{n,i}, t_{q_n+1} + sh) L_j(s)\,ds \right).$$

The matrices $B_{n,\ell} \in L(\mathbb{R}^m)$ coincide with the ones in (5.8).

5.3.2. Optimal orders of convergence

Suppose that a given delay integral with (vanishing) proportional delay is solved by collocation in $S_{m-1}^{(-1)}(I_h)$, with the underlying mesh I_h being a *uniform* one, and assume that the functions defining the functional equation have arbitrarily high degree of regularity on their respective domains (implying, as we have seen in Theorem 4.4, that the exact solution has the same regularity). What can be said, as $h \to 0$, $N \to \infty$ ($Nh = T$) about the optimal values of p and p^* in the estimates

$$\|y - u_h\|_\infty := \max\{|y(t) - u_h(t)| : t \in I\} \le C(q) h^p \qquad (5.27)$$

and

$$\|y - u_h\|_{I_h,\infty} := \max\{|y(t) - u_h(t)| : t \in I_h \setminus \{0\}\} \le C(q) h^{p^*}? \qquad (5.28)$$

Do higher values of p and p^* result for the *iterated* collocation solution u_h^{it} based on u_h?

It turns out that we have $p = m$ in the global estimate (5.27) if the set $\{c_i\}$ is arbitrary; that is, the results that hold for 'classical' VIEs of the second kind remain valid. However, the question regarding the optimal value of p^* in (5.28) (attainable order of *local superconvergence* on $I_h \setminus \{0\}$) has an answer that differs sharply from the one for non-delay VIEs. Moreover, we shall see that it is not yet known under what conditions on the collocation parameters $\{c_i\}$ the collocation solutions u_h to the *first-kind* delay VIE (5.27) converge uniformly on I to the exact solution y (see also Section 5.2.1).

Theorem 5.2. Consider the second-kind delay VIE (4.6) and assume that the given functions g, K_1, K_2 are at least m times continuously differentiable on their respective domains. Then, for all sufficiently small $h > 0$ (so that the difference equations (5.11), (5.16), (5.20) possess unique solutions), we have $p = m$ in (5.27) for arbitrary $\{c_i\}$ in X_h and for all delay functions $\theta(t) = qt$ with $0 < q < 1$. The constant $C(q)$ depends on the $\{c_i\}$ and on q but not on h.

The proof of this result is an adaptation of the one for classical second-kind VIEs and delay VIEs with non-vanishing delays: it consists in showing that in each of the Phases I–III the ℓ^1-norms of the vectors $\mathbf{U}_n$ defined by (5.11), (5.16), and (5.20) satisfy a discrete Gronwall inequality. An elementary induction argument then leads to the assertion, observing the local representation (5.5) of u_h and the fact that, by the assumption on the regularity of the given data, $y \in C^m(I)$. The detailed arguments can be found in Chapter 5 (Section 5.3) of Brunner (2004b); see also Zhang (1998) and Brunner and Zhang (1999) for related results.

We now turn to the question of global and local *superconvergence* on I and I_h, respectively: are there collocation parameters $\{c_i\}$ for which $p = m$ in (5.27) can be replaced by $p^* > m$, and how large can p^* become in (5.28)?

Assume that the solution of the second-kind delay VIE (4.6) is approximated by the collocation solution $u_h \in S_{m-1}^{(-1)}(I_h)$ and the corresponding *iterated collocation solution*,

$$u_h^{it}(t) := g(t) + (\mathcal{V}u_h)(t) + (\mathcal{V}_\theta u_h)(t), \quad t \in I. \tag{5.29}$$

Recall that $u_h^{it}(t) = u_h(t)$ whenever $t \in X_h$; in particular, if $c_m = 1$ then $t_n \in X_h$ ($n = 1, \ldots, N$) and hence $u_h^{it}(t_n) = u_h(t_n)$.

It is well known (see, e.g., Brunner and van der Houwen (1986, Chapter 5)) that if $\mathcal{V}_\theta = 0$ (classical Volterra integral equation) and if the $\{c_i\}$ are the *Gauss (–Legendre) points* (given by the zeros of the shifted Legendre polynomial $P_m(2s - 1)$), then we only attain $p^* = m$ in (5.28): local superconvergence of order $p^* = 2m$ is only possible if u_h is replaced by the iterated collocation solution u_h^{it}. For the general delay VIE (4.6) with $\theta(t) = qt$ ($0 < q < 1$) this is no longer true (see Brunner (1997a), Takama et al. (2000), Muroya et al. (2002), Brunner (2004a)): for *arbitrary* $q \in (0, 1)$ we have $p^* < 2m$ when $m \geq 3$. The results of Theorems 5.3 and 5.5 have recently been established in Brunner and Hu (2003); see also Brunner (2004a, 2004b). Theorem 5.5 disproves a conjecture in Brunner (1997a) and Brunner, Hu and Lin (2001a) for $m > 2$.

Theorem 5.3. Let the collocation parameters $\{c_i\}$ satisfy the orthogonality condition

$$J_0 := \int_0^1 \prod_{i=1}^m (s - c_i) \, ds = 0. \tag{5.30}$$

Then

$$\|y - u_h^{it}\|_\infty \leq C(q) h^{m+1},$$

where u_h^{it} is the iterated collocation approximation (5.29) corresponding to the collocation solution $u_h \in S_{m-1}^{(-1)}(I_h)$ for (4.4). Here, $m + 1$ cannot, in general, be replaced by $m + 2$.

We note that the most prominent set of parameters $\{c_i\}$ satisfying the above orthogonality condition are the *Gauss points*.

Proof. We will sketch the principal steps leading to the above global superconvergence result by using an approach different from the one in Brunner and Hu (2003). For ease of notation we will do this for (4.4) with $\mathcal{V} = 0$, that is,

$$y(t) = g(t) + (\mathcal{V}_\theta y)(t), \quad t \in I.$$

In this case, the collocation error $e_h := y - u_h$ satisfies the integral equation

$$e_h(t) = \delta_h(t) + (\mathcal{V}_\theta e_h)(t), \quad t \in I, \tag{5.31}$$

where the defect δ_h vanishes on X_h and inherits (piecewise, on each subinterval σ_n) the regularity of g and K_2. Moreover, we have $e_h^{it} := y - u_h^{it} = e_h - \delta_h$. Hence, it follows from the solution representation (4.5) in Theorem 4.2 that

$$e_h^{it}(t) = \sum_{j=1}^{\infty} \int_0^{q^j t} K_{2,j}(t,s)\delta_h(s)\,ds, \quad t \in I. \tag{5.32}$$

Here, $K_{2,j}(t,s)$ denotes the jth iterated kernel of the given kernel $K_2(t,s)$ in $\mathcal{V}_\theta$ (cf. Theorem 4.2)), and the infinite series converges uniformly on I.

Now let $t = t_n + vh$, $v \in [0,1]$ be given and define, as in (5.1),

$$q_{j,n}(v) := \lfloor q^j(n+v) \rfloor, \qquad \gamma_{j,n}(v) := q^j(n+v) - q_{j,n}(v) \in [0,1), \quad j \in \mathbb{N}.$$

This allows us to write (5.32) in the form

$$e_h^{it}(t) = \sum_{j=1}^{\infty} \left(\int_0^{t_{q_{j,n}(v)}} K_{2,j}(t,s)\delta_h(s)\,ds \right. \tag{5.33}$$

$$\left. + h \int_0^{\gamma_{j,n}(v)} K_{2,j}(t, t_{q_{j,n}(v)} + sh)\delta_h(t_{q_{j,n}(v)} + sh)\,ds \right).$$

The assertion in Theorem 5.3 now follows from the following observations.

(i) Since $0 < q < 1$, we have $q_{j,n}(v) < N$ for all $n \le N-1$ and $v \in [0,1]$. Upon writing

$$\int_0^{t_{q_{j,n}(v)}} K_{2,j}(t,s)\delta_h(s)\,ds = h \sum_{\ell=0}^{q_{j,n}(v)-1} \int_0^1 K_{2,j}(t, t_\ell + sh)\delta_h(t_\ell + sh)\,ds,$$

we can again resort to the classical 'quadrature error argument' of Sections 3.4.2 and 3.4.3, consisting in replacing each of the above integrals by the sum of the interpolatory m-point quadrature approximation based on the points $\{t_\ell + c_i h\}$ and the corresponding quadrature error. As $\delta_h(t_\ell + c_i h) = 0$, and $Nh = T$, it follows that the absolute values of the integrals are bounded by $C_Q h^{m+1}$, because the orthogonality condition (5.30) implies that the quadrature formulas all possess a degree of precision of (at least) m.

(ii) The global convergence estimate given by Theorem 5.2 can be used in (5.31) to obtain the estimate

$$\|\delta_h\|_\infty \le (1 + \|\mathcal{V}_\theta\|)C(q)h^m,$$

where

$$\|\mathcal{V}_\theta\| := \max_{t \in D_\theta} \int_0^{\theta(t)} |K_2(t,s)|\,ds.$$

(iii) The iterated kernels $K_{2,j}(t,s)$ satisfy

$$|K_{2,j}(t,s)| \leq \frac{q^{j(j-1)/2}}{(j-1)!} T^{j-1} \bar{K}_\theta^j, \quad t \in D_\theta^{(j)}$$

(*cf.* Lemma 4.3), where

$$\bar{K}_\theta := \max_{(D_\theta)} |K_2(t,s)| \quad \text{and} \quad D_\theta^{(j)} := \{(t,s) : 0 \leq s \leq q^j t, \, t \in I\}.$$

This result implies that the infinite series involving the second terms on the right-hand side of (5.33) converges uniformly and is $\mathcal{O}(h^{m+1})$. □

In contrast to this result (the analogue of the global superconvergence result for ODEs), collocation at the Gauss points no longer leads to *local superconvergence* of order $2m$ on I_h when $m \geq 3$, as the following theorem shows. (Observe that for $m = 2$ we have $2m = m+2$.) A first hint that this is so may be divined from the following result comparing the collocation solution for the special pantograph equation (4.2) with the iterated collocation solution (based on the same collocation parameters) for its integrated form,

$$y(t) = y_0 + \int_0^{qt} (b/q) y(s) \, ds, \quad t \in I.$$

Theorem 5.4. Let the following conditions hold.

(a) $u_h \in S_{m-1}^{(-1)}(I_h)$ is the collocation solution (with respect to X_h) to the integrated form of $y'(t) = by(qt)$, $y(0) = y_0$ ($b \neq 0$, $y_0 \neq 0$), and u_h^{it} denotes the corresponding iterated collocation solution.

(b) $v_h \in S_m^{(0)}(I_h)$ is the collocation solution (also with respect to X_h) to $y'(t) = by(qt)$, $y(0) = y_0$.

Then we have, for all $q \in (0,1)$,

$$u_h^{it}(t) \neq v_h(t) \quad \text{whenever } t \in I_h \setminus \{0\}.$$

Remark. We remind the reader that for $q = 1$ we obtain $u_h^{it}(t) = v_h(t)$ for all $t \in I$: the 'indirect' collocation approximation u_h^{it} has the same (super-)convergence properties as the 'direct' one for the original ODE.

The proof of the following theorem is based on interpolatory projection techniques and can be found in Brunner and Hu (2003).

Theorem 5.5. Assume that g, K_1 and K_2 in (4.6) are at least $d \geq m+2$ times continuously differentiable on their respective domains. Let $u_h \in S_{m-1}^{(-1)}(I_h)$ be the collocation solution corresponding to the Gauss points $\{c_i\}$, and let the iterated collocation solution be defined by (5.29). Then for any $q \in (0,1)$ the order p^* in the local estimate

$$\|y(t) - u_h^{it}(t)\|_{h,\infty} \leq C(q) h^{p^*}$$

cannot exceed $m + 2$, regardless of the value of d. More precisely, the following is true.

(i) If $q = 1/2$ then

$$\|y - u_h^{it}\|_{h,\infty} \leq C(q) \begin{cases} h^{m+2}, & \text{if } m \text{ is even,} \\ h^{m+1}, & \text{if } m \text{ is odd.} \end{cases}$$

(ii) For $q \in (0,1) \setminus \{1/2\}$, we have

$$\|y - u_h^{it}\|_{h,\infty} \leq C(q) h^{m+1},$$

for all $m \geq 2$.

Remarks. (1) It is intuitively clear that the error constants $C(q)$ in the above order estimates will change their values dramatically as $q \to 1^-$. Thus it is a challenging problem to find this dependence on q explicitly: even insight obtained from a simple (linear) 'toy problem' would be valuable.

(2) For the more general second-kind DVIE (4.9), which we write now as

$$y(t) = g(t) + b(t)y(\theta(t)) + (\mathcal{W}_\theta y)(t), \quad t \in I, \quad \theta(t) = qt, \quad 0 < q < 1,$$

the existence of a unique collocation solution $u_h \in S_{m-1}^{(-1)}(I_h)$ is no longer guaranteed in Phases I and II for all sufficiently small mesh diameters $h > 0$. This is due to the presence of the terms $b(t_{n,i})u_h(\theta(t_{n,i}))$ which, using the local representation of u_h on $\sigma_{qn,i}$ (cf. (5.5)), assume the form $b(t_{n,i}) \sum_{j=1}^m L_j(\gamma_{n,i}) U_{qn,i,j}$ ($i = 1, \ldots, m$). Thus, the matrix describing the left-hand side of the difference equation (5.11) (Phase I) now has the form $I_m - \mathcal{B}_n - h(B_n + B_n^I(q))$, with

$$\mathcal{B}_n := \text{diag}(b(t_{n,i})) \begin{pmatrix} L_j(\gamma_{n,i}) \\ (i,j = 1, \ldots, m) \end{pmatrix},$$

and hence its inverse will no longer exist for all sufficiently small $h > 0$. (Compare also Liu (1995b), where this problem is studied for the case $\mathcal{W}_\theta = 0$, $m = 1$, and $c_1 = 1$.) The superconvergence analysis for this more general Volterra integral equation with proportional delay is yet to be established.

5.4. Proportional delay VIEs of the first kind

Turning to Volterra's first-kind VIE with proportional delay $\theta(t) = qt$ ($0 < q < 1$) of 1897,

$$(\mathcal{W}_\theta y)(t) = g(t), \quad t \in I := [0, T], \tag{5.34}$$

the linear algebraic systems whose solutions $\mathbf{U}_n \in \mathbb{R}^m$ define the local representations (5.5) of its collocation solution $u_h \in S_{m-1}^{(-1)}(I_h)$ can be obtained from (5.24)–(5.26): they are, respectively,

$$\bar{B}_n^I(q)\mathbf{U}_n = h^{-1}\mathbf{g}_n, \quad 0 \le n < q^I, \tag{5.35}$$

$$\bar{B}_n^{II}(q)\mathbf{U}_n = h^{-1}\mathbf{g}_n - \bar{S}_{n-1}^{II}(q)\mathbf{U}_{n-1}, \quad q^I \le n < q^{II}, \tag{5.36}$$

and

$$B_n\mathbf{U}_n = h^{-1}\mathbf{g}_n - [\bar{S}_{q_n}^{III}(q)\mathbf{U}_{q_n} \tag{5.37}$$

$$+ \sum_{\ell=q_n+1}^{n-1} B_{n,\ell}\mathbf{U}_\ell + S_{q_n+1}^{III}(q)\mathbf{U}_{q_n+1}, \quad q^{II} \le n \le N-1.$$

Example 5.2. Consider the first-kind delay VIE (5.34), and assume that its collocation solution is to be in the collocation space of Example 5.1. Thus, for $0 \le n < q^I$ the collocation equation at $t = t_{n,1} = t_n + c_1 h$ may be written in the form

$$\left(\int_{\gamma_{n,1}}^{c_1} K(t_{n,1}, t_n + sh) \, ds \right) y_{n+1} = h^{-1}g(t_{n,1}).$$

If $n \ge q^I = q^{II}$ (see Example 5.1) we have, by (5.22),

$$(\mathcal{W}_\theta u_h)(t_{n,1}) = \int_{qt_{n,1}}^{t_n} K(t_{n,1}, s) u_h(s) \, ds + \int_{t_n}^{t_{n,1}} K(t_{n,1}, s) u_h(s) \, ds,$$

and this can be written as

$$\left(\int_0^{c_1} K(t_{n,1}, t_n + sh) \, ds \right) y_{n+1}$$

$$= h^{-1}g(t_{n,1}) - \left(\int_{\gamma_{n,1}}^1 K(t_{n,1}, t_{q_{n,1}} + sh) \, ds \right) y_{q_{n,1}+1}$$

$$- \sum_{\ell=q_{n,1}+1}^{n-1} \left(\int_0^1 K(t_{n,1}, t_\ell + sh) \, ds \right) y_{\ell+1}.$$

Setting

$$A_n := \int_0^{c_1} K(t_{n,1}, t_n + sh) \, ds,$$

$$B_{n,\ell} := \int_0^1 K(t_{n,1}, t_\ell + sh) \, ds \quad (q_{n,1} + 1 \le \ell \le n - 1),$$

and

$$C_{n,q_n} := \int_{\gamma_{n,1}}^1 K(t_{n,1}, t_{q_{n,1}} + sh) \, ds,$$

with $q_{n,1} := \lfloor q(n + c_1) \rfloor$ and $\gamma_{n,1} := q(n + c_1) - q_{n,1}$, the above difference equation for $\{y_{n+1}\}$ becomes

$$A_n y_{n+1} + \sum_{\ell=q_{n,1}+1}^{n-1} B_{n,\ell} y_{\ell+1} + C_{n,q_n} y_{q_{n,1}+1} = h^{-1} g(t_{n,1}). \qquad (5.38)$$

If $K(t,s) \equiv 1$ the delay integral equation (5.34) reduces to

$$\int_{qt}^{t} y(s)\,ds = g(t), \quad t \in I, \quad g(0) = 0,$$

and this is equivalent to the functional equation

$$y(t) - qy(qt) = g'(t), \quad t \in I.$$

The corresponding collocation solution $u_h \in S_0^{(-1)}(I_h)$ is thus determined by the solution of the difference equation

$$c_1 y_{n+1} + \sum_{\ell=q_{n,1}+1}^{n-1} y_{\ell+1} + (1-\gamma_{n,1}) y_{q_{n,1}+1} = h^{-1} g(t_n + c_1 h), \quad n \geq 0. \qquad (5.39)$$

In order to obtain the difference equation corresponding to collocation at the Gauss points ($c_1 = 1/2$) recall Tables 5.3 and 5.4 of Example 5.1 for the values of $q_{n,1}$ and $\gamma_{n,1}$.

What can be said about the attainable orders of global and local (super-) convergence: what do the analogues of Theorems 5.2, 5.3 and 5.4 look like? As we have briefly mentioned before, the analysis leading to answers for these questions remains open. We only know from numerical evidence that the condition

$$-1 \leq \rho_m := (-1)^m \prod_{i=1}^{m} \frac{1-c_i}{c_i} \leq 1$$

(which guarantees uniform convergence when $q = 0$ in $\mathcal{W}_\theta$ with $\theta(t) = qt$) is no longer sufficient for this to be true. In particular, it is not yet known for which values of $c_1 \in (0,1]$ the solution of the simple difference equation (5.38) remains uniformly bounded as $N \to \infty$ ($h \to 0$, $Nh = T$) when $q \in (0,1)$.

5.5. Volterra integro-differential equations with proportional delays

The collocation solution $u_h \in S_m^{(0)}(I_h)$ for the DVIDE (4.12) is determined by

$$u_h'(t) = a(t)u_h(t) + b(t)u_h(\theta(t)) + (\mathcal{W}_\theta u_h)(t), \quad t \in X_h, \qquad (5.40)$$

with initial condition $u_h(0) = y_0$. Since the collocation space is now a subspace of $C(I)$ the system of difference equations arising in the computation of u_h has a somewhat more complex structure than the ones we encountered in the previous section. To be somewhat more precise, the local representation of $u_h \in S_m^{(0)}(I_h)$ is now

$$u_h(t_n + vh) = y_n + h \sum_{j=1}^{m} \beta_j(v) Y_{n,j}, \quad v \in [0,1], \quad 0 \le n \le N-1, \quad (5.41)$$

where $y_n := u_h(t_n)$, $Y_{n,j} := u_h'(t_{n,j})$, and $\beta_j(v) := \int_0^1 L_j(s)\,ds$. Hence, the key ingredients in the derivation of the difference equations are essentially the same as in Section 5.2, that is, the terms of $(\mathcal{W}_\theta u_h)(t_{n,i})$ corresponding to Phases I, II, and III and containing the vectors $\mathbf{Y}_n$ are described by matrices similar to those in (5.24)–(5.26), except that instead of $L_j(s)$ their integrands contain the integrated Lagrange polynomials $\beta_j(s)$. However, there are now also additional terms reflecting the continuity constraint of u_h at the interior mesh points $t_1, \ldots, t_{N-1}$. We leave the details to the reader; compare also Chapter 5 in Brunner (2004b). Instead, we present an illustration from which the difference equations for arbitrary $m \ge 2$ can readily be deduced.

Example 5.3. Consider (5.40) and suppose that $u_h \in S_1^{(0)}(I_h)$ ($m = 1$), with $0 < c_1 \le 1$. The collocation equation defining this collocation solution u_h,

$$u_h'(t_{n,1}) = a(t_{n,1}) u_h(t_{n,1}) + b(t_{n,1}) u_h(qt_{n,1}) + (\mathcal{W}_\theta u_h)(t_{n,1}),$$

where, by (5.41), $u_h(t_n + vh) = y_n + hv Y_{n,1}$ ($v \in [0,1]$), can be written as

$$Y_{n,1} = a(t_{n,1})\{y_n + hc_1 Y_{n,1}\} + b(t_{n,1})\{y_{q_{n,1}} + h\gamma_{n,1} Y_{q_{n,1}}\} + (\mathcal{W}_\theta u_h)(t_{n,1}),$$

with

$$(\mathcal{W}_\theta u_h)(t_{n,1}) = \int_{t_{q_{n,1}}}^{t_{q_{n,1}+1}} K(t_{n,1}, s) u_h(s)\,ds + \int_{t_{q_{n,1}+1}}^{t_n} K(t_{n,1}, s) u_h(s)\,ds$$

$$+ h \int_0^{c_1} K(t_{n,1}, t_n + sh)\{y_n + hs Y_{n,1}\}\,ds.$$

Hence, the resulting difference equation is

$$\left(1 - ha(t_{n,1})c_1 - h^2 \int_0^{c_1} K(t_{n,1}, t_n + sh) s\,ds\right) Y_{n,1} \qquad (5.42)$$

$$= \left[hb(t_{n,1})\gamma_{n,1} + h^2 \left(\int_{\gamma_{n,1}}^{1} K(t_{n,1}, t_{q_{n,1}} + sh) s\,ds\right)\right] Y_{q_{n,1}}$$

$$+ h^2 \sum_{\ell=q_{n,1}+1}^{n-1} \left(\int_0^1 K(t_{n,1}, t_\ell + sh) s\,ds\right) Y_{\ell,1} + \rho_n,$$

where

$$\rho_n := a(t_{n,1})y_n + \left(b(t_{n,1}) + h\int_{\gamma_{n,1}}^{1} K(t_{n,1}, t_{q_{n,1}} + sh)\,ds\right)y_{q_{n,1}}$$

$$+ h\sum_{\ell=q_{n,1}+1}^{n-1}\left(\int_{0}^{1} K(t_{n,1}, t_\ell + sh)\,ds\right)y_\ell$$

$$+ h\left(\int_{0}^{c_1} K(t_{n,1}, t_n + sh)\,ds\right)y_n.$$

The values of $q_{n,1}$ and $\gamma_{n,1}$ can be found in Example 5.1. Observe also that the above difference equation (5.42) can be reformulated as a difference equation for $\{y_n\}$, by setting $Y_{n,1} = (y_{n+1} - y_n)/h$ (cf. (5.41) with $m = 1$ and $v = 1$).

The problem of asymptotic stability for u_h in special case corresponding to $\mathcal{W} = 0$ and constant coefficients a, b, i.e., the pantograph equation (4.1), was studied in Buhmann et al. (1993) and Iserles (1994a) for $q = 1/2$. For pantograph-type VIDEs the problem is completely open (see also Section 5.7).

5.5.1. Optimal convergence estimates

The first theorem shows that the classical global order of convergence also remains valid for delay VIDEs with vanishing proportional delay. The proof is again based on a 'Phase I–III' Gronwall-type inductive argument; the details can be found in Brunner (2004b, Section 5.5). See also Zhang (1998) and Zhang and Brunner (1998) for some related results.

Theorem 5.6. Consider the delay VIDE (4.12) (which includes (4.14) and the pantograph equation (4.1) as special cases) and assume that the given functions a, b, K_1 and K_2 are at least m times continuously differentiable on their respective domains. Then, for any $\theta(t) = qt$ with $q \in (0,1)$ and for all sufficiently small $h > 0$, the unique collocation solution $u_h \in S_m^{(0)}(I_h)$ given by (5.40) satisfies

$$\|y^{(\nu)} - u_h^{(\nu)}\|_\infty := \sup\{|y^{(\nu)}(t) - u_h^{(\nu)}(t)| : t \in I\} \leq C_\nu(q)h^m, \quad \nu = 0, 1,$$

and this holds for any set $\{c_i\}$ defining the collocation points X_h. The constants $C_\nu(q)$ depend on the $\{c_i\}$ and on q but not on h.

As for delay VIEs of the second kind with proportional delays, *local superconvergence results* for analogous delay VIDEs differ sharply from the classical results. However, the *global order* $p = m + 1$ is possible for the collocation solution $u_h \in S_m^{(0)}(I_h)$. We summarize this results, and a conjecture, below. Note, incidentally, that these delay VIDEs include the pantograph equation and its counterpart with variable coefficients a and b as special cases.

Theorem 5.7. Let $u_h \in S_m^{(0)}(I_h)$ be the collocation solution to the proportional delay VIDE (4.12). If the given functions possess continuous derivatives of at least order $m+1$ on their respective domains, and if the $\{c_i\}$ satisfy the orthogonality condition (5.30) of Theorem 5.3, then

$$\|y - u_h\|_\infty \leq C(q) h^{m+1}$$

holds for all $q \in (0,1)$.

Proof. We write the error equation for (5.40),

$$e_h'(t) = a(t) e_h(t) + b(t) e_h(qt) + \delta_h(t) + (\mathcal{V}e_h)(t) + (\mathcal{V}_\theta e_h)(t),$$
$$t \in I, \quad e_h(0) = 0,$$

in integrated form. The analysis in Brunner and Hu (2003) (or, if $a(t) \equiv 0$, $\mathcal{V} = 0$, the one employed in the proof of Theorem 5.3) can then be applied to the resulting delay integral equation for e_h. We leave the details to the reader. □

Conjecture 5.8. The order of local superconvergence of $u_h \in S_m^{(0)}(I_h)$ on I_h cannot exceed $p^* = m + 2$. If the $\{c_i\}$ defining the collocation points X_h are the Gauss points, then we have $p^* = m + 2$ for any $q \in (0,1)$ and all $m \geq 2$. The same is true for the general pantograph equation corresponding to $\mathcal{W}_\theta = 0$ in (4.14).

5.6. Collocation on geometric meshes

The special form of the delay function $\theta(t) = qt$ ($0 < q < 1$) suggests that u_h^{it} might possibly attain the classical optimal order of superconvergence $p^* = 2m$ on I_h if I_h is a suitable *geometric mesh* and if collocation is at the Gauss points. That this is (almost) so was verified in the paper by Brunner et al. (2001a). We briefly describe this result and sketch its proof.

Assume that I_h is a *geometric mesh* defined by

$$I_h := \{t_n : t_n = \gamma^{N-n} T, \ n = 0, 1, \ldots, N; \ \gamma \in (0,1)\}. \tag{5.43}$$

The mesh parameter γ will depend, as is made precise below, on N (but not on n), on q, and on m. This mesh possesses the following obvious properties.

(i) $h_n := t_{n+1} - t_n = \gamma^{N-n-1}(1-\gamma) T \quad (n = 0, 1, \ldots, N-1)$.

(ii) $\max_{(n)} h_n = h_{N-1} = (1-\gamma) T$ (for any $N \in \mathbb{N}$). Hence, $\gamma = \gamma(N)$ will have to be chosen so that $\gamma \to 1^-$, as $N \to \infty$, for all $q \in (0,1)$.

Let $\rho \in \mathbb{N}$ be defined by

$$\rho := \left\lfloor \frac{\ln(q)}{\ln\left(1 - \frac{2m \ln(N)}{(m+1)N}\right)} \right\rfloor. \tag{5.44}$$

It is the largest integer for which

$$q^{1/\rho} \leq 1 - \frac{2m \cdot \ln(N)}{(m+1)N}.$$

Theorem 5.9 will reveal the motivation for introducing this integer ρ. Observe that for given (fixed) $q \in (0,1)$ and $m \geq 1$, we have $\rho > 1$ for all sufficiently large N. This is true because

$$1 - \frac{2m \cdot \ln(N)}{(m+1)N} \longrightarrow 1^{-}, \quad \text{as } N \to \infty,$$

for any $m \in \mathbb{N}$.

Theorem 5.9. Let the following be satisfied.

(a) $g \in C^{2m}(I)$, $K_1 \in C^{2m}(D)$, $K_2 \in C^{2m}(D_\theta)$.
(b) I_h is the geometric mesh described by (5.43) and (5.44), with $\gamma = q^{1/\rho}$.
(c) $u_h \in S_{m-1}^{(-1)}(I_h)$ is the collocation solution to the delay VIE (4.6), with the $\{c_i\}$ given by the Gauss points, and u_h^{it} denotes the corresponding iterated collocation solution.

Then, for all sufficiently large N, the resulting local order of convergence of u_h^{it} is given by

$$\max_{t \in I_h \setminus \{0\}} |y(t) - u_h^{it}(t)| \leq C(q) N^{-(2m - \varepsilon_N)},$$

where

$$\varepsilon_N := \log_N \left(\frac{(2m \cdot \ln(N))^{2m}}{(2m+1)(m+1)^{2m}} \right)$$

satisfies

$$\lim_{N \to \infty} \varepsilon_N = 0.$$

Proof. Since the proof is technically quite complex (using interpolatory projection techniques), we will only exhibit one of its key ingredients. □

Lemma 5.10. Let I_h be the geometric mesh (5.43), (5.44), with $\gamma = q^{1/\rho}$. Then:

(i) $h_0 \leq CN^{-2m/(m+1)}$;
(ii) $\sum_{n=1}^{N-1} h_n^{2m+1} \leq CN^{-(2m - \varepsilon_N)}$;
(iii) for $\rho + 1 \leq n \leq N$ we have $qt_n = t_{n-\rho} \in I_h \setminus \{0\}$.

Note, incidentally, that proposition (iii) may be viewed as generalized θ-invariance for this geometric mesh I_h.

Remarks. (1) Geometric meshes similar to the ones employed here were introduced by Hu (1998) for piecewise polynomial collocation methods applied to VIDEs with weakly singular kernels, to obtain local superconvergence of the collocation solution on I_h.

(2) It is clear that analogous superconvergence results can be derived for collocation solutions in $S_m^{(0)}(I_h)$, with suitable geometric mesh I_h, for the VIDE (4.12) with proportional delays. However, since this has not yet been worked out in detail, the reader is invited to take up the challenge.

5.7. Equations with nonlinear vanishing delays

Suppose that the linear delay function $\theta(t) = qt$ $(0 < q < 1)$ is replaced by a *nonlinear* function θ satisfying the following conditions:

(ND1) $\theta \in C^1(I)$, with $\theta(0) = 0$ and $\theta(t) < t$ for $t > 0$;
(ND2) $\min_{t \in I} \theta'(t) = q_0 > 0$.

The (linear) proportional delay function $\theta(t) = qt$ $(0 < q < 1)$ of course satisfies (ND1) and (ND2), with $\theta'(t) = q =: q_0$ for all t. Similar nonlinear vanishing delays were considered by Denisov and Korovin (1992) and Denisov and Lorenzi (1995); see also Bellen et al. (2002).

While it seems clear from the foregoing convergence analyses that the (super-) convergence results for collocation solutions remain valid for (3.8) and (3.45) with such nonlinear vanishing delays, the details have yet to be worked out.

5.8. Open problems

Our previous discussion of collocation methods has made it clear that even for delay VFIEs and VFIDEs with the simple linear lag function $\theta(t) = qt$ $(0 < q < 1;\ t \in [0,T])$ we have a long way to go to understand the dynamics of collocation solutions. We list below a selection of open problems whose solution would significantly advance our understanding of such functional equations.

(1) Superconvergence analysis of collocation solutions in $S_m^{(0)}(I_h)$, with uniform mesh I_h, for *nonlinear* Volterra functional integro-differential equations of Hammerstein type,

$$y'(t) = f(y(t), y(qt), y'(qt)) + \int_{qt}^{t} k(t-s) G(y(s), y'(s))\,ds, \quad 0 < q < 1,$$

in particular if f has one of the (Riccati, or rational) forms considered in Iserles (1994*b*).

(2) Long-time integration of pantograph-type functional equations: how do the error constants associated with the collocation solutions for DVIEs and DVIDEs depend on q and grow as $q \uparrow 1^-$?

(3) For which continuous convolution kernels k_1 and k_2 in

$$y(t) = 1 + \int_0^t k_1(t-s)y(s)\,ds + \int_0^{qt} k_2(t-s)y(s)\,ds, \quad t \geq 0, \qquad (5.45)$$

does

$$\lim_{t \to \infty} y(t) = 0$$

hold?

The answer to the analogous question for

$$y'(t) = ay(t) + by(qt) + \int_0^t k_1(t-s)y(s)\,ds + \int_0^{qt} k_2(t-s)y(s)\,ds, \quad t \geq 0, \qquad (5.46)$$

where $y(0) = y_0 \neq 0$, is also open. In particular, for which continuous convolution kernels k do we obtain asymptotic stability in (5.43) and (5.44) when the sum of the two integral operators is replaced by the Volterra operator corresponding to $\mathcal{W}_\theta$,

$$(\mathcal{W}_q \phi)(t) := \int_{qt}^t k(t-s)\phi(s)\,ds?$$

(4) Suppose that the given functions in (6.1) and (6.2) are such that the solutions of these pantograph-type Volterra equations are asymptotically stable. For which collocation parameters $\{c_i\}$ do the corresponding collocation solutions possess the same asymptotic behaviour?

6. Summary and outlook

6.1. Delay VIEs with weakly singular kernels

Owing to limitation of space we can only briefly touch upon the convergence properties of collocation solutions to delay VIEs of the form

$$y(t) = g(t) + (\mathcal{V}_\alpha y)(t) + (\mathcal{V}_{\theta,\alpha} y)(t), \quad t \in (t_0, T], \qquad (6.1)$$

where the Volterra integral operators are based on kernels containing a weak (i.e., unbounded but integrable) singularity:

$$(\mathcal{V}_\alpha y)(t) := \int_{t_0}^t (t-s)^{-\alpha} K_1(t,s) y(s)\,ds,$$

$$(\mathcal{V}_{\theta,\alpha} y)(t) := \int_{t_0}^{\theta(t)} (t-s)^{-\alpha} K_2(t,s) y(s)\,ds,$$

with $0 < \alpha < 1$, smooth K_i, and $K_1(t,t) \neq 0$ on I. The generalization of $\mathcal{W}_\theta$ in (3.22) is thus given by

$$(\mathcal{W}_{\theta,\alpha} y)(t) := \int_{\theta(t)}^t (t-s)^{-\alpha} K(t,s) y(s) \, ds,$$

with $K(t,t) \neq 0$ on I.

It follows from the theory of classical (non-delay) VIEs with weakly singular kernels (Brunner, Pedas and Vainikko 1999) that on *uniform meshes* the collocation solution $u_h \in S_{m-1}^{(-1)}(I_h)$ to (6.1) has global order of convergence of order $p = 1 - \alpha$, regardless of the regularity of the given functions or the degree of the approximating piecewise polynomial u_h. This is due to the low regularity of y at $t = t_0^+$: its first derivative behaves like $C(t-t_0)^{-\alpha}$ near $t = t_0^+$. An analogous result holds for VIDEs with weakly singular kernels (see Brunner, Pedas and Vainikko (2001b)): here, it is y'' that is similarly unbounded at $t = t_0^+$.

The optimal order of global convergence, $p = m$, can only be restored if the mesh I_h is suitably *graded*, i.e., if it is given by

$$I_h = \left\{ t_n := t_0 + \left(\frac{n}{N}\right)^r (T - t_0) : n = 0, 1, \ldots, N \right\},$$

with grading exponent $r \geq m/(1-\alpha)$ in the case of (6.1) with $\mathcal{V}_{\theta,\alpha} = 0$. For the analogous VIDE we must have $r \geq (m+1-\alpha)/(2-\alpha)$ (see Brunner (1985), Brunner, Pedas and Vainikko (1999, 2001)).

When solving weakly singular VIEs with *non-vanishing delays*, the mesh I_h will obviously be the constrained mesh (3.3) where the local meshes are now graded ones:

$$I_h^{(\mu)} := \left\{ t_n^{(\mu)} := \xi_\mu + \left(\frac{n}{N}\right)^{r_\mu} (\xi_{\mu+1} - \xi_\mu) : n = 0, 1, \ldots, N \right\},$$

with (local) grading exponents $r_\mu > 1$ depending on α and μ. Note that we have $\xi_{\mu+1} - \xi_\mu = \tau(\xi_{\mu+1}) \geq \tau_0 > 0$.

Lemma 6.1. *Let I_h be a θ-invariant mesh defined by (3.3) and (3.4), and assume that the first local mesh $I_h^{(0)}$ is optimally graded, that is,*

$$I_h^{(0)} := \left\{ t_n^{(0)} = \left(\frac{n}{N}\right)^{r_0} (\xi_1 - t_0) : n = 0, 1, \ldots, N \right\},$$

with $r_0 = m/(1-\alpha)$.

(i) *If the lag function θ is linear, then the optimal grading is inherited by the macro-meshes $I_h^{(1)}, \ldots, I_h^{(M)}$, with $r_\mu = r_0$ for all μ.*

(ii) *If θ is nonlinear, the (optimal) grading is lost for $I_h^{(\mu)}$ ($\mu \geq 1$).*

This result tells us that for the weakly singular Volterra integral or integro-differential equations (2.30) and (2.31) with non-vanishing delays, collocation on θ-invariant meshes I_h with optimally graded initial mesh $I_h^{(0)}$ will exhibit the classical optimal global and local convergence orders if the *delay function* θ is *linear*. For *nonlinear* θ, this will no longer be true.

Similar (positive and negative) results hold for piecewise polynomial collocation solutions applied to *neutral* VFIDEs with weakly singular kernels,

$$\frac{d}{dt}(a_0 y(t) - (\mathcal{T}_{\theta,\alpha} y)(t)) = F(t, y(t), y(\theta(t)), y'(\theta(t))), \quad (6.2)$$

with $\mathcal{T}_{\theta,\alpha}$ denoting one of the Volterra delay operators $\mathcal{V}_{\theta,\alpha}$ or $\mathcal{W}_{\theta,\alpha}$ ($0 < \alpha < 1$), when $a_0 = 1$. An alternative numerical approach to such functional integro-differential equations can be found in Ito and Turi (1991): it is based on the semigroup framework underlying functional equations of this kind.

For VFIDEs of the first kind, corresponding to $a_0 = 0$ in (6.2), e.g.,

$$\frac{d}{dt}((\mathcal{W}_{\theta,\alpha} y)(t)) = f(t), \quad t \in (t_0, T],$$

the convergence analysis of collocation solutions is much more difficult, and many questions remain to be answered. This is not so surprising when we recall that collocation solutions in $S_{m-1}^{(-1)}(I_h)$ for classical first-kind VIEs,

$$(\mathcal{V} y)(t) = g(t), \quad t \in I := [0, T], \quad g(0) = 0,$$

converge to y uniformly on I only if the collocation parameters $\{c_i\} \subset (0, 1]$ satisfy the stability condition

$$-1 \leq \rho_m := (-1)^m \prod_{i=1}^{m} \frac{1 - c_i}{c_i} \leq 1$$

(see Brunner and van der Houwen (1986) and Kauthen and Brunner (1997)). For the weakly singular version,

$$(\mathcal{V}_\alpha y)(t) = g(t), \quad t \in I, \quad 0 < \alpha < 1,$$

it is not known for which sets $\{c_i\}$ the collocation solution is convergent (see Brunner (1999b)). We note that Ito and Turi (1991) use their semigroup-based method to solve NVFIDEs of the form (6.2) however, the question regarding the attainable order of convergence on graded meshes remains open.

6.2. Integral-algebraic Volterra equations with non-vanishing delays

The numerical solution of differential-algebraic equations (DAEs) with constant delays by Runge–Kutta and collocation methods is studied in Ascher and Petzold (1995) and in Hauber (1997). However, as we mentioned in the discussion following equation (1.10), it is not clear how to obtain a *numerically properly formulated* form of a delay DAE, or of an analogous system of integral-algebraic or integro-differential-algebraic Volterra equations with delay arguments. This is an important issue for the understanding of the quantitative and – especially – the dynamical properties of collocation solutions to functional equations like (1.10). Although some partial answers are known for index-1 IAEs and IDAEs (see Chapter 8 in Brunner (2004b)), the numerical analysis of DDAEs and analogous delay Volterra IAEs of higher index is much more challenging. As März (2002a) has shown, the key to our understanding may possibly be found in a suitable reformulation of integral-algebraic equations as abstract (infinite-dimensional) DAEs, to which the elegant analysis of März (1992, 2002b) can be extended.

6.3. VFIEs with state-dependent delays

The numerical analysis of DDEs with state-dependent delays is now well understood. The papers by Feldstein and Neves (1984), Neves and Thompson (1992), Hartung and Turi (1995), Hartung, Herdman and Turi (1997), Györi, Hartung and Turi (1998), and the monograph by Bellen and Zennaro (2003) convey a fairly complete picture of its state of the arts.

For Volterra functional integro-differential equations with state-dependent delays we have the substantial work by Tavernini (1978) on general one-step methods. Cahlon and Nachman (1985) and Cahlon (1992) deal with a class of numerical methods for solving analogous Volterra functional integral equations. However, except for the results in Cryer and Tavernini (1972) (Euler's method may be viewed as a simple collocation method) the general (super-) convergence analysis for piecewise collocation methods is still outstanding. For example, we do not know if the collocation solution $u_h \in S_1^{(0)}(I_h)$ for DVIDEs of the form

$$y'(t) = g(t) + \int_{t-\tau(y(t))}^{t} k(t-s)G(y(s))\,ds$$

(*i.e.*, the analogue of (1.8)) exhibits $\mathcal{O}(h^2)$-superconvergence if collocation is based on the Gauss point $c_1 = 1/2$. We are similarly ignorant about the optimal order of convergence on I_h for u_h^{it} corresponding to the collocation solution $u_h \in S_0^{(-1)}(I_h)$ for Bélair's state-dependent delay integral equation (1.8).

Acknowledgements

A substantial part of this research was supported by a research grant from the Natural Sciences and Engineering Research Council of Canada (NSERC). The author would also like to acknowledge the many inspiring discussions he has had in recent years with Alfredo Bellen, Marino Zennaro, Lucio

Torelli, Nicola Guglielmi and Stefano Maset (University of Trieste), Rossana Vermiglio (University of Udine), and Arieh Iserles (DAMTP, University of Cambridge) on various aspects of delay differential and integral equations.

REFERENCES

G. Andreoli (1913), 'Sulle equazioni integrali', *Rend. R. Accad. Naz. Lincei Cl. Sci. Fis. Mat.* **22**, 776–781.

G. Andreoli (1914), 'Sulle equazioni integrali', *Rend. Circ. Mat. Palermo* **37**, 76–112.

H. Arndt and C. T. H. Baker (1988), 'Runge–Kutta formulae applied to Volterra functional equations with fixed delay', in *Numerical Treatment of Differential Equations, Halle-Wittenberg 1987*, Vol. 104 of *Teubner-Texte Math.*, Teubner, Leipzig, pp. 19–30.

U. Ascher and L. R. Petzold (1995), 'The numerical solution of delay-differential-algebraic equations of retarded and neutral type', *SIAM J. Numer. Anal.* **32**, 1635–1657.

N. Baddour and H. Brunner (1993), 'Continuous Volterra–Runge–Kutta methods for integral equations with pure delay', *Computing* **50**, 213–227.

C. T. H. Baker (1997), 'Numerical analysis of Volterra functional and integral equations', in *The State of the Art in Numerical Analysis* (I. S. Duff and G. A. Watson, eds), Clarendon Press, Oxford, pp. 193–222.

C. T. H. Baker (2000), 'A perspective on the numerical treatment of Volterra equations', *J. Comput. Appl. Math.* **125**, 217–249.

C. T. H. Baker and M. S. Derakhshan (1990), 'R-K formulae applied to Volterra equations with delay', *J. Comput. Appl. Math.* **29**, 293–310.

C. T. H. Baker and C. A. H. Paul (1997), 'Pitfalls in parameter estimation for delay differential equations', *SIAM J. Sci. Comput.* **18**, 305–314.

C. T. H. Baker, C. A. H. Paul and D. R. Willé (1995), 'Issues in the numerical solution of evolutionary delay differential equations', *Adv. Comput. Math.* **3**, 171–196.

C. T. H. Baker and A. Tang (1997), 'Stability analysis of continuous implicit Runge–Kutta methods for Volterra integro-differential systems with unbounded delays', *Appl. Numer. Math.* **24**, 153–173.

C. T. H. Baker and A. Tang (2000), 'Generalized Halanay inequalities for Volterra functional differential equations and discretized versions', in Corduneanu and Sandberg (2000), pp. 39–55.

H. T. Banks and F. Kappel (1979), 'Spline approximations to functional differential equations', *J. Differential Equations* **34**, 496–522.

J. Bélair (1981), 'Sur une équation différentielle fonctionnelle analytique', *Canad. Math. Bull.* **24**, 43–46.

J. Bélair (1991), 'Population models with state-dependent delays', in *Mathematical Population Dynamics* (O. Arino, D. E. Axelrod and M. Kimmel, eds), Marcel Dekker, New York, pp. 165–176.

A. Bellen (1984), 'One-step collocation for delay differential equations', *J. Comput. Appl. Math.* **10**, 275–283.

A. Bellen (1985), 'Constrained mesh methods for functional-differential equations', in *Delay Equations, Approximation and Application* (G. Meinardus and G. Nürnberger, eds), Vol. 74 of *International Series of Numerical Mathematics*, Birkhäuser, Basel/Boston, pp. 52–70.

A. Bellen (1997), 'Contractivity of continuous Runge–Kutta methods for delay differential equations', *Appl. Numer. Math.* **24**, 219–232.

A. Bellen (2001), 'Preservation of superconvergence in the numerical integration of delay differential equations with proportional delay', *IMA J. Numer. Anal.* **22**, 529–536.

A. Bellen, H. Brunner, S. Maset and L. Torelli (2002), 'Superconvergence of collocation solutions on quasi-geometric meshes for Volterra integro-differential equations with vanishing delays', preprint, Dipartimento di Scienze Matematiche, University of Trieste.

A. Bellen, N. Guglielmi and L. Torelli (1997), 'Asymptotic stability properties of θ-methods for the pantograph equation', *Appl. Numer. Math.* **24**, 275–293.

A. Bellen, N. Guglielmi and M. Zennaro (1999), 'On the contractivity and asymptotic stability of systems of delay differential equations of neutral type', *BIT* **39**, 1–24.

A. Bellen, Z. Jackiewicz, R. Vermiglio and M. Zennaro (1989), 'Natural continuous extensions of Runge–Kutta methods for Volterra integral equations of the second kind and their applications', *Math. Comp.* **52**, 49–63.

A. Bellen and S. Maset (2000), 'Numerical solution of constant coefficient linear delay differential equations as abstract Cauchy problems', *Numer. Math.* **84**, 351–374.

A. Bellen and M. Zennaro (1985), 'Numerical solution of delay differential equations by uniform corrections to an implicit Runge–Kutta method', *Numer. Math.* **47**, 301–316.

A. Bellen and M. Zennaro (2003), *Numerical Methods for Delay Differential Equations*, Oxford University Press.

R. Bellman (1963), 'On the computational solution of differential-difference equations', *J. Math. Anal. Appl.* **2**, 108–110.

R. Bellman and K. L. Cooke (1963), *Differential–Difference Equations*, Academic Press, New York.

L. Berg and M. Krüppel (1998a), 'On the solution of an integral-functional equation with a parameter', *Z. Anal. Anwendungen* **17**, 159–181.

L. Berg and M. Krüppel (1998b), 'Cantor sets and integral-functional equations', *Z. Anal. Anwendungen* **17**, 997–1020.

G. A. Bocharov and F. A. Rihan (2000), 'Numerical modelling in biosciences using delay differential equations', *J. Comput. Appl. Math.* **125**, 183–199.

J. M. Bownds, J. M. Cushing and R. Schutte (1976), 'Existence, uniqueness, and extendibility of solutions to Volterra integral systems with multiple variable delays', *Funkcial. Ekvac.* **19**, 101–111.

F. Brauer and C. Castillo-Chávez (2001), *Mathematical Models in Population Biology and Epidemiology*, Springer, New York.

F. Brauer and P. van den Driessche (2003), Some directions for mathematical epidemiology, in *Dynamical Systems and Their Applications in Biology, Cape*

Breton 2001 (S. Ruan, G. S. K. Wolkowicz and J. Wu, eds), Vol. 36 of *Fields Institute Communications*, AMS, Providence, RI, pp. 95–112.

H. Brezis and F. E. Browder (1975), 'Existence theorems for nonlinear integral equations of Hammerstein type', *Bull. Amer. Math. Soc.* **81**, 73–78.

H. Brunner (1985), 'The numerical solution of weakly singular Volterra integral equations by collocation on graded meshes', *Math. Comp.* **45**, 417–437.

H. Brunner (1992), 'Implicitly linear collocation methods for nonlinear Volterra integral equations', *Appl. Numer. Math.* **9**, 235–247.

H. Brunner (1994a), 'Iterated collocation methods for Volterra integral equations with delay arguments', *Math. Comp.* **62**, 581–599.

H. Brunner (1994b), 'The numerical solution of neutral Volterra integro-differential equations with delay arguments', *Ann. Numer. Math.* **1**, 309–322.

H. Brunner (1997a), 'On the discretization of differential and Volterra integral equations with variable delay', *BIT* **37**, 1–12.

H. Brunner (1997b), 1896–1996: One hundred years of Volterra integral equations of the first kind, *Appl. Numer. Math.* **24**, 83–93.

H. Brunner (1999a), 'The discretization of neutral functional integro-differential equations by collocation methods', *Z. Anal. Anwendungen* **18**, 393–406.

H. Brunner (1999b), 'The numerical solution of weakly singular first-kind Volterra integral equations with delay arguments', *Proc. Estonian Acad. Sci. Phys. Math.* **48**, 90–100.

H. Brunner (2004a), 'The discretization of Volterra functional integral equations with proportional delays', in *Difference and Differential Equations, Changsha 2002* (S. Elaydi, G. Lada, J. Wu and X. Zou, eds), Vol. 42 of *Fields Institute Communications*, AMS, Providence, RI. To appear.

H. Brunner (2004b), *Collocation Methods for Volterra Integral and Related Functional Differential Equations*, Cambridge Monographs on Applied and Computational Mathematics, Cambridge University Press. To appear.

H. Brunner and P. J. van der Houwen (1986), *The Numerical Solution of Volterra Equations*, Vol. 3 of *CWI Monographs*, North-Holland, Amsterdam.

H. Brunner and Q.-Y. Hu (2003), 'Superconvergence orders of iterated collocation solutions for Volterra integral equations with variable delays', preprint.

H. Brunner, Q.-Y. Hu and Q. Lin (2001a), 'Geometric meshes in collocation methods for Volterra integral equations with proportional delays', *IMA J. Numer. Anal.* **21**, 783–798.

H. Brunner and J. Ma (2004), 'Primary discontinuities in solutions of neutral Volterra functional integro-differential equations with weakly singular kernels', to appear.

H. Brunner, A. Pedas and G. Vainikko (1999), 'The piecewise polynomial collocation method for nonlinear weakly singular Volterra equations', *Math. Comp.* **68**, 1079–1095.

H. Brunner, A. Pedas and G. Vainikko (2001b), 'Piecewise polynomial collocation methods for linear Volterra integro-differential equations with weakly singular kernels', *SIAM J. Numer. Anal.* **39**, 957–982.

H. Brunner and R. Vermiglio (2003), 'Stability of solutions of neutral functional integro-differential equations and their discretizations', *Computing* **71**, 229–245.

H. Brunner and W. Zhang (1999), 'Primary discontinuities in solutions for delay integro-differential equations', *Methods Appl. Anal.* **6**, 525–533.

M. Buhmann and A. Iserles (1991), 'Numerical analysis of functional differential equations with a variable delay', in *Numerical Analysis, Dundee 1991* (D. F. Griffiths and G. A. Watson, eds), Vol. 260 of *Pitman Research Notes in Mathematics Series*, Longman, Harlow, pp. 17–33.

M. Buhmann and A. Iserles (1992), 'On the dynamics of a discretized neutral equation', *IMA J. Numer. Anal.* **12**, 339–363.

M. Buhmann and A. Iserles (1993), 'Stability of the discretized pantograph differential equation', *Math. Comp.* **60**, 575–589.

M. Buhmann, A. Iserles and S. P. Nørsett (1993), 'Runge–Kutta methods for neutral differential equations', in *Contributions in Numerical Mathematics, Singapore 1993* (R. P. Agarwal, ed.), World Scientific, River Edge, NJ, pp. 85–98.

A. Burgstaller (1993), *Kollokationsverfahren für Anfangswertprobleme*, Dissertation, Fakultät für Mathematik, Ludwig-Maximilians-Universität München.

A. Burgstaller (2000), 'A modified collocation method for Volterra delay integro-differential equations with multiple delays', in *Integral and Integrodifferential Equations: Theory, Methods and Applications* (R. P. Agarwal and D. O'Regan, eds), Gordon and Breach, Amsterdam, pp. 39–53.

J. A. Burns, E. M. Cliff and T. L. Herdman (1983a), 'A state-space model for an aeroelastic system', *22nd IEEE Conference on Decision and Control*, Vol. 3, pp. 1074–1077.

J. A. Burns, E. M. Cliff and T. L. Herdman (1983b), 'On integral transforms appearing in the derivation of the equations of an aeroelastic system', in Lakshmikantham (1987), pp. 89–98.

J. A. Burns, T. L. Herdman and H. W. Stech (1983c), 'Linear functional differential equations as semigroups on product spaces', *SIAM J. Math. Anal.* **14**, 98–116.

J. A. Burns, T. L. Herdman and J. Turi (1987), 'Nonatomic neutral functional differential equations', in Lakshmikantham (1987), pp. 635–646.

J. A. Burns, T. L. Herdman and J. Turi (1990), 'Neutral functional integro-differential equations with weakly singular kernels', *J. Math. Anal. Appl.* **145**, 371–401.

T. A. Burton (1983), *Volterra Integral and Differential Equations*, Academic Press, New York.

S. Busenberg and K. L. Cooke (1980), 'The effect of integral conditions in certain equations modelling epidemics and population growth', *J. Math. Biol.* **10**, 13–32.

B. Cahlon (1990), 'On the numerical stability of Volterra integral equations with delay arguments', *J. Comput. Appl. Math.* **33**, 97–104.

B. Cahlon (1992), 'Numerical solutions for functional integral equations with state-dependent delay', *Appl. Numer. Math.* **9**, 291–305.

B. Cahlon (1995), 'On the stability of Volterra integral equations with a lagging argument', *BIT* **35**, 19–29.

B. Cahlon and L. J. Nachman (1985), 'Numerical solutions of Volterra integral equations with a solution dependent delay', *J. Math. Anal. Appl.* **112**, 541–562.

B. Cahlon, L. J. Nachman and D. Schmidt (1984), 'Numerical solution of Volterra integral equations with delay arguments', *J. Integral Equations* **7**, 191–208.

B. Cahlon and D. Schmidt (1997), 'Stability criteria for certain delay integral equations of Volterra type', *J. Comput. Appl. Math.* **84**, 161–188.

A. Cañada and A. Zertiti (1994), 'Methods of upper and lower solutions for nonlinear delay integral equations modelling epidemics and population growth', *Math. Models Methods Appl. Sci.* **4**, 107–119.

J. Carr and J. Dyson (1976), 'The functional differential equation $y'(x) = ay(\lambda x) + by(x)$', *Proc. Roy. Soc. Edinburgh Sect. A* **74**, 5–22.

L. A. V. Carvalho and K. L. Cooke (1998), 'Collapsible backward continuation and numerical approximations in a functional differential equation', *J. Differential Equations* **143**, 96–109.

J. Cerha (1976), 'On some linear Volterra delay equations', *Časopis Pěst. Mat.* **101**, 111–123.

Ll. G. Chambers (1990), 'Some properties of the functional equation $\phi(x) = f(x) + \int_0^{\lambda x} g(x, y, f(y))dy$', *Internat. J. Math. Math. Sci.* **14**, 27–44.

P. Clément, W. Desch and K. W. Homan (2002), 'An analytic semigroup setting for a class of Volterra equations', *J. Integral Equations Appl.* **14**, 239–281.

K. L. Cooke (1976), 'An epidemic equation with immigration', *Math. Biosci.* **29**, 135–158.

K. L. Cooke and J. L. Kaplan (1976), 'A periodicity threshold theorem for epidemics and population growth', *Math. Biosci.* **31**, 87–104.

K. L. Cooke and J. A. Yorke (1973), 'Some equations modelling growth processes and epidemics', *Math. Biosci.* **16**, 75–101.

C. Corduneanu and V. Lakshmikantham (1980), 'Equations with unbounded delay: a survey', *Nonlinear Anal.* **4**, 831–877.

C. Corduneanu and I. W. Sandberg, eds (2000), *Volterra Equations and Applications*, Vol. 10 of *Stability Control, Theory, Methods Appl.*, Gordon and Breach, Amsterdam.

C. W. Cryer (1972), 'Numerical methods for functional differential equations', in Schmitt (1972), pp. 17–101.

C. W. Cryer and L. Tavernini (1972), 'The numerical solution of Volterra functional differential equations by Euler's method', *SIAM J. Numer. Anal.* **9**, 105–129.

J. M. Cushing (1977), *Integro-Differential Equations and Delay Models in Population Dynamics*, Vol. 20 of *Lecture Notes in Biomathematics*, Springer, Berlin/Heidelberg/New York.

A. M. Denisov and S. V. Korovin (1992), 'On Volterra's integral equation of the first kind', *Moscow Univ. Comput. Math. Cybernet.* **3**, 19–24.

A. M. Denisov and A. Lorenzi (1995), 'On a special Volterra integral equation of the first kind', *Boll. Un. Mat. Ital. B* **9**, 443–457.

A. M. Denisov and A. Lorenzi (1997), 'Existence results and regularization techniques for severely ill-posed integrofunctional equations', *Boll. Un. Mat. Ital. B* **11**, 713–732.

G. A. Derfel (1990), 'Kato problem for functional-differential equations and difference Schrödinger operators', in *Order, Disorder and Chaos in Quantum Systems, Dubna 1989* (P. Exner and H. Neidhardt, eds), *Oper. Theory Adv. Appl.* **46**, 319–321.

G. A. Derfel (1991), 'Functional differential equations with linearly transformed arguments and their applications', in *Proc. EQUADIFF 91, Barcelona 1991* (C. Perelló et al., eds), World Scientific, River Edge, NJ, pp. 421–423.

G. A. Derfel and S. A. Molchanov (1990), 'Spectral methods in the theory of differential-functional equations', *Math. Notes Acad. Sci. USSR* **47**, 254–260.

G. A. Derfel and F. Vogl (1996), 'On the asymptotics of solutions of a class of linear functional-differential equations', *Europ. J. Appl. Math.* **7**, 511–518.

O. Diekmann, S. A. van Gils, S. M. Verduyn Lunel and H.-O. Walther (1995), *Delay Equations: Functional, Complex, and Nonlinear Analysis*, Springer, New York.

Á. Elbert (1992), 'Asymptotic behaviour of the analytic solution of the delay differential equation $y'(t) + y(qt) = 0$ as $q \to 1^-$', *J. Comput. Appl. Math.* **41**, 5–22.

L. E. El'sgol'ts and S. B. Norkin (1973), *Introduction to the Theory and Application of Differential Equations with Deviating Arguments*, Academic Press, New York.

K. Engelborghs and E. Doedel (2002), 'Stability of piecewise polynomial collocation for computing periodic solutions of delay differential equations', *Numer. Math.* **91**, 627–648.

K. Engelborghs, T. Luzyanina, K. J. in 't Hout and D. Roose (2000), 'Collocation methods for the computation of periodic solutions of delay differential equations', *SIAM J. Sci. Comput.* **22**, 1593–1609.

W. H. Enright and H. Hayashi (1998), 'Convergence analysis of the solution of retarded and neutral delay differential equations by continuous numerical methods', *SIAM J. Numer. Anal.* **35**, 572–585.

W. H. Enright and M. Hu (1997), 'Continuous Runge–Kutta methods for neutral Volterra integro-differential equations with delay', *Appl. Numer. Math.* **24**, 175–190.

R. Esser (1976), *Numerische Lösung einer verallgemeinerten Volterra'schen Integralgleichung zweiter Art*, PhD Dissertation, Math.-Naturwiss. Fakultät, Universität zu Köln.

R. Esser (1978), 'Numerische Behandlung einer Volterraschen Integralgleichung', *Computing* **19**, 269–284.

A. Feldstein, A. Iserles and D. Levin (1995), 'Embedding of delay equations into an infinite-dimensional ODE system', *J. Differential Equations* **117**, 127–150.

A. Feldstein and Y. Liu (1998), 'On neutral functional-differential equations with variable time delays', *Math. Proc. Cambridge Phil. Soc.* **124**, 371–384.

A. Feldstein and K. W. Neves (1984), 'High order methods for state-dependent delay differential equations with nonsmooth solutions', *SIAM J. Numer. Anal.* **21**, 844–863.

S. Fenyö and H. W. Stolle (1984), *Theory und Praxis der Linearen Integralgleichungen*, Band 3, VEB Deutscher Verlag der Wissenschaften, Berlin, and Birkhäuser, Basel/Boston.

L. Fox, D. F. Mayers, J. R. Ockendon and A. B. Tayler (1971), 'On a functional differential equation', *J. Inst. Math. Appl.* **8**, 271–307.

P. O. Frederickson (1971), 'Dirichlet solutions for certain functional differential equations', in *Japan-United States Seminar on Ordinary Differential and Functional Equations, Kyoto 1971* (M. Urabe, ed.), Vol. 243 of *Lecture Notes in Mathematics*, Springer, Berlin/Heidelberg, pp. 249–251.

M. de Gee (1985), 'Smoothness of solutions of functional differential equations', *J. Math. Anal. Appl.* **107**, 103–121.

S. I. Grossman and R. K. Miller (1970), 'Perturbation theory for Volterra integro-differential systems', *J. Differential Equations* **8**, 457–474.

N. Guglielmi (1998), 'Delay dependent stability regions of θ-methods for delay differential equations', *IMA J. Numer. Anal.* **18**, 399–418.

N. Guglielmi and E. Hairer (2001*a*), 'Geometric proofs of numerical stability for delay differential equations', *IMA J. Numer. Anal.* **21**, 439–450.

N. Guglielmi and E. Hairer (2001*b*), 'Implementing Radau IIA methods for stiff delay differential equations', *Computing* **67**, 1–12.

N. Guglielmi and M. Zennaro (2003), 'Stability of one-leg θ-methods for the variable coefficient pantograph equation on the quasi-geometric mesh', *IMA J. Numer. Anal.* **23**, 421–438.

I. Györi and F. Hartung (2002), 'Numerical approximation of neutral differential equations on infinite intervals', *J. Differential Equations Appl.* **8**, 983–999.

I. Györi, F. Hartung and J. Turi (1995), 'Numerical approximations for a class of differential equations with time- and state-dependent delays', *Appl. Math. Lett.* **8**, 19–24.

I. Györi, F. Hartung and J. Turi (1998), 'Preservation of stability in delay equations under delay perturbations', *J. Math. Anal. Appl.* **220**, 290–312.

I. Györi and G. Ladas (1991), *Oscillation Theory of Delay Differential Equations*, Clarendon Press, Oxford.

A. Halanay and J. A. Yorke (1971), 'Some new results and problems in the theory of functional-differential equations, *SIAM Review* **13**, 5–80.

J. K. Hale (1977), *Theory of Functional Differential Equations*, Springer, New York.

J. K. Hale and S. M. Verduyn Lunel (1993), *Introduction to Functional Differential Equations*, Springer, New York.

F. Hartung, T. L. Herdman and J. Turi (1997), 'On existence, uniqueness and numerical approximation for neutral equations with state-dependent delay', *Appl. Numer. Math.* **24**, 393–409.

F. Hartung and J. Turi (1995), 'On the asymptotic behavior of the solutions of a state-dependent differential equation', *Differential Integral Equations* **8**, 1867–1872.

R. Hauber (1997), 'Numerical treatment of retarded differential-algebraic equations by collocation methods', *Adv. Comput. Math.* **7**, 573–592.

T. L. Herdman and J. A. Burns (1979), 'Functional differential equations with discontinuous right-hand side', in *Volterra Equations, Otaniemi 1978* (S. O. Londen and O. J. Staffans, eds), Springer, Berlin, pp. 99–106.

T. L. Herdman and J. Turi (1991), 'An application of finite Hilbert transforms in the derivation of a state space model for an aeroelastic system', *J. Integral Equations Appl.* **3**, 271–287.

H. W. Hethcote and P. van den Driessche (1995), 'An SIS epidemic model with variable population size and a delay', *J. Math. Biol.* **34**, 177–194.

H. W. Hethcote and P. van den Driessche (2000), 'Two SIS epidemiologic models with delays', *J. Math. Biol.* **40**, 3–26.

H. W. Hethcote, M. A. Lewis and P. van den Driessche (1989), 'An epidemiological model with a delay and a nonlinear incidence rate', *J. Math. Biol.* **27**, 49–64.

H. W. Hethcote and D. W. Tudor (1980), 'Integral equation models for endemic infectious diseases', *J. Math. Biol.* **9**, 37–47.

K. J. in 't Hout (1992), 'A new interpolation procedure for adapting Runge–Kutta methods to delay differential equations', *BIT* **32**, 634–649.

N. Hritonenko and Y. Yatsenko (1996), *Modeling and Optimization of the Lifetime of Technologies*, Kluwer, Dordrecht.

Q.-Y. Hu (1997), 'Stepwise collocation methods based on the high-order interpolation for Volterra integral equations with multiple delays' (in Chinese), *Math. Numer. Sinica* **19**, 353–358.

Q.-Y. Hu (1998), 'Geometric meshes and their application to Volterra integro-differential equations with singularities', *IMA J. Numer. Anal.* **18**, 151–164.

Q.-Y. Hu (1999), 'Multilevel correction for discrete collocation solutions of Volterra integral equations with delay arguments', *Appl. Numer. Math.* **31**, 159–171.

Q.-Y. Hu and L. Peng (1999), 'Multilevel correction for collocation solutions of Volterra delay integro-differential equations' (in Chinese), *Systems Sci. Math. Sci.* **19**, 134–141.

A. Iserles (1993), 'On the generalized pantograph functional differential equation', *Europ. J. Appl. Math.* **4**, 1–38.

A. Iserles (1994a), 'Numerical analysis of delay differential equations with variable delays', *Ann. Numer. Math.* **1**, 133–152.

A. Iserles (1994b), 'On nonlinear delay-differential equations', *Trans. Amer. Math. Soc.* **344**, 441–477.

A. Iserles (1994c), 'The asymptotic behaviour of certain difference equations with proportional delay', *Comput. Math. Appl.* **28**, 141–152.

A. Iserles (1997a), 'Beyond the classical theory of computational ODEs', in *The State of the Art in Numerical Analysis, York 1996* (I. S. Duff and G. A. Watson, eds), Clarendon Press, Oxford, pp. 171–192.

A. Iserles (1997b), 'Exact and discretized stability of the pantograph equation', *Appl. Numer. Math.* **24**, 295–308.

A. Iserles and Y. Liu (1994), 'On pantograph integro-differential equations', *J. Integral Equations Appl.* **6**, 213–237.

A. Iserles and Y. Liu (1997), 'On neutral functional-differential equations with proportional delays', *J. Math. Anal. Appl.* **207**, 73–95.

A. Iserles and J. Terjéki (1995), 'Stability and asymptotic stability of functional-differential equations', *J. London Math. Soc.* **51**, 559–572.

E. Ishiwata (2000), 'On the attainable order of collocation methods for the neutral functional-differential equations with proportional delays', *Computing* **64**, 207–222.

K. Ito and F. Kappel (1989), 'Approximation of infinite delay and Volterra type equations', *Numer. Math.* **54**, 405–444.

K. Ito and F. Kappel (1991), 'On integro-differential equations with weakly singular kernels', in *Differential Equations with Applications* (J. A. Goldstein et al., eds), Vol. 133 of *Lecture Notes in Pure and Applied Mathematics*, Marcel Dekker, New York, pp. 209–218.

K. Ito and F. Kappel (2002), *Evolution Equations and Approximations*, World Scientific, Singapore.

K. Ito, F. Kappel and J. Turi (1996), 'On well-posedness of singular neutral equations in the state space C', *J. Differential Equations* **125**, 40–72.

K. Ito and J. Turi (1991), 'Numerical methods for a class of singular integro-differential equations based on semigroup approximation', *SIAM J. Numer. Anal.* **28**, 1698–1722.

Z. Jackiewicz (1984), 'One-step methods of any order for neutral functional differential equations', *SIAM J. Numer. Anal.* **21**, 486–511.

Z. Jackiewicz and M. Kwapisz (1991), 'The numerical solution of functional differential equations', *Mat. Stos.* **33**, 57–78.

F. Kappel and K. Kunisch (1981), 'Spline approximations for neutral functional differential equations', *SIAM J. Numer. Anal.* **18**, 1058–1080.

F. Kappel and K. Kunisch (1987), 'Invariance results for delay and Volterra equations in fractional order Sobolev space', *Trans. Amer. Math. Soc.* **304**, 1–51.

F. Kappel and K. P. Zhang (1986), 'On neutral functional differential equations with nonatomic difference operator', *J. Math. Anal. Appl.* **113**, 311–343.

A. Karoui and R. Vaillancourt (1994), 'Computer solutions of state-dependent delay differential equations', *Comput. Math. Appl.* **27**, 37–51.

T. Kato (1972), 'Asymptotic behaviour of solutions of the functional-differential equations $y'(x) = ay(\lambda x) + by(x)$', in Schmitt (1972), pp. 197–217.

T. Kato and J. B. McLeod (1971), 'The functional-differential equation $y'(x) = ay(\lambda x) + by(x)$', *Bull. Amer. Math. Soc.* **77**, 891–937.

J.-P. Kauthen and H. Brunner (1997), 'Continuous collocation approximations to solutions of first kind Volterra equations', *Math. Comp.* **66**, 1441–1459.

N. G. Kazakova and D. D. Bainov (1990), 'An approximate solution of the initial value problem for integro-differential equations with a deviating argument', *Math. J. Toyama Univ.* **13**, 9–27.

V. Kolmanovskii and A. Myshkis (1992), *Applied Theory of Functional Differential Equations*, Kluwer, Dordrecht.

T. Koto (1999), 'Stability of Runge–Kutta methods for the generalized pantograph equation', *Numer. Math.* **84**, 233–247.

T. Koto (2002), 'Stability of Runge–Kutta methods for delay integro-differential equations', *J. Comput. Appl. Math.* **145**, 483–492.

M. A. Krasnosel'skii and P. P. Zabreiko (1984) *Geometric Methods of Nonlinear Analysis*, Springer, Berlin/Heidelberg/New York.

Y. Kuang and A. Feldstein (1990), 'Monotonic and oscillatory solutions of a linear neutral delay equation with infinite lag', *SIAM J. Math. Anal.* **21**, 1633–1641.

S. Kumar and I. H. Sloan (1987), 'A new collocation-type method for Hammerstein integral equations', *Math. Comp.* **48**, 585–593.

V. Lakshmikantham, ed. (1987), *Nonlinear Analysis and Applications*, Vol. 109 of Lecture Notes in Pure and Applied Mathematics, Marcel Dekker, New York.

V. Lakshmikantham, L. Wen and B. Zhang (1994), *Theory of Differential Equations with Unbounded Delay*, Kluwer, Dordrecht.

T. Lalesco (1908), *Sur l'équation de Volterra*, Thèse de doctorat, Gauthier-Villars, Paris; *J. de Math.* **4**, 125–202.

T. Lalesco (1911), 'Sur une équation intégrale du type Volterra', *CR Acad. Sci. Paris* **52**, 579–580.

T. Lalesco (1912), *Introduction à la théorie des équations intégrales*, Hermann and Fils, Paris.

J. J. Levin and J. A. Nohel (1964), 'On a nonlinear delay equation', *J. Math. Anal. Appl.* **8**, 31–44.

D.-S. Li and M.-Z. Liu (1999), 'Asymptotic stability of numerical solution of pantograph delay differential equations' (in Chinese), *J. Harbin Inst. Tech.* **31**, 57–59.

J. Liang and M. Liu (1996), 'Numerical stability of θ-methods for pantograph delay differential equations' (in Chinese), *J. Numer. Methods Comput. Appl.* **12**, 271–278.

J. Liang, S. Qiu and M. Liu (1996), 'The stability of θ-methods for pantograph delay differential equations', *Numer. Math. (Engl. Ser.)* **5**, 80–85.

Q. Lin (1963), 'Comparison theorems for difference-differential equations', *Sci. Sinica* **12**, 449.

W. J. Liu and J. C. Clements (2002), 'On solutions of evolution equations with proportional time delay', *Int. J. Differ. Equ. Appl.* **4**, 229–254.

Y. Liu (1995a), 'Stability analysis of θ-methods for neutral functional-differential equations', *Numer. Math.* **70**, 473–485.

Y. Liu (1995b), 'The linear q-difference equation', *Appl. Math. Lett.* **8**, 15–18.

Y. Liu (1996a), 'Asymptotic behaviour of functional-differential equations with proportional time delays', *Europ. J. Appl. Math.* **7**, 11–30.

Y. Liu (1996b), 'On θ-methods for delay differential equations with infinite lag', *J. Comput. Appl. Math.* **71**, 177–190.

Y. Liu (1997), 'Numerical investigation of the pantograph equation', *Appl. Numer. Math.* **24**, 309–317.

Y. Liu (1999a), 'Numerical solution of implicit neutral functional differential equations', *SIAM J. Numer. Anal.* **36**, 516–528.

Y. Liu (1999b), 'Runge–Kutta-collocation methods for systems of functional-differential and functional equations', *Adv. Comput. Math.* **11**, 315–329.

A. Makroglou (1983), 'A block-by-block method for the numerical solution of Volterra delay integro-differential equations', *Computing* **30**, 49–62.

J. E. Marshall (1979), *Control of Time-Delay Systems*, Peregrinus, London.

R. März (1992), 'Numerical methods for differential-algebraic equations', in *Acta Numerica*, Vol. 1, Cambridge University Press, pp. 141–198.

R. März (2002a), 'Differential algebraic equations anew', *Appl. Numer. Math.* **42**, 315–335.

R. März (2002b), 'The index of linear differential-algebraic equations with properly stated leading terms', *Resultate Math.* **42**, 308–338.

S. Maset (1999), 'Asymptotic stability in the numerical solution of linear pure delay differential equations as abstract Cauchy problems', *J. Comput. Appl. Math.* **111**, 163–172.

S. Maset (2003), 'Numerical solution of retarded functional differential equations as abstract Cauchy problems', *J. Comput. Appl. Math.* **161**, 259–282.

S. Maset, L. Torelli and R. Vermiglio (2002), 'Runge–Kutta methods for general retarded functional differential equations', preprint.

T. Meis (1976), 'Eine spezielle Integralgleichung erster Art', in *Numerical Treatment of Differential Equations, Oberwolfach 1976* (R. Bulirsch, R. D. Grigorieff and J. Schröder, eds), Vol. 631 of *Lecture Notes in Mathematics*, Springer, Berlin/Heidelberg, pp. 107–120.

J. A. J. Metz and O. Diekmann, eds (1986), *The Dynamics of Physiologically Structured Populations*, Vol. 68 of *Lecture Notes in Biomathematics*, Springer, Berlin/Heidelberg.

R. K. Miller (1971), *Nonlinear Volterra Integral Equations*, Benjamin, Menlo Park, CA.

G. R. Morris, A. Feldstein and E. W. Bowen (1972), 'The Phragmén–Lindelöf principle and a class of functional differential equations', in *Ordinary Differential Equations, Washington, DC, 1971* (L. Weiss, ed.), Academic Press, New York, pp. 513–540.

V. Mureşan (1984), 'Die Methode der sukzessiven Approximationen für eine Integralgleichung vom Typ Volterra–Sobolev', *Mathematica (Cluj)* **26**, 129–136.

V. Mureşan (1999), 'On a class of Volterra integral equations with deviating argument', *Studia Univ. Babeş-Bolyai Math.* **XLIV**, 47–54.

Y. Muroya, E. Ishiwata and H. Brunner (2002), 'On the attainable order of collocation methods for pantograph integro-differential equations', *J. Comput. Appl. Math.* **152**, 347–366.

A. D. Myshkis (1972), *Linear Differential Equations with Retarded Argument* (in Russian), revised 2nd edn, Izdat. Nauka, Moscow.

K. W. Neves and A. Feldstein (1976), 'Characterization of jump discontinuities for state dependent delay differential equations', *J. Math. Anal. Appl.* **56**, 689–707.

K. W. Neves and S. Thompson (1992), 'Software for the numerical solution of systems of functional differential equations with state-dependent delays', *Appl. Numer. Math.* **9**, 385–401.

H. J. Oberle and H. J. Pesch (1981), 'Numerical treatment of delay differential equations by Hermite interpolation', *Numer. Math.* **37**, 235–255.

J. R. Ockendon and A. B. Tayler (1971), 'The dynamics of a current collection system for an electric locomotive', *Proc. Roy. Soc. London Ser. A* **322**, 447–468.

J. Piila (1996), 'Characterization of the membrane theory of a clamped shell: The hyperbolic case', *Math. Methods Appl. Sci.* **6**, 169–194.

J. Piila and J. Pitkäranta (1996), 'On the integral equation $f(x) - (c/L(x)) \int_{L(x)}^{x} f(y)\,dy = g(x)$, where $L(x) = \min\{ax, 1\}$, $a > 1$', *J. Integral Equations Appl.* **8**, 363–378.

T. P. Pukhnacheva (1990), 'A functional equation with contracting argument', *Siberian Math. J.* **31**, 365–367.

J. Reverdy (1981), *Sur l'approximation d'équations d'évolution linéaires du premier ordre à retard par des méthodes de type Runge–Kutta*, Thèse de doctorat, Université Paul Sabbatier de Toulouse.

J. Reverdy (1990), 'Sur la B-stabilité pour une équation différentielle à retard', *CR Acad. Sci. Paris Sér. I Math.* **310**, 461–463.

S. Ruan and G. S. K. Wolkowicz (1996), 'Bifurcation analysis of a chemostat model with a distributed delay', *J. Math. Anal. Appl.* **204**, 786–812.

S. Ruan, G. Wolkowicz and J. Wu, eds (2003), *Dynamical Systems and Their Application in Biology, Cape Breton, NS, 2001*, AMS, Providence, RI.

S. Ruan and J. Wu (1994), 'Reaction-diffusion equations with infinite delay', *Canad. Appl. Math. Quart.* **2**, 485–550.

K. Schmitt, ed. (1972), *Delay and Functional Differential Equations and Their Applications*, Academic Press, New York.

J.-G. Si (2000), 'Analytic solutions of a nonlinear functional differential equation with proportional delays', *Demonstratio Math.* **33**, 747–752.

J.-G. Si and S. S. Cheng (2002), 'Analytic solutions of a functional differential equation with proportional delays', *Bull. Korean Math. Soc.* **39**, 225–236.

H. L. Smith (1977), 'On periodic solutions of a delay integral equation modelling epidemics', *J. Math. Biol.* **4**, 69–80.

M. N. Spijker (1997), 'Numerical stability, resolvent conditions and delay differential equations', *Appl. Numer. Math.* **24**, 233–246.

O. J. Staffans (1985a), 'Extended initial and forcing function semigroups generated by a functional equation', *SIAM J. Math. Anal.* **16**, 1034–1048.

O. J. Staffans (1985b), 'Some well-posed functional equations which generate semigroups', *J. Differential Equations* **58**, 157–191.

N. Takama, Y. Muroya and E. Ishiwata (2000), 'On the attainable order of collocation methods for the delay differential equations with proportional delay', *BIT* **40**, 374–394.

L. Tavernini (1971), 'One-step methods for the numerical solution of Volterra functional differential equations', *SIAM J. Numer. Anal.* **8**, 786–795.

L. Tavernini (1973), 'Linear multistep methods for the numerical solution of Volterra functional differential equations', *Appl. Anal.* **3**, 169–185.

L. Tavernini (1978), 'The approximate solution of Volterra differential systems with state dependent time lags', *SIAM J. Numer. Anal.* **15**, 1039–1052.

A. B. Tayler (1986), *Mathematical Models in Applied Mechanics*, Clarendon Press, Oxford.

J. Terjéki (1995), 'Representation of the solutions to linear pantograph equations', *Acta Sci. Math. (Szeged)* **60**, 705–713.

H. R. Thieme and X.-Q. Zhao (2003), 'Asymptotic speeds of spread and traveling waves for integral equations and delayed reaction-diffusion models', *J. Differential Equations* **195**, 430–470.

R. J. Thompson (1968), 'On some functional differential equations: existence of solutions and difference approximations', *SIAM J. Numer. Anal.* **5**, 475–487.

H.-J. Tian and J.-X. Kuang (1995), 'Numerical stability analysis of numerical methods for Volterra integral equations with delay argument', *Appl. Math. Mech. (Engl. Edn)* **16**, 485–491.

L. Torelli (1989), 'Stability of numerical methods for delay differential equations', *J. Comput. Appl. Math.* **25**, 15–26.

L. Torelli and R. Vermiglio (2003), 'A numerical approach for implicit non-linear neutral delay differential equations and its stability analysis', *BIT* **43**, 195–215.

J. Turi and W. Desch (1993), 'A neutral functional differential equation with an unbounded kernel', *J. Integral Equations Appl.* **5**, 569–582.

A. Tychonoff (1938), 'Sur les équations fonctionnelles de Volterra et leurs applications à certains problémes de la physique mathématique', *Bull. Univ. d'État de Moscou Sér. Internat. Sér. A Math. Méchan.* **1**, 1–25.

P. Vâţă (1978), 'Convergence theorems of some numerical approximation scheme for the class of nonlinear integral equation', *Bul. Univ. Galaţi Fasc. II Mat. Fiz. Mec. Teoret.* **1**, 25–33.

R. Vermiglio (1985), 'A one-step subregion method for delay differential equations', *Calcolo* **22**, 429–455.

R. Vermiglio (1988), 'Natural continuous extension of Runge–Kutta methods for Volterra integrodifferential equations', *Numer. Math.* **53**, 439–458.

R. Vermiglio (1992), 'On the stability of Runge–Kutta methods for delay integral equations', *Numer. Math.* **61**, 561–577.

R. Vermiglio and L. Torelli (1998), 'A stable numerical approach for implicit nonlinear neutral delay differential equations', *BIT* **43**, 195–215.

R. Vermiglio and M. Zennaro (1993), 'Multistep natural continuous extensions of Runge–Kutta methods: The potential for stable interpolation', *Appl. Numer. Math.* **12**, 521–546.

Th. Vogel (1965), *Théorie des systèmes évolutifs, Traité de Physique Théorique et de Physique Mathématique*, XXII, Gauthier-Villars, Paris. (See also MR 32, #8546.)

V. Volterra (1896), 'Sulla inversione degli integrali definiti', *Atti R. Accad. Sci. Torino* **31**, 311–323 (Nota I); 400–408 (Nota II); 557–567 (Nota III); 693–708 (Nota IV).

V. Volterra (1897), 'Sopra alcune questioni di inversione di integrali definite', *Ann. Mat. Pura Appl.* **25**, 139–178.

V. Volterra (1909), 'Sulle equazioni integro-differenziali, *Rend. R. Accad. Lincei*, **18**, 167–174.

V. Volterra (1912), 'Sur les équations intégro-différentielles et leurs applications', *Acta Math.* **35**, 295–356.

V. Volterra (1913), *Leçons sur les équations intégrales et les équations intégro-différentielles*, Gauthier-Villars, Paris. (VIEs with proportional delays are discussed on pp. 92–101.)

V. Volterra (1927), 'Variazioni e fluttuazioni del numero d'individui in specie animali conviventi', *Memorie del R. Comitato talassografico italiano* **CXXXI**; also Volterra (1956), Vol. V, 1–111.

V. Volterra (1928), 'Sur la théorie mathématique des phénomèmes héréditaires', *J. Math. Pures Appl.* **7**, 249–298.

V. Volterra (1931), *Leçons sur la théorie mathématique de la lutte pour la vie*, Gauthier-Villars, Paris; also Éditions Jacques Gabay, Sceaux (1990).

V. Volterra (1934), 'Remarques sur la Note de M. Régnier et Mlle Lambin', *CR Acad. Sci.* **199**, 1684–1686; also: Volterra (1956), Vol. V, 390–391.

V. Volterra (1939), 'The general equations of biological strife in the case of historical actions', *Proc. Edinburgh Math. Soc.* **6**, 4–10.

V. Volterra (1956), *Opere Matematiche, Vol. I–V* (1956–1962), Accademia Nazionale dei Lincei, Roma.

V. Volterra (1959), *Theory of Functionals and of Integral and Integro-Differential Equations*, Dover, New York.

V. Volterra and U. d'Ancona (1935), *Les Associations Biologiques au Point de Vue Mathématique*, Hermann, Paris.

P. Waltman (1974), *Deterministic Threshold models in the Theory of Epidemics*, Vol. 1 of *Lecture Notes in Biomathematics*, Springer, Berlin/Heidelberg.

G. F. Webb (1985), *Theory of Nonlinear Age-Dependent Population Dynamics*, Marcel Dekker, New York.

D. R. Willé and C. T. H. Baker (1992), 'The tracking of derivative discontinuities in systems of delay differential equations', *Appl. Numer. Math.* **9**, 209–222.

J. Wolff (1982), 'Numerische Lösung Volterrascher Integralgleichungen zweiter Art mit Nacheilung unter Verwendung kubischer Splines', *Wiss. Z. Pädag. Hochschule 'Liselotte Herrmann' Güstrow Math. Nat.-Wiss. Fak.* **20**, 225–244.

J. Wu (1996), *Theory and Applications of Partial Functional Differential Equations*, Springer, New York.

Y. Yatsenko (1995), 'Volterra integral equations with unknown delay time', *Methods Appl. Anal.* **2**, 408–419.

T. Yoshizawa and J. Kato, eds (1991), *Functional Differential Equations, Kyoto 1990*, World Scientific, Singapore.

M. Zennaro (1986), 'Natural continuous extensions of Runge–Kutta methods', *Math. Comp.* **46**, 119–133.

M. Zennaro (1993), 'Contractivity of Runge–Kutta methods with respect to forcing terms', *Appl. Numer. Math.* **10**, 321–345.

M. Zennaro (1995), 'Delay differential equations: Theory and numerics', in *Theory and Numerics of Ordinary and Partial Differential Equations, Leicester 1994* (M. Ainsworth et al., eds), Vol. 4 of *Advances in Numerical Analysis*, Clarendon Press, Oxford, pp. 291–333.

M. Zennaro (1997), 'Asymptotic stability analysis of Runge–Kutta methods for nonlinear systems of delay differential equations', *Numer. Math.* **77**, 549–563.

C. Zhang and S. Vandewalle (2004), 'Stability analysis of Runge–Kutta methods for nonlinear Volterra delay-integro-differential equations', *IMA J. Numer. Anal.*, to appear.

C. J. Zhang and X. X. Liao (2002), 'Stability of BDF methods for nonlinear Volterra integral equations with delay', *Comput. Math. Appl.* **43**, 95–102.

W. Zhang (1998), *Numerical Analysis of Delay Differential and Integro-Differential Equations*, PhD Dissertation, Department of Mathematics and Statistics, Memorial University of Newfoundland, St. John's, NL.

W. Zhang and H. Brunner (1998), 'Collocation approximations for second-order differential equations and Volterra integro-differential equations with variable delays', *Canad. Appl. Math. Quart.* **6**, 269–285.

X.-Q. Zhao (2003), *Dynamical Systems in Population Biology*, CMS Books in Mathematics, Springer, New York.

B. Zubik-Kowal (1999), 'Stability in the numerical solution of linear parabolic equations with a delay term', *BIT* **41**, 191–206.

B. Zubik-Kowal and S. Vandewalle (1999), 'Waveform relaxation for functional-differential equations', *SIAM J. Sci. Comput.* **21**, 207–226.

Sparse grids

Hans-Joachim Bungartz
IPVS, Universität Stuttgart,
Universitätsstraße 38, D-70569 Stuttgart, Germany
E-mail: `bungartz@informatik.uni-stuttgart.de`

Michael Griebel
Institut für Numerische Simulation, Universität Bonn,
Wegelerstraße 6, D-53113 Bonn, Germany
E-mail: `griebel@iam.uni-bonn.de`

We present a survey of the fundamentals and the applications of sparse grids, with a focus on the solution of partial differential equations (PDEs). The sparse grid approach, introduced in Zenger (1991), is based on a higher-dimensional multiscale basis, which is derived from a one-dimensional multiscale basis by a tensor product construction. Discretizations on sparse grids involve $O(N \cdot (\log N)^{d-1})$ degrees of freedom only, where d denotes the underlying problem's dimensionality and where N is the number of grid points in one coordinate direction at the boundary. The accuracy obtained with piecewise linear basis functions, for example, is $O(N^{-2} \cdot (\log N)^{d-1})$ with respect to the L_2- and L_∞-norm, if the solution has bounded second mixed derivatives. This way, the curse of dimensionality, *i.e.*, the exponential dependence $O(N^d)$ of conventional approaches, is overcome to some extent. For the energy norm, only $O(N)$ degrees of freedom are needed to give an accuracy of $O(N^{-1})$. That is why sparse grids are especially well-suited for problems of very high dimensionality.

The sparse grid approach can be extended to nonsmooth solutions by adaptive refinement methods. Furthermore, it can be generalized from piecewise linear to higher-order polynomials. Also, more sophisticated basis functions like interpolets, prewavelets, or wavelets can be used in a straightforward way.

We describe the basic features of sparse grids and report the results of various numerical experiments for the solution of elliptic PDEs as well as for other selected problems such as numerical quadrature and data mining.

CONTENTS

1	Introduction	148
2	Breaking the curse of dimensionality	151
3	Piecewise linear interpolation on sparse grids	154
4	Generalizations, related concepts, applications	188
5	Numerical experiments	219
6	Concluding remarks	255
	References	256

1. Introduction

The discretization of PDEs by conventional methods is limited to problems with up to three or four dimensions, due to storage requirements and computational complexity. The reason is the so-called *curse of dimensionality*, a term coined in Bellmann (1961). Here, the cost of computing and representing an approximation with a prescribed accuracy ε depends exponentially on the dimensionality d of the considered problem. We encounter complexities of the order $O(\varepsilon^{-\alpha d})$ with $\alpha > 0$ depending on the respective approach, the smoothness of the function under consideration, and the details of the implementation. If we consider simple uniform grids with piecewise d-polynomial functions over a bounded domain in a finite element or finite difference approach, for instance, this complexity estimate translates to $O(N^d)$ grid points or degrees of freedom for which approximation accuracies of the order $O(N^{-\alpha'})$ are achieved, where α' depends on the smoothness of the function under consideration and the polynomial degree of the approximating functions.[1] Thus, the computational cost and storage requirements grow exponentially with the dimensionality of the problem, which is the reason for the dimensional restrictions mentioned above, even on the most powerful machines presently available.

The curse of dimensionality can be circumvented to some extent by restricting the class of functions under consideration. If we make a stronger assumption on the smoothness of the solution such that the order of accuracy depends on d in a *negative* exponential way, *i.e.*, it behaves like $O(N^{-\beta/d})$, we directly see that the cost–benefit ratio is independent of d and that it is of the order $O(N^d \cdot N^{-\beta/d}) = O(N^{-\beta})$, with some β independent of d. This way, the curse of dimensionality can be broken easily. However, such an assumption is somewhat unrealistic.

---

[1] If the solution is not smooth but possesses singularities, the order α of accuracy deteriorates. Adaptive refinement/nonlinear approximation is employed with success. In the best case, the cost–benefit ratio of a smooth solution can be recovered.

In the sparse grid method, in principle, we follow this approach, but we only assume the functions to live in spaces of functions with bounded mixed derivatives instead. First, we need a 1D multilevel basis, preferably an H^1- and L_2-stable one. Then, if we express a 1D function as usual as a linear combination of these basis functions, the corresponding coefficients decrease from level to level with a rate which depends on the smoothness of the function and on the chosen set of basis functions. From this, by a simple tensor product construction, we obtain a multilevel basis for the higher-dimensional case. Note that here 1D bases living on different levels are used in the product construction, *i.e.*, we obtain basis functions with anisotropic support in the higher-dimensional case. Now, if the function to be represented has bounded second mixed derivatives and if we use a piecewise linear 1D basis as a starting point, it can be shown that the corresponding coefficients decrease with a factor proportional to $2^{-2|\mathbf{l}|_1}$ where the multi-index $\mathbf{l} = (l_1, \ldots, l_d)$ denotes the different levels involved. If we then omit coefficients whose absolute values are smaller than a prescribed tolerance, we obtain sparse grids. It turns out that the number of degrees of freedom needed for some prescribed accuracy no longer depends (up to logarithmic factors) on d exponentially. This allows substantially faster solution of moderate-dimensional problems and can enable the solution of higher-dimensional problems.[2]

Of course, this sparse grid approach is not restricted to the piecewise linear case. Extensions of the standard piecewise linear hierarchical basis to general polynomial degree p as well as interpolets (see Section 4.3) or (pre-) wavelets (see Section 4.4) have been successfully used as the univariate ingredient for tensor product construction. Finally, the spectrum of applications of sparse grids ranges from numerical quadrature, via the discretization of PDEs, to fields such as data mining.

The remainder of this paper is organized as follows. In Section 2, we briefly discuss the breaking of the curse of dimensionality from the theoretical point of view. Here, we resort to harmonic analysis and collect known approaches for escaping the exponential dependence on d. It turns out that one example of such a method simply corresponds to the assumption of using a space of bounded mixed first variation, *i.e.*, an anisotropic multivariate smoothness assumption. Except for the degree of the derivative, this resembles the explicit assumption of Section 3, where we introduce the principles of the sparse grid technique and derive the interpolation properties of the resulting sparse grid spaces. As a starting point, we use the

---

[2] The constant in the corresponding complexity estimates, however, still depends exponentially on d. This still limits the approach for PDEs to moderate-dimensional problems. At present we are able to deal with 18D PDEs.

standard piecewise linear multiscale basis in one dimension, *i.e.*, the Faber basis (Faber 1909, Yserentant 1986), as the simplest example of a multiscale series expansion which involves interpolation by piecewise linears. To attain higher dimensions, we resort to a tensor product construction. Then, for functions with bounded mixed second derivatives, sparse grid approximation schemes are derived which exhibit cost complexities of the order $O(N(\log N)^{d-1})$ and give an error of $O(N^{-2}(\log N)^{d-1})$ in the L_2-norm. We then show that, for our smoothness assumption, the sparse grid discretization scheme can also be formally derived by solving an optimization problem which is closely related to n-term approximation (DeVore 1998). Finally, we consider the energy norm, for which optimality leads us to an energy-based sparse grid with cost complexity $O(N)$ and accuracy $O(N^{-1})$. Thus the exponential dependence of the logarithmic terms on d is completely removed (although present in the constants).

In Section 4 we discuss generalizations of the above approach based on piecewise linear hierarchical bases. First, instead of these, higher-order polynomial hierarchical bases can be employed. Here we describe the construction of multilevel polynomial bases by means of a hierarchical Lagrangian interpolation scheme and analyse the resulting sparse grid approximation properties. A further example of an extension is based on the interpolets due to Deslauriers and Dubuc (1989). Such higher-order approaches allow us to take advantage of higher regularity assumptions concerning the solution, resulting in better approximation rates. However, all these hierarchical multiscale bases have a crucial drawback when $d > 1$. They are not stable, in the sense that the error can be estimated from above by a multilevel norm with constants independent of the level, but not from below. Lower bounds can be obtained by using wavelet-type multiscale bases, semi-orthogonal prewavelets, or related biorthogonal multiscale bases instead. Again, then, the tensor product construction and successive optimization lead to sparse grids. Finally, we close this section with a short overview of the state of the art of sparse grid research. Here, the focus is put on the numerical treatment of PDEs based on different discretization approaches, and we include a short discussion of adaptive grid refinement and fast solvers in the sparse grid context.

In Section 5 we present numerical results of selected experiments. First, to show the properties of the sparse grid approximation, we discuss some PDE model problems in two or three dimensions. Second, we apply sparse grids to the solution of flow problems via the Navier–Stokes equations. Finally, we present results for two non-PDE problem classes of high dimensionality: numerical quadrature and data mining.

The concluding remarks of Section 6 close this discussion of sparse grid methods.

2. Breaking the curse of dimensionality

Classical approximation schemes exhibit the curse of dimensionality (Bellmann 1961) mentioned above. We have

$$\|f - f_n\| = O(n^{-r/d}),$$

where r and d denote the isotropic smoothness of the function f and the problem's dimensionality, respectively. This is one of the main obstacles in the treatment of high-dimensional problems. Therefore, the question is whether we can find situations, *i.e.*, either function spaces or error norms, for which the curse of dimensionality can be broken. At first glance, there is an easy way out: if we make a stronger assumption on the smoothness of the function f such that $r = O(d)$, then, we directly obtain $\|f - f_n\| = O(n^{-c})$ with constant $c > 0$. Of course, such an assumption is completely unrealistic.

However, about ten years ago, Barron (1993) found an interesting result. Denote by $\mathcal{F}L_1$ the class of functions with Fourier transforms in L_1. Then, consider the class of functions of $\mathbb{R}^d$ with

$$\nabla f \in \mathcal{F}L_1.$$

We expect an approximation rate

$$\|f - f_n\| = O(n^{-1/d})$$

since $\nabla f \in \mathcal{F}L_1 \approx r = 1$. However, Barron was able to show

$$\|f - f_n\| = O(n^{-1/2})$$

independent of d. Meanwhile, other function classes are known with such properties. These comprise certain radial basis schemes, stochastic sampling techniques, and approaches that work with spaces of functions with bounded mixed derivatives.

A better understanding of these results is possible with the help of harmonic analysis (Donoho 2000). Here we resort to the approach of the L_1-combination of L_∞-atoms; see also Triebel (1992) and DeVore (1998). Consider the class of functions $\mathcal{F}(M)$ with integral representation

$$f(x) = \int A(x,t)\,d|\mu|(t)$$

with

$$\int d|\mu|(t) \leq M, \tag{2.1}$$

where, for fixed t, we call $A(x,t) = A_t(x)$ an L_∞-atom if $|A_t(x)| \leq 1$ holds. Then there are results from Maurey for Banach spaces and Stechkin in

Fourier analysis which state that there exists an n-term sum

$$f_n(x) = \sum_{j=1}^{n} a_j A_{t_j}(x)$$

where

$$\|f - f_n\|_\infty \leq C \cdot n^{-1/2}$$

with C independent of d. In the following, we consider examples of such spaces.

Example 1. (Radial basis schemes) Consider superpositions of Gaussian bumps. These resemble the space $\mathcal{F}(M, \text{Gaussians})$ with $t := (x_0, s)$ and Gaussian atoms $A(x,t) = \exp(-\|x - x_0\|^2/s^2)$, where $\|\cdot\|$ denotes the Euclidean norm. Now, if the sum of the height of all Gaussians is bounded by M, Niyogi and Girosi (1998) showed that the resulting approximation rate is independent of d for the corresponding radial basis schemes. There is no further condition on the widths or positions of the bumps. Note that this corresponds to a ball in Besov space $B^d_{1,1}(\mathbb{R}^d)$ which is just Meyer's bump algebra (Meyer 1992). Thus, we have a restriction to smoother functions in higher dimensions such that the ratio r/d stays constant and, consequently, $n^{-r/d}$ does not grow with d.

Example 2. (Orthant scheme I) Another class of functions with an approximation rate independent of d is $\mathcal{F}(M, \text{Orthant})$. Now $t = (x_0, k)$, and k is the orthant indicator. Furthermore, $A(x,t)$ is the indicator of orthant k with apex at x_0. Again, if the integral (2.1) is at most M, the resulting approximation rate is of order $O(n^{-1/2})$ independent of d. A typical and well-known example for such a construction is the cumulative distribution function in $\mathbb{R}^d$. This simply results in the Monte Carlo method.

Example 3. (Orthant scheme II) A further interesting function class are the functions which are formed by any superposition of 2^d functions, each orthant-wise monotone for a different orthant. Now, the condition $\int d|\mu|(t) \leq 1$ is the same as

$$\frac{\partial^d f}{\partial x_1 \cdots \partial x_d} \in L_1,$$

i.e., we obtain the space of bounded mixed first variation. Again, this means considering only functions that get smoother as the dimensionality grows, but, in contrast to the examples mentioned above, only an anisotropic smoothness assumption is involved. Note that this is just the prerequisite for sparse grids with the piecewise constant hierarchical basis.

Further results on high-dimensional (and even infinite-dimensional) problems and their tractability were recently given by Woźniakowski, Sloan, and others (Wasilkovski and Woźniakowski 1995, Sloan and Woźniakowski 1998, Wasilkovski and Woźniakowski 1999, Sloan 2001, Hickernell, Sloan and Wasilkovski 2003, Dick, Sloan, Wang and Woźniakowski 2003). Here, especially in the context of numerical integration, they introduce the notion of *weighted* Sobolev spaces. Having observed that for some problems the integrand becomes less and less variable in successive coordinate directions, they introduce a sequence of positive weights $\{\gamma_j\}$ with decreasing values, with the weight γ_j being associated with coordinate direction j. They are able to show that the integration problem in a particular Sobolev space setting becomes *strongly tractable* (Traub and Woźniakowski 1980, Traub, Wasilkowski and Woźniakowski 1983, 1988), *i.e.*, that the worst-case error for all functions in the unit ball of the weighted Sobolev space is bounded independently of d and goes polynomially to zero if and only if the sum of the weights is asymptotically bounded from above. This corresponds to the decay of the kernel contributions in a reproducing kernel Hilbert space with rising d. The original paper (Sloan and Woźniakowski 1998) assumes that the integrand belongs to a Sobolev space of functions with square-integrable mixed first derivatives with the weights built into the definition of the associated inner product. Note that this assumption is closely related to that of Example 3 above. Since then, more general assumptions on the weights, and thus on the induced weighted function spaces, have been found (Dick *et al.* 2003, Hickernell *et al.* 2003).

In any case, we observe that a certain smoothness assumption on the function under consideration changes with d to obtain approximation rates which no longer depend exponentially on d. This raises the question of the very meaning of smoothness as the dimension changes and tends to ∞.

To this end, let us finally note an interesting aspect, namely the concentration of measure phenomenon (Milman 1988, Talagrand 1995, Gromov 1999) for probabilities in normed spaces of high dimension (also known as the geometric law of large numbers). This is an important development in modern analysis and geometry, manifesting itself across a wide range of mathematical sciences, particularly geometric functional analysis, probability theory, graph theory, diverse fields of computer science, and statistical physics. In the statistical setting it states the following. Let f be a Lipschitz function, Lipschitz constant L, on the d-sphere. Let P be normalized Lebesgue measure on the sphere and let X be uniformly distributed with respect to P. Then

$$P(|f(X) - Ef(X)| > t) \leq c_1 \exp(-c_2 t^2 / L^2),$$

with constants c_1 and c_2 independent of f and d (see Milman and Schechtman (1986) or Baxter and Iserles (2003, Section 2.3)). In its simplest form, the concentration of measure phenomenon states that every Lipschitz function on a sufficiently high-dimensional domain Ω is well approximated by a constant function (Hegland and Pestov 1999). Thus, there is some chance of treating high-dimensional problems despite the curse of dimensionality.

3. Piecewise linear interpolation on sparse grids

As a first approach to sparse grids and their underlying hierarchical multilevel setting, we discuss the problem of *interpolating* smooth functions with the help of piecewise d-linear hierarchical bases. For that, we introduce a tensor product-based subspace splitting and study the resulting subspaces. Starting from their properties, sparse grids are defined via an optimization process in a cost–benefit spirit closely related to the notion of n-term approximation. Out of the variety of norms with respect to which such an optimized discretization scheme can be derived, we restrict ourselves to the L_2-, the L_∞- and the energy norm, and thus to the respective types of sparse grids. After presenting the most important approximation properties of the latter, a short digression into recurrences and complexity will demonstrate their asymptotic characteristics and, consequently, their potential for problems of high dimensionality.

3.1. Hierarchical multilevel subspace splitting

In this section, the basic ingredients for our tensor product-based hierarchical setting are provided.

Subspace decomposition

Let us start with some notation and with the preliminaries that are necessary for a detailed discussion of sparse grids for purposes of interpolation or approximation, respectively. On the d-dimensional unit interval $\bar\Omega := [0,1]^d$, we consider multivariate functions u, $u(\mathbf{x}) \in \mathbb{R}$, $\mathbf{x} := (x_1, \ldots, x_d) \in \bar\Omega$, with (in some sense) bounded weak mixed derivatives

$$D^{\boldsymbol\alpha} u := \frac{\partial^{|\boldsymbol\alpha|_1} u}{\partial x_1^{\alpha_1} \cdots \partial x_d^{\alpha_d}} \qquad (3.1)$$

up to some given order $r \in \mathbb{N}_0$. Here, $\boldsymbol\alpha \in \mathbb{N}_0^d$ denotes a d-dimensional multi-index with the two norms

$$|\boldsymbol\alpha|_1 := \sum_{j=1}^d \alpha_j \quad \text{and} \quad |\boldsymbol\alpha|_\infty := \max_{1 \le j \le d} \alpha_j.$$

In the context of multi-indices, we use component-wise arithmetic operations, for example

$$\boldsymbol\alpha \cdot \boldsymbol\beta := (\alpha_1 \beta_1, \ldots, \alpha_d \beta_d), \qquad \gamma \cdot \boldsymbol\alpha := (\gamma \alpha_1, \ldots, \gamma \alpha_d),$$

or
$$2^{\boldsymbol{\alpha}} := (2^{\alpha_1}, \ldots, 2^{\alpha_d}),$$

the relational operators
$$\boldsymbol{\alpha} \leq \boldsymbol{\beta} \iff \forall_{1 \leq j \leq d} \; \alpha_j \leq \beta_j$$

and
$$\boldsymbol{\alpha} < \boldsymbol{\beta} \iff \boldsymbol{\alpha} \leq \boldsymbol{\beta} \text{ and } \boldsymbol{\alpha} \neq \boldsymbol{\beta},$$

and, finally, special multi-indices such as
$$\mathbf{0} := (0, \ldots, 0) \quad \text{or} \quad \mathbf{1} := (1, \ldots, 1),$$

and so on. In the following, for $q \in \{2, \infty\}$ and $r \in \mathbb{N}_0$, we study the spaces
$$X^{q,r}(\bar{\Omega}) := \{u : \bar{\Omega} \to \mathbb{R} : D^{\boldsymbol{\alpha}} u \in L_q(\Omega), |\boldsymbol{\alpha}|_\infty \leq r\}, \tag{3.2}$$
$$X_0^{q,r}(\bar{\Omega}) := \{u \in X^r(\bar{\Omega}) : u|_{\partial \Omega} = 0\}.$$

Thus, $X^{q,r}(\bar{\Omega})$ denotes the space of all functions of bounded (with respect to the L_q-norm) mixed derivatives up to order r, and $X_0^{q,r}(\bar{\Omega})$ will be the subspace of $X^{q,r}(\bar{\Omega})$ consisting of those $u \in X^r(\bar{\Omega})$ vanishing on the boundary $\partial \Omega$. Note that, for the theoretical considerations, we shall restrict ourselves to the case of homogeneous boundary conditions, i.e., to $X_0^{q,r}(\bar{\Omega})$. Furthermore, note that we omit the ambient dimension d when clear from the context. Concerning the smoothness parameter $r \in \mathbb{N}_0$, we need $r = 2$ for the case of piecewise linear approximations which, for the moment, will be in the centre of interest. Finally, for functions $u \in X_0^{q,r}(\bar{\Omega})$ and multi-indices $\boldsymbol{\alpha}$ with $|\boldsymbol{\alpha}|_\infty \leq r$, we introduce the seminorms
$$|u|_{\boldsymbol{\alpha}, \infty} := \|D^{\boldsymbol{\alpha}} u\|_\infty, \tag{3.3}$$
$$|u|_{\boldsymbol{\alpha}, 2} := \|D^{\boldsymbol{\alpha}} u\|_2 = \left(\int_{\bar{\Omega}} |D^{\boldsymbol{\alpha}} u|^2 \, d\mathbf{x} \right)^{1/2}.$$

Now, with the multi-index $\mathbf{l} = (l_1, \ldots, l_d) \in \mathbb{N}^d$, which indicates the level in a multivariate sense, we consider the family of d-dimensional standard rectangular grids
$$\{\Omega_\mathbf{l} : \mathbf{l} \in \mathbb{N}^d\} \tag{3.4}$$

on $\bar{\Omega}$ with mesh size
$$\mathbf{h}_\mathbf{l} := (h_{l_1}, \ldots, h_{l_d}) := 2^{-\mathbf{l}}. \tag{3.5}$$

That is, the grid $\Omega_\mathbf{l}$ is equidistant with respect to each individual coordinate direction, but, in general, may have different mesh sizes in the different coordinate directions. The grid points $\mathbf{x}_{\mathbf{l},\mathbf{i}}$ of grid $\Omega_\mathbf{l}$ are just the points
$$\mathbf{x}_{\mathbf{l},\mathbf{i}} := (x_{l_1, i_1}, \ldots, x_{l_d, i_d}) := \mathbf{i} \cdot \mathbf{h}_\mathbf{l}, \quad \mathbf{0} \leq \mathbf{i} \leq 2^\mathbf{l}. \tag{3.6}$$

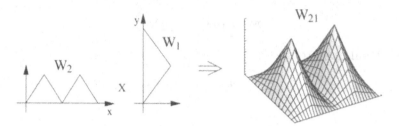

Figure 3.1. Tensor product approach for piecewise bilinear basis functions.

Thus, here and in the following, the multi-index **l** indicates the level (of a grid, a point, or, later on, a basis function, respectively), whereas the multi-index **i** denotes the location of a given grid point $\mathbf{x}_{\mathbf{l},\mathbf{i}}$ in the respective grid $\Omega_{\mathbf{l}}$.

Next, we have to define discrete approximation spaces and sets of basis functions that span those discrete spaces. In a piecewise linear setting, the simplest choice of a 1D basis function is the standard hat function $\phi(x)$,

$$\phi(x) := \begin{cases} 1 - |x|, & \text{if } x \in [-1, 1], \\ 0, & \text{otherwise.} \end{cases} \quad (3.7)$$

This mother of all piecewise linear basis functions can be used to generate an arbitrary $\phi_{l_j,i_j}(x_j)$ with support $[x_{l_j,i_j} - h_{l_j}, x_{l_j,i_j} + h_{l_j}] = [(i_j-1)h_{l_j}, (i_j+1)h_{l_j}]$ by dilation and translation, that is,

$$\phi_{l_j,i_j}(x_j) := \phi\left(\frac{x_j - i_j \cdot h_{l_j}}{h_{l_j}}\right). \quad (3.8)$$

The resulting 1D basis functions are the input of the tensor product construction which provides a suitable piecewise d-linear basis function in each grid point $\mathbf{x}_{\mathbf{l},\mathbf{i}}$ (see Figure 3.1):

$$\phi_{\mathbf{l},\mathbf{i}}(\mathbf{x}) := \prod_{j=1}^{d} \phi_{l_j,i_j}(x_j). \quad (3.9)$$

Since we deal with homogeneous boundary conditions (*i.e.*, with $X_0^{q,2}(\bar{\Omega})$), only those $\phi_{\mathbf{l},\mathbf{i}}(\mathbf{x})$ that correspond to *inner* grid points of $\Omega_{\mathbf{l}}$ are taken into account for the definition of

$$V_{\mathbf{l}} := \text{span}\{\phi_{\mathbf{l},\mathbf{i}} : \mathbf{1} \leq \mathbf{i} \leq 2^{\mathbf{l}} - \mathbf{1}\}, \quad (3.10)$$

the space of piecewise d-linear functions with respect to the interior of $\Omega_{\mathbf{l}}$. Obviously, the $\phi_{\mathbf{l},\mathbf{i}}$ form a basis of $V_{\mathbf{l}}$, with one basis function $\phi_{\mathbf{l},\mathbf{i}}$ of a support of the fixed size $2 \cdot \mathbf{h}_{\mathbf{l}}$ for each inner grid point $\mathbf{x}_{\mathbf{l},\mathbf{i}}$ of $\Omega_{\mathbf{l}}$, and this basis $\{\phi_{\mathbf{l},\mathbf{i}}\}$ is just the standard *nodal point basis* of the finite-dimensional space $V_{\mathbf{l}}$.

Additionally, we introduce the hierarchical increments $W_\mathbf{l}$,

$$W_\mathbf{l} := \text{span}\{\phi_{\mathbf{l},\mathbf{i}} : \mathbf{1} \leq \mathbf{i} \leq 2^\mathbf{l} - \mathbf{1},\ i_j \text{ odd for all } 1 \leq j \leq d\}, \qquad (3.11)$$

for which the relation

$$V_\mathbf{l} = \bigoplus_{\mathbf{k} \leq \mathbf{l}} W_\mathbf{k} \qquad (3.12)$$

can be easily seen. Note that the supports of all basis functions $\phi_{\mathbf{l},\mathbf{i}}$ spanning $W_\mathbf{l}$ are mutually disjoint. Thus, with the index set

$$\mathbf{I}_\mathbf{l} := \{\mathbf{i} \in \mathbb{N}^d_{\bullet} : \mathbf{1} \leq \mathbf{i} \leq 2^\mathbf{l} - \mathbf{1},\ i_j \text{ odd for all } 1 \leq j \leq d\}, \qquad (3.13)$$

we get another basis of $V_\mathbf{l}$, the *hierarchical basis*

$$\{\phi_{\mathbf{k},\mathbf{i}} : \mathbf{i} \in \mathbf{I}_\mathbf{k}, \mathbf{k} \leq \mathbf{l}\}, \qquad (3.14)$$

which generalizes the well-known 1D basis shown in Figure 3.2 to the d-dimensional case by means of a tensor product approach. With these hierarchical difference spaces $W_\mathbf{l}$, we can define

$$V := \sum_{l_1=1}^{\infty} \cdots \sum_{l_d=1}^{\infty} W_{(l_1,\ldots,l_d)} = \bigoplus_{\mathbf{l} \in \mathbb{N}^d} W_\mathbf{l} \qquad (3.15)$$

with its natural hierarchical basis

$$\{\phi_{\mathbf{l},\mathbf{i}} : \mathbf{i} \in \mathbf{I}_\mathbf{l}, \mathbf{l} \in \mathbb{N}^d\}. \qquad (3.16)$$

Except for completion with respect to the H^1-norm, V is simply the underlying Sobolev space $H^1_0(\bar{\Omega})$, i.e., $\bar{V} = H^1_0(\bar{\Omega})$.

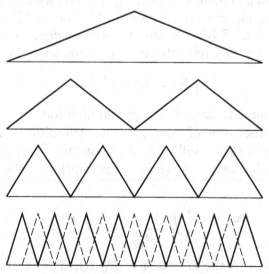

Figure 3.2. Piecewise linear hierarchical basis (solid) vs. nodal point basis (dashed).

Later we shall deal with finite-dimensional subspaces of V. Note that, for instance, with the discrete spaces

$$V_n^{(\infty)} := \bigoplus_{|\mathbf{l}|_\infty \leq n} W_\mathbf{l}, \qquad (3.17)$$

the limit

$$\lim_{n \to \infty} V_n^{(\infty)} = \lim_{n \to \infty} \bigoplus_{|\mathbf{l}|_\infty \leq n} W_\mathbf{l} := \bigcup_{n=1}^{\infty} V_n^{(\infty)} = V \qquad (3.18)$$

exists due to $V_n^{(\infty)} \subset V_{n+1}^{(\infty)}$. Hence, any function $u \in H_0^1(\bar{\Omega})$ and, consequently, any $u \in X_0^{q,2}(\bar{\Omega})$ can be uniquely split by

$$u(\mathbf{x}) = \sum_\mathbf{l} u_\mathbf{l}(\mathbf{x}), \qquad u_\mathbf{l}(\mathbf{x}) = \sum_{i \in I_\mathbf{l}} v_{\mathbf{l},i} \cdot \phi_{\mathbf{l},i}(\mathbf{x}) \in W_\mathbf{l}, \qquad (3.19)$$

where the $v_{\mathbf{l},i} \in \mathbb{R}$ are the coefficient values of the hierarchical product basis representation of u also called *hierarchical surplus*.

Before we turn to finite-dimensional approximation spaces for $X_0^{q,2}(\bar{\Omega})$, we summarize the most important properties of the hierarchical subspaces $W_\mathbf{l}$ according to Bungartz (1992b) and Bungartz and Griebel (1999).

Basic properties of the subspaces
Concerning the subspaces $W_\mathbf{l}$, the crucial questions are how important $W_\mathbf{l}$ is for the interpolation of some given $u \in X_0^{q,2}(\bar{\Omega})$ and what computational and storage cost come along with it. From (3.11) and (3.13), we immediately learn the dimension of $W_\mathbf{l}$, i.e., the number of degrees of freedom (grid points or basis functions, respectively) associated with $W_\mathbf{l}$:

$$|W_\mathbf{l}| = |I_\mathbf{l}| = 2^{|\mathbf{l}-\mathbf{1}|_1}. \qquad (3.20)$$

Equation (3.20) already answers the second question.

The following discussion of a subspace's contribution to the overall interpolant according to (3.19) will be based upon three norms: the maximum norm $\|\cdot\|_\infty$, the L_p-norm $\|\cdot\|_p$ ($p = 2$ in general), and the energy norm

$$\|u\|_E := \left(\int_\Omega \sum_{j=1}^d \left(\frac{\partial u(\mathbf{x})}{\partial x_j} \right)^2 d\mathbf{x} \right)^{1/2}, \qquad (3.21)$$

which is equivalent to the H^1-norm in $H_0^1(\bar{\Omega})$. For the Laplacian, (3.21) indeed indicates the energy norm in finite element terminology. First we look at the different hierarchical basis functions $\phi_{\mathbf{l},i}(\mathbf{x})$.

Lemma 3.1. For any piecewise d-linear basis function $\phi_{\mathbf{l},\mathbf{i}}(\mathbf{x})$, the following equations hold:

$$\|\phi_{\mathbf{l},\mathbf{i}}\|_\infty = 1, \qquad (3.22)$$

$$\|\phi_{\mathbf{l},\mathbf{i}}\|_p = \left(\frac{2}{p+1}\right)^{d/p} \cdot 2^{-|\mathbf{l}|_1/p}, \qquad (p \geq 1),$$

$$\|\phi_{\mathbf{l},\mathbf{i}}\|_E = \sqrt{2} \cdot \left(\frac{2}{3}\right)^{(d-1)/2} \cdot 2^{-|\mathbf{l}|_1/2} \cdot \left(\sum_{j=1}^{d} 2^{2l_j}\right)^{1/2}.$$

Proof. All equalities result from straightforward calculations based on the definition of $\phi_{\mathbf{l},\mathbf{i}}(\mathbf{x})$ (see (3.7) to (3.9), and Bungartz (1992b) and Bungartz and Griebel (1999), for example). □

Next, we consider the hierarchical coefficient values $v_{\mathbf{l},\mathbf{i}}$ in more detail. They can be computed from the function values $u(\mathbf{x}_{\mathbf{l},\mathbf{i}})$ in the following way:

$$v_{\mathbf{l},\mathbf{i}} = \left(\prod_{j=1}^{d}[\,-\tfrac{1}{2}\ \ 1\ \ -\tfrac{1}{2}\,]_{x_{l_j,i_j},l_j}\right) u =: \left(\prod_{j=1}^{d} I_{x_{l_j,i_j},l_j}\right) u \qquad (3.23)$$

$$=: I_{\mathbf{x}_{\mathbf{l},\mathbf{i}},\mathbf{l}}\, u.$$

This is due to the definition of the spaces $W_\mathbf{l}$ and their basis functions (3.11), whose supports are mutually disjoint and do not contain coarse grid points $\mathbf{x}_{\mathbf{k},\mathbf{j}}$, $\mathbf{k} < \mathbf{l}$, in their interior. Definition (3.23) illustrates why $v_{\mathbf{l},\mathbf{i}}$ is also called *hierarchical surplus*. In (3.23), as usual in multigrid terminology (*cf.* Hackbusch (1985, 1986), for instance), $I_{\mathbf{x}_{\mathbf{l},\mathbf{i}},\mathbf{l}}$ denotes a d-dimensional stencil which gives the coefficients for a linear combination of nodal values of its argument u. This operator-based representation of the hierarchical coefficients $v_{\mathbf{l},\mathbf{i}}$ leads to an integral representation of $v_{\mathbf{l},\mathbf{i}}$, as follows.

Lemma 3.2. Let $\psi_{l_j,i_j}(x_j) := -2^{-(l_j+1)} \cdot \phi_{l_j,i_j}(x_j)$. Further, let $\psi_{\mathbf{l},\mathbf{i}}(\mathbf{x}) := \prod_{j=1}^{d} \psi_{l_j,i_j}(x_j)$. For any coefficient value $v_{\mathbf{l},\mathbf{i}}$ of the hierarchical representation (3.19) of $u \in X_0^{q,2}(\bar{\Omega})$, the following relation holds:

$$v_{\mathbf{l},\mathbf{i}} = \int_\Omega \psi_{\mathbf{l},\mathbf{i}}(\mathbf{x}) \cdot D^2 u(\mathbf{x})\, d\mathbf{x}. \qquad (3.24)$$

Proof. First we look at the simplest case $d = 1$, where we can omit the index j for clarity. Partial integration provides

$$\int_\Omega \psi_{l,i}(x) \cdot \frac{\partial^2 u(x)}{\partial x^2}\, dx$$

$$= \int_{x_{l,i}-h_l}^{x_{l,i}+h_l} \psi_{l,i}(x) \cdot \frac{\partial^2 u(x)}{\partial x^2}\, dx$$

$$= \left[\psi_{l,i}(x) \cdot \frac{\partial u(x)}{\partial x}\right]_{x_{l,i}-h_l}^{x_{l,i}+h_l} - \int_{x_{l,i}-h_l}^{x_{l,i}+h_l} \frac{\partial \psi_{l,i}(x)}{\partial x} \cdot \frac{\partial u(x)}{\partial x} \, dx$$

$$= \int_{x_{l,i}-h_l}^{x_{l,i}} \frac{1}{2} \cdot \frac{\partial u(x)}{\partial x} \, dx - \int_{x_{l,i}}^{x_{l,i}+h_l} \frac{1}{2} \cdot \frac{\partial u(x)}{\partial x} \, dx$$

$$= I_{x_{l,i},l} \, u,$$

since $\psi_{l,i}(x_{l,i} - h_l) = \psi_{l,i}(x_{l,i} + h_l) = 0$ and since $\partial \psi_{l,i}(x)/\partial x \in \{\frac{1}{2}, -\frac{1}{2}\}$ due to the construction of $\psi_{l,i}$ and $\phi_{l,i}$. Finally, the tensor product approach according to the operator product given in (3.23) leads to a straightforward generalization to $d > 1$. □

The above lemma and its proof show the close relations of our hierarchical basis approach to integral transforms like wavelet transforms. Applying successive partial integration to (3.24), twice for $d = 1$ and $2d$ times for general dimensionality, we get

$$v_{\mathbf{l},\mathbf{i}} = \int_\Omega \psi_{\mathbf{l},\mathbf{i}}(\mathbf{x}) \cdot D^2 u(\mathbf{x}) \, d\mathbf{x} = \int_\Omega \hat{\psi}_{\mathbf{l},\mathbf{i}}(\mathbf{x}) \cdot u(\mathbf{x}) \, d\mathbf{x}, \qquad (3.25)$$

where $\hat{\psi}_{\mathbf{l},\mathbf{i}}(\mathbf{x})$ equals $D^2 \psi_{\mathbf{l},\mathbf{i}}(\mathbf{x})$ in a weak sense (i.e., in the sense of distributions) and is a linear combination of 3^d Dirac pulses of alternating sign. Thus, the hierarchical surplus $v_{\mathbf{l},\mathbf{i}}$ can be interpreted as the coefficient resulting from an integral transform with respect to a function $\hat{\psi}_{\mathbf{l},\mathbf{i}}(\mathbf{x})$ of an oscillating structure.

Starting from (3.24), we are now able to give bounds for the hierarchical coefficients with respect to the different seminorms introduced in (3.3).

Lemma 3.3. Let $u \in X_0^{q,2}(\bar{\Omega})$ be given in its hierarchical representation. Then, the following estimates for the hierarchical coefficients $v_{\mathbf{l},\mathbf{i}}$ hold:

$$|v_{\mathbf{l},\mathbf{i}}| \leq 2^{-d} \cdot 2^{-2 \cdot |\mathbf{l}|_1} \cdot |u|_{2,\infty}, \qquad (3.26)$$

$$|v_{\mathbf{l},\mathbf{i}}| \leq 2^{-d} \cdot \left(\frac{2}{3}\right)^{d/2} \cdot 2^{-(3/2) \cdot |\mathbf{l}|_1} \cdot \left. |u| \right._{\mathrm{supp}(\phi_{\mathbf{l},\mathbf{i}})}|_{2,2},$$

where $\mathrm{supp}(\phi_{\mathbf{l},\mathbf{i}})$ denotes the support of $\phi_{\mathbf{l},\mathbf{i}}(\mathbf{x})$.

Proof. With (3.22), (3.24), and with the definition of $\psi_{\mathbf{l},\mathbf{i}}$, we get

$$|v_{\mathbf{l},\mathbf{i}}| = \left|\int_\Omega \psi_{\mathbf{l},\mathbf{i}}(\mathbf{x}) \cdot D^2 u(\mathbf{x}) \, d\mathbf{x}\right| \leq \|\psi_{\mathbf{l},\mathbf{i}}\|_1 \cdot \|D^2 u \, |_{\mathrm{supp}(\phi_{\mathbf{l},\mathbf{i}})}\|_\infty$$

$$= 2^{-d} \cdot 2^{-|\mathbf{l}|_1} \cdot \|\phi_{\mathbf{l},\mathbf{i}}\|_1 \cdot |u \, |_{\mathrm{supp}(\phi_{\mathbf{l},\mathbf{i}})}|_{2,\infty}$$

$$\leq 2^{-d} \cdot 2^{-2 \cdot |\mathbf{l}|_1} \cdot |u|_{2,\infty}.$$

For $|\cdot|_{2,2}$, the Cauchy–Schwarz inequality provides

$$|v_{\mathbf{l},\mathbf{i}}| = \left| \int_\Omega \psi_{\mathbf{l},\mathbf{i}}(\mathbf{x}) \cdot D^2 u(\mathbf{x})\, d\mathbf{x} \right| \leq \|\psi_{\mathbf{l},\mathbf{i}}\|_2 \cdot \|D^2 u\,|_{\mathrm{supp}(\phi_{\mathbf{l},\mathbf{i}})}\|_2$$

$$= 2^{-d} \cdot 2^{-|\mathbf{l}|_1} \cdot \|\phi_{\mathbf{l},\mathbf{i}}\|_2 \cdot |u\,|_{\mathrm{supp}(\phi_{\mathbf{l},\mathbf{i}})}|_{2,2},$$

which, with (3.22), is the desired result. □

Finally, the results from the previous three lemmata lead to bounds for the contribution $u_{\mathbf{l}} \in W_{\mathbf{l}}$ of a subspace $W_{\mathbf{l}}$ to the hierarchical representation (3.19) of a given $u \in X_0^{q,2}(\bar{\Omega})$.

Lemma 3.4. Let $u \in X_0^{q,2}(\bar{\Omega})$ be given in its hierarchical representation (3.19). Then, the following estimates for its components $u_{\mathbf{l}} \in W_{\mathbf{l}}$ hold:

$$\|u_{\mathbf{l}}\|_\infty \leq 2^{-d} \cdot 2^{-2\cdot|\mathbf{l}|_1} \cdot |u|_{2,\infty}, \qquad (3.27)$$

$$\|u_{\mathbf{l}}\|_2 \leq 3^{-d} \cdot 2^{-2\cdot|\mathbf{l}|_1} \cdot |u|_{2,2},$$

$$\|u_{\mathbf{l}}\|_E \leq \frac{1}{2 \cdot 12^{(d-1)/2}} \cdot 2^{-2\cdot|\mathbf{l}|_1} \cdot \left(\sum_{j=1}^d 2^{2\cdot l_j} \right)^{1/2} \cdot |u|_{2,\infty},$$

$$\|u_{\mathbf{l}}\|_E \leq \sqrt{3} \cdot 3^{-d} \cdot 2^{-2\cdot|\mathbf{l}|_1} \cdot \left(\sum_{j=1}^d 2^{2\cdot l_j} \right)^{1/2} \cdot |u|_{2,2}.$$

Proof. Since the supports of all $\phi_{\mathbf{l},\mathbf{i}}$ contributing to $u_{\mathbf{l}}$ according to (3.19) are mutually disjoint, the first estimate follows immediately from the respective statements in (3.22) and (3.26). For the estimate concerning the L_2-norm, we get with the same argument of disjoint supports and with (3.22) and (3.26)

$$\|u_{\mathbf{l}}\|_2^2 = \left\| \sum_{\mathbf{i} \in I_{\mathbf{l}}} v_{\mathbf{l},\mathbf{i}} \cdot \phi_{\mathbf{l},\mathbf{i}} \right\|_2^2 = \sum_{\mathbf{i} \in I_{\mathbf{l}}} |v_{\mathbf{l},\mathbf{i}}|^2 \cdot \|\phi_{\mathbf{l},\mathbf{i}}\|_2^2$$

$$\leq \sum_{\mathbf{i} \in I_{\mathbf{l}}} \frac{1}{6^d} \cdot 2^{-3\cdot|\mathbf{l}|_1} \cdot |u\,|_{\mathrm{supp}(\phi_{\mathbf{l},\mathbf{i}})}|_{2,2}^2 \cdot \left(\frac{2}{3}\right)^d \cdot 2^{-|\mathbf{l}|_1}$$

$$= 9^{-d} \cdot 2^{-4\cdot|\mathbf{l}|_1} \cdot |u|_{2,2}^2.$$

Finally, an analogous argument provides

$$\|u_{\mathbf{l}}\|_E^2 = \left\| \sum_{\mathbf{i} \in I_{\mathbf{l}}} v_{\mathbf{l},\mathbf{i}} \cdot \phi_{\mathbf{l},\mathbf{i}} \right\|_E^2$$

$$= \sum_{\mathbf{i} \in I_{\mathbf{l}}} |v_{\mathbf{l},\mathbf{i}}|^2 \cdot \|\phi_{\mathbf{l},\mathbf{i}}\|_E^2$$

$$\leq \sum_{\mathbf{i} \in I_{\mathbf{l}}} \frac{1}{4^d} \cdot 2^{-4 \cdot |\mathbf{l}|_1} \cdot |u|_{2,\infty}^2 \cdot 2 \cdot \left(\frac{2}{3}\right)^{d-1} \cdot 2^{-|\mathbf{l}|_1} \cdot \left(\sum_{j=1}^{d} 2^{2 \cdot l_j}\right)$$

$$= \frac{1}{2 \cdot 6^{d-1}} \cdot 2^{-5 \cdot |\mathbf{l}|_1} \cdot \left(\sum_{j=1}^{d} 2^{2 \cdot l_j}\right) \cdot \sum_{\mathbf{i} \in I_{\mathbf{l}}} |u|_{2,\infty}^2$$

$$= \frac{1}{4 \cdot 12^{d-1}} \cdot 2^{-4 \cdot |\mathbf{l}|_1} \cdot \left(\sum_{j=1}^{d} 2^{2 \cdot l_j}\right) \cdot |u|_{2,\infty}^2$$

as well as the second estimate for $\|u_\mathbf{l}\|_E$. □

In the next section, the information gathered above will be used to construct finite-dimensional approximation spaces U for V or $X_0^{q,2}(\bar{\Omega})$, respectively. Such a U shall be based on a subspace selection $\mathbf{I} \subset \mathbb{N}^d$,

$$U := \bigoplus_{\mathbf{l} \in \mathbf{I}} W_\mathbf{l}, \tag{3.28}$$

with corresponding interpolants or approximants

$$u_U := \sum_{\mathbf{l} \in \mathbf{I}} u_\mathbf{l}, \quad u_\mathbf{l} \in W_\mathbf{l}. \tag{3.29}$$

The estimate

$$\|u - u_U\| = \left\|\sum_{\mathbf{l}} u_\mathbf{l} - \sum_{\mathbf{l} \in \mathbf{I}} u_\mathbf{l}\right\| \leq \sum_{\mathbf{l} \notin \mathbf{I}} \|u_\mathbf{l}\| \leq \sum_{\mathbf{l} \notin \mathbf{I}} b(\mathbf{l}) \cdot |u| \tag{3.30}$$

will allow the evaluation of the approximation space U with respect to a norm $\|\cdot\|$ and a corresponding seminorm $|\cdot|$ on the basis of the bounds from above indicating the *benefit* $b(\mathbf{l})$ of $W_\mathbf{l}$.

3.2. Sparse grids

Interpolation in finite-dimensional spaces

The hierarchical multilevel splitting introduced in the previous section brings along a whole family of hierarchical subspaces $W_\mathbf{l}$ of V. However, for discretization purposes, we are more interested in decompositions of finite-dimensional subspaces of V or $X_0^{q,2}(\bar{\Omega})$ than in the splitting of V itself. Therefore, we now turn to finite sets $\mathbf{I}$ of active levels $\mathbf{l}$ in the summation (3.19). For some $n \in \mathbb{N}$, for instance, one possibility $V_n^{(\infty)}$ has already been mentioned in (3.17). The finite-dimensional $V_n^{(\infty)}$ is just the usual space of piecewise d-linear functions on the rectangular grid $\Omega_{n \cdot \mathbf{1}} = \Omega_{(n,\ldots,n)}$ with equidistant mesh size $h_n = 2^{-n}$ in each coordinate direction. In a scheme

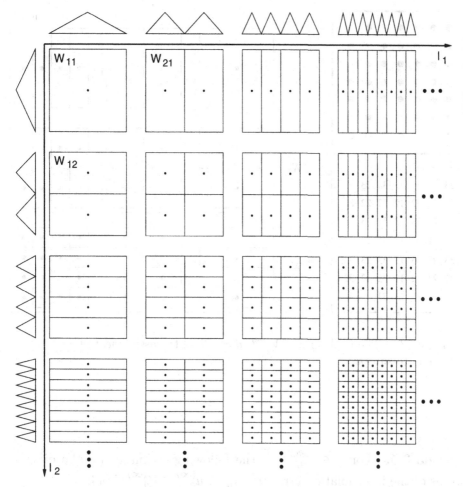

Figure 3.3. Scheme of subspaces for $d = 2$. Each square represents one subspace W_l with its associated grid points. The supports of the corresponding basis functions have the same mesh size $\mathbf{h_l}$ and cover the domain Ω.

of subspaces W_l as shown in Figure 3.3 for the 2D case, $V_n^{(\infty)}$ corresponds to a square sector of subspaces: see Figure 3.4.

Obviously, the dimension of $V_n^{(\infty)}$ (i.e., the number of inner grid points in the underlying grid) is

$$|V_n^{(\infty)}| = (2^n - 1)^d = O(2^{d \cdot n}) = O(h_n^{-d}). \qquad (3.31)$$

For the error $u - u_n^{(\infty)}$ of the interpolant $u_n^{(\infty)} \in V_n^{(\infty)}$ of a given function $u \in X_0^{q,2}(\bar{\Omega})$ with respect to the different norms we are interested in, the following lemma states the respective results.

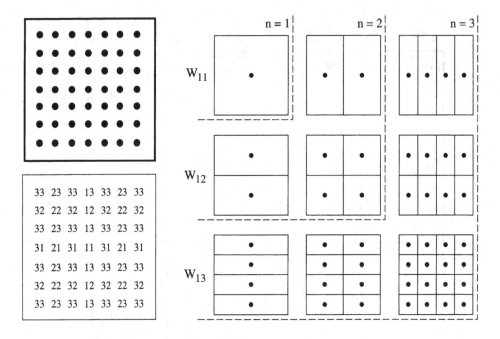

Figure 3.4. The (full) grid of $V_3^{(\infty)}$, $d = 2$, and the assignment of grid points to subspaces.

Lemma 3.5. For $u \in X_0^{q,2}(\bar{\Omega})$, the following estimates for the different norms of the interpolation error $u - u_n^{(\infty)}$, $u_n^{(\infty)} \in V_n^{(\infty)}$, hold:

$$\|u - u_n^{(\infty)}\|_\infty \leq \frac{d}{6^d} \cdot 2^{-2n} \cdot |u|_{2,\infty} \qquad = O(h_n^2), \qquad (3.32)$$

$$\|u - u_n^{(\infty)}\|_2 \leq \frac{d}{9^d} \cdot 2^{-2n} \cdot |u|_{2,2} \qquad = O(h_n^2),$$

$$\|u - u_n^{(\infty)}\|_E \leq \frac{d^{3/2}}{2 \cdot 3^{(d-1)/2} \cdot 6^{d-1}} \cdot 2^{-n} \cdot |u|_{2,\infty} = O(h_n),$$

$$\|u - u_n^{(\infty)}\|_E \leq \frac{d^{3/2}}{\sqrt{3} \cdot 9^{d-1}} \cdot 2^{-n} \cdot |u|_{2,2} \qquad = O(h_n).$$

Proof. For the L_∞-norm, (3.27) provides

$$\|u - u_n^{(\infty)}\|_\infty \leq \sum_{|\mathbf{l}|_\infty > n} \|u_\mathbf{l}\|_\infty \leq \frac{1}{2^d} \cdot |u|_{2,\infty} \cdot \sum_{|\mathbf{l}|_\infty > n} 2^{-2 \cdot |\mathbf{l}|_1},$$

from which we get

$$\|u - u_n^{(\infty)}\|_\infty \leq \frac{1}{2^d} \cdot |u|_{2,\infty} \cdot \left(\sum_{\mathbf{l}} 4^{-|\mathbf{l}|_1} - \sum_{|\mathbf{l}|_\infty \leq n} 4^{-|\mathbf{l}|_1}\right)$$

$$= \frac{1}{2^d} \cdot |u|_{2,\infty} \cdot \left(\left(\frac{1}{3}\right)^d - \left(\sum_{i=1}^n 4^{-i}\right)^d\right)$$

$$= \frac{1}{2^d} \cdot |u|_{2,\infty} \cdot \left(\frac{1}{3}\right)^d \cdot (1 - (1 - 4^{-n})^d)$$

$$\leq \frac{d}{6^d} \cdot |u|_{2,\infty} \cdot 4^{-n}.$$

The respective result for the L_2-norm can be obtained in exactly the same way. For the error with respect to the energy norm, (3.27) leads to

$$\|u - u_n^{(\infty)}\|_E \leq \frac{d^{1/2}}{2 \cdot 12^{(d-1)/2}} \cdot |u|_{2,\infty} \cdot \sum_{|\mathbf{l}|_\infty > n} 2^{-2\cdot|\mathbf{l}|_1} \cdot \max_{1 \leq j \leq d} 2^{l_j}$$

$$\leq \frac{d^{3/2}}{2 \cdot 12^{(d-1)/2}} \cdot |u|_{2,\infty} \cdot \sum_{|\mathbf{l}|_\infty = l_1 > n} 2^{-2\cdot|\mathbf{l}|_1} \cdot 2^{l_1}$$

$$= \frac{d^{3/2}}{2 \cdot 12^{(d-1)/2}} \cdot |u|_{2,\infty} \cdot \sum_{l_1 > n} 2^{-l_1} \cdot \left(\sum_{l_j=1}^{l_1} 4^{-l_j}\right)^{d-1}$$

$$\leq \frac{d^{3/2}}{2 \cdot 12^{(d-1)/2}} \cdot |u|_{2,\infty} \cdot \frac{1}{3^{d-1}} \cdot 2^{-n},$$

and an analogous argument provides the second estimate. $\square$

It is important to note that we get the same order of accuracy as in standard approximation theory, although our regularity assumptions differ from those normally used there.

Equations (3.31) and (3.32) clearly reveal the crucial drawback of $V_n^{(\infty)}$, the *curse of dimensionality* discussed in detail in Section 2. With d increasing, the number of degrees of freedom that are necessary to achieve an accuracy of $O(h)$ or $O(h^2)$, respectively, grows exponentially. Therefore, we ask how to construct discrete approximation spaces that are better than $V_n^{(\infty)}$ in the sense that the same number of invested grid points leads to a higher order of accuracy. Hence, in the following, we look for an optimum $V^{(\text{opt})}$ by solving a restricted optimization problem of the type

$$\max_{u \in X_0^{q,2}: |u|=1} \|u - u_{V^{(\text{opt})}}\| = \min_{U \subset V: |U|=w} \max_{u \in X_0^{q,2}: |u|=1} \|u - u_U\| \quad (3.33)$$

for some prescribed cost or work count w. The aim is to profit from a given work count as much as possible. Note that an optimization the other way round could be done as well. Prescribe some desired accuracy ε and look for the discrete approximation scheme that achieves this with the smallest work count possible. This is in fact the point of view of computational complexity. Of course, any potential solution $V^{(\text{opt})}$ of (3.33) has to be expected to depend on the norm $\|\cdot\|$ as well as on the seminorm $|\cdot|$ used to measure the error of u's interpolant $u_U \in U$ or the smoothness of u, respectively. According to our hierarchical setting, we will allow discrete spaces of the type $U := \bigoplus_{\mathbf{l} \in \mathbf{I}} W_{\mathbf{l}}$ for an arbitrary finite index set $\mathbf{I} \subset \mathbb{N}^d$ as candidates for the optimization process only.

An approach such as (3.33) selects certain $W_{\mathbf{l}}$ due to their importance, and thus selects the respective underlying grid points. Depending on the invested work count w, we can expect to get some kind of regular structure or grid patterns. However, in contrast to adaptive grid refinement, which is highly problem-dependent, such a proceeding simply depends on the problem class (*i.e.*, on the space u has to belong to, here $X_0^{q,2}(\bar{\Omega})$), but not on u itself. Although such *a priori optimization* strategies are not very widespread in the context of PDEs, there is a long tradition in approximation theory and numerical quadrature. For example, think of the Gauss quadrature rules where the grid points are chosen as the roots of certain classes of orthogonal polynomials; *cf.* Krommer and Ueberhuber (1994), for instance. Compared with equidistant quadrature rules based on polynomial interpolants with the same number n of grid points, the degree of accuracy, *i.e.*, the maximum polynomial degree up to which a numerical quadrature rule provides exact results, can be augmented from at most n to $2n-1$.

Another nice example of the usefulness of an *a priori* grid optimization in numerical quadrature is provided by the *Koksma–Hlawka inequality* (Hlawka 1961, Krommer and Ueberhuber 1994), which says that, for every quadrature formula Q_n based on a simple averaging of samples,

$$Q_n u := \frac{1}{n} \cdot \sum_{i=1}^{n} u(\mathbf{x}_i), \qquad (3.34)$$

that is used to get an approximation to $Iu := \int_{\bar{\Omega}} u(\mathbf{x})\,d\mathbf{x}$, a sharp error bound is given by

$$|Q_n u - Iu| \leq V(u) \cdot D_n^*(\mathbf{x}_1, \ldots, \mathbf{x}_n). \qquad (3.35)$$

Here, $V(u)$ is the so-called *variation of u in the sense of Hardy and Krause*, a property of u indicating the global smoothness of u: the smoother u is on $\bar{\Omega}$, the smaller the values of $V(u)$. $D_n^*(\mathbf{x}_1, \ldots, \mathbf{x}_n)$ denotes the so-called *star discrepancy* of the grid $(\mathbf{x}_1, \ldots, \mathbf{x}_n)$, which measures the deviation of a finite part of the sequence $\mathbf{x}_1, \mathbf{x}_2, \ldots$ from the uniform distribution, and is defined by

$$D_n^*(\mathbf{x}_1, \ldots, \mathbf{x}_n) := \sup_{E \in \mathcal{E}} \left| \frac{1}{n} \cdot \sum_{i=1}^n \chi_E(\mathbf{x}_i) - \int_{\bar{\Omega}} \chi_E(\mathbf{x}) \, d\mathbf{x} \right|, \quad (3.36)$$

where $\mathcal{E} := \{[0, e_1[\times \cdots \times [0, e_d[\subset \bar{\Omega}\}$ is the set of all subcubes of $\bar{\Omega}$ with $\mathbf{0}$ as a corner and χ_E is the characteristic function of $E \in \mathcal{E}$. Although we do not want to go into detail here, we emphasize the crucial point of (3.35). The quadrature error $|Q_n u - I u|$ divides into two parts: a *problem-dependent* one (the variation of u) unaffected by the grid, and a *grid-dependent* one (the star discrepancy) uninfluenced by the actual problem (*i.e.*, u). This clearly shows the benefit of the two optimization strategies mentioned above: the construction of grids of low discrepancy (*low-discrepancy formulas*) reduces the second factor in (3.35), whereas adaptive grid refinement can help to concentrate further grid points on subregions of $\bar{\Omega}$ with a (locally) high variation. After this digression into numerical quadrature, we return to our actual topic, the solution of (3.33) in our hierarchical subspace setting.

Formal derivation and properties of sparse grids
As already mentioned, the candidates for $V^{(\text{opt})}$ are finite sets of $W_{\mathbf{l}}$. Therefore, spaces $U := \bigoplus_{\mathbf{l} \in \mathbf{I}} W_{\mathbf{l}}$, the respective grids, and the underlying index sets $\mathbf{I} \subset \mathbb{N}^d$ have to be identified. There are two obvious ways to tackle such problems: a continuous one based on an analytical approach where the multi-index $\mathbf{l}$ is generalized to a nonnegative real one, and a discrete one which uses techniques known from combinatorial optimization; for details, we refer the reader to Bungartz (1998).

Continuous optimization
For the following, a grid and its representation $\mathbf{I}$ – formerly a finite set of multi-indices – is nothing but a bounded subset of $\mathbb{R}_+^d$, and a hierarchical subspace $W_{\mathbf{l}}$ just corresponds to a point $\mathbf{l} \in \mathbb{R}_+^d$.

First we have to formulate the optimization problem (3.33). To this end, and inspired by (3.20), the *local cost function* $c(\mathbf{l})$ is defined as a straightforward generalization of the number of degrees of freedom involved:

$$c(\mathbf{l}) := 2^{|\mathbf{l}|_1 - d} = 2^{l_1 + \cdots + l_d - d}. \quad (3.37)$$

For the *local benefit function* $b(\mathbf{l})$, we use the squared upper bounds for $\|u_{\mathbf{l}}\|$ according to (3.27). At the moment, we do not fix the norm to be used here. Obviously, the search for an optimal $\mathbf{I} \subset \mathbb{R}_+^d$ can be restricted to $\mathbf{I} \subset \mathbf{I}^{(\max)} := [0, N]^d$ for a sufficiently large N without loss of generality. Based on the two local quantities $c(\mathbf{l})$ and $b(\mathbf{l})$, the *global cost* $C(\mathbf{I})$ and the *global benefit* $B(\mathbf{I})$ of a grid $\mathbf{I}$ are defined by

$$C(I) := \int_I c(\mathbf{l})\,\mathrm{d}\mathbf{l}, \qquad B(I) := \int_I b(\mathbf{l})\,\mathrm{d}\mathbf{l}. \tag{3.38}$$

This leads to the desired restricted optimization problem according to (3.33):

$$\max_{C(\mathbf{I})=w} B(\mathbf{I}). \tag{3.39}$$

For the solution of (3.39), we start from an arbitrary $\mathbf{I} \subset \mathbf{I}^{(\mathrm{max})}$ that has a sufficiently smooth boundary $\partial \mathbf{I}$. With a sufficiently smooth mapping τ,

$$\tau: \mathbb{R}_+^d \to \mathbb{R}_+^d, \qquad \tau(\mathbf{l}) = 0 \quad \text{for } \mathbf{l} \in \partial \mathbb{R}_+^d, \tag{3.40}$$

we define a small disturbance $\varphi_{\varepsilon,\tau}$ of the grid $\mathbf{I}$:

$$\varphi_{\varepsilon,\tau}: \mathbf{I} \to \mathbf{I}_{\varepsilon,\tau} \subset \mathbf{I}^{(\mathrm{max})}, \qquad \varphi_{\varepsilon,\tau}(\mathbf{l}) := \mathbf{l} + \varepsilon \cdot \tau(\mathbf{l}), \qquad \varepsilon \in \mathbb{R}. \tag{3.41}$$

For the global cost of the disturbed grid $\mathbf{I}_{\varepsilon,\tau}$, we get

$$C(\mathbf{I}_{\varepsilon,\tau}) = \int_{\mathbf{I}_{\varepsilon,\tau}} c(\mathbf{k})\,\mathrm{d}\mathbf{k} = \int_{\mathbf{I}} c(\mathbf{l} + \varepsilon \cdot \tau(\mathbf{l})) \cdot |\det D\varphi_{\varepsilon,\tau}|\,\mathrm{d}\mathbf{l}. \tag{3.42}$$

Taylor expansion of $c(\mathbf{l} + \varepsilon \cdot \tau(\mathbf{l}))$ in $\varepsilon = 0$ provides

$$c(\mathbf{l} + \varepsilon \cdot \tau(\mathbf{l})) = c(\mathbf{l}) + \varepsilon \cdot \nabla c(\mathbf{l}) \cdot \tau(\mathbf{l}) + O(\varepsilon^2), \tag{3.43}$$

where $\nabla c(\mathbf{l}) \cdot \tau(\mathbf{l})$ denotes the scalar product. Furthermore, a straightforward calculation shows

$$|\det D\varphi_{\varepsilon,\tau}| = 1 + \varepsilon \cdot \mathrm{div}\,\tau + O(\varepsilon^2). \tag{3.44}$$

Thus, since $\mathbf{I} \subset \mathbf{I}^{(\mathrm{max})}$ with $\mathbf{I}^{(\mathrm{max})}$ bounded, Gauss's theorem leads to

$$C(\mathbf{I}_{\varepsilon,\tau}) = C(\mathbf{I}) + \varepsilon \cdot \int_{\partial \mathbf{I}} c(\mathbf{l}) \cdot \tau(\mathbf{l})\,\mathrm{d}\vec{S} + O(\varepsilon^2). \tag{3.45}$$

Consequently, for the derivative with respect to ε, we get

$$\left.\frac{\partial C(\mathbf{I}_{\varepsilon,\tau})}{\partial \varepsilon}\right|_{\varepsilon=0} = \lim_{\varepsilon \to 0} \frac{C(\mathbf{I}_{\varepsilon,\tau}) - C(\mathbf{I})}{\varepsilon} = \int_{\partial \mathbf{I}} c(\mathbf{l}) \cdot \tau(\mathbf{l})\,\mathrm{d}\vec{S}. \tag{3.46}$$

Similar arguments hold for the global benefit $B(\mathbf{I})$ and result in

$$\left.\frac{\partial B(\mathbf{I}_{\varepsilon,\tau})}{\partial \varepsilon}\right|_{\varepsilon=0} = \lim_{\varepsilon \to 0} \frac{B(\mathbf{I}_{\varepsilon,\tau}) - B(\mathbf{I})}{\varepsilon} = \int_{\partial \mathbf{I}} b(\mathbf{l}) \cdot \tau(\mathbf{l})\,\mathrm{d}\vec{S}. \tag{3.47}$$

Now, starting from the *optimal* grid $\mathbf{I}^{(\mathrm{opt})}$, Lagrange's principle for the optimization under a constraint can be applied, and we get

$$\lambda \cdot \int_{\partial \mathbf{I}^{(\mathrm{opt})}} c(\mathbf{l}) \cdot \tau(\mathbf{l})\,\mathrm{d}\vec{S} = \int_{\partial \mathbf{I}^{(\mathrm{opt})}} b(\mathbf{l}) \cdot \tau(\mathbf{l})\,\mathrm{d}\vec{S}. \tag{3.48}$$

Since τ vanishes on the boundary of $\mathbb{R}_+^d$, i.e., $\tau(\mathbf{l}) = 0$ when any component of $\mathbf{l}$ vanishes, (3.48) is equivalent to

$$\lambda \cdot \int_{\partial \mathbf{I}^{(\mathrm{opt})} \setminus \partial \mathbb{R}_+^d} c(\mathbf{l}) \cdot \tau(\mathbf{l}) \, \mathrm{d}\vec{S} = \int_{\partial \mathbf{I}^{(\mathrm{opt})} \setminus \partial \mathbb{R}_+^d} b(\mathbf{l}) \cdot \tau(\mathbf{l}) \, \mathrm{d}\vec{S}. \tag{3.49}$$

Finally, since (3.49) is valid for all appropriate smooth disturbances τ,

$$\lambda \cdot c(\mathbf{l}) = b(\mathbf{l}) \tag{3.50}$$

holds for all $\mathbf{l} \in \partial \mathbf{I}^{(\mathrm{opt})} \setminus \partial \mathbb{R}_+^d$.

This is a quite interesting result, because (3.50) says that the ratio of the local benefit $b(\mathbf{l})$ to the local cost $c(\mathbf{l})$ is constant on the boundary $\partial \mathbf{I}^{(\mathrm{opt})} \setminus \partial \mathbb{R}_+^d$ of any grid $\mathbf{I}^{(\mathrm{opt})}$ that is optimal in our sense. This means that the *global* optimization process (3.33) or (3.39), respectively, in which we look for an optimal grid can be reduced to studying the *local* cost-benefit ratios $b(\mathbf{l})/c(\mathbf{l})$ of the subspaces associated with $\mathbf{l}$. Therefore, if we come back to real hierarchical subspaces $W_{\mathbf{l}}$ and to indices $\mathbf{l} \in \mathbb{N}^d$, all one has to do is to identify sets of subspaces $W_{\mathbf{l}}$ with constant cost-benefit ratio in the subspace scheme of Figure 3.3. The grid $\mathbf{I}^{(\mathrm{opt})}$, then, contains the region where the cost-benefit ratio is bigger than or equal to the constant value on the boundary $\partial \mathbf{I}^{(\mathrm{opt})} \setminus \partial \mathbb{R}_+^d$.

Discrete optimization

Since the above continuous optimization process with its roundabout way of generalizing integer multi-indices to real ones is a bit unnatural, (3.33) is now formulated as a discrete optimization problem.

First of all, we redefine the local functions $c(\mathbf{l})$ and $b(\mathbf{l})$, now for multi-indices $\mathbf{l} \in \mathbb{N}^d$ only. According to (3.20), the local cost $c(\mathbf{l})$ is defined by

$$c(\mathbf{l}) := |W_{\mathbf{l}}| = 2^{|\mathbf{l}-\mathbf{1}|_1}, \tag{3.51}$$

which is exactly the same as (3.37) restricted to $\mathbf{l} \in \mathbb{N}^d$. Obviously, $c(\mathbf{l}) \in \mathbb{N}$ holds for all $\mathbf{l} \in \mathbb{N}^d$. Concerning the local benefit function, we define

$$b(\mathbf{l}) := \gamma \cdot \beta(\mathbf{l}), \tag{3.52}$$

where $\beta(\mathbf{l})$ is an upper bound for $\|u_{\mathbf{l}}\|^2$ according to (3.27), and γ is a factor depending on the problem's dimensionality d and on the smoothness of the data, i.e., of u, but constant with respect to $\mathbf{l}$, such that $b(\mathbf{l}) \in \mathbb{N}$. The respective bounds in (3.27) show that such a choice of γ is possible for each of the three norms that are of interest in our context. Note that, as in the continuous case, we do not make any decision concerning the actual choice of norm to be used for $b(\mathbf{l})$ for the moment.

Again, the search for an optimal grid $\mathbf{I} \subset \mathbb{N}^d$ can be restricted to all $\mathbf{I} \subset \mathbf{I}^{(\mathrm{max})} := \{1, \ldots, N\}^d$ for a sufficiently large N without loss of generality.

Next, the global cost benefit functions are redefined as well. For $C(\mathbf{I})$, we define

$$C(\mathbf{I}) := \sum_{\mathbf{l} \in \mathbf{I}} c(\mathbf{l}) = \sum_{\mathbf{l} \in \mathbf{I}^{(\max)}} x(\mathbf{l}) \cdot c(\mathbf{l}), \qquad (3.53)$$

where

$$x(\mathbf{l}) := \begin{cases} 0, & \mathbf{l} \notin \mathbf{I}, \\ 1, & \mathbf{l} \in \mathbf{I}. \end{cases} \qquad (3.54)$$

The interpolant to u on a grid $\mathbf{I}$ provides the global benefit $B(\mathbf{I})$:

$$\left\| u - \sum_{\mathbf{l} \in \mathbf{I}} u_{\mathbf{l}} \right\|^2 \approx \left\| \sum_{\mathbf{l} \in \mathbf{I}^{(\max)}} u_{\mathbf{l}} - \sum_{\mathbf{l} \in \mathbf{I}} u_{\mathbf{l}} \right\|^2 \qquad (3.55)$$

$$\leq \sum_{\mathbf{l} \in \mathbf{I}^{(\max)} \setminus \mathbf{I}} \| u_{\mathbf{l}} \|^2$$

$$\leq \sum_{\mathbf{l} \in \mathbf{I}^{(\max)}} (1 - x(\mathbf{l})) \cdot \gamma \cdot \beta(\mathbf{l})$$

$$= \sum_{\mathbf{l} \in \mathbf{I}^{(\max)}} \gamma \cdot \beta(\mathbf{l}) - \sum_{\mathbf{l} \in \mathbf{I}^{(\max)}} x(\mathbf{l}) \cdot \gamma \cdot \beta(\mathbf{l})$$

$$=: \sum_{\mathbf{l} \in \mathbf{I}^{(\max)}} \gamma \cdot \beta(\mathbf{l}) - B(\mathbf{I}).$$

Of course, (3.55) gives only an upper bound for an approximation to the (squared) interpolation error, because it does not take into account all $\mathbf{l} \notin \mathbf{I}^{(\max)}$. However, since N and, consequently, $\mathbf{I}^{(\max)}$ can be chosen to be as big as appropriate, this is not a serious restriction. Altogether, we get the following reformulation of (3.33):

$$\max_{\mathbf{I} \subset \mathbf{I}^{(\max)}} \sum_{\mathbf{l} \in \mathbf{I}^{(\max)}} x(\mathbf{l}) \cdot \gamma \cdot \beta(\mathbf{l}) \quad \text{with} \quad \sum_{\mathbf{l} \in \mathbf{I}^{(\max)}} x(\mathbf{l}) \cdot c(\mathbf{l}) = w. \qquad (3.56)$$

If we arrange the $\mathbf{l} \in \mathbf{I}^{(\max)}$ in some linear order (e.g., a lexicographical one) with local cost c_i and benefit b_i, $i = 1, \ldots, N^d =: M$, (3.56) reads as

$$\max_{\mathbf{x}} \mathbf{b}^T \mathbf{x} \quad \text{with} \quad \mathbf{c}^T \mathbf{x} = w, \qquad (3.57)$$

where $\mathbf{b} \in \mathbb{N}^M$, $\mathbf{c} \in \mathbb{N}^M$, $\mathbf{x} \in \{0,1\}^M$, and, without loss of generality, $w \in \mathbb{N}$. In combinatorial optimization, a problem like (3.57) is called a *binary knapsack problem* (Martello and Toth 1990), which is known to be an NP-hard one. However, a slight change makes things much easier. If *rational* solutions, i.e., $\mathbf{x} \in \left([0,1] \cap \mathbb{Q} \right)^M$, are allowed too, there exists a very simple algorithm that provides an optimal solution vector $\mathbf{x} \in ([0,1] \cap \mathbb{Q})^M$:

(1) rearrange the order such that $\frac{b_1}{c_1} \geq \frac{b_2}{c_2} \cdots \geq \frac{b_M}{c_M}$;
(2) let $r := \max\{j : \sum_{i=1}^{j} c_i \leq w\}$;
(3) $x_1 := \cdots := x_r := 1$,
$x_{r+1} := (w - \sum_{i=1}^{r} c_i)/c_{r+1}$,
$x_{r+2} := \cdots := x_M := 0$.

Although there is only one potential non-binary coefficient x_{r+1}, the rational solution vector $\mathbf{x}$, generally, has nothing to do with its binary counterpart. But fortunately our knapsack is of variable size, since the global work count w is an arbitrarily chosen natural number. Therefore, it is possible to force the solution of the *rational* problem to be a *binary* one which is, of course, also a solution of the corresponding *binary* problem. Consequently, as in the continuous case before, the *global* optimization problem (3.33) or (3.57), respectively, can be reduced to the discussion of the *local* cost–benefit ratios b_i/c_i or $b(\mathbf{l})/c(\mathbf{l})$ of the underlying subspaces $W_\mathbf{l}$. Those subspaces with the best cost–benefit ratios are taken into account first, and the smaller these ratios become, the more negligible the underlying subspaces turn out to be. This is in the same spirit as n-term approximation (DeVore 1998).

L_2-based sparse grids
Owing to (3.27), the L_2- and L_∞-norm of $W_\mathbf{l}$'s contribution $u_\mathbf{l}$ to the hierarchical representation (3.19) of $u \in X_0^{q,2}(\bar{\Omega})$ are of the same order of magnitude. Therefore there are no differences in the character of the cost–benefit ratio, and the same optimal grids $\mathbf{I}^{(\mathrm{opt})}$ will result from the optimization process described above. According to (3.20) and (3.27), we define

$$\mathrm{cbr}_\infty(\mathbf{l}) := \frac{b_\infty(\mathbf{l})}{c(\mathbf{l})} := \frac{2^{-4 \cdot |\mathbf{l}|_1} \cdot |u|^2_{2,\infty}}{4^d \cdot 2^{|\mathbf{l}-1|_1}} = \frac{1}{2^d} \cdot 2^{-5 \cdot |\mathbf{l}|_1} \cdot |u|^2_{2,\infty}, \quad (3.58)$$

$$\mathrm{cbr}_2(\mathbf{l}) := \frac{b_2(\mathbf{l})}{c(\mathbf{l})} := \frac{2^{-4 \cdot |\mathbf{l}|_1} \cdot |u|^2_{2,2}}{9^d \cdot 2^{|\mathbf{l}-1|_1}} = \left(\frac{2}{9}\right)^d \cdot 2^{-5 \cdot |\mathbf{l}|_1} \cdot |u|^2_{2,2}$$

as the local cost–benefit ratios. Note that we use bounds for the squared norms of $u_\mathbf{l}$ for reasons of simplicity, but without loss of generality An optimal grid $\mathbf{I}^{(\mathrm{opt})}$ will consist of all multi-indices $\mathbf{l}$ or their corresponding subspaces $W_\mathbf{l}$ where $\mathrm{cbr}_\infty(\mathbf{l})$ or $\mathrm{cbr}_2(\mathbf{l})$ is bigger than some prescribed threshold $\sigma_\infty(n)$ or $\sigma_2(n)$, respectively. We choose those thresholds to be of the order of $\mathrm{cbr}_\infty(\bar{\mathbf{l}})$ or $\mathrm{cbr}_2(\bar{\mathbf{l}})$ with $\bar{\mathbf{l}} := (n, 1, \ldots, 1)$:

$$\sigma_\infty(n) := \mathrm{cbr}_\infty(\bar{\mathbf{l}}) = \frac{1}{2^d} \cdot 2^{-5 \cdot (n+d-1)} \cdot |u|^2_{2,\infty}, \quad (3.59)$$

$$\sigma_2(n) := \mathrm{cbr}_2(\bar{\mathbf{l}}) = \left(\frac{2}{9}\right)^d \cdot 2^{-5 \cdot (n+d-1)} \cdot |u|^2_{2,2}.$$

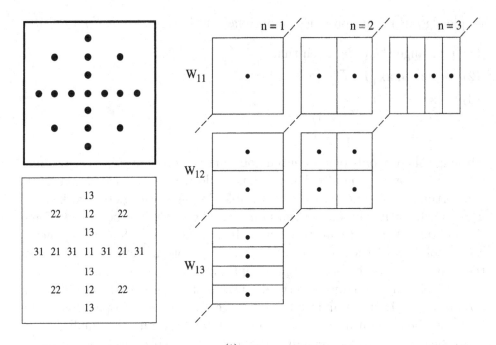

Figure 3.5. The sparse grid of $V_3^{(1)}$, $d = 2$, and the assignment of grid points to subspaces.

That is, we fix d subspaces on the axes in the subspace scheme of Figure 3.3 and search for all $W_\mathbf{l}$ whose cost–benefit ratio is equal or better. Thus, applying the criterion $\mathrm{cbr}_\infty(\mathbf{1}) \geq \sigma_\infty(n)$ or $\mathrm{cbr}_2(\mathbf{1}) \geq \sigma_2(n)$, respectively, we get the relation

$$|\mathbf{l}|_1 \leq n + d - 1 \qquad (3.60)$$

that qualifies a subspace $W_\mathbf{l}$ to be taken into account. This result leads us to the definition of a new discrete approximation space $V_n^{(1)}$,

$$V_n^{(1)} := \bigoplus_{|\mathbf{l}|_1 \leq n+d-1} W_\mathbf{l}, \qquad (3.61)$$

which is L_∞- and L_2-optimal with respect to our cost–benefit setting. The grids that correspond to the spaces $V_n^{(1)}$ are just the standard *sparse grids* as were introduced in Zenger (1991), studied in detail in Bungartz (1992b), and discussed in a variety of other papers for different applications. In comparison with the standard *full grid* space $V_n^{(\infty)}$, we now have triangular or simplicial sectors of subspaces in the scheme of Figure 3.3: see Figure 3.5. Figure 3.6, finally, gives two examples of sparse grids: a regular 2D and an adaptively refined 3D one.

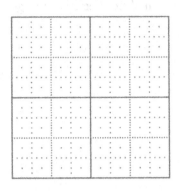 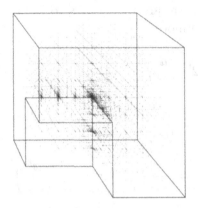

Figure 3.6. Sparse grids: regular example (left) and adaptive one (right).

Now, let us turn to the basic properties of the sparse grid approximation spaces $V_n^{(1)}$.

Lemma 3.6. The dimension of the space $V_n^{(1)}$, i.e., the number of degrees of freedom or inner grid points, is given by

$$|V_n^{(1)}| = \sum_{i=0}^{n-1} 2^i \cdot \binom{d-1+i}{d-1} \qquad (3.62)$$

$$= (-1)^d + 2^n \cdot \sum_{i=0}^{d-1} \binom{n+d-1}{i} \cdot (-2)^{d-1-i}$$

$$= 2^n \cdot \left(\frac{n^{d-1}}{(d-1)!} + O(n^{d-2}) \right).$$

Thus, we have

$$|V_n^{(1)}| = O(h_n^{-1} \cdot |\log_2 h_n|^{d-1}). \qquad (3.63)$$

Proof. With (3.20) and (3.61), we get

$$|V_n^{(1)}| = \left| \bigoplus_{|\mathbf{l}|_1 \leq n+d-1} W_\mathbf{l} \right| = \sum_{|\mathbf{l}|_1 \leq n+d-1} 2^{|\mathbf{l}-\mathbf{1}|_1} = \sum_{i=d}^{n+d-1} 2^{i-d} \cdot \sum_{|\mathbf{l}|_1 = i} 1$$

$$= \sum_{i=d}^{n+d-1} 2^{i-d} \cdot \binom{i-1}{d-1}$$

$$= \sum_{i=0}^{n-1} 2^i \cdot \binom{d-1+i}{d-1},$$

since there are $\binom{i-1}{d-1}$ ways to form the sum i with d nonnegative integers. Furthermore,

$$\sum_{i=0}^{n-1} 2^i \cdot \binom{d-1+i}{d-1}$$

$$= \frac{1}{(d-1)!} \cdot \sum_{i=0}^{n-1} (x^{i+d-1})^{(d-1)}\bigg|_{x=2}$$

$$= \frac{1}{(d-1)!} \cdot \left(x^{d-1} \cdot \frac{1-x^n}{1-x}\right)^{(d-1)}\bigg|_{x=2}$$

$$= \frac{1}{(d-1)!} \cdot \sum_{i=0}^{d-1} \binom{d-1}{i} \cdot (x^{d-1} - x^{n+d-1})^{(i)} \cdot \left(\frac{1}{1-x}\right)^{(d-1-i)}\bigg|_{x=2}$$

$$= (-1)^d + 2^n \cdot \sum_{i=0}^{d-1} \binom{n+d-1}{i} \cdot (-2)^{d-1-i},$$

from which the result concerning the order and the leading coefficient follows immediately. $\square$

The above lemma shows the order $O(2^n \cdot n^{d-1})$ or, with $h_n = 2^{-n}$, $O(h_n^{-1} \cdot |\log_2 h_n|^{d-1})$, which is a significant reduction of the number of degrees of freedom and, thus, of the computational and storage requirement compared with the order $O(h_n^{-d})$ of $V_n^{(\infty)}$.

The other question to be discussed concerns the interpolation accuracy that can be obtained on sparse grids. For that, we look at the interpolation error $u - u_n^{(1)}$ of the sparse grid interpolant $u_n^{(1)} \in V_n^{(1)}$ which, due to (3.19) and (3.61), can be written as

$$u - u_n^{(1)} = \sum_{\mathbf{l}} u_\mathbf{l} - \sum_{|\mathbf{l}|_1 \leq n+d-1} u_\mathbf{l} = \sum_{|\mathbf{l}|_1 > n+d-1} u_\mathbf{l}.$$

Therefore, for any norm $\|\cdot\|$, we have

$$\|u - u_n^{(1)}\| \leq \sum_{|\mathbf{l}|_1 > n+d-1} \|u_\mathbf{l}\|. \tag{3.64}$$

The following lemma provides a prerequisite for the estimates of the interpolation error with respect to the different norms we are interested in. For $d, n \in \mathbb{N}$, we define

$$A(d,n) := \sum_{k=0}^{d-1} \binom{n+d-1}{k} = \frac{n^{d-1}}{(d-1)!} + O(n^{d-2}). \tag{3.65}$$

SPARSE GRIDS

Lemma 3.7. For purposes of summation over all grid points $x_{l,i}$ with corresponding basis functions $\phi_{l,i} \notin V_n^{(1)}$, we obtain for arbitrary $s \in \mathbb{N}$

$$\sum_{|l|_1 > n+d-1} 2^{-s \cdot |l|_1} = 2^{-s \cdot n} \cdot 2^{-s \cdot d} \cdot \sum_{i=0}^{\infty} 2^{-s \cdot i} \cdot \binom{n+i+d-1}{d-1} \quad (3.66)$$

$$\leq 2^{-s \cdot n} \cdot 2^{-s \cdot d} \cdot 2 \cdot A(d, n).$$

Proof. As for the proof of the previous lemma, we get

$$\sum_{|l|_1 > n+d-1} 2^{-s \cdot |l|_1} = \sum_{i=n+d}^{\infty} 2^{-s \cdot i} \cdot \sum_{|l|_1 = i} 1$$

$$= \sum_{i=n+d}^{\infty} 2^{-s \cdot i} \cdot \binom{i-1}{d-1}$$

$$= 2^{-s \cdot n} \cdot 2^{-s \cdot d} \cdot \sum_{i=0}^{\infty} 2^{-s \cdot i} \cdot \binom{n+i+d-1}{d-1}.$$

Since

$$\sum_{i=0}^{\infty} x^i \cdot \binom{n+i+d-1}{d-1} = \frac{x^{-n}}{(d-1)!} \cdot \left(\sum_{i=0}^{\infty} x^{n+i+d-1} \right)^{(d-1)} \quad (3.67)$$

$$= \frac{x^{-n}}{(d-1)!} \cdot \left(x^{n+d-1} \cdot \frac{1}{1-x} \right)^{(d-1)}$$

$$= \frac{x^{-n}}{(d-1)!} \cdot \sum_{k=0}^{d-1} \binom{d-1}{k} \cdot (x^{n+d-1})^{(k)} \cdot \left(\frac{1}{1-x} \right)^{(d-1-k)}$$

$$= \sum_{k=0}^{d-1} \binom{n+d-1}{k} \cdot \left(\frac{x}{1-x} \right)^{d-1-k} \cdot \frac{1}{1-x},$$

we get with $x := 2^{-s}$:

$$\sum_{i=0}^{\infty} 2^{-s \cdot i} \cdot \binom{n+i+d-1}{d-1} \leq 2 \cdot \sum_{k=0}^{d-1} \binom{n+d-1}{k} = 2 \cdot A(d, n). \quad \square$$

With the above lemma, we obtain the desired result concerning the interpolation quality of standard sparse grid spaces $V_n^{(1)}$.

Theorem 3.8. For the L_∞-, the L_2-, and the energy norm, we have the following upper bounds for the interpolation error of a function $u \in X_0^{q,2}(\bar{\Omega})$ in the sparse grid space $V_n^{(1)}$:

$$\|u - u_n^{(1)}\|_\infty \leq \frac{2 \cdot |u|_{2,\infty}}{8^d} \cdot 2^{-2n} \cdot A(d,n) = O(h_n^2 \cdot n^{d-1}), \qquad (3.68)$$

$$\|u - u_n^{(1)}\|_2 \leq \frac{2 \cdot |u|_{2,2}}{12^d} \cdot 2^{-2n} \cdot A(d,n) = O(h_n^2 \cdot n^{d-1}),$$

$$\|u - u_n^{(1)}\|_E \leq \frac{d \cdot |u|_{2,\infty}}{2 \cdot 3^{(d-1)/2} \cdot 4^{d-1}} \cdot 2^{-n} = O(h_n),$$

$$\|u - u_n^{(1)}\|_E \leq \frac{d \cdot |u|_{2,2}}{\sqrt{3} \cdot 6^{d-1}} \cdot 2^{-n} = O(h_n).$$

Proof. With (3.27), (3.64), and (3.66) for $s = 2$, we get

$$\|u - u_n^{(1)}\|_\infty \leq \sum_{|\mathbf{l}|_1 > n+d-1} \|u_\mathbf{l}\|_\infty \leq \frac{|u|_{2,\infty}}{2^d} \cdot \sum_{|\mathbf{l}|_1 > n+d-1} 2^{-2 \cdot |\mathbf{l}|_1}$$

$$\leq \frac{2 \cdot |u|_{2,\infty}}{8^d} \cdot 2^{-2n} \cdot A(d,n)$$

and, analogously, the corresponding result for the L_2-norm. Concerning the first bound with respect to the energy norm, we have

$$\|u - u_n^{(1)}\|_E \leq \sum_{|\mathbf{l}|_1 > n+d-1} \|u_\mathbf{l}\|_E$$

$$\leq \frac{|u|_{2,\infty}}{2 \cdot 12^{(d-1)/2}} \cdot \sum_{|\mathbf{l}|_1 > n+d-1} 4^{-|\mathbf{l}|_1} \cdot \left(\sum_{j=1}^d 4^{l_j}\right)^{1/2}$$

$$= \frac{|u|_{2,\infty}}{2 \cdot 12^{(d-1)/2}} \cdot \sum_{i=n+d}^\infty 4^{-i} \cdot \sum_{|\mathbf{l}|_1 = i} \left(\sum_{j=1}^d 4^{l_j}\right)^{1/2}$$

$$\leq \frac{|u|_{2,\infty}}{2 \cdot 12^{(d-1)/2}} \cdot \sum_{i=n+d}^\infty d \cdot 2^{-i}$$

$$= \frac{d \cdot |u|_{2,\infty}}{2 \cdot 3^{(d-1)/2} \cdot 4^{d-1}} \cdot 2^{-n},$$

because

$$\sum_{|\mathbf{l}|_1 = i} \left(\sum_{j=1}^d 4^{l_j}\right)^{1/2} \leq d \cdot 2^i,$$

which can be shown by complete induction with respect to d. The last estimate can be obtained with analogous arguments. □

This theorem shows the crucial improvement of the sparse grid space $V_n^{(1)}$ in comparison with $V_n^{(\infty)}$. The number of degrees of freedom is reduced significantly, whereas the accuracy is only slightly deteriorated – for the L_∞- and the L_2-norm – or even stays of the same order if the error is measured in the energy norm. This lessens the curse of dimensionality, but it does not overcome it completely. Since this result is optimal with respect to both the L_∞- and the L_2-norm, a further improvement can only be expected if we change the setting. Therefore, in the following, we study the optimization process with respect to the energy norm.

Energy-based sparse grids

Now, we base our cost–benefit approach on the energy norm. According to (3.20) and (3.27), we define

$$\mathrm{cbr}_E(\mathbf{l}) := \frac{b_E(\mathbf{l})}{c(\mathbf{l})} := \frac{2^{-4\cdot|\mathbf{l}|_1} \cdot |u|_{2,\infty}^2}{4 \cdot 12^{d-1} \cdot 2^{|\mathbf{l}-\mathbf{1}|_1}} \cdot \sum_{j=1}^{d} 4^{l_j} \qquad (3.69)$$

$$= \frac{3}{6^d} \cdot 2^{-5\cdot|\mathbf{l}|_1} \cdot \sum_{j=1}^{d} 4^{l_j} \cdot |u|_{2,\infty}^2$$

as the local cost–benefit ratio. Again, instead of $\|u_\mathbf{l}\|_E$ itself, only an upper bound for the squared energy norm of $u_\mathbf{l}$ is used. The resulting optimal grid $\mathbf{I}^{(\mathrm{opt})}$ will consist of all those multi-indices $\mathbf{l}$ or their respective hierarchical subspaces $W_\mathbf{l}$ that fulfil $\mathrm{cbr}_E(\mathbf{l}) \geq \sigma_E(n)$ for some given constant threshold $\sigma_E(n)$. As before, $\sigma_E(n)$ is defined via the cost–benefit ratio of $W_{\bar{\mathbf{1}}}$ with $\bar{\mathbf{1}} := (n, 1, \ldots, 1)$:

$$\sigma_E(n) := \mathrm{cbr}_E(\bar{\mathbf{1}}) = \frac{3}{6^d} \cdot 2^{-5\cdot(n+d-1)} \cdot \left(4^n + 4\cdot(d-1)\right) \cdot |u|_{2,\infty}^2. \qquad (3.70)$$

Thus, applying the criterion $\mathrm{cbr}_E(\mathbf{l}) \geq \sigma_E(n)$, we come to an alternative sparse grid approximation space $V_n^{(E)}$, which is based on the energy norm:

$$V_n^{(E)} := \bigoplus_{|\mathbf{l}|_1 - \frac{1}{5}\cdot\log_2\left(\sum_{j=1}^d 4^{l_j}\right) \leq (n+d-1) - \frac{1}{5}\cdot\log_2\left(4^n + 4d - 4\right)} W_\mathbf{l}. \qquad (3.71)$$

First, we look at the number of grid points of the underlying sparse grids.

Lemma 3.9. *The energy-based sparse grid space $V_n^{(E)}$ is a subspace of $V_n^{(1)}$, and its dimension fulfils*

$$|V_n^{(E)}| \leq 2^n \cdot \frac{d}{2} \cdot e^d = O(h_n^{-1}). \qquad (3.72)$$

Proof. For subspaces W_l with $|l|_1 = n + d - 1 + i$, $i \in \mathbb{N}$, we have

$$|l|_1 - \frac{1}{5} \cdot \log_2\left(\sum_{j=1}^{d} 4^{l_j}\right) \geq n + d - 1 + i - \frac{1}{5} \cdot \log_2(4^{n+i} + 4d - 4)$$

$$\geq n + d - 1 + i - \frac{1}{5} \cdot \log_2(4^i(4^n + 4d - 4))$$

$$> n + d - 1 - \frac{1}{5} \cdot \log_2(4^n + 4d - 4).$$

Therefore, no W_l with $|l|_1 > n + d - 1$ can belong to $V_n^{(E)}$. Consequently, $V_n^{(E)}$ is a subspace of $V_n^{(1)}$ and $|V_n^{(E)}| \leq |V_n^{(1)}|$ for all $n \in \mathbb{N}$. Starting from that, (3.20) provides

$$|V_n^{(E)}| = \sum_{i=0}^{n-1} \sum_{|l|_1 = n+d-1-i, \ \sum_{j=1}^{d} 4^{l_j} \geq \frac{4^n + 4d - 4}{32^i}} |W_l|$$

$$= 2^n \cdot \frac{1}{2} \cdot \sum_{i=0}^{n-1} 2^{-i} \cdot \sum_{|l|_1 = n+d-1-i, \ \sum_{j=1}^{d} 4^{l_j} \geq \frac{4^n + 4d - 4}{32^i}} 1$$

$$\leq 2^n \cdot \frac{1}{2} \cdot \lim_{n \to \infty} \sum_{i=0}^{n-1} 2^{-i} \cdot \sum_{|l|_1 = n+d-1-i, \ \sum_{j=1}^{d} 4^{l_j} \geq \frac{4^n + 4d - 4}{32^i}} 1$$

$$= 2^n \cdot \frac{1}{2} \cdot \lim_{n \to \infty} \sum_{i=0}^{n-1} 2^{-i} \cdot d \cdot \binom{d - 1 - \lfloor 1.5i \rfloor}{d - 1},$$

since it can be shown that, for $n \to \infty$, our energy-based sparse grid and the grid resulting from the second condition $|l|_\infty \geq n - \lfloor 2.5i \rfloor$ for the inner

Table 3.1. Dimension of $V_n^{(\infty)}$, $V_n^{(1)}$, and $V_n^{(E)}$ for different values of d and n.

	$d = 2$		$d = 3$		$d = 4$	
	$n = 10$	$n = 20$	$n = 10$	$n = 20$	$n = 10$	$n = 20$
$V_n^{(\infty)}$	$1.05 \cdot 10^6$	$1.10 \cdot 10^{12}$	$1.07 \cdot 10^9$	$1.15 \cdot 10^{18}$	$1.10 \cdot 10^{12}$	$1.21 \cdot 10^{24}$
$V_n^{(1)}$	9217	$1.99 \cdot 10^7$	47103	$2.00 \cdot 10^8$	$1.78 \cdot 10^5$	$1.41 \cdot 10^9$
$V_n^{(E)}$	3841	$4.72 \cdot 10^6$	10495	$1.68 \cdot 10^7$	24321	$5.27 \cdot 10^7$

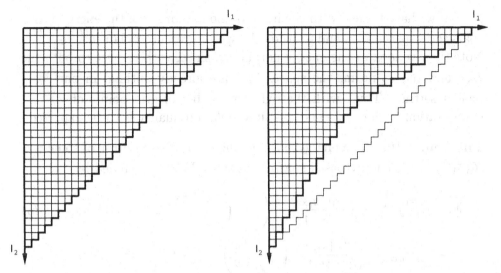

Figure 3.7. Scheme of subspaces for $V_{30}^{(1)}$ (left) and $V_{30}^{(E)}$ (right), $d = 2$.

sum instead of
$$\sum_{j=1}^{d} 4^{l_j} \geq \frac{4^n + 4d - 4}{32^i}$$
are the same, and since there exist
$$\binom{d - 1 + \lfloor 1.5i \rfloor}{d - 1}$$
such subspaces W_l with $|l|_\infty = l_1$. Consequently, we obtain
$$|V_n^{(E)}| \leq 2^n \cdot \frac{d}{2} \cdot \sum_{i=0}^{\infty} 2^{-\frac{2}{3}i} \cdot \binom{d - 1 + i}{d - 1}$$
$$= 2^n \cdot \frac{d}{2} \cdot \left(1 - 2^{-\frac{2}{3}}\right)^{-d}$$
$$\leq 2^n \cdot \frac{d}{2} \cdot e^d,$$
since $\sum_{i=0}^{\infty} x^i \cdot \binom{k+i}{k} = (1-x)^{-k-1}$ for $k \in \mathbb{N}_0$ and $0 < x < 1$. □

Table 3.1 compares the dimensions of the standard full grid approximation space $V_n^{(\infty)}$ with both sparse grid spaces $V_n^{(1)}$ and $V_n^{(E)}$ for different dimensionalities $d \in \{2, 3, 4\}$ and for the two resolutions $n = 10$ and $n = 20$. The sparse grid effect is already obvious for $V_n^{(1)}$. Especially for larger d, the advantages of $V_n^{(E)}$ become evident. For a comparison of the underlying subspace schemes of $V_n^{(1)}$ and $V_n^{(E)}$ in 2D, see Figure 3.7.

Next we have to deal with the interpolation accuracy of the energy-based sparse grid spaces $V_n^{(E)}$ and to study the sparse grid interpolant $u_n^{(E)} \in V_n^{(E)}$. Note that we look at the energy norm only, since, with respect to both the L_∞- and the L_2-norm, the spaces $V_n^{(1)}$ are already optimal in our cost–benefit setting. Thus, with the reduced number of grid points of $V_n^{(E)}$, a deterioration of the (L_∞- and L_2-) interpolation quality is to be expected.

Theorem 3.10. The energy norm of the interpolation error of some $u \in X_0^{q,2}(\bar\Omega)$ in the energy-based sparse grid space $V_n^{(E)}$ is bounded by

$$\|u - u_n^{(E)}\|_E \leq \frac{d \cdot |u|_{2,\infty}}{3^{(d-1)/2} \cdot 4^{d-1}} \cdot \left(\frac{1}{2} + \left(\frac{5}{2}\right)^{d-1}\right) \cdot 2^{-n} = O(h_n), \quad (3.73)$$

$$\|u - u_n^{(E)}\|_E \leq \frac{2 \cdot d \cdot |u|_{2,2}}{\sqrt{3} \cdot 6^{d-1}} \cdot \left(\frac{1}{2} + \left(\frac{5}{2}\right)^{d-1}\right) \cdot 2^{-n} = O(h_n).$$

Proof. First, since

$$\|u - u_n^{(E)}\|_E \leq \|u - u_n^{(1)}\|_E + \|u_n^{(1)} - u_n^{(E)}\|_E,$$

and since we already know that $\|u - u_n^{(1)}\|_E$ is of the order $O(h_n)$, we can restrict ourselves to $\|u_n^{(1)} - u_n^{(E)}\|_E$. For that, it can be shown that, for $i \in \mathbb{N}_0$, each $W_\mathbf{l}$ with $|\mathbf{l}|_1 = n + d - 1 - i$ and $|\mathbf{l}|_\infty \geq n - 2.5i$ is a subspace of $V_n^{(E)}$. Therefore, we obtain with (3.27)

$$\|u_n^{(1)} - u_n^{(E)}\|_E$$

$$\leq \sum_{W_\mathbf{l} \subseteq V_n^{(1)} \ominus V_n^{(E)}} \|u_\mathbf{l}\|_E$$

$$\leq \sum_{i=0}^{i^*} \sum_{|\mathbf{l}|_1 = n+d-1-i,\ |\mathbf{l}|_\infty < n-2.5i} \|u_\mathbf{l}\|_E$$

$$\leq \frac{|u|_\infty}{2 \cdot 12^{(d-1)/2}} \cdot \sum_{i=0}^{i^*} \sum_{|\mathbf{l}|_1=n+d-1-i,\ |\mathbf{l}|_\infty<n-2.5i} 4^{-|\mathbf{l}|_1} \cdot \left(\sum_{j=1}^d 4^{l_j}\right)^{1/2}$$

$$\leq \frac{|u|_\infty}{2 \cdot 12^{(d-1)/2}} \cdot 4^{-n-d+1} \cdot \sum_{i=0}^{i^*} 4^i \cdot \sum_{|\mathbf{l}|_1=n+d-1-i,\ |\mathbf{l}|_\infty<n-2.5i} \left(\sum_{j=1}^d 2^{l_j}\right)$$

$$\leq \frac{|u|_\infty}{2 \cdot 12^{(d-1)/2}} \cdot 4^{-n-d+1} \cdot \sum_{i=0}^{i^*} 4^i \cdot \sum_{j=1}^{n-1-\lfloor 2.5i \rfloor} d \cdot \binom{n+d-2-i-j}{d-2} \cdot 2^j$$

$$= \frac{|u|_\infty}{2 \cdot 12^{(d-1)/2}} 4^{-n-d+1} \sum_{i=0}^{i^*} 4^i \sum_{k=1}^{n-1-\lfloor 2.5i \rfloor} d \binom{d-2+\lfloor 1.5i \rfloor + k}{d-2} 2^{n-\lfloor 2.5i \rfloor - k}$$

$$= \frac{d \cdot |u|_\infty}{2 \cdot 12^{(d-1)/2}} 4^{-(d-1)} 2^{-n} \sum_{i=0}^{i^*} 2^{-\lfloor \frac{i}{2} \rfloor} \sum_{k=1}^{n-1-\lfloor 2.5i \rfloor} \binom{d-2+\lfloor 1.5i \rfloor + k}{d-2} 2^{-k}$$

$$\leq \frac{d \cdot |u|_\infty}{2 \cdot 12^{(d-1)/2}} \cdot 4^{-(d-1)} \cdot 2^{-n} \cdot 2 \cdot 5^{d-1}$$

$$= \frac{d \cdot |u|_\infty}{3^{(d-1)/2} \cdot 4^{d-1}} \cdot \left(\frac{5}{2}\right)^{d-1} \cdot 2^{-n},$$

where $0 \leq i^* \leq n-1$ is the maximum value of i for which the set of indices **l** with $|\mathbf{l}|_1 = n+d-1-i$ and $|\mathbf{l}|_\infty < n-2.5i$ is not empty. Together with (3.68), we get the first result and, in a completely analogous way, the second one, too. □

Though we have only derived upper bounds for the energy norm of the interpolation error, it is helpful to compare the respective results (3.32), (3.68) and (3.73) for the three approximation spaces $V_n^{(\infty)}$, $V_n^{(1)}$ and $V_n^{(E)}$. Table 3.2 shows that there is no asymptotic growth with respect to d, either for the full grid case or for our two sparse grid spaces.

The crucial result of this section is that, with the energy-based sparse grid spaces $V_n^{(E)}$, the curse of dimensionality can be overcome. In both (3.72) and (3.73), the n-dependent terms are free of any d-dependencies. There is an order of $O(2^n)$ for the dimension and $O(2^{-n})$ for the interpolation error. Especially, there is no longer any polynomial term in n like n^{d-1}. That is, apart from the factors that are constant with respect to n, there is no d-dependence of both $|V_n^{(E)}|$ and $\|u - u_n^{(E)}\|_E$, and thus no deterioration in complexity for higher-dimensional problems. Furthermore, the growth of the d-dependent terms in d is not too serious, since we have a factor

Table 3.2. d-depending constants in the bounds for $\|u - u_n^{(\cdot)}\|_E$ (multiply with $|u|_{2,\infty} \cdot 2^{-n}$ (first row) or $|u|_{2,2} \cdot 2^{-n}$ (second row) to get the respective bounds).

$V_n^{(\infty)}$	$V_n^{(1)}$	$V_n^{(E)}$
$\dfrac{d^{3/2}}{2 \cdot 3^{(d-1)/2} \cdot 6^{d-1}}$	$\dfrac{d}{2 \cdot 3^{(d-1)/2} \cdot 4^{d-1}}$	$\dfrac{d}{3^{(d-1)/2} \cdot 4^{d-1}} \cdot \left(\dfrac{1}{2} + \left(\dfrac{5}{2}\right)^{d-1}\right)$
$\dfrac{d^{3/2}}{\sqrt{3} \cdot 9^{d-1}}$	$\dfrac{d}{\sqrt{3} \cdot 6^{d-1}}$	$\dfrac{2 \cdot d}{\sqrt{3} \cdot 6^{d-1}} \cdot \left(\dfrac{1}{2} + \left(\dfrac{5}{2}\right)^{d-1}\right)$

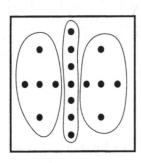 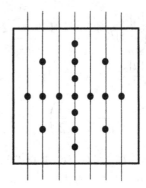

Figure 3.8. Recursive structure of $V_3^{(1)}$ for $d = 2$.

of $\frac{d}{2} \cdot e^d$ in the upper bound of $|V_n^{(E)}|$ and (in the best case; see Table 3.2) $\frac{d}{3(d-1)/2 \cdot 4^{d-1}} \cdot \left(\frac{1}{2} + \left(\frac{5}{2}\right)^{d-1}\right)$ in the upper bound of $\|u - u_n^{(E)}\|_E$.

3.3. Recurrences and complexity

In this section, we make a short digression into recurrence formulas for sparse grids and into sparse grid complexity, in order to learn more about their asymptotic behaviour. We present the most interesting results only and refer the reader to Bungartz (1998) for details or proofs.

Recurrence formulas

For the following, we restrict ourselves to the sparse grid spaces $V_n^{(1)}$. In (3.61), they were introduced with the help of an explicit formula. Now we study their recursive character to obtain further results concerning their complexity and asymptotic properties. First, starting from (3.62), one can show a recurrence relation for $|V_n^{(1)}|$, the number of (inner) grid points of a sparse grid. Note that $|V_n^{(1)}|$ depends on two parameters: the *dimensionality* d and the *resolution* n. Defining

$$a_{n,d} := |V_n^{(1)}|, \quad (3.74)$$

we get

$$a_{n,d} = a_{n,d-1} + 2 \cdot a_{n-1,d}. \quad (3.75)$$

That is, the d-dimensional sparse grid of resolution (or depth) n consists of a $(d-1)$-dimensional one of depth $n-1$ (separator) and of two d-dimensional sparse grids of depth $n-1$ (*cf.* Figure 3.8, left). If we continue with this decomposition in a recursive way, we finally obtain a full-history version of (3.75) with respect to n,

$$a_{n,d} = \sum_{i=0}^{n-1} 2^i \cdot a_{n-i,d-1}, \quad (3.76)$$

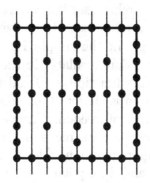

Figure 3.9. Recursive structure of $b_{3,2}$.

since $a_{1,d} = 1$ for all $d \in \mathbb{N}$ due to (3.74). Thus, a sparse grid $V_n^{(1)}$ of dimensionality d and depth n can be completely reduced to sparse grids of dimensionality $d-1$ and depth k, $k = 1, \ldots, n$ (cf. Figure 3.8, right).

In addition to $a_{n,d}$, we shall now deal with sparse grids with grid points on the boundary. To this end, let $b_{n,d}$ be the number of overall grid points of the L_2-based sparse grid of parameters d and n, i.e., in the interior *and* on the boundary of $\bar{\Omega}$. On $\partial\bar{\Omega}$, we assume sparse grids of the same resolution n, but of a reduced dimensionality. Since the boundary of the d-dimensional unit interval $\bar{\Omega}$ consists of $\binom{d}{j} \cdot 2^{d-j}$ j-dimensional unit intervals ($j = 0, \ldots, d$), we get

$$b_{n,d} := \sum_{j=0}^{d} \binom{d}{j} \cdot 2^{d-j} \cdot a_{n,j}, \qquad (3.77)$$

where $a_{n,0} := 1$ for all $n \in \mathbb{N}$. With the help of (3.75) and (3.77), the following recurrence relation for the $b_{n,d}$ can be derived:

$$b_{n,d} = 2 \cdot b_{n-1,d} + 3 \cdot b_{n,d-1} - 4 \cdot b_{n-1,d-1} \qquad (3.78)$$

with its full-history version with respect to n

$$b_{n,d} = 3 \cdot b_{n,d-1} + \sum_{i=1}^{n-1} 2^i \cdot b_{n-i,d-1} \qquad (3.79)$$

$$= 2 \cdot b_{n,d-1} + \sum_{i=0}^{n-1} 2^i \cdot b_{n-i,d-1},$$

where the first term stands for the boundary faces $x_d \in \{0, 1\}$, whereas the sum denotes the inner part of the grid with respect to direction x_d. Figure 3.9 illustrates the recursive structure of $b_{n,d}$ in the 2D case.

Finally, a third quantity $c_{p,d}$ shall be introduced that motivates the sparse grid pattern from a perspective of approximation with polynomials, thus anticipating, to some extent, the higher-order approaches to be discussed later. Starting from our hierarchical setting, we are looking for the minimum number of grid points that are necessary to realize a polynomial approximation of a certain degree p. In the simple 1D case, things are evident. First, with one degree of freedom available in the left point of the boundary ($x = 0$), a constant basis function is placed there, while on the right side ($x = 1$), a linear function is chosen, thus allowing linear interpolation ($p = 1$) in $\bar{\Omega}$ with two degrees of freedom. Afterwards, on each level $l := 1$ of inner grid points, we raise the degree p of the basis functions by one, and consequently get an approximation order of degree $p = l + 1$ for level l. Note that there is no overall interpolant of degree $p > 2$ on $\bar{\Omega}$, but, owing to the hierarchical subspace decomposition, there exists an interpolant, continuous on $\bar{\Omega}$, that is piecewise polynomial of degree p with respect to level $l - 1 = p - 2$.

For $d > 1$, things are a little bit more complicated. We discuss the definition of $c_{p,d}$ for $d = 2$. For constant interpolation, just one degree of freedom is necessary. Linear interpolation, i.e., $p = 1$, can be obtained with three grid points, e.g., with the three degrees of freedom 1, x, and y living in three of the unit square's four corners. This proceeding is completely consistent with the tensor product approach: for $x = 0$, the 1D basis function is constant with respect to x. Thus, we need linear interpolation with respect to y (two degrees of freedom for $x = 0$). On the right-hand side of the unit square ($x = 1$), we are linear with respect to x and thus need only one degree of freedom with respect to y (see Figure 3.10). For a quadratic approximation, six degrees of freedom are necessary, and so on. This economic use of degrees of freedom leads to a certain asymmetry of the respective grid patterns and to a delayed generation of inner grid points: when the first inner grid point appears, the maximum depth on the boundary is already three in the 2D case. Nevertheless, the resulting grids for a given polynomial degree p are very closely related to the standard sparse grids described by $b_{n,d}$, since, obviously, the grid patterns resulting in the interior are just the standard patterns of $V_n^{(1)}$. Hence, here, the maximum degree p of the basis functions used takes the part of the resolution n as the second parameter besides d.

The principles of construction for the grids resulting from this polynomial approach lead us to a recurrence relation for $c_{p,d}$,

$$c_{p,d} = c_{p,d-1} + c_{p-1,d-1} + \sum_{i=0}^{p-2} 2^i \cdot c_{p-2-i,d-1}, \qquad (3.80)$$

from which we get

$$c_{p,d} = 2 \cdot c_{p-1,d} + c_{p,d-1} - c_{p-1,d-1} - c_{p-2,d-1}. \qquad (3.81)$$

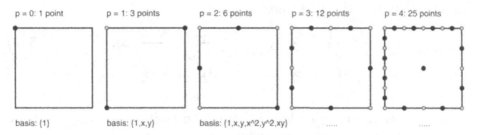

Figure 3.10. Minimum number of grid points for piecewise polynomial degree $p = 3$ in the hierarchical setting ($d = 2$) and corresponding local bases.

In the following, we present some properties of the $a_{n,d}$, $b_{n,d}$, and $c_{p,d}$. Though we do not want to go into detail here, note that, in contrast to many recurrences studied for the analysis of algorithms (*cf.* Graham, Knuth and Patashnik (1994), Sedgewick and Flajolet (1996), for example), these quantities, and thus the overall storage requirement and computational cost connected with sparse grids, depend on two parameters. Let $d \in \mathbb{N}$, $n \in \mathbb{N}$, and $p \in \mathbb{N}_0$. The initial conditions

$$a_{1,d} = 1 \quad \forall d \in \mathbb{N}, \qquad a_{n,1} = 2^n - 1 \quad \forall n \in \mathbb{N}, \qquad (3.82)$$
$$b_{1,d} = 3^d \quad \forall d \in \mathbb{N}, \qquad b_{n,1} = 2^n + 1 \quad \forall n \in \mathbb{N},$$
$$c_{0,d} = 1 \quad \forall d \in \mathbb{N}, \qquad c_{p,1} = 2^{p-1} + 1 \quad \forall p \in \mathbb{N},$$
$$c_{1,d} = d + 1 \quad \forall d \in \mathbb{N}$$

follow immediately from the semantics of $a_{n,d}$, $b_{n,d}$, and $c_{p,d}$. Owing to (3.76), (3.79), (3.80), and the initial conditions (3.82), all $a_{n,d}$, $b_{n,d}$, and $c_{p,d}$ are natural numbers. Next, we study the behaviour of our three quantities for increasing d, n, or p, respectively. The following lemma summarizes the asymptotic behaviour of $a_{n,d}$, $b_{n,d}$, and $c_{p,d}$.

Lemma 3.11. The following relations are valid for the n- and d-asymptotic behaviour of $a_{n,d}$ and $b_{n,d}$:

$$\frac{a_{n,d+1}}{a_{n,d}} \xrightarrow{d \to \infty} 1, \qquad \frac{a_{n,d+1}}{a_{n,d}} \xrightarrow{n \to \infty} \infty, \qquad (3.83)$$

$$\frac{a_{n+1,d}}{a_{n,d}} \xrightarrow{n \to \infty} 2, \qquad \frac{a_{n+1,d}}{a_{n,d}} \xrightarrow{d \to \infty} \infty,$$

$$\frac{b_{n,d+1}}{b_{n,d}} \xrightarrow{d \to \infty} 3, \qquad \frac{b_{n,d+1}}{b_{n,d}} \xrightarrow{n \to \infty} \infty,$$

$$\frac{b_{n+1,d}}{b_{n,d}} \xrightarrow{n \to \infty} 2, \qquad \frac{b_{n+1,d}}{b_{n,d}} \xrightarrow{d \to \infty} \infty.$$

For $c_{p,d}$, the respective limits are

$$\frac{c_{p,d+1}}{c_{p,d}} \xrightarrow{d \to \infty} 1, \qquad \frac{c_{p,d+1}}{c_{p,d}} \xrightarrow{p \to \infty} \infty, \qquad (3.84)$$

$$\frac{c_{p+1,d}}{c_{p,d}} \xrightarrow{p \to \infty} 2, \qquad \frac{c_{p+1,d}}{c_{p,d}} \xrightarrow{d \to \infty} \infty.$$

For $a_{n,d}$, $b_{n,d}$, and $c_{p,d}$, the results for the limits $n \to \infty$ are not too surprising. In the case of $a_{n,d} = |V_n^{(d)}|$, for instance, they can be derived from the order terms $2^n \cdot n^{d-1}$ with respect to n according to (3.63). However, the statements concerning the limits $d \to \infty$ provide some new and interesting information. First, the ratio $a_{n+1,d}/a_{n,d}$ of two sparse grids of resolution $n+1$ and n, for example, is not bounded for increasing d. That is, increasing the resolution does not entail only a certain bounded factor of increasing cost, but the relative increase in cost switching from resolution n to $n+1$ becomes bigger and bigger for higher dimensionality and is not bounded. Second, for a fixed resolution n, the relative difference between two sparse grids of dimensionality d and $d+1$ becomes more and more negligible for increasing d, which is, in fact, somewhat surprising. Note that this is a hard statement and not just one dealing with orders of magnitude.

However, after all those asymptotic considerations, it is important to note that, often, such limits are not excessively useful for numerical purposes, since practical computations do not always reach the region of asymptotic behaviour. Therefore, usually, the constant factors play a more predominant part than an asymptotic point of view suggests. Furthermore, it certainly makes sense to increase resolution during the numerical solution of a given problem, but a variable dimensionality is, of course, more of a theoretical interest.

ε-complexity

An approach that is closely related to the cost–benefit setting used for the derivation of the sparse grid approximation spaces is the concept of the ε-complexity (Traub and Woźniakowski 1980, Traub, Wasilkowski and Woźniakowski 1983, Traub, Wasilkowski and Woźniakowski 1988, Woźniakowski 1985). The ε-complexity of a numerical method or algorithm indicates the computational work that is necessary to produce an approximate solution of some prescribed accuracy ε. In particular, for the complexity of general multivariate tensor product problems, see Wasilkovski and Woźniakowski (1995). We consider the ε-complexity of the different discrete approximation spaces $V_n^{(\infty)}$, $V_n^{(1)}$, and $V_n^{(E)}$ for the problem of *representing* a function $u \in X_0^{q,2}(\bar{\Omega})$ on a grid, *i.e.*, the problem of constructing the interpolant $u_n^{(\infty)} \in V_n^{(\infty)}$, $u_n^{(1)} \in V_n^{(1)}$, or $u_n^{(E)} \in V_n^{(E)}$, respectively. To this end, the overall computational cost caused by the interpolation in one of the three above discrete spaces will be estimated by the number of degrees of freedom (*i.e.*, grid points), or by an upper bound for this number. This does not constitute a restriction, since there are algorithms that can calculate the interpolant $u_n^{(\cdot)}$ in $O(|V_n^{(\cdot)}|)$ arithmetic operations, of course. Furthermore, as a measure for the accuracy, we use the upper bounds for the interpolation error $\|u - u_n^{(\infty)}\|$, $\|u - u_n^{(1)}\|$, and $\|u - u_n^{(E)}\|$ with respect to the different norms, as provided by (3.32), (3.68), and (3.73).

First, we deal with the well-known case of the regular full grid space $V_n^{(\infty)}$, where the curse of dimensionality is predominant and causes the problem to be *intractable* in the sense of Traub et al. (1988): the computational cost of obtaining an approximate solution of some given accuracy ε grows exponentially in the problem's dimensionality d. Note that, in the following, all occurring order terms have to be read with respect to ε or N, respectively, *i.e.*, for arbitrary, but fixed d.

Lemma 3.12. For the ε-complexities $N_\infty(\varepsilon)$, $N_2(\varepsilon)$, and $N_E(\varepsilon)$ of the problem of computing the interpolant $u_n^{(\infty)} \in V_n^{(\infty)}$ with respect to the L_∞-, the L_2-, and the energy norm for some prescribed accuracy ε, the following relations hold:

$$N_\infty(\varepsilon) = O(\varepsilon^{-\frac{d}{2}}), \qquad N_2(\varepsilon) = O(\varepsilon^{-\frac{d}{2}}), \qquad (3.85)$$

$$N_E(\varepsilon) = O(\varepsilon^{-d}).$$

Conversely, given a number N of grid points, the following accuracies can be obtained with respect to the different norms:

$$\varepsilon_{L_\infty}(N) = O(N^{-\frac{2}{d}}), \qquad \varepsilon_{L_2}(N) = O(N^{-\frac{2}{d}}), \qquad (3.86)$$

$$\varepsilon_E(N) = O(N^{-\frac{1}{d}}).$$

Proof. The statements follow directly from (3.31) and (3.32). □

Next we turn to the L_2-based sparse grid space $V_n^{(1)}$. As we have already seen, $V_n^{(1)}$ lessens the curse of dimensionality, but does not yet overcome it.

Lemma 3.13. For the ε-complexities $N_\infty(\varepsilon)$, $N_2(\varepsilon)$, and $N_E(\varepsilon)$ of the problem of computing the interpolant $u_n^{(1)} \in V_n^{(1)}$ with respect to the L_∞-, the L_2-, and the energy norm for some prescribed accuracy ε, the following relations hold:

$$N_\infty(\varepsilon), N_2(\varepsilon) = O(\varepsilon^{-\frac{1}{2}} \cdot |\log_2 \varepsilon|^{\frac{3}{2} \cdot (d-1)}), \qquad (3.87)$$

$$N_E(\varepsilon) = O(\varepsilon^{-1} \cdot |\log_2 \varepsilon|^{d-1}).$$

Conversely, given a number N of grid points, the following accuracies can be obtained with respect to the different norms:

$$\varepsilon_\infty(N), \varepsilon_2(N) = O(N^{-2} \cdot |\log_2 N|^{3 \cdot (d-1)}), \tag{3.88}$$

$$\varepsilon_E(N) = O(N^{-1} \cdot |\log_2 N|^{d-1}).$$

Proof. See Bungartz (1998), for example. □

Finally, for the energy-based sparse grid space $V_n^{(E)}$, the situation is evident and gratifying: the curse of dimensionality has disappeared.

Lemma 3.14. For the ε-complexity $N_E(\varepsilon)$ of the problem of computing the interpolant $u_n^{(E)} \in V_n^{(E)}$ with respect to the energy norm for some prescribed accuracy ε, the following relation holds:

$$N_E(\varepsilon) = O(\varepsilon^{-1}). \tag{3.89}$$

Thus, for a fixed number of N of grid points, the following accuracy can be obtained:

$$\varepsilon_E(N) = O(N^{-1}). \tag{3.90}$$

Proof. Both results are obvious consequences of (3.72) and (3.73). □

With the above remarks on the ε-complexity of the problem of interpolating functions $u \in X_0^{q,2}(\bar{\Omega})$ in our two sparse grid spaces $V_n^{(1)}$ and $V_n^{(E)}$, we close the discussion of the piecewise d-linear case.

4. Generalizations, related concepts, applications

In the previous section we presented the main ideas of the sparse grid approach starting from the 1D piecewise linear hierarchical basis, which was then extended to the general piecewise d-linear hierarchical basis by the discussed tensor product construction. However, the discretization on sparse grids is, of course, not limited to this explanatory example, but can be directly generalized to other multiscale bases such as p-type hierarchical bases, prewavelets, or wavelets, for instance. To this end, another 1D multiscale basis must be chosen. Then the tensor product construction as well as the cut-off of the resulting series expansion will lead to closely related sparse grids.

In the following, we first give a short survey of the historic background of sparse grids. Then we generalize the piecewise linear hierarchical basis to hierarchical polynomial bases of higher order, and discuss hierarchical Lagrangian interpolation of Bungartz (1998) and the so-called interpolets of Deslauriers and Dubuc (1989). After that, we discuss the use of prewavelets and wavelets in a sparse grid context. A short overview of the current state concerning sparse grid applications, with a focus on the discretization of PDEs including some remarks on adaptive refinement and fast solvers, will close this section.

4.1. Ancestors

The discussion of hierarchical finite elements (Peano 1976, Zienkiewicz, Kelly, Gago and Babuška 1982) and, in particular, the series of articles by Yserentant (1986, 1990, 1992) introducing the use of hierarchical bases for the numerical solution of PDEs, both for purposes of an explicit discretization and for the construction of preconditioners, was the starting point of Zenger's sparse grid concept (Zenger 1991). The generalization of Yserentant's hierarchical bases to a strict tensor product approach with its underlying hierarchical subspace splitting discussed in Section 3 allowed the *a priori* identification of more and of less important subspaces and grid points. As we have seen, it is this characterization of subspaces that the definition of sparse grids is based on. With the sparse grid approach, for the first time, *a priori* optimized and fully structured grid patterns were integrated into existing and well-established discretization schemes for PDEs such as finite elements, and were combined with a very straightforward access to adaptive grid refinement.

Even if the sparse grid concept was new for the context of PDEs, very closely related techniques had been studied for purposes of approximation, recovery, or numerical integration of smooth functions, before. For instance, the generalization of Archimedes' well-known hierarchical quadrature of $1 - x^2$ on $[-1, 1]$ to the d-dimensional case via Cavalieri's principle (see Figure 4.1) is a very prominent example of an (indeed early) hierarchical tensor product approach. Much later, Faber (1909) discussed the hierarchical representation of functions.

Finally, once more, the Russian literature turns out to be helpful for exploring the roots of a new approach in numerical mathematics. Two

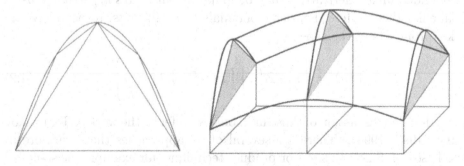

Figure 4.1. Hierarchical quadrature according to Archimedes (left) and application of Cavalieri's principle (right).

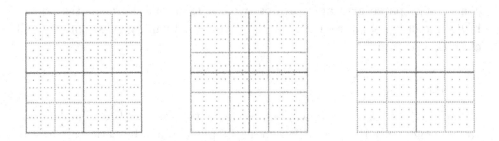

Figure 4.2. Smolyak quadrature patterns based on the trapezoidal rule as the one-dimensional algorithm: $p = 2$ and $n = 8$ (left), $p = 3$ and $n = 5$ (centre), $p = 4$ and $n = 4$ (right).

names that have to be mentioned here are those of Smolyak and Babenko. Smolyak (1963) studied classes of quadrature formulas of the type

$$Q_n^{(d)} f := \left(\sum_{i=0}^{n} (Q_i^{(1)} - Q_{i-1}^{(1)}) \otimes Q_{n-i}^{(d-1)} \right) f \qquad (4.1)$$

that are based on a tensor product $\otimes$ of lower-dimensional operators. In (4.1), $Q_n^{(d)}$ denotes a d-dimensional quadrature formula based on the 1D rule $Q_n^{(1)}$ that is, for $n \in \mathbb{N}$, usually chosen to be the compound formula resulting from the application of some simple formula Q on p^n subintervals $[i/p^n, (i+1)/p^n]$, $i = 0, \ldots, p^n - 1$, of $[0,1]$ for some natural number $p \geq 2$. Furthermore, the midpoint rule is usually taken as $Q_0^{(1)}$, and $Q_{-1}^{(1)} \equiv 0$ (see Frank and Heinrich (1996), Novak and Ritter (1996), and Smolyak (1963)). Figure 4.2 shows several examples of grids resulting from the application of 2D Smolyak quadrature for different n and p. Functions suitable for the Smolyak approach typically live in spaces of bounded (L_p-integrable) mixed derivatives which are closely related to our choice of $X_0^{q,r}(\bar{\Omega})$ in (3.2). A similar approach to the approximation of periodic multivariate functions of bounded mixed derivatives may be found in Babenko's *hyperbolic crosses* (Babenko 1960). Here, Fourier monomials or coefficients, respectively, are taken from sets of the type

$$\Gamma(n) := \left\{ \mathbf{k} \in \mathbb{Z}^d : \prod_{j=1}^{d} \max\{|k_j|, 1\} \leq n \right\}. \qquad (4.2)$$

For a detailed discussion of those methods, we refer to the work of Temlyakov (1989, 1993, 1994). There are several other approaches that are more or less based on Smolyak's tensor product technique, for example the so-called *Boolean methods* of Delvos et al. (Delvos 1982, Delvos and Schempp 1989, Delvos 1990, Baszenski and Delvos 1993) or the *discrete blending methods* of Baszenski, Delvos and Jester (1992) going back to work of Gordon (Gordon 1969, 1971, Gordon and Hall 1973). For the general analysis of such tensor product methods, see also Wasilkovski and Woźniakowski (1995). Concerning the computational cost, Smolyak's tensor product approach and its derivatives are characterized by terms of the order $O(N \cdot (\log_2 N)^{d-1})$, where N denotes the 1D cost, as we have shown them for the sparse grid spaces $V_n^{(1)}$ in (3.63). Furthermore, the tensor product approach of (4.1) itself, obviously, calls to mind the definition (3.61) of the L_2-based sparse grid spaces $V_n^{(1)}$, which can be written as

$$V_n^{(d,1)} := V_n^{(1)} = \sum_{|\mathbf{l}|_1 \leq n+d-1} W_{\mathbf{l}}^{(d)} = \sum_{l=1}^{n} W_l^{(1)} \otimes V_{n+d-1-l}^{(d-1,1)}, \qquad (4.3)$$

that is, in a tensor product form, too.

Finally, there are of course close relations of hierarchical bases and sparse grids on the one hand and wavelets on the other hand. These will be discussed in Section 4.4.

4.2. Higher-order polynomials

In this section, we discuss how to generalize the piecewise linear sparse grid method to higher-order basis functions.

The hierarchical Lagrangian interpolation

The first approach, the so-called hierarchical Lagrangian interpolation introduced in Bungartz (1998), uses a hierarchical basis of piecewise polynomials of arbitrary degree p, still working with just one degree of freedom per node. Before we discuss the underlying p-hierarchy, note that the piecewise constant case, *i.e.*, $p = 0$, as illustrated in Figure 4.3, is the natural starting point of such a hierarchy.

In accordance with the tensor product approach (3.9), we define basis functions of the (generalized) degree $\mathbf{p} := (p_1, \ldots, p_d) \in \mathbb{N}^d$ as products

$$\phi_{\mathbf{l},\mathbf{i}}^{(\mathbf{p})}(\mathbf{x}) := \prod_{j=1}^{d} \phi_{l_j,i_j}^{(p_j)}(x_j) \qquad (4.4)$$

of d 1D basis polynomials of degree p_j with the respective supports $[x_{l_j,i_j} - h_{l_j}, x_{l_j,i_j} + h_{l_j}]$. Note that, since there is no change in the underlying grids $\Omega_{\mathbf{l}}$, the grid points $\mathbf{x}_{\mathbf{l},\mathbf{i}}$ or x_{l_j,i_j} and the mesh widths $\mathbf{h}_{\mathbf{l}}$ or h_{l_j} are defined exactly as in the linear case. This, again, allows the restriction to the 1D case. For reasons of clarity, we will omit the index j. As already mentioned, we want to preserve the 'character' of the elements, *i.e.*, we want to keep to C^0-elements and manage without increasing the number of degrees of

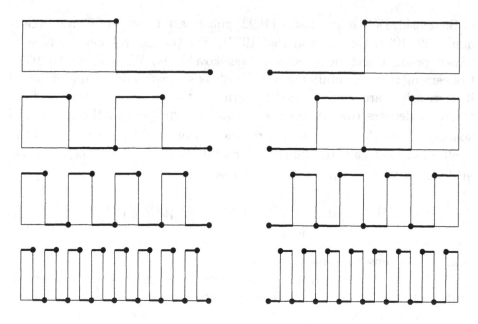

Figure 4.3. Piecewise constant hierarchical bases: continuous from the left (left) or continuous from the right (right).

freedom per element or per grid point for a higher p. However, to determine a polynomial $u^{(p)}(x)$ of degree p uniquely on $[x_{l,i} - h_l, x_{l,i} + h_l]$, we need $p+1$ conditions $u^{(p)}(x)$ has to fulfil. In the linear case, the interpolant resulting from the hierarchically higher levels is defined by its values in the two boundary points $x_{l,i} \pm h_l$. For $p \geq 2$, these two conditions are no longer sufficient. Therefore, we profit from the hierarchical history of $x_{l,i}$. Figure 4.4 shows the hierarchical relations of the grid points according to the hierarchical subspace splitting of Section 3.1. Apart from $x_{l,i} \pm h_l$, which mark the boundary of the support of $\phi_{l,i}^{(p)}$, $x_{l,i}$ may have hierarchical ancestors that are all located outside this support. Consequently, for the definition of such a local interpolant $u^{(p)}(x)$, it is reasonable and, for the construction of a hierarchical basis, essential to take the values of u in $x_{l,i} \pm h_l$ (as in the linear case) and, in addition, in a sufficient number of hierarchically next ancestors of $x_{l,i}$. These considerations lead us to the following definition.

Let $u \in C^{p+1}([0,1])$, $1 \leq p \leq l$, and let Ω_l denote the 1D grid of mesh width $h_l = 2^{-l}$ with grid points $x_{l,i}$ according to (3.4)–(3.6). Then, the *hierarchical Lagrangian interpolant* $u^{(p)}(x)$ of degree p of $u(x)$ with respect to Ω_l is defined on $[x_{l,i} - h_l, x_{l,i} + h_l]$, i odd, as the polynomial interpolant of $(x_k, u(x_k))$, $k = 1, \ldots, p+1$, where the x_k are just $x_{l,i} \pm h_l$ and the $p-1$ next hierarchical ancestors of $x_{l,i}$. Note that $u^{(p)}(x)$ is continuous on $\bar{\Omega}$, piecewise of polynomial degree p with respect to the grid Ω_{l-1}, and it interpolates $u(x)$

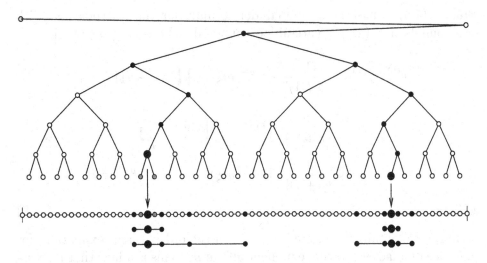

Figure 4.4. Ancestors (here: boundary points of the respective basis function's support and two more ($p = 4$); solid) and descendants (dotted) of two grid points.

on Ω_{l-1}. Nevertheless, it is defined *locally* on $[x_{l,i} - h_l, x_{l,i} + h_l]$. Thus, the width of the interval taken for the local *definition* of $u^{(p)}(x)$ is $2^p \cdot h_l$, but for any kind of further calculations, $u^{(p)}(x)$ is living only on the local interval of size $2 \cdot h_l$. The restriction $l \geq p$ is due to the fact that, with degree p, we need at least $p + 1$ ancestors for the interpolation. Now we study the local approximation properties of the hierarchical Lagrangian interpolation, that is, the interpolation error $u(x_{l,i}) - u^{(p)}(x_{l,i})$ or hierarchical surplus (*cf.* (3.19)).

Lemma 4.1. Let $u \in C^{p+1}([0,1])$, $1 \leq p \leq l$, and let $x_1 < \cdots < x_{p+1}$ be the ancestors of $x_{l,i}$ on level l, i odd, taken for the construction of the hierarchical Lagrangian interpolant $u^{(p)}$ of u in $[x_{l,i} - h_l, x_{l,i} + h_l]$. Then the hierarchical surplus $v^{(p)}(x_{l,i})$ in $x_{l,i}$ fulfils

$$v^{(p)}(x_{l,i}) := u(x_{l,i}) - u^{(p)}(x_{l,i}) = \frac{1}{(p+1)!} \cdot D^{p+1}u(\xi) \cdot \prod_{k=1}^{p+1}(x_{l,i} - x_k) \quad (4.5)$$

for some $\xi \in [x_1, x_{p+1}]$. Moreover, the order of approximation is given by

$$|v^{(p)}(x_{l,i})| \leq \frac{1}{(p+1)!} \cdot |D^{p+1}u(\xi)| \cdot h_l^{p+1} \cdot 2^{p \cdot (p+1)/2 - 1}. \quad (4.6)$$

Proof. (4.5) is a standard remainder formula for the interpolation with polynomials. For (4.6), a careful look at the distances $x_{l,i} - x_k$ provides

$$\left|v^{(p)}(x_{l,i})\right| = \frac{1}{(p+1)!} \cdot \left|D^{p+1}u(\xi)\right| \cdot \prod_{k=1}^{p+1} |x_{l,i} - x_k|$$

$$\leq \frac{1}{(p+1)!} \cdot \left|D^{p+1}u(\xi)\right| \cdot h_l^{p+1} \cdot \prod_{k=1}^{p} (2^k - 1)$$

$$\leq \frac{1}{(p+1)!} \cdot \left|D^{p+1}u(\xi)\right| \cdot h_l^{p+1} \cdot 2^{p \cdot (p+1)/2 - 1},$$

which is (4.6). □

Hence, we obtain the desired increase in the order of approximation, for the price of a factor growing exponentially in p^2. This is a hint that increasing p on each new level will not be the best strategy.

As an analogue to (3.24), an integral representation can be shown for the general order hierarchical surplus, too. For that, define $s_{l,i}^{(p)}(t)$ as the minimum support spline with respect to $x_{l,i}$ and its $p+1$ direct hierarchical ancestors (renamed $x_0, \ldots, x_{p+1}$ in increasing order):

$$s_{l,i}^{(p)}(t) := [x_0, \ldots, x_{p+1}](x-t)_+^p = \sum_{k=0}^{p+1} \frac{(x_k - t)_+^p}{w'_{l,i}(x_k)}. \tag{4.7}$$

Here, $[x_0, \ldots, x_{p+1}]f(x)$ just denotes the divided differences of order p with respect to x,

$$(x-t)_+^p := \begin{cases} (x-t)^p, & \text{for } x-t \geq 0, \\ 0, & \text{otherwise,} \end{cases} \tag{4.8}$$

and

$$w_{l,i}(x) := \prod_{j=0}^{p+1}(x - x_j). \tag{4.9}$$

Lemma 4.2. *With the above definitions, we get the following integral representation for the hierarchical surplus $v^{(p)}(x_{l,i})$:*

$$v^{(p)}(x_{l,i}) = \frac{w'_{l,i}(x_{l,i})}{p!} \cdot \int_{-\infty}^{\infty} s_{l,i}^{(p)}(t) \cdot D^{p+1}u(t)\, dt. \tag{4.10}$$

Proof. See Bungartz (1998). □

An immediate consequence of (4.10) is the relation

$$\int_{x_0}^{x_{p+1}} D^{p-1} s_{l,i}^{(p)}(t) \cdot f(t)\, dt = 0 \tag{4.11}$$

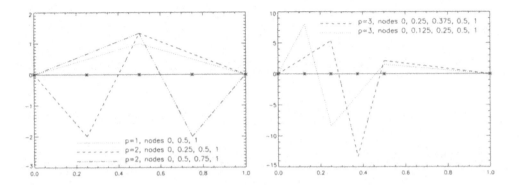

Figure 4.5. $D^{p-1}s_{l,i}^{(p)}(t)$ for $p = 1$ and the two constellations of $p = 2$ (left) and for two of the four cubic constellations (right).

for all $f \in \mathbb{P}_{p-2}$. Owing to (4.7),

$$D^{p-1}s_{l,i}^{(p)}(t) = (-1)^{p-1} \cdot \sum_{k=0}^{p+1} \frac{(x_k - t)_+}{w'_{l,i}(x_k)} \qquad (4.12)$$

is piecewise linear and continuous. In wavelet terminology, (4.11) means that the $p - 1$ first moments of the $D^{p-1}s_{l,i}^{(p)}(t)$ vanish. Therefore we could construct $s_{l,i}^{(p)}(t)$ and, hence, the hierarchical surplus $v^{(p)}(x_{l,i})$ of degree p the other way round, too. Starting from $x_0, \ldots, x_{p+1}$, look for the piecewise linear and continuous function $\sigma^{(p)}(t)$ with $\sigma^{(p)}(x_0) = \sigma^{(p)}(x_{p+1}) = 0$ and with vanishing first $p - 1$ moments, which is determined up to a constant factor, and integrate the resulting $\sigma^{(p)}(t)$ $p-1$ times. The remaining degree of freedom can be fixed by forcing the coefficient of $u(x_m)$ to be 1 (as it is in $v^{(p)}(x_m)$). The left-hand part of Figure 4.5 shows this $(p-1)$st derivative of $s_{l,i}^{(p)}(t)$ for the possible constellations of ancestors for $p = 1, 2$. In the linear case, we have no vanishing moment; for $p = 2$, we get one. The cubic case is illustrated in the right-hand part of Figure 4.5.

Note that Figure 4.5 suggests that the higher-order case may be obtained without explicitly working with polynomials of arbitrary degree, but just by a superposition of the linear approximations in different ancestors. In fact, the spline's $(p - 1)$st derivative, and thus the calculation of the hierarchical surplus $v^{(p)}(x_{l,i})$ of degree p, can be reduced to the linear case.

Lemma 4.3. Let p, l, and i be defined as before. Furthermore, let x_m denote the current grid point $x_{l,i}$, and let x_n be its hierarchical father. For any $u \in C^{p+1}([0,1])$, the hierarchical surplus $v^{(p)}(x_m)$ of degree p in x_m can be calculated with the help of $v^{(p-1)}(x_m)$ and of $v^{(p-1)}(x_n)$:

$$v^{(p)}(x_m) = v^{(p-1)}(x_m) - \alpha(x_0, \ldots, x_{p+1}) \cdot v^{(p-1)}(x_n), \qquad (4.13)$$

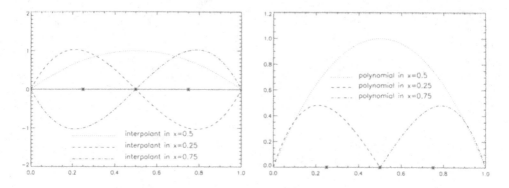

Figure 4.6. Hierarchical basis polynomials for $p = 2$ and $p = 3$ (different scaling for clarity): construction via hierarchical Lagrangian interpolation (left) and restriction to the respective hierarchical support (right).

where α depends on the relative position of x_m's ancestors, but not on the interpolated values.

Proof. See Bungartz (1998). □

Having introduced the hierarchical Lagrangian interpolation we now discuss the corresponding hierarchical basis polynomials of degree p. Such a $\phi_{l,i}^{(p)}$ in $x_{l,i}$ with i odd, $l \geq p - 1$, and $p \geq 2$ is uniquely defined on $[x_{l,i} - h_l, x_{l,i} + h_l]$ by the following $p + 1$ conditions:

$$\phi_{l,i}^{(p)}(x_{l,i}) := 1, \qquad \phi_{l,i}^{(p)}(x_k) := 0, \qquad (4.14)$$

where the x_k are just $x_{l,i} \pm h_l$ and the $p - 2$ next hierarchical ancestors of $x_{l,i}$. Additionally, we force $\phi_{l,i}^{(p)}$ to vanish outside $[x_{l,i} - h_l, x_{l,i} + h_l]$:

$$\phi_{l,i}^{(p)}(x) := 0 \quad \text{for} \quad x \notin [x_{l,i} - h_l, x_{l,i} + h_l]. \qquad (4.15)$$

Note that the restriction $p \geq 2$ is due to the fact that $p = 1$ does not fit into this scheme. Since the typical basis function used for linear approximation is a *piecewise* linear one only, there are three degrees of freedom to determine it uniquely on $[x_{l,i} - h_l, x_{l,i} + h_l]$, as in the quadratic case.

At this point, three things are important to realize. First, this definition is fully consistent with the hierarchical Lagrangian interpolation: a global interpolant $u^{(p)}(x)$ built up with the help of these $\phi_{l,i}^{(p)}$ fulfils all requirements. Second, though the definition of the $\phi_{l,i}^{(p)}$ is based on points outside $[x_{l,i} - h_l, x_{l,i} + h_l]$ (for $p > 2$, at least, to be precise), we use only $[x_{l,i} - h_l, x_{l,i} + h_l]$ as support of $\phi_{l,i}^{(p)}$ according to (4.15). Thus, the support of the basis polynomials does not change in comparison with the piecewise

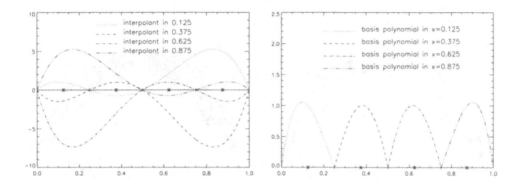

Figure 4.7. Hierarchical basis polynomials for $p = 4$: construction via hierarchical Lagrangian interpolation (left) and restriction to the respective hierarchical support (right).

linear case. Finally, since we need $p-2$ ancestors outside $[x_{l,i} - h_l, x_{l,i} + h_l]$, a basis polynomial of degree p can not be used earlier than on level $p-1$.

The quadratic basis polynomial is uniquely defined by its values in $x_{l,i}$ and $x_{l,i} \pm h_l$. For $p = 3$, however, the shape of the resulting polynomial depends on where the third zero outside $\phi_{l,i}^{(p)}$'s support is located. Owing to the hierarchical relations illustrated in Figure 4.4, there are two possibilities for this ancestor's position: to the left of $x_{l,i}$, i.e., in $x = x_{l,i} - 3 \cdot h_l$, or to the right in $x = x_{l,i} + 3 \cdot h_l$. Thus we get two different types of cubic basis polynomials. Figure 4.6 illustrates this for the highest two levels $l \in \{1, 2\}$. In $x = 0.5$ (i.e., $l = 1$), only $p = 2$ is possible owing to the lack of ancestors outside $\bar{\Omega}$. On the next level, in $x = 0.25$, the third ancestor $x = 1.0$ is situated to the right of 0.5. In $x = 0.75$, the third ancestor is $x = 0.0$, to the left of $x = 0.75$. Of course, both cubic functions are symmetric with respect to $x = 0.5$. If we continue with cubic basis polynomials on the lower levels $l > 2$, no new relationships will occur. Thus we can manage with these two types of cubic polynomials. Figure 4.7 shows the four types of quartic basis polynomials. Owing to the underlying hierarchy, these four (pairwise symmetric) polynomials cover all possible constellations. Obviously, in the general case of an arbitrary polynomial degree $p > 1$, our approach leads to 2^{p-2} different types of basis functions.

Moreover, Figure 4.7 illustrates that the four quartic basis polynomials do not differ that much. This effect even holds for all basis polynomials of arbitrary p. Figure 4.8 shows this similarity for all different types of basis polynomials up to degree 7 (all now plotted with respect to the common support $[-1, 1]$). This similarity is also reflected in the analytical properties of the basis polynomials. For details, we refer the reader to Bungartz (1998).

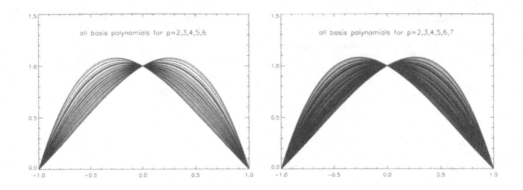

Figure 4.8. All hierarchical basis polynomials for $p \leq 6$ (31 different types; left) and $p \leq 7$ (63 different types; right) with respect to the support $[-1, 1]$.

Higher-order approximation on sparse grids

Finally, we deal with the generalization of the piecewise linear sparse grid approximation spaces $V_n^{(1)}$ and $V_n^{(E)}$ to the case of piecewise polynomials of degree $p \geq 2$ or $\mathbf{p} \geq \mathbf{2}$ according to the hierarchical Lagrangian interpolation. For the discussion of the approximation properties of the resulting sparse grid spaces of higher order, we restrict ourselves to the *p-regular* case. For some fixed maximum degree $p := p_{\max}$, the polynomials are chosen as described in the previous section until we reach level $p - 1$ or $(p - 1) \cdot \mathbf{1}$. On the further levels, we use this p. Note that the degree $\mathbf{p}$ of a basis function or of a subspace, respectively, has to fulfil the condition

$$\mathbf{2} \leq \mathbf{p} = \min\{p \cdot \mathbf{1}, \mathbf{l} + \mathbf{1}\}, \tag{4.16}$$

where the minimum is taken component-wise. The lower bound reflects the fact that there are no linear hierarchical Lagrangian basis functions; the upper bound is caused by our choice of a maximum degree p and by the lack of basis polynomials of degree $\mathbf{p}$ earlier than on level $\mathbf{p} - \mathbf{1}$.

The starting point is the space $X_0^{p+1,q}$ of functions of (in some sense) bounded weak mixed derivatives of the order less than or equal to $p + 1$ in each direction: *cf.* (3.2). For the approximation of some $u \in X_0^{p+1,q}$, we use product-type basis functions $\phi_{\mathbf{l},\mathbf{i}}^{(\mathbf{p})}(\mathbf{x})$ according to (4.4). Without introducing nodal subspaces $V_{\mathbf{l}}^{(\mathbf{p})}$, we directly turn to the hierarchical subspaces $W_{\mathbf{l}}^{(\mathbf{p})}$:

$$W_{\mathbf{l}}^{(\mathbf{p})} := \text{span}\{\phi_{\mathbf{l},\mathbf{i}}^{(\mathbf{p})} : \mathbf{i} \in \mathbf{I}_{\mathbf{l}}\}. \tag{4.17}$$

The completion of the sum of all $W_{\mathbf{l}}^{(\mathbf{p})}$ with respect to the energy norm contains $X_0^{p+1,q}$. Thus, analogously to (3.19), the hierarchical subspace decomposition leads to a hierarchical representation of degree p of $u \in X_0^{p+1,q}$:

$$u(\mathbf{x}) = \sum_{\mathbf{l}} u_{\mathbf{l}}^{(\mathbf{p})}(\mathbf{x}), \quad u_{\mathbf{l}}^{(\mathbf{p})}(\mathbf{x}) = \sum_{\mathbf{i} \in I_{\mathbf{l}}} v_{\mathbf{l},\mathbf{i}}^{(\mathbf{p})} \cdot \phi_{\mathbf{l},\mathbf{i}}^{(\mathbf{p})}(\mathbf{x}) \in u_{\mathbf{l}}^{(\mathbf{p})} \in W_{\mathbf{l}}^{(\mathbf{p})}, \quad (4.18)$$

where $v_{\mathbf{l},\mathbf{i}}^{(\mathbf{p})} \in \mathbb{R}$ is just the hierarchical coefficient or surplus. Note that, in (4.18), the degree $\mathbf{p}$ is not constant because of (4.16).

Concerning the cost and benefit of the $W_{\mathbf{l}}^{(\mathbf{p})}$, remember that the number of degrees of freedom induced by $W_{\mathbf{l}}^{(\mathbf{p})}$ does not increase in comparison to the linear situation. Thus we still have

$$|W_{\mathbf{l}}^{(\mathbf{p})}| = |W_{\mathbf{l}}| = 2^{|\mathbf{l}-\mathbf{1}|_1}. \quad (4.19)$$

On the other hand, the discussion of $W_{\mathbf{l}}^{(\mathbf{p})}$'s benefit, *i.e.*, of its contribution to the overall interpolant $u_{\mathbf{l}}^{(\mathbf{p})} \in W_{\mathbf{l}}^{(\mathbf{p})}$ of some $u \in X_0^{p+1,q}$, requires studying the different norms of our hierarchical Lagrangian basis polynomials $\phi_{\mathbf{l},\mathbf{i}}^{(\mathbf{p})}(\mathbf{x})$.

Lemma 4.4. *For any d-dimensional hierarchical Lagrangian basis polynomial $\phi_{\mathbf{l},\mathbf{i}}^{(\mathbf{p})}(\mathbf{x})$ according to the above discussion, the following relations hold:*

$$\|\phi_{\mathbf{l},\mathbf{i}}^{(\mathbf{p})}\|_\infty \leq 1.117^d, \quad (4.20)$$

$$\|\phi_{\mathbf{l},\mathbf{i}}^{(\mathbf{p})}\|_q \leq 1.117^d \cdot 2^{d/q} \cdot 2^{-|\mathbf{l}|_1/q}, \quad q \geq 1,$$

$$\|\phi_{\mathbf{l},\mathbf{i}}^{(\mathbf{p})}\|_E \leq 3.257 \cdot \left(\frac{5}{2}\right)^{d/2} \cdot 2^{-|\mathbf{l}|_1/2} \cdot \left(\sum_{j=1}^{d} 2^{2l_j}\right)^{1/2}.$$

Proof. The statements follow immediately from analytical properties of the basis polynomials: see Bungartz (1998) for details. □

Next we consider $v_{\mathbf{l},\mathbf{i}}^{(\mathbf{p})}$, the d-dimensional hierarchical surplus of degree $\mathbf{p}$. Inspired by the integral representation (4.10) of the 1D surplus, we define

$$\sigma_{\mathbf{l},\mathbf{i}}^{(\mathbf{p})}(\mathbf{x}) := \prod_{j=1}^{d} \frac{w'_{l_j,i_j}(x_{l_j,i_j})}{p_j!} \cdot s_{l_j,i_j}^{(p_j)}(x_j), \quad (4.21)$$

where $w_{l_j,i_j}(x_j)$ and $s_{l_j,i_j}^{(p_j)}(x_j)$ are defined exactly as in (4.9) and (4.7), but now for direction x_j and based on the respective hierarchical ancestors. With the help of (4.21) we obtain an integral representation analogous to (3.24).

Lemma 4.5. *For $u \in X_0^{p+1,q}$, the hierarchical surplus $v_{\mathbf{l},\mathbf{i}}^{(\mathbf{p})}$ fulfils*

$$v_{\mathbf{l},\mathbf{i}}^{(\mathbf{p})} = \int_\Omega \sigma_{\mathbf{l},\mathbf{i}}^{(\mathbf{p})}(\mathbf{x}) \cdot D^{\mathbf{p}+\mathbf{1}} u(\mathbf{x}) \, d\mathbf{x}. \quad (4.22)$$

Proof. Analogously to (3.23) for $p = 1$, the d-dimensional interpolation operator or stencil can be written as an operator product of the d univariate operators. Thus we can proceed as in the proof of (3.24) and do the integration with respect to the d coordinate directions one after the other. According to (4.10) and (4.21), the 1D integral with respect to x_j leads to the 1D surplus with respect to x_j. Consequently, the d-dimensional integral equals the d-dimensional surplus, as asserted in (4.22). □

Again, note the close relations between the hierarchical approach and integral transforms. Applying successive partial integration to (4.22), we get

$$v_{\mathbf{l},\mathbf{i}}^{(\mathbf{p})} = \int_\Omega \sigma_{\mathbf{l},\mathbf{i}}^{(\mathbf{p})}(\mathbf{x}) D^{\mathbf{p}+1} u(\mathbf{x}) \, d\mathbf{x} = (-1)^{|\mathbf{p}+1|_1} \int_\Omega \hat{\sigma}_{\mathbf{l},\mathbf{i}}^{(\mathbf{p})}(\mathbf{x}) u(\mathbf{x}) \, d\mathbf{x}, \quad (4.23)$$

where $\hat{\sigma}_{\mathbf{l},\mathbf{i}}^{(\mathbf{p})}(\mathbf{x})$ equals $D^{\mathbf{p}+1}\sigma_{\mathbf{l},\mathbf{i}}^{(\mathbf{p})}(\mathbf{x})$ in a weak sense and is a linear combination of $(p+2)^d$ Dirac pulses of alternating sign. Thus, again, the surplus can be interpreted as the coefficient resulting from an integral transform based on the function $\hat{\sigma}_{\mathbf{l},\mathbf{i}}^{(\mathbf{p})}(\mathbf{x})$.

Next, (4.22) leads us to upper bounds for the d-dimensional surplus of degree $\mathbf{p}$ and for $W_{\mathbf{l}}^{(\mathbf{p})}$'s contribution $u_{\mathbf{l}}^{(\mathbf{p})}$ to the hierarchical representation of u.

Lemma 4.6. *Let* $u \in X_0^{\mathbf{p}+1,q}$ *be given in its hierarchical representation* (4.18). *Then, with*

$$c(\mathbf{p}) := \prod_{j=1}^d \frac{2^{p_j \cdot (p_j+1)/2}}{(p_j+1)!}, \quad (4.24)$$

and with the seminorms $|u|_{\alpha,\infty}$ *and* $|u|_{\alpha,2}$ *defined in* (3.3), *the following estimates hold for the d-dimensional hierarchical surplus* $v_{\mathbf{l},\mathbf{i}}^{(\mathbf{p})}$:

$$|v_{\mathbf{l},\mathbf{i}}^{(\mathbf{p})}| \leq \left(\frac{1}{2}\right)^d \cdot c(\mathbf{p}) \cdot 2^{-|\mathbf{l}\cdot(\mathbf{p}+1)|_1} \cdot |u|_{\mathbf{p}+1,\infty}, \quad (4.25)$$

$$|v_{\mathbf{l},\mathbf{i}}^{(\mathbf{p})}| \leq \left(\frac{1}{6}\right)^{d/2} \cdot c(\mathbf{p}) \cdot 2^{-|\mathbf{l}\cdot(\mathbf{p}+1)|_1} \cdot 2^{|\mathbf{l}|_1/2} \cdot \left\| D^{\mathbf{p}+1} u \right\|_{\text{supp}(\phi_{\mathbf{l},\mathbf{i}})}\Big\|_2.$$

Proof. Owing to (4.21), (4.22), and $s_{l_j,i_j}^{(p_j)}(x_j) \geq 0$, we have

$$|v_{\mathbf{l},\mathbf{i}}^{(\mathbf{p})}| = \left| \int_\Omega \sigma_{\mathbf{l},\mathbf{i}}^{(\mathbf{p})}(\mathbf{x}) \cdot D^{\mathbf{p}+1} u(\mathbf{x}) \, d\mathbf{x} \right|$$

$$= \left| \int_\Omega \prod_{j=1}^d \frac{w'_{l_j,i_j}(x_{l_j,i_j})}{p_j!} \cdot s_{l_j,i_j}^{(p_j)}(x_j) \cdot D^{\mathbf{p}+1} u(\mathbf{x}) \, d\mathbf{x} \right|$$

$$\leq \left| \prod_{j=1}^{d} \frac{w'_{l_j,i_j}(x_{l_j,i_j})}{p_j!} \cdot \int_{[0,1]} s^{(p_j)}_{l_j,i_j}(x_j) \, \mathrm{d}x_j \right| \cdot \|D^{\mathbf{p}+\mathbf{1}}u\|_{\infty}$$

$$= \prod_{j=1}^{d} \left| \frac{w'_{l_j,i_j}(x_{l_j,i_j})}{p_j!} \cdot \int_{[0,1]} s^{(p_j)}_{l_j,i_j}(x_j) \, \mathrm{d}x_j \right| \cdot |u|_{\mathbf{p}+\mathbf{1},\infty}.$$

Because of (4.10), each of the d factors in the above product is the absolute value of the hierarchical surplus of $x_j^{p_j+1}/(p_j+1)!$, for $j = 1, \ldots, d$, which is bounded by

$$\frac{1}{(p_j+1)!} \cdot h_{l_j}^{p_j+1} \cdot 2^{p_j \cdot (p_j+1)/2 - 1}$$

because of (4.6). Thus we obtain

$$|v_{\mathbf{l},\mathbf{i}}^{(\mathbf{p})}| \leq \prod_{j=1}^{d} \frac{1}{(p_j+1)!} \cdot h_{l_j}^{p_j+1} \cdot 2^{p_j \cdot (p_j+1)/2 - 1} \cdot |u|_{\mathbf{p}+\mathbf{1},\infty}$$

$$= 2^{-d} \cdot c(\mathbf{p}) \cdot 2^{-|\mathbf{l} \cdot (\mathbf{p}+\mathbf{1})|_1} \cdot |u|_{\mathbf{p}+\mathbf{1},\infty}.$$

For the bound with respect to the L_2-norm, we start from

$$|v_{\mathbf{l},\mathbf{i}}^{(\mathbf{p})}| = \left| \int_{\Omega} \sigma_{\mathbf{l},\mathbf{i}}^{(\mathbf{p})}(\mathbf{x}) \cdot D^{\mathbf{p}+\mathbf{1}}u(\mathbf{x}) \, \mathrm{d}\mathbf{x} \right| \leq \|\sigma_{\mathbf{l},\mathbf{i}}^{(\mathbf{p})}\|_2 \cdot \|D^{\mathbf{p}+\mathbf{1}}u\, |_{\mathrm{supp}(\phi_{\mathbf{l},\mathbf{i}})}\|_2.$$

According to (4.21), and since $s^{(p_j)}_{l_j,i_j}(x_j) \geq 0$,

$$\|\sigma_{\mathbf{l},\mathbf{i}}^{(\mathbf{p})}\|_2 = \prod_{j=1}^{d} \left\| \frac{w'_{l_j,i_j}(x_{l_j,i_j})}{p_j!} \cdot s^{(p_j)}_{l_j,i_j} \right\|_2$$

$$= \prod_{j=1}^{d} \left(\int_{[0,1]} \left| \frac{w'_{l_j,i_j}(x_{l_j,i_j})}{p_j!} \cdot s^{(p_j)}_{l_j,i_j}(x_j) \right|^2 \mathrm{d}x_j \right)^{1/2}$$

$$\leq \prod_{j=1}^{d} \left(\max_{x_j \in [0,1]} \left| \frac{w'_{l_j,i_j}(x_{l_j,i_j})}{p_j!} \cdot s^{(p_j)}_{l_j,i_j}(x_j) \right| \cdot \frac{1}{(p_j+1)!} h_{l_j}^{p_j+1} 2^{p_j \cdot (p_j+1)/2 - 1} \right)^{1/2}$$

$$\leq \prod_{j=1}^{d} \left(\max_{x_j \in [0,1]} |s^{(p_j)}_{l_j,i_j}(x_j)| \cdot \frac{1}{p_j! \cdot (p_j+1)!} h_{l_j}^{2 \cdot (p_j+1)} 2^{p_j \cdot (p_j+1) - 2} \right)^{1/2}$$

holds because of (4.6) and (4.10). In (4.10), choose u such that

$$D^{p_j+1}u(x_j) = \frac{w'_{l_j,i_j}(x_{l_j,i_j})}{p_j!} \cdot s^{(p_j)}_{l_j,i_j}(x_j)$$

and apply (4.6). Finally, since

$$|s^{(p_j)}_{l_j,i_j}(x_j)| \leq \frac{2}{3} \cdot \frac{1}{(p_j+1) \cdot h_{l_j}}.$$

can be shown for all $x_j \in [0,1]$ and for all possible l_j, i_j, and p_j, we get

$$\|\sigma^{(\mathbf{p})}_{\mathbf{l},\mathbf{i}}\|_2 \leq 2^{-d} \cdot \left(\frac{2}{3}\right)^{d/2} \cdot c(\mathbf{p}) \cdot 2^{-|\mathbf{l} \cdot (\mathbf{p}+1)|_1} \cdot 2^{|\mathbf{l}|_1/2},$$

$$|v^{(\mathbf{p})}_{\mathbf{l},\mathbf{i}}| \leq \left(\frac{1}{6}\right)^{d/2} \cdot c(\mathbf{p}) \cdot 2^{-|\mathbf{l} \cdot (\mathbf{p}+1)|_1} \cdot 2^{|\mathbf{l}|_1/2} \cdot \|D^{\mathbf{p}+1} u\,|_{\mathrm{supp}(\phi_{\mathbf{l},\mathbf{i}})}\|_2. \qquad \Box$$

Lemma 4.7. Let $u \in X^{p+1,q}_0$ be given in its representation (4.18). Then the following upper bounds hold for the contributions $u^{(\mathbf{p})}_{\mathbf{l}} \in W^{(\mathbf{p})}_{\mathbf{l}}$:

$$\|u^{(\mathbf{p})}_{\mathbf{l}}\|_\infty \leq 0.5585^d c(\mathbf{p}) \cdot 2^{-|\mathbf{l} \cdot (\mathbf{p}+1)|_1} \cdot |u|_{\mathbf{p}+1,\infty}, \qquad (4.26)$$

$$\|u^{(\mathbf{p})}_{\mathbf{l}}\|_2 \leq 1.117^d \cdot \left(\frac{1}{3}\right)^{d/2} c(\mathbf{p}) \cdot 2^{-|\mathbf{l} \cdot (\mathbf{p}+1)|_1} \cdot |u|_{\mathbf{p}+1,2},$$

$$\|u^{(\mathbf{p})}_{\mathbf{l}}\|_E \leq 3.257 \cdot \left(\frac{5}{8}\right)^{d/2} c(\mathbf{p}) \cdot 2^{-|\mathbf{l} \cdot (\mathbf{p}+1)|_1} \cdot \left(\sum_{j=1}^d 2^{2l_j}\right)^{1/2} \cdot |u|_{\mathbf{p}+1,\infty},$$

$$\|u^{(\mathbf{p})}_{\mathbf{l}}\|_E \leq 3.257 \cdot \left(\frac{5}{12}\right)^{d/2} c(\mathbf{p}) \cdot 2^{-|\mathbf{l} \cdot (\mathbf{p}+1)|_1} \cdot \left(\sum_{j=1}^d 2^{2l_j}\right)^{1/2} \cdot |u|_{\mathbf{p}+1,2}.$$

Proof. All results follow from the previous lemmata, with arguments completely analogous to the piecewise linear case. $\Box$

Now, we are ready for the optimization process studied in detail for the piecewise linear case in Section 3. For the p-regular scenario, a slight simplification allows us to use the diagonal subspace pattern again, that is,

$$V^{(1,1)}_n := V^{(1)}_n, \qquad V^{(p,1)}_n := \bigoplus_{|\mathbf{l}|_1 \leq n+d-1} W^{(\mathbf{p})}_{\mathbf{l}} \quad \text{for } p > 1. \qquad (4.27)$$

As before, note that $\mathbf{p}$ is not constant, owing to (4.16). The following theorem deals with the approximation quality of these sparse grid spaces $V^{(p,1)}_n$.

Theorem 4.8. For the L_∞-, L_2-, and energy norm, the following bounds for the error of the interpolant $u^{(p,1)}_n \in V^{(p,1)}_n$ of $u \in X^{p+1,q}_0$ hold:

$$\|u - u^{(p,1)}_n\|_\infty \leq \left(\frac{0.5585}{2^{p+1}}\right)^d \cdot c(p \cdot \mathbf{1}) \cdot |u|_{(p+1)\cdot\mathbf{1},\infty} \cdot A(d,n) \cdot h^{p+1}_n + O(h^{p+1}_n)$$

$$= O(h^{p+1}_n \cdot n^{d-1}), \qquad (4.28)$$

$$\|u - u_n^{(p,1)}\|_2 \leq \left(\frac{1.117}{\sqrt{3} \cdot 2^{p+1}}\right)^d \cdot c(p \cdot \mathbf{1}) \cdot |u|_{(p+1) \cdot \mathbf{1}, 2} \cdot A(d, n) \cdot h_n^{p+1} + O(h_n^{p+1})$$

$$= O(h_n^{p+1} \cdot n^{d-1}),$$

$$\|u - u_n^{(p,1)}\|_E \leq 3.257 \cdot \left(\frac{\sqrt{5}}{\sqrt{2} \cdot 2^{p+1}}\right)^d \cdot c(p \cdot \mathbf{1}) \cdot \frac{d \cdot |u|_{(p+1) \cdot \mathbf{1}, \infty}}{1 - 2^{-p}} \cdot h_n^p + O(h_n^p)$$

$$= O(h_n^p),$$

$$\|u - u_n^{(p,1)}\|_E \leq 3.257 \cdot \left(\frac{\sqrt{5}}{\sqrt{3} \cdot 2^{p+1}}\right)^d \cdot c(p \cdot \mathbf{1}) \cdot \frac{d \cdot |u|_{(p+1) \cdot \mathbf{1}, 2}}{1 - 2^{-p}} \cdot h_n^p + O(h_n^p)$$

$$= O(h_n^p).$$

Proof. Actually, there is just one major difference compared to the proof of (3.68) dealing with the piecewise linear case. Now, owing to (4.16), the polynomial degree $\mathbf{p}$ is not constant for all subspaces $W_\mathbf{l}^{(\mathbf{p})}$ neglected in $V_n^{(p,1)}$, but depends on the respective level $\mathbf{l}$. However, the influence of all subspaces with $\mathbf{p} < p \cdot \mathbf{1}$ can be collected in a term of the order $O(h_n^{p+1})$ with respect to the L_∞- and L_2-norm or of the order $O(h_n^p)$ with respect to the energy norm, if $n \geq p - 1$: for sufficiently large n, each of those subspaces involves at least one coordinate direction x_j with $l_j = O(n)$ and $p_j = p$.

Therefore we can proceed as in the proof of (3.68) and assume a constant degree $\mathbf{p} = p \cdot \mathbf{1}$. With (4.26) and (3.66) for $s = p + 1$, we get

$$\|u - u_n^{(p,1)}\|_\infty \leq \sum_{|\mathbf{l}|_1 > n+d-1} \|u_\mathbf{l}^{(\mathbf{p})}\|_\infty$$

$$\leq \sum_{|\mathbf{l}|_1 > n+d-1} 0.5585^d \cdot c(p \cdot \mathbf{1}) \cdot 2^{-(p+1) \cdot |\mathbf{l}|_1} \cdot |u|_{(p+1) \cdot \mathbf{1}, \infty} + O(h_n^{p+1})$$

$$= 0.5585^d \cdot c(p \cdot \mathbf{1}) \cdot |u|_{(p+1) \cdot \mathbf{1}, \infty} \cdot \sum_{|\mathbf{l}|_1 > n+d-1} 2^{-(p+1) \cdot |\mathbf{l}|_1} + O(h_n^{p+1})$$

$$\leq 0.5585^d \cdot c(p \cdot \mathbf{1}) \cdot |u|_{(p+1) \cdot \mathbf{1}, \infty} \cdot 2^{-(p+1) \cdot (n+d)} \cdot A(d, n) + O(h_n^{p+1})$$

$$= \left(\frac{0.5585}{2^{p+1}}\right)^d \cdot c(p \cdot \mathbf{1}) \cdot |u|_{(p+1) \cdot \mathbf{1}, \infty} \cdot A(d, n) \cdot h_n^{p+1} + O(h_n^{p+1})$$

$$= O(h_n^{p+1} \cdot n^{d-1})$$

owing to the definition of $A(d, n)$ in (3.65), and, by analogy, the corresponding result for the L_2-norm. Concerning the energy norm, we have

$$\left\|u - u_n^{(p,1)}\right\|_E \leq \sum_{|\mathbf{l}|_1 > n+d-1} \left\|u_\mathbf{l}^{(\mathbf{p})}\right\|_E$$

$$\leq 3.257 \left(\frac{5}{8}\right)^{d/2} c(\mathbf{p} \cdot \mathbf{1}) |u|_{(\mathbf{p}+\mathbf{1}) \cdot \mathbf{1}, \infty} \cdot \sum_{|\mathbf{l}|_1 > n+d-1} 2^{-(p+1) \cdot |\mathbf{l}|_1} \cdot \left(\sum_{j=1}^d 4^{l_j}\right)^{1/2}$$

$$+ O(h_n^p)$$

and, as for the linear case,

$$\left\|u - u_n^{(p,1)}\right\|_E$$

$$\leq 3.257 \left(\frac{5}{8}\right)^{d/2} c(\mathbf{p} \cdot \mathbf{1}) |u|_{(\mathbf{p}+\mathbf{1}) \cdot \mathbf{1}, \infty} \cdot \sum_{i=n+d}^{\infty} 2^{-(p+1) \cdot i} \cdot \sum_{|\mathbf{l}|_1 = i} \left(\sum_{j=1}^d 4^{l_j}\right)^{1/2}$$

$$+ O(h_n^p)$$

$$\leq 3.257 \left(\frac{5}{8}\right)^{d/2} c(\mathbf{p} \cdot \mathbf{1}) |u|_{(\mathbf{p}+\mathbf{1}) \cdot \mathbf{1}, \infty} \cdot \sum_{i=n+d}^{\infty} d \cdot 2^{-p \cdot i} + O(h_n^p)$$

$$\leq 3.257 \left(\frac{\sqrt{5}}{\sqrt{2} \cdot 2^{p+1}}\right)^d c(\mathbf{p} \cdot \mathbf{1}) \frac{d \cdot |u|_{(\mathbf{p}+\mathbf{1}) \cdot \mathbf{1}, \infty}}{1 - 2^{-p}} \cdot h_n^p + O(h_n^p)$$

$$= O(h_n^p),$$

because $\sum_{|\mathbf{l}|_1 = i} (\sum_{j=1}^d 4^{l_j})^{1/2} \leq d \cdot 2^i$ as in the proof of (3.68). The second energy estimate can be obtained in an analogous way. □

This theorem shows that our approach, indeed, leads to a sparse grid approximation of higher order. For the space $V_n^{(p,1)}$, i.e., for a maximum degree of p in each direction, we get an interpolation error of the order $O(h_n^{p+1} \cdot |\log_2 h_n|^{d-1})$ with respect to both the L_∞- and the L_2-norm. For the energy norm, the result is an error of the order $O(h_n^p)$.

Of course, the above optimization process can be based on the energy norm, too. For that, we start from (4.26) and define the local cost–benefit ratio, now with respect to the energy norm (cf. (3.69)):

$$\operatorname{cbr}_E(\mathbf{l}) := \frac{b_E(\mathbf{l})}{c(\mathbf{l})} \tag{4.29}$$

$$= 10.608 \cdot \left(\frac{5}{4}\right)^d \cdot c^2(\mathbf{p}) \cdot 2^{-|2 \cdot \mathbf{1} \cdot \mathbf{p} + 3 \cdot \mathbf{1}|_1} \cdot \left(\sum_{j=1}^d 4^{l_j}\right) \cdot |u|_{\mathbf{p}+\mathbf{1}, \infty}^2.$$

Although we do not want to study the energy-based sparse grid spaces of higher order $V_n^{(p,E)}$ resulting from (4.29) in detail, we shall nevertheless mention their most important properties. First, we can show that

$$V_n^{(E)} \subseteq V_n^{(p,E)} \subseteq V_n^{(1)} \tag{4.30}$$

is valid for arbitrary n. Second, as long as $V_n^{(p,E)}$ is not the same space as $V_n^{(1)}$, we are rid of the log-factors, again – both concerning the overall cost (which is of the order $O(h_n^{-1})$) and the approximation quality with respect to the energy norm (which is of the order $O(h_n^p)$).

We close this discussion by briefly returning to the notion of ε-complexity. As in the discussion of the piecewise linear counterpart, all occurring order terms have to be read with respect to ε or N. That is, for the following, both d and p are supposed to be arbitrary, but fixed.

Lemma 4.9. For the ε-complexities $N_\infty^{(p)}(\varepsilon)$, $N_2^{(p)}(\varepsilon)$, and $N_E^{(p)}(\varepsilon)$ of the problem of computing the interpolant $u_n^{(p,1)} \in V_n^{(p,1)}$ with respect to the L_∞-, L_2-, and the energy norm for given accuracy ε, the following relations hold:

$$N_\infty^{(p)}(\varepsilon) = O\big(\varepsilon^{-\frac{1}{p+1}} \cdot |\log_2 \varepsilon|^{\frac{p+2}{p+1}\cdot(d-1)}\big), \tag{4.31}$$

$$N_2^{(p)}(\varepsilon) = O\big(\varepsilon^{-\frac{1}{p+1}} \cdot |\log_2 \varepsilon|^{\frac{p+2}{p+1}\cdot(d-1)}\big),$$

$$N_E^{(p)}(\varepsilon) = O\big(\varepsilon^{-\frac{1}{p}} \cdot |\log_2 \varepsilon|^{d-1}\big).$$

Conversely, given a number of N of grid points, the following accuracies can be obtained with respect to the different norms:

$$\varepsilon_\infty^{(p)}(N) = O\big(N^{-(p+1)} \cdot |\log_2 N|^{(p+2)\cdot(d-1)}\big), \tag{4.32}$$

$$\varepsilon_2^{(p)}(N) = O\big(N^{-(p+1)} \cdot |\log_2 N|^{(p+2)\cdot(d-1)}\big),$$

$$\varepsilon_E^{(p)}(N) = O\big(N^{-p} \cdot |\log_2 N|^{p\cdot(d-1)}\big).$$

Finally, for the same problem tackled by the energy-based sparse grid approximation space $V_n^{(p,E)}$, we get

$$N_E^{(p)}(\varepsilon) = O\big(\varepsilon^{-\frac{1}{p}}\big) \quad \text{and} \quad \varepsilon_E^{(p)}(N) = O(N^{-p}). \tag{4.33}$$

Proof. The proof follows exactly the argumentation of the linear case. □

In comparison with the full grid case, where we get $N_\infty^{(p)}(\varepsilon) = O(\varepsilon^{-d/(p+1)})$ and $\varepsilon_\infty^{(p)}(N) = O(N^{-(p+1)/d})$ with respect to the L_∞- or L_2-norm and $N_E^{(p)}(\varepsilon) = O(\varepsilon^{-d/p})$ and $\varepsilon_E^{(p)}(N) = O(N^{-p/d})$ with respect to the energy norm, as before, the sparse grid space $V_n^{(p,1)}$ lessens the curse of dimensionality in a significant manner; $V_n^{(p,E)}$, however, completely overcomes it.

4.3. Interpolets

Another hierarchical multiscale basis with higher-order functions is given by the interpolet family (Deslauriers and Dubuc 1989, Donoho and Yu 1999). These functions are obtained from a simple but powerful interpolation process. For given data $y(s)$, $s \in \mathbb{Z}$, we seek an interpolating function $y : \mathbb{R} \to \mathbb{R}$ which is as smooth as possible. To this end, in a first step, the interpolated values in $\mathbb{Z} + \frac{1}{2}$ are determined. Here, for the determination of $y(s + \frac{1}{2})$, the Lagrangian polynomial $p(x)$ of degree $2n - 1$ is calculated which interpolates the data in $s - n + 1, \ldots, s + n$. Then we set $y(s + \frac{1}{2}) := p(s + \frac{1}{2})$. The parameter n later determines the smoothness of the interpolant and the degree of polynomial exactness. Since the values $y(\mathbb{Z})$ and $y(\mathbb{Z} + \frac{1}{2})$ are now known, we can compute the values $y(\mathbb{Z} + \frac{1}{4})$ and $y(\mathbb{Z} + \frac{3}{4})$ using the same scheme. Here, for example, the values $y(s + \frac{1}{2}(-n + 1)), \ldots, y(s + \frac{n}{2})$ are used for $y(s + \frac{1}{4})$. This way, interpolation values for y can be found on a set which is dense in $\mathbb{R}$. Since the interpolant depends linearly on the data, there exists a fundamental function F with

$$y(x) = \sum_{s \in \mathbb{Z}} y(s) F(x - s).$$

F is the interpolant for the data $y(s) = \delta_{0,s}, s \in \mathbb{Z}$. Now, the interpolet mother function of the order $2n$ is just $\phi := F$. A hierarchical multiscale basis is formed from that ϕ by dilation and translation as in (3.8). The function ϕ has the following properties.

Scaling equation. ϕ is the solution of

$$\phi(x) = \sum_{s \in \mathbb{Z}} h_s \phi(2x - s). \tag{4.34}$$

The mask coefficients $\mathbf{h} := \{h_s\}_{s \in \mathbb{Z}}$ are given by $h_0 = 1$, $h_s = h_{-s}$ ($s \in \mathbb{Z}$), and

$$\begin{pmatrix} 1^0 & 3^0 & \cdots & (2n-1)^0 \\ 1^2 & 3^2 & \cdots & (2n-1)^2 \\ \vdots & \vdots & \vdots & \vdots \\ 1^{2n-2} & 3^{2n-2} & \cdots & (2n-1)^{2n-2} \end{pmatrix} \cdot \begin{pmatrix} h_1 \\ h_3 \\ \vdots \\ h_{2n-1} \end{pmatrix} = \begin{pmatrix} 1/2 \\ 0 \\ \vdots \\ 0 \end{pmatrix}. \tag{4.35}$$

All other h_s are zero.

Compact support.

$$\operatorname{supp} \phi = [-2n + 1, 2n - 1]. \tag{4.36}$$

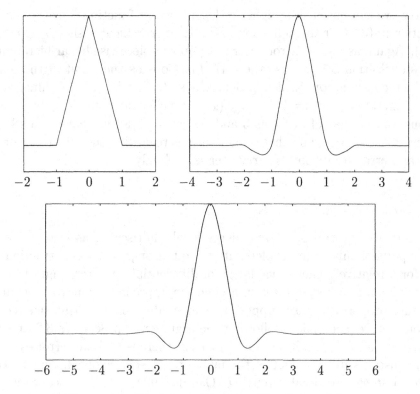

Figure 4.9. The interpolets ϕ for $N = 2, 4$ and 6.

Polynomial exactness. In a pointwise sense, polynomials of degree *less* than $N = 2n$ can be written as linear combinations of translates of ϕ, *e.g.*,

$$\forall\, 0 \leq i < N \; : \; \forall x \in \mathbb{R} \; : \; x^i = \sum_{s \in \mathbb{Z}} s^i \phi(x - s). \qquad (4.37)$$

Interpolation property. With $\delta(x)$ denoting the Dirac functional, we get

$$\forall s \in \mathbb{Z} \; : \; \phi(s) = \delta(s). \qquad (4.38)$$

This is the main property which distinguishes interpolating multiscale bases from other non-interpolating multiscale bases. It allows particularly simple multilevel algorithms for the evaluation of nonlinear terms.

The functions ϕ for different values of the parameter $N = 2n$ are given in Figure 4.9. Note that there is not much difference between the interpolets with $N = 4$ and $N = 6$. This behaviour is quite similar to that of the higher-order polynomials of Section 4.2, which were created by hierarchical Lagrangian interpolation. Note, furthermore, that whilst interpolets are defined on the whole of $\mathbb{R}$, their construction can easily be adapted to a bounded interval; see Koster (2002) for the respective construction.

Now, we can use such a higher-order interpolet function ϕ as the mother function in (3.7) for the definition of the 1D hierarchical basis $\{\phi_{l_j,i_j}(x_j)\}$ in (3.8). Again, as in our introductory example of piecewise linear hierarchical bases of Section 3.1, we use these 1D functions as the input of the tensor product construction, which provides a suitable piecewise d-dimensional basis function $\phi_{\mathbf{l},\mathbf{i}}(\mathbf{x}) := \prod_{j=1}^{d} \phi_{l_j,i_j}(x_j)$ in each grid point $\mathbf{x}_{\mathbf{l},\mathbf{i}}$. In an analogous way to that of Sections 3 and 4.2, we can derive sparse grids based on interpolets. Also, all other considerations regarding the estimates for the cost and error complexities carry over accordingly.

4.4. Prewavelets and wavelets

Note again that the presented hierarchical multiscale bases with higher-order polynomials or interpolets, after the tensor product construction, allow for a relatively cheap evaluation of differential operators and discretization schemes on sparse grids in a Galerkin approach owing to their interpolating properties. Also, upper error estimates can be easily derived for various sparse grid spaces following the arguments in Sections 3.2 and 4.2. However, they all exhibit a main drawback: there is no lower error estimate with constants independent of the number of levels involved, that is, they form no *stable* multiscale splitting (Oswald 1994). The consequences are twofold. First, the (absolute) value of the hierarchical coefficient is just a local error indicator and no true error estimator. We obtain sufficiently refined sparse grids (compare the associated numerical experiments in Section 5) on which the error is properly reduced, but it may happen that too many grid points are employed for a prescribed error tolerance in an adaptive procedure. Efficiency is thus not guaranteed. Second, the condition number of the linear system which results from a symmetric elliptic partial differential operator in multiscale basis representation is, after diagonal scaling, in general not independent of the finest mesh size involved.[3] To obtain a fast mesh independent solver, additional lifting tricks (Sweldens 1997, Daubechies and Sweldens 1998, Koster 2002) or multigrid-type extensions (Oswald 1994, Griebel 1994a, Griebel and Oswald 1994, Griebel and Oswald 1995a, Griebel and Oswald 1995b) are necessary. These difficulties are avoided if we employ stable multiscale splittings (Oswald 1994, Dahmen and Kunoth 1992, Carnicer, Dahmen and Pena 1996, Cohen 2003) and the respective L_2- and H_1-stable multiscale bases, for which two-sided estimates exist.

---

[3] This is the case for stable splittings with wavelets. Then a simple fast solver results from the diagonal scaling preconditioner: see Dahmen and Kunoth (1992) and Oswald (1994).

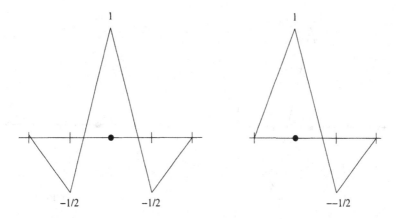

Figure 4.10. Non-orthogonal linear spline wavelet (left: interior; right: boundary).

Since Daubechies' fundamental discovery of orthogonal multiscale bases with local support, *i.e.*, the classical wavelets (Daubechies 1988), an enormous literature has arisen. At present, there exists a whole zoo of such bases from the wavelet family. Here, we stick to simple hierarchical spline-like function systems. Within this class, we have orthonormal wavelets (Lemarié 1988), biorthogonal spline wavelets (Cohen, Daubechies and Feauveau 1992), semi-orthogonal spline wavelets (Chui and Wang 1992), fine grid correction spline wavelets (Cohen and Daubechies 1996, Lorentz and Oswald 1998) and multiwavelets derived from splines. The construction principles of such functions are highly developed, and there is abundant literature on them. Most theory can be found in Daubechies (1992), Chui (1992) and Meyer (1992), and the references cited therein. Further reading is Daubechies (1988, 1993), and Cohen *et al.* (1992, 2001). A nice introduction, similar in spirit to this paper, is given in Cohen (2003).

In the following, let us briefly mention the simplest but also cheapest mother wavelets which are made up from piecewise linears by stable completion procedures (Carnicer *et al.* 1996). They are sufficient to be used for a second-order PDE within the Galerkin method.

These are the so-called linear pre-prewavelets, which are a special case of non-orthogonal spline wavelets (Cohen and Daubechies 1996, Lorentz and Oswald 1998, Stevenson 1996) of linear order. The corresponding mother function ϕ is shown in Figure 4.10 (left). Again, by translation and dilation in an analogous way to (3.8), we get a 1D multilevel basis from this ϕ. Note that a modification is necessary near the boundary. For homogeneous Dirichlet conditions, the scaled and dilated function in Figure 4.10 (right) can be used in points next to the boundary.

Another, still reasonably cheap approach involves linear prewavelets, a special case of semi-orthogonal spline wavelets (Chui and Wang 1992, Griebel

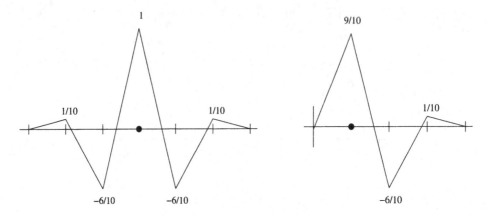

Figure 4.11. Semi-orthogonal linear spline wavelet (left: interior; right: boundary).

and Oswald 1995b). The corresponding mother function ϕ is shown in Figure 4.11 (left). Again, a modification is necessary near the boundary. For homogeneous Dirichlet conditions, this is shown in Figure 4.11 (right). A construction for Neumann conditions can be found in Griebel and Oswald (1995b).

Finally, let us mention the so-called lifting wavelets. They are the result of the application of Sweldens' lifting scheme (Carnicer et al. 1996, Sweldens 1997) to the interpolet basis $\{\phi_{l_j,i_j}(x_j)\}$ or other interpolatory multiscale bases, for example. Here, the hierarchical (i.e., if we consider odd indices only; see (3.11)) lifting wavelets are defined by

$$\hat{\phi}_{l_j,2i_j+1} := \phi_{(l_j,2i_j+1)} + \sum_{s_j} Q^{l_j}_{s_j,2i_j+1}\phi_{(l_j-1,s_j)}$$

on finer levels $l_j > 0$. The basic idea is to choose the weights $Q^{l_j}_{s,i_j}$ in this linear combination in such a way that $\hat{\phi}_{l_j,i_j}$ has more vanishing moments than ϕ_{l_j,i_j}, and thus to obtain a stabilization effect. If we apply this approach to the hierarchical interpolet basis of Section 4.3 so that we achieve two vanishing moments in the lifting wavelet basis, we end up with

$$\hat{\phi}_{l_j,2i_j+1} = \phi_{(l_j,2i_j+1)} - \frac{1}{4}(\phi_{(l_j-1,i_j)} + \phi_{(l_j-1,i_j+1)})$$

for the odd indices $2i_j + 1$. The corresponding mother function $\hat{\phi}$ is shown in Figure 4.12.

Again, as in our introductory example of piecewise linear hierarchical bases of Section 3.1, we can use these 1D multilevel basis functions as the input of the tensor product construction which provides a suitable piecewise d-dimensional basis function $\phi_{\mathbf{l},\mathbf{i}}(\mathbf{x})$ in each grid point $\mathbf{x}_{\mathbf{l},\mathbf{i}}$. As in Sections 3

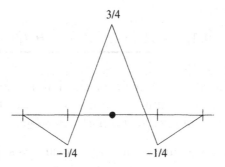

Figure 4.12. Lifting interpolet with $N = 2$ and two vanishing moments.

and 4.2, we can derive sparse grids based on wavelets. Also, most other considerations regarding the estimates for the cost and error complexities carry over accordingly. Further information on wavelets and sparse grids can be found in Griebel and Oswald (1995b), Hochmuth (1999), DeVore, Konyagin and Temlyakov (1998), Koster (2002), Knapek (2000a), Griebel and Knapek (2000) and the references cited therein.

At the end of this subsection, let us dwell a bit into theory, and let us give an argument why there is a difference between hierarchical polynomials and interpolets on the one hand and prewavelets and pre-prewavelets on the other hand.

For all mother functions and the resulting product multiscale bases and sparse grid subspaces, we can, in principle, follow the arguments and proofs of Sections 3 and 4.2, respectively, to obtain cost complexity estimates and upper error bounds with relatively sharp estimates for the order constants. The main tools in the proofs are the simple triangle inequality and geometric series arguments. However, as already mentioned, a lower bound for the error can not be obtained so easily. An alternative approach is that of Griebel and Oswald (1995b) and Knapek (2000a), where we developed a technique for which two-sided error norm estimates for the 1D situation can be carried over to the higher-dimensional case. The approach is based on the representation of Sobolev spaces $H^s([0,1]^d), s \geq 0$, as

$$H^s([0,1]^d) = \bigcap_{i=1}^{d} \underbrace{L_2([0,1]) \otimes \cdots \otimes L_2([0,1])}_{(i-1) \text{ times}} \otimes H^s([0,1]) \otimes \underbrace{L_2([0,1]) \otimes \cdots \otimes L_2([0,1])}_{(d-i) \text{ times}}. \qquad (4.39)$$

The Sobolev space $H^s_{\text{mix}}([0,1]^d), s \geq 0$, on the other hand is defined as the simple tensor product

$$H^s_{\text{mix}}([0,1]^d) = \underbrace{H^s([0,1]) \otimes \cdots \otimes H^s([0,1])}_{d \text{ times}}.$$

Now, for the different components of the intersection, we obtain two-sided norm estimates, if the univariate multiscale functions ϕ_{l_j,i_j} of (3.8) allow two-sided norm estimates for both H^s and L_2.

Theorem 4.10. Let the univariate multiscale basis $\{\phi_{l_j,i_j}\}$ satisfy the norm equivalence

$$\|u\|^2_{H^s} \sim \sum_{l_j} 2^{-l_j} \sum_{i_j} 2^{2l_j s} |u_{l_j,i_j}|^2, \qquad (4.40)$$

where $u(x_j) = \sum_{l_j,i_j} u_{l_j,i_j} \phi_{l_j,i_j}(x_j)$, and for $-\gamma_1 < s < \gamma_2$, $\gamma_1, \gamma_2 > 0$. Then, the multivariate basis functions $\{\phi_{\mathbf{l},\mathbf{i}}(\mathbf{x})\}$ fulfil the norm equivalences

$$\|u\|^2_{H^s} = \left\| \sum_{\mathbf{l},\mathbf{i}} u_{\mathbf{l},\mathbf{i}} \phi_{\mathbf{l},\mathbf{i}} \right\|^2_{H^s} \sim \sum_{\mathbf{l},\mathbf{i}} 2^{2s|\mathbf{l}|_\infty} |u_{\mathbf{l},\mathbf{i}}|^2 2^{-|\mathbf{l}|_1},$$

$$\|u\|^2_{H^s_{\text{mix}}} = \left\| \sum_{\mathbf{l},\mathbf{i}} u_{\mathbf{l},\mathbf{i}} \phi_{\mathbf{l},\mathbf{i}} \right\|^2_{H^s_{\text{mix}}} \sim \sum_{\mathbf{l},\mathbf{i}} 2^{2s|\mathbf{l}|_1} |u_{\mathbf{l},\mathbf{i}}|^2 2^{-|\mathbf{l}|_1},$$

where $u(\mathbf{x}) = \sum_{\mathbf{l},\mathbf{i}} u_{\mathbf{l},\mathbf{i}} \phi_{\mathbf{l},\mathbf{i}}(\mathbf{x})$.

Here, $\sim$ denotes a two-sided equivalence, *i.e.*, $a \sim b$ means that there exist positive constants c_1, c_2 such that $c_1 \cdot b \leq a \leq c_2 \cdot b$. Note the distinct difference in the quality of the two estimates. It is exactly the difference between $|\mathbf{l}|_\infty$ and $|\mathbf{l}|_1$ which leads to a significant reduction in ε-complexity, *i.e.*, it allows us to use substantially fewer degrees of freedom to reach the same truncation error in the H^s_{mix}-norm than in the H^s-norm.

For the proof and further details, see Griebel and Oswald (1995b), Knapek (2000a) and Koster (2002). Here, the bounds γ_1, γ_2 for the range of the regularity parameter s depend on the specific choice of the mother function ϕ. The value γ_2 is determined by the Sobolev regularity of ϕ, *i.e.*, $\gamma_2 = \sup\{s : \phi \in H^s\}$. The theory here works with biorthogonality arguments and dual spaces: our univariate multiscale basis $\{\phi_{l_j,i_j}\}$ possesses a *dual basis* $\{\tilde{\phi}_{l_j,i_j}\}$, in the sense that (for our hierarchical functions[4]) the

---

[4] In the general setting, there is a multiresolution analysis with spaces spanned by scaling functions and difference spaces spanned by wavelets. Then, biorthogonality conditions must be fulfilled between all these function systems. In the hierarchical approach, certain scaling functions are simply wavelets, and the biorthogonality conditions are reduced.

biorthogonality relations (Cohen *et al.* 1992)

$$\langle \tilde{\phi}_{l_j,i_j}, \phi_{l_j,k_j} \rangle = \delta_{i_j,k_j}, \qquad \langle \tilde{\phi}_{l_j-1,i_j}, \phi_{l_j,k_j} \rangle = \delta_{i_j,k_j}.$$

If the primal and dual functions are at least in L_2, then $\langle .,. \rangle$ is the usual L_2-inner product. The value of γ_1 is here just the Sobolev regularity of the dual mother function $\tilde{\phi}$.

Note that, with $\gamma_1, \gamma_2 > 0$, the 1D norm equivalence for $s = 0$, and thus the L_2-stability of the 1D hierarchical basis $\{\phi_{l_j,i_j}\}$, is a prerequisite in Theorem 4.10. However, the linear hierarchical bases of Section 3, the higher-order Lagrangian interpolation functions of Section 4.2 and the interpolets of Section 4.3 do not fulfil this prerequisite, so they are not L_2-stable.[5] Then we obtain simply one-sided estimates. For these function systems, we get the same upper estimates with our simple tools as the triangle inequality and the geometric series arguments, but we can pinpoint, additionally, quite sharp bounds for the constants. The one-sided estimates still allow us to use the hierarchical coefficients $|u_{\mathbf{l},\mathbf{i}}|$ (*e.g.*, after suitable weighting) as error indicators in a refinement procedure, but give us no true error estimator. To this end, grids are constructed in an adaptation process steered by the indicator such that a prescribed global error tolerance is reached. This aim, however, might be reached by using more points than necessary. For the function systems built from the wavelet-type mother functions, the prerequisite of L_2-stability is fulfilled, and we obtain two-sided estimates. Thus, the wavelet coefficients $|u_{\mathbf{l},\mathbf{i}}|$ can serve as local error estimators, and the lower estimate part then gives us efficiency of the corresponding adaptive scheme, which means that only the necessary number of points (up to a constant) to reach a prescribed error tolerance is employed in a grid adaptation process. Furthermore, a fast solver can now be gained with level-independent convergence rate by simple diagonal scaling as a preconditioner in the multiscale system.

Note that from relation (4.39) it also becomes clear that the constants in the multivariate norm equivalence are more or less given by the dth power of the constants in the univariate equivalence. This causes rather large constants for higher dimensionality d.

4.5. Sparse grid applications

PDE discretization techniques

Since its introduction, the sparse grid concept has been applied to most of the relevant discretization schemes for PDEs. These are finite element

---

[5] The reason is that the dual 'functions' are Dirac functionals. Then $\gamma_1 = -\frac{1}{2}$ and the norm equivalence (4.40) is only valid for $s \in]\frac{1}{2}, \gamma_2[$, *i.e.*, it does not hold for $s = 0$.

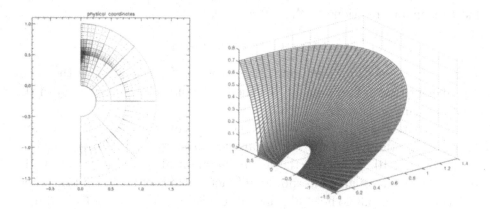

Figure 4.13. Laplace equation with a singularity on the boundary: adaptive sparse grid (left) and corresponding solution (right).

methods and Galerkin techniques, finite differences, finite volumes, spectral methods, and splitting extrapolation, which leads to the combination technique.

The first main focus of the development of sparse grid discretization methods was on piecewise linear *finite elements*. In the pioneering work of Zenger (1991) and Griebel (1991b), the foundations for adaptive refinement, multilevel solvers, and parallel algorithms for sparse grids were laid. Subsequent studies included the solution of the 3D Poisson equation Bungartz (1992a, 1992b), the generalization to arbitrary dimensionality d (Balder 1994) and to more general equations (the Helmholtz equation (Balder and Zenger 1996), parabolic problems using a time–space discretization (Balder, Rüde, Schneider and Zenger 1994), the biharmonic equation (Störtkuhl 1995), and general linear elliptic operators of second order in 2D (Pflaum 1996, Dornseifer and Pflaum 1996). As a next step, the solution of general linear elliptic differential equations and, via mapping techniques, the treatment of more general geometries was implemented (Bungartz and Dornseifer 1998, Dornseifer 1997) (see Figure 4.13). Since then, algorithmic improvements for the general linear elliptic operator of second order have been studied in detail (Schneider 2000, Achatz 2003a, Achatz 2003b).

In the first experiments with polynomial bases of higher-order, bicubic Hermite bases for the biharmonic equation (Störtkuhl 1995) and hierarchical Lagrange bases of degree two (Bungartz 1996) as well as of arbitrary degree (Bungartz and Dornseifer 1998, Bungartz 1997) were studied. The general concept of the hierarchical Lagrangian interpolation on sparse grids was introduced in Bungartz (1998). Afterwards, the higher-order concept was combined with a duality-based adaptivity approach (Schneider 2000) and with operator extension (Achatz 2003a). In addition to that, aspects of complexity, especially for the solution of the Poisson equation (Bungartz and Griebel 1999), were discussed with regard to Werschulz (1995). Furthermore, besides the classical hierarchical bases according to Section 3, prewavelet bases (Oswald 1994, Griebel and Oswald 1995b) and wavelet decompositions (Sprengel 1997a) were considered. Also, the parallelization of the sparse grid algorithms was pursued: see Pfaffinger (1997), Hahn (1990), Griebel (1991a, 1991b).

Another development in the discretization of PDEs on sparse grids using second-order *finite differences* was derived and implemented in Griebel (1998), Griebel and Schiekofer (1999), Schiekofer (1998), and Schiekofer and Zumbusch (1998). There, based on Taylor expansions and sufficient regularity assumptions, the same orders of consistency can be shown as they are known from the comparable full grid operators. This means, in particular, that the typical log-term does not occur. Generalizations to higher-order finite difference schemes using conventional Taylor expansion (Schiekofer 1998) and interpolets (Koster 2002) as discussed in Section 4.3 were also developed. The conceptual simplicity of finite difference schemes allowed for straightforward adaptive refinement strategies based on hashing (Griebel 1998, Schiekofer 1998), for the application of sparse grids to elliptic problems with general coefficient functions, and for the handling of general domains and nonlinear systems of parabolic PDEs of reaction–diffusion type (Schiekofer 1998). This opened the way to a first solution of the Navier–Stokes equations on sparse grids: see Schiekofer (1998), Griebel and Koster (2000), Koster (2002) and Griebel and Koster (2003).

Furthermore, the so-called *combination technique* (Griebel, Schneider and Zenger 1992), an extrapolation-type sparse grid variant, has been discussed in several papers (Bungartz, Griebel, Röschke and Zenger 1994b, 1994c, Bungartz, Griebel and Rüde 1994a, Pflaum and Zhou 1999). Because a sparse grid can be represented as a superposition of several (much coarser) full grids Ω_l (see Figure 4.14 for the 2D case), a sparse grid solution for some PDEs can be obtained from the linear combination of solutions u_l computed on the respective coarse grids. Since the latter ones are regular full grids, existing solvers can be used without any need for an explicit discretization on a sparse grid. For the 2D case, we get

$$u_n^{(c)}(\mathbf{x}) := \sum_{|\mathbf{l}|_1 = n+1} u_\mathbf{l}(\mathbf{x}) - \sum_{|\mathbf{l}|_1 = n} u_\mathbf{l}(\mathbf{x}), \qquad (4.41)$$

and the 3D combination formula is given by

$$u_n^{(c)}(\mathbf{x}) := \sum_{|\mathbf{l}|_1 = n+2} u_\mathbf{l}(\mathbf{x}) - 2 \cdot \sum_{|\mathbf{l}|_1 = n+1} u_\mathbf{l}(\mathbf{x}) + \sum_{|\mathbf{l}|_1 = n} u_\mathbf{l}(\mathbf{x}). \qquad (4.42)$$

Note that, in contrast to (3.19), where $u_\mathbf{l} \in W_\mathbf{l}$, $u_\mathbf{l}$ denotes the approximate solution of the underlying PDE on the regular full grid $\Omega_\mathbf{l}$. Compared with the direct discretization on sparse grids, the combination technique

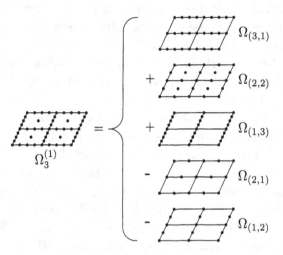

Figure 4.14. Combination technique for $d = 2$: combine coarse full grids $\Omega_{\mathbf{l}}$, $|\mathbf{l}|_1 \in \{n, n+1\}$, with mesh widths 2^{-l_1} and 2^{-l_2} to get a sparse grid $\Omega_n^{(1)}$ corresponding to $V_n^{(1)}$.

turns out to be advantageous in two respects. First, the possibility of using existing codes allows the straightforward application of the combination technique to complicated problems. Second, since the different subproblems can be solved fully in parallel, there is a very elegant and efficient inherent coarse-grain parallelism that makes the combination technique perfectly suited to modern high-performance computers (Griebel 1992, 1993). This has been shown in a series of papers dealing with problems from computational fluid dynamics, including turbulence simulation (Bungartz and Huber 1995, Huber 1996b, Griebel and Thurner 1993, Griebel, Huber and Zenger 1993a, Griebel 1993, Griebel and Thurner 1995, Griebel and Huber 1995, Griebel, Huber and Zenger 1996, Huber 1996a). Furthermore, note that there are close relations to the above-mentioned Boolean and blending methods (Delvos 1982, Delvos and Schempp 1989, Gordon 1969, Gordon and Hall 1973, Baszenski et al. 1992, Hennart and Mund 1988) as well as to the so-called *splitting extrapolation method* (Liem, Lu and Shih 1995).

Concerning alternative discretization schemes, there are ongoing investigations on sparse grid finite *volume methods* for the Euler equations (Hemker 1995, Hemker and de Zeeuw 1996, Koren, Hemker and de Zeeuw 1996). In this context, a variant – the so-called *sets of grids* or *grids of grids* – has been developed (Hemker and Pflaum 1997, Hemker, Koren and Noordmans 1998). This approach goes back to the notion of semi-coarsening (Mulder 1989, Naik and van Rosendale 1993). Furthermore, *spectral methods* have been implemented on sparse grids and studied for PDEs with periodic boundary conditions (Kupka 1997).

Adaptivity and fast solvers
Especially in the context of PDE, two ingredients are essential for each kind of discretization scheme: adaptive mesh refinement and fast linear solvers.

Concerning the first, grid adaptation based on local error estimators (typically closely related to the hierarchical surplus, *i.e.*, the hierarchical basis coefficients) is a very widely used approach (Zenger 1991, Bungartz 1992*a*, 1992*b*, Griebel 1998, Koster 2002, Griebel and Koster 2000). However, in Schneider (2000), the *dual problem approach* (Becker and Rannacher 1996, Eriksson, Estep, Hansbo and Johnson 1996) has been successfully applied to sparse grids, allowing for global error estimation too. As a result, the error control can be governed by some appropriate linear error functional, and the respective refinement strategies include l_2-driven as well as point-driven adaptation.

Concerning the efficient multilevel solution of the resulting linear systems, a number of contributions using additive and multiplicative subspace correction schemes have been made (Griebel 1991*a*, 1991*b*, Pflaum 1992, Griebel, Zenger and Zimmer 1993*b*, Griebel 1994*b*, Griebel and Oswald 1994, 1995*a*, 1995*b*, Bungartz 1997, Pflaum 1998), showing the availability of fast algorithms for the solution of the linear systems resulting from sparse grid discretizations.

Applications
Flow problems were the first focus for the use of sparse grid PDE solvers. Now, however, sparse grids are also used for problems from quantum mechanics (Garcke 1998, Garcke and Griebel 2000, Yserentant 2004, Hackbusch 2001), for problems in the context of stochastic differential equations (Schwab and Todor 2002, 2003), or for the discretization of differential forms arising from Maxwell's equations (Hiptmair and Gradinaru 2003).

Aside from the field of PDEs, sparse grids are being applied to a variety of problems that will not be forgotten here. Among these problems are integral equations (Frank, Heinrich and Pereverzev 1996, Griebel, Oswald and Schiekofer 1999, Knapek and Koster 2002, Knapek 2000*b*), general operator equations (Griebel and Knapek 2000, Knapek 2000*a*, Hochmuth, Knapek and Zumbusch 2000), eigenvalue problems (Garcke 1998, Garcke and Griebel 2000), periodic interpolation (Pöplau and Sprengel 1997), interpolation on Gauss–Chebyshev grids (Sprengel 1997*b*), Fourier transforms (Hallatschek 1992), tabulation of reduced chemical systems (Heroth 1997), digital elevation models and terrain representation Gerstner (1995, 1999), audio and image compression (Frank 1995, Paul 1995) (see Figure 4.15), and possibly others not listed here.

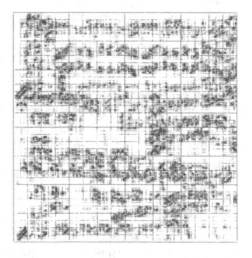

Figure 4.15. Data compression with sparse grids: Mozart's autograph of the fourth movement of KV 525 'Eine kleine Nachtmusik'. Top left: original image. Top right: compressed version. Right: corresponding adaptive sparse grid.

Of course, owing to the historic background described in Section 4.1, numerical quadrature has always been a hot topic in sparse grid research. Starting from the explicit use of the piecewise linear basis functions to calculate integrals Bonk (1994b, 1994a), quadrature formulas based on the midpoint rule (Baszenski and Delvos 1993), the rectangle rule (Paskov 1993), the trapezoidal rule (Bonk 1994a), the Clenshaw–Curtis rule (Cools and Maerten 1997, Novak and Ritter 1998), the Gauss rules (Novak and Ritter 1997), and the Gauss–Patterson rules (Gerstner and Griebel 1998, Petras 2000) were used as 1D input for Smolyak's principle according to (4.1). Furthermore, the higher-order basis polynomials from Section 4.2 were used for an adaptive quadrature algorithm (Bungartz and Dirnstorfer 2003). These techniques can be used advantageously for the computation of stiffness matrix entries in conventional finite element methods and especially in partition of unity methods (Griebel and Schweitzer 2002, Schweitzer 2003). Gerstner and Griebel (2003) developed a generalization of the conventional sparse grid approach which is able to adaptively assess the dimensions according to their importance. For further information on numerical quadrature on sparse grids and on data mining, another interesting field of application, we refer to the respective numerical examples in Section 5.3.

5. Numerical experiments

In this section, we present a collection of numerical results for different problems solved on sparse grids. We start with the discussion of the basic features of sparse grid methods for PDEs, applied to simpler 2D and 3D model problems. Then we turn to the solution of the Navier–Stokes equations on sparse grids. Finally, we illustrate the potential of sparse grids for problems of a higher dimensionality, here in the context of numerical quadrature and data mining.

5.1. PDE model problems

We start with some PDE model problems to demonstrate the interpolation properties of the L_2-based sparse grid spaces presented as well as their behaviour concerning the approximation of a PDE's solution. We omit examples with energy-based sparse grids, since the situation with respect to the energy norm in a finite element context is completely clear owing to Céa's lemma, whereas the L_2-quality of the finite element solution is still an open question.

Hence we choose a standard Ritz–Galerkin finite element setting and use the hierarchical Lagrangian basis polynomials from Section 4.2 in the p-regular way, *i.e.*, choosing the local degree as high as possible according to (4.16) up to some given maximum value for p. Note that a Petrov–Galerkin approach with the higher-order basis functions appearing in the approximation space only leads to the same approximation properties. We will not discuss the algorithmic scheme which is applied to ensure the linear complexity of the matrix–vector product, the so-called *unidirectional principle* (Bungartz 1998), allowing us to treat the single dimensions separately. However, note that, especially in the case of more general operators, this is actually the challenging part of a sparse grid finite element implementation; see Dornseifer (1997), Bungartz and Dornseifer (1998), Schneider (2000), and Achatz (2003a, 2003b). Concerning the iterative solution of the resulting linear systems, several multilevel schemes are available, most of them based upon the semi-coarsening approach of Mulder (1989) and Naik and van Rosendale (1993) and its sparse grid extension (Griebel 1991b, 1991a, Griebel *et al.* 1993b, Griebel and Oswald 1995b, Pflaum 1992, 1998, Bungartz 1997). In the context of solvers, it is important that the influence

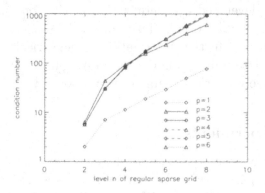

Figure 5.1. Condition numbers of the (diagonally preconditioned) stiffness matrix for the sparse grid spaces $V_n^{(p,1)}$.

of the polynomial degree starting from $p = 2$ be only moderate, which is in contrast to the behaviour known from hierarchical polynomials in a p- or h-p-version context (*cf.* Zumbusch (1996), for example). Figure 5.1 hints this, showing spectral condition numbers of the stiffness matrix.

Our model problems cover regular and adaptively refined sparse grids in two and three dimensions, mainly using the Laplacian as the operator, but also the second-order linear operator with more general coefficients. For measuring the error, we consider the error's discrete maximum norm, the discrete l_2-norm, and the error in the centre point $P = (0.5, \ldots, 0.5)$. Where grid adaptation is used, it is based on a sparse grid extension of the dual problem approach (Becker and Rannacher 1996, Eriksson *et al.* 1996); see Schneider (2000).

The 2D Laplace equation
On $\bar{\Omega} = [0,1]^2$, let

$$\Delta u(\mathbf{x}) = 0 \qquad \text{in } \Omega, \qquad (5.1)$$

$$u(\mathbf{x}) = \sin(\pi x_1) \cdot \frac{\sinh(\pi x_2)}{\sinh(\pi)} \qquad \text{on } \partial\Omega,$$

Figure 5.2 shows the smooth solution $u(\mathbf{x})$ of (5.1). We study the accuracy of the hierarchical Lagrangian approach for the regular sparse grid spaces $V_n^{(p,1)}$ with polynomial degrees $p \in \{1, \ldots, 6\}$. Figure 5.3 illustrates the maximum and the l_2-error. The effect of the higher-order approximation can be seen clearly. For $p = 6$, we already get troubles with machine accuracy because of the use of standard double precision floating point numbers.

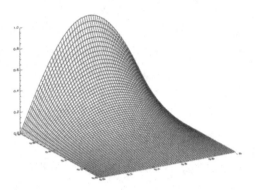

Figure 5.2. Solution $u(\mathbf{x}) = \sin(\pi x_1) \cdot \frac{\sinh(\pi x_2)}{\sinh(\pi)}$ of (5.1).

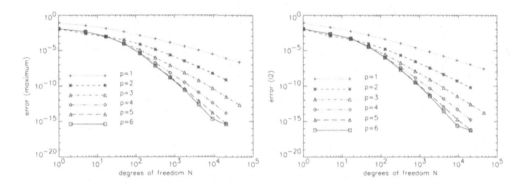

Figure 5.3. Example (5.1): maximum and l_2-error for regular sparse grid spaces $V_n^{(p,1)}$, $p \in \{1, \ldots, 6\}$.

In Figure 5.4, a summary of the convergence behaviour with respect to all error norms regarded and for $p \in \{1, \ldots, 6\}$ is provided. In addition to the error plots, we show the curves of expected sparse grid convergence (sgc) due to the results (3.88) and (4.32), stating

$$\varepsilon_\infty(N) = O\big(N^{-2} \cdot |\log_2 N|^{3 \cdot (d-1)}\big), \tag{5.2}$$
$$\varepsilon_\infty^{(p)}(N) = O\big(N^{-(p+1)} \cdot |\log_2 N|^{(p+2) \cdot (d-1)}\big).$$

These curves indicate the (asymptotic) behaviour of the ε-complexity of sparse grids with respect to the problem of interpolating a given function. For all polynomial degrees presented, we observe a rather striking correspondence of theoretical results concerning the mere quality of interpolation and experimental results for the accuracy of calculating approximate solutions of PDEs. Note that this correspondence is still an open question with respect to the L_2- or maximum norm, in contrast to the energy norm for which the answer is provided by Céa's lemma.

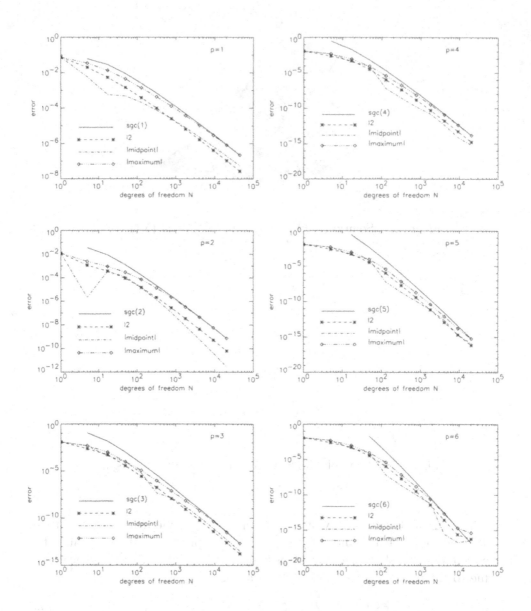

Figure 5.4. Example (5.1): convergence on regular sparse grid spaces $V_n^{(p,1)}$ for $p \in \{1,2,3\}$ (left) and $p \in \{4,5,6\}$ (right); the solid lines indicate the respective expected sparse grid convergence (sgc) due to the interpolation accuracy.

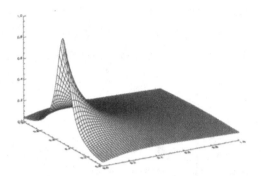

Figure 5.5. Solution $u(\mathbf{x}) = \frac{0.1 \cdot (x_1+0.1)}{(x_1+0.1)^2+(x_2-0.5)^2}$ of example (5.3).

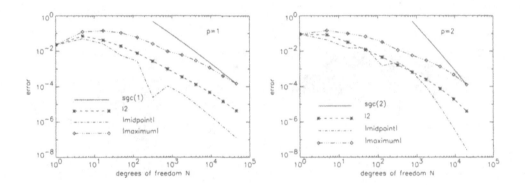

Figure 5.6. Example (5.3): convergence on $V_n^{(p,1)}$ for $p = 1$ (left) and $p = 2$ (right); the solid lines indicate the respective expected sparse grid convergence (sgc; position of curve chosen for clarity).

Next, we turn to adaptively refined sparse grids. On $\bar{\Omega} = [0,1]^2$, let

$$\Delta u(\mathbf{x}) = 0 \qquad \text{in } \Omega, \tag{5.3}$$

$$u(\mathbf{x}) = \frac{0.1 \cdot (x_1+0.1)}{(x_1+0.1)^2 + (x_2-0.5)^2} \qquad \text{on } \partial\Omega.$$

In $\mathbf{x} = (-0.1, 0.5)$, $u(\mathbf{x})$ has a singularity which is outside $\bar{\Omega}$, but close to the boundary $\partial\Omega$. Figure 5.5 shows the solution $u(\mathbf{x})$.

Figure 5.6 illustrates the case of the regular sparse grid spaces $V_n^{(1,1)}$ (left) and $V_n^{(2,1)}$ (right). Obviously, things are not too convincing for reasonable N. However, decisive progress can be made if we apply our l_2-adaptivity.

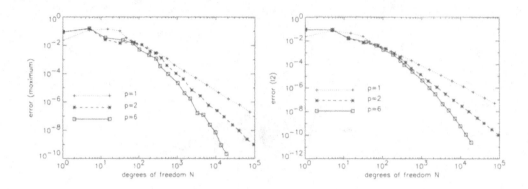

Figure 5.7. Example (5.3): maximum and l_2-error for adaptive refinement, $p \in \{1, 2, 6\}$.

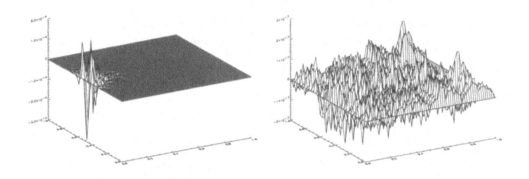

Figure 5.8. Example (5.3): error on regular (left) and adaptive sparse grid (right).

Figure 5.7 shows a gain in accuracy with higher p that is comparable to the smooth situation of (5.1).

In Figure 5.8, we compare the error on a regular sparse grid and on an adaptively refined one. As expected, the l_2-adaptation process reduces the error equally over the whole domain. In contrast to that, regular sparse grids show large errors near the singularity.

As for (5.1), the achieved accuracy will be compared to the theoretical results concerning the ε-complexity of interpolation on sparse grids. In Figure 5.9, again for $p = 1$, the correspondence is striking. For $p = 6$, it seems that the asymptotic behaviour needs bigger values of N to appear. This was to be expected, since our hierarchical Lagrangian basis polynomials of degree $p = 6$ need at least level 5 to enter the game. However, the adaptation process causes a delayed creation of new grid points in the

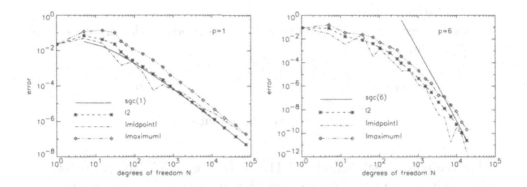

Figure 5.9. Example (5.3): convergence with adaptive mesh refinement for $p = 1$ (left) and $p = 6$ (right); the solid lines indicate the respective expected sparse grid convergence (sgc; position of curve chosen for clarity).

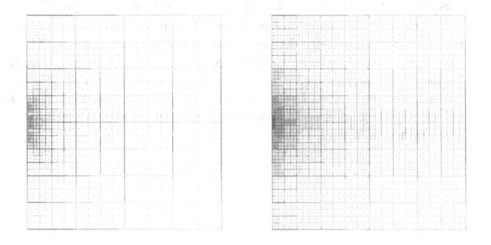

Figure 5.10. Example (5.3): adaptive grids for $p = 1$ (left) and $p = 6$ (right).

smooth areas. Nevertheless, with the adaptive refinement advancing, the higher-order accuracy comes to fruition.

Finally, to get an impression of the adaptation process, Figure 5.10 shows two adaptively refined grids with 7 641 grid points ($p = 1$, left), and 10 965 grid points ($p = 6$, right).

With these results supporting the efficiency of the combination of higher-order approximation and adaptive mesh refinement on sparse grids, we close the discussion of the 2D Laplacian and turn to 3D problems.

The 3D Laplace equation

For the 3D case, we restrict ourselves to a smooth model problem. On $\bar{\Omega} = [0,1]^3$, let

$$\Delta u(\mathbf{x}) = 0 \qquad \text{in } \Omega, \qquad (5.4)$$

$$u(\mathbf{x}) = \frac{\sinh(\sqrt{2}\pi x_1)}{\sinh(\sqrt{2}\pi)} \cdot \sin(\pi x_2) \cdot \sin(\pi x_3) \qquad \text{on } \partial\Omega.$$

For the polynomial degrees $p \in \{1, \ldots, 4\}$, Figure 5.11 compares the accuracy with respect to the error's maximum norm or its l_2-norm, respectively. Again, the effects of the improved approximation properties of our hierarchical polynomial bases are evident.

Figure 5.12 shows that we do already come quite close to the asymptotic behaviour predicted for the quality of mere interpolation. This is true in spite of the fact that up to 100 000 degrees of freedom are not excessive for a 3D problem. Remember that, although the curse of dimensionality is lessened significantly by the L_2- or L_∞-based sparse grid spaces $V_n^{(1)}$ and $V_n^{(p,1)}$, there is still some d-dependence. Evaluating the respective order terms of the ε-complexity derived before and given once more in (5.2) for the 3D case, we observe an exponent of $(p+2) \cdot (d-1) = 2p + 4$ in the log-factor, *i.e.*, an exponent of 12 for polynomial degree $p = 4$, for example. Nevertheless, as in the 2D case of (5.1) and (5.3), the benefit caused by higher polynomial degrees in combination with sparse grid discretization is again rather impressive.

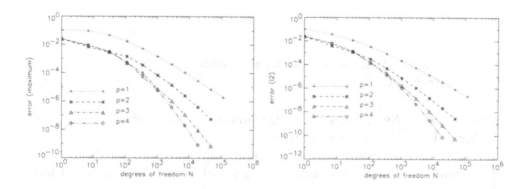

Figure 5.11. Example (5.4): maximum and l_2-error for regular sparse grid spaces $V_n^{(p,1)}$, $p \in \{1, 2, 3, 4\}$.

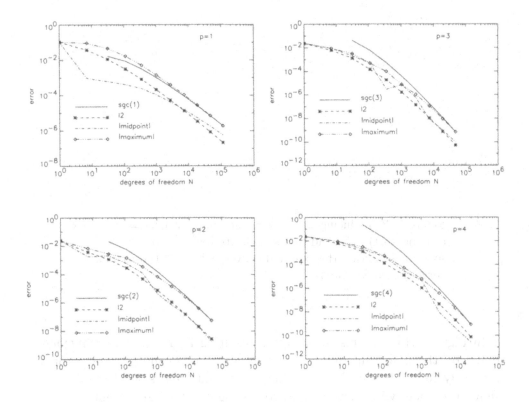

Figure 5.12. Example (5.4): convergence on $V_n^{(p,1)}$, $p \in \{1,2\}$ (left) and $p \in \{3,4\}$ (right); the solid lines indicate the respective expected sparse grid convergence (sgc; position of curve chosen for clarity).

Towards greater generality

Figure 5.13 illustrates the effect of using the point functional with respect to the midpoint P of Ω for the dual problem-based control of adaptive refinement instead of the l_2-adaptation regarded up to now. It can be shown that the resulting error estimator is strongly related to the energy of the basis function living in P. Since we have an $O(h^p)$-behaviour of the energy norm, the energy decreases of the order $O(h^{2p})$. Consequently, a convergence behaviour of $O(h^{2p})$ for the error in the midpoint can be expected. Actually, for the quadratic case, this can be seen from Figure 5.13 for both example (5.1) and even for the root singularity $\text{Re}\,(z^{1/2})$ on $\Omega := \,]0,1[\,\times\,]-\frac{1}{2},\frac{1}{2}[$. Note that the solid line now indicates an N^{-4}-behaviour. In addition to (5.3), the results for the root singularity show that we can tackle really singular problems too.

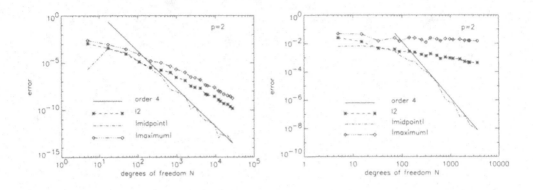

Figure 5.13. Reducing the error in the midpoint for $p = 2$: example (5.1) (left) and root singularity $\operatorname{Re}(z^{1/2})$ (right); the solid lines indicate the expected convergence N^{-4} (position of curve chosen for clarity).

In Figure 5.14, we present the error resulting from adaptive refinement based on the point functional for example (5.1). Obviously, the error is primarily reduced near the midpoint. The right-hand side of Figure 5.14 shows the underlying adaptive grid consisting of 12 767 grid points.

Since all examples have so far been treated with a p-regular approach, we want to present at least one result for a p-asymptotic proceeding, *i.e.*, for the scenario of increasing p for each new level l of grid points occurring. On $\Omega = \,]0,1[^2$, let

$$-\Delta u(\mathbf{x}) = -2 \cdot \frac{x_1 \cdot (x_1^2 x_2^2 - 3x_2^2 + x_1^4 + 2x_1^2 + 1)}{(x_1^2 + 1)^3} \quad \text{in } \Omega, \qquad (5.5)$$

$$u(\mathbf{x}) = \frac{x_1 x_2^2}{x_1^2 + 1} \qquad \text{on } \partial\Omega.$$

According to the p-asymptotic strategy, in Figure 5.15, the polynomial degree is chosen as $\mathbf{p} := l + 1$ up to $p = 12$. In order to avoid trouble with machine accuracy, we switch to quadruple precision. Of course, striving for twenty-five or more correct digits is not particularly useful. Nevertheless, we obtain a convincing diagram that shows the sub-exponential behaviour of the p-asymptotic strategy which, by the way, performs better than theory had suggested.

The last finite element example of this section demonstrates that our approach is not limited to simple operators such as the Laplacian. Here we turn to a variable diffusion coefficient. For a detailed discussion of the general linear elliptic operator of second order, see Achatz (2003b), for instance.

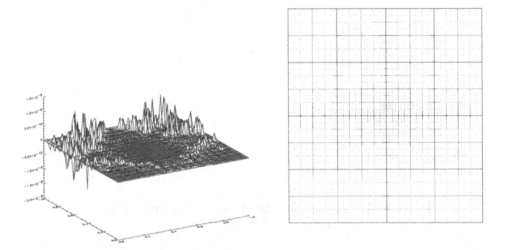

Figure 5.14. Example (5.1): reducing the error in the midpoint for $p = 2$; error (left) and adaptively refined grid (right).

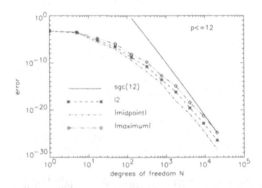

Figure 5.15. Example (5.5): regular sparse grid spaces $V_n^{(p,1)}$, p-asymptotic proceeding up to $p = 12$ with quadruple precision.

On the unit square, let

$$-\nabla \cdot \big(A(\mathbf{x})\nabla u(\mathbf{x})\big) = f(\mathbf{x}) \quad \text{in } \Omega, \tag{5.6}$$

where

$$A(\mathbf{x}) := \begin{pmatrix} 4 + \sin(2\pi x_1) + \sin(2\pi x_2) & 1 + 4x_2 \cdot (1 - x_2) \cdot \sin(2\pi x_1) \\ 1 + 4x_2 \cdot (1 - x_2) \cdot \sin(2\pi x_1) & 4 + \sin(2\pi x_1 x_2) \end{pmatrix},$$

and where $f(\mathbf{x})$ and the Dirichlet boundary conditions are chosen such that

$$u(\mathbf{x}) := \sin(\pi x_1) \cdot \sin(\pi x_2) \tag{5.7}$$

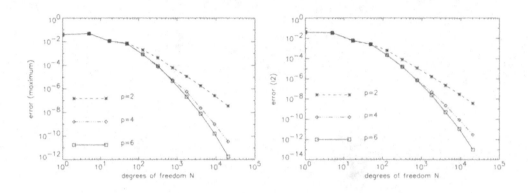

Figure 5.16. Example (5.6): maximum and l_2-error for the regular sparse grid spaces $V_n^{(p,1)}$, $p \in \{2,4,6\}$.

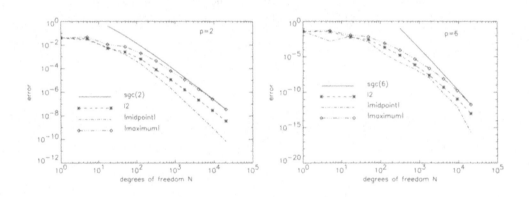

Figure 5.17. Example (5.6): convergence on $V_n^{(p,1)}$ for $p = 2$ (left) and $p = 6$ (right); the solid lines indicate the respective expected sparse grid convergence (sgc; position of curve chosen for clarity).

is the solution. We present results for $p \in \{2, 4, 6\}$. Note that, owing to the smoothness of $u(\mathbf{x})$, we restrict ourselves to the regular sparse grid spaces $V_n^{(p,1)}$. Figure 5.16 shows the maximum and the l_2-error.

Obviously, the higher-order approximation of our hierarchical Lagrangian basis polynomials comes to fruition in the more general situation of example (5.6) too. Figure 5.17 illustrates that we come close to the expected asymptotic behaviour already for moderate values of N.

So far, all problems have been tackled with a finite element approach. However, finite differences have been successfully used for the numerical solution of PDEs too (see Griebel and Schiekofer (1999), Schiekofer (1998), Schiekofer and Zumbusch (1998), for example). Hence we want to present one model problem that has been solved with finite differences on sparse grids, *i.e.*, on $V_n^{(1)}$, to be precise. On the unit cube, define the matrix

$$\hat{A}(\mathbf{x}) := \begin{pmatrix} x_1^2 \exp(\pi x_2 - 1) & x_1^2 & \pi^{-3} \\ 0 & \pi & 0 \\ 0 & \sin(x_1) & 0.1 \cdot (x_1^2 + 2x_2) \end{pmatrix},$$

the vector

$$\hat{b}(\mathbf{x}) := \begin{pmatrix} \hat{b}_1(\mathbf{x}) \\ \hat{b}_2(\mathbf{x}) \\ \hat{b}_3(\mathbf{x}) \end{pmatrix} := \begin{pmatrix} x_1/(x_1 x_3 + 0.1) \\ \cos(\exp(x_1 + x_2 x_3)) \\ x_1 x_2^2 \end{pmatrix},$$

and the scalar function

$$\hat{c}(\mathbf{x}) := \pi \cdot (x_1 + x_2 + x_3).$$

Furthermore, choose the right-hand side $f(\mathbf{x})$ such that

$$u(\mathbf{x}) = \arctan\left(100\left(\frac{x_1 + x_2 + x_3}{\sqrt{3}} - \frac{4}{5}\right)\prod_{i=1}^{3}(x_i - x_i^2)\right) \quad (5.8)$$

solves

$$-\nabla \cdot (\hat{A}(\mathbf{x})\nabla u(\mathbf{x})) + \hat{b}(\mathbf{x}) \cdot \nabla u(\mathbf{x}) + \hat{c}(\mathbf{x})u(\mathbf{x}) = f(\mathbf{x}) \quad (5.9)$$

on the unit cube with Dirichlet boundary conditions. Figure 5.19 (overleaf) illustrates the solution $u(\mathbf{x})$ due to (5.8) for the two x_3-values 0.5 and 0.875. Numerical results are presented in Figure 5.18.

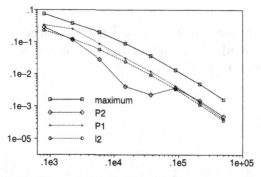

Figure 5.18. Example (5.9), finite differences, regular sparse grid spaces $V_n^{(1)}$: maximum l_2-error as well as errors in $P_1 = (0.5, 0.5, 0.5)$ and $P_2 = (1/\pi, 1/\pi, 1/\pi)$.

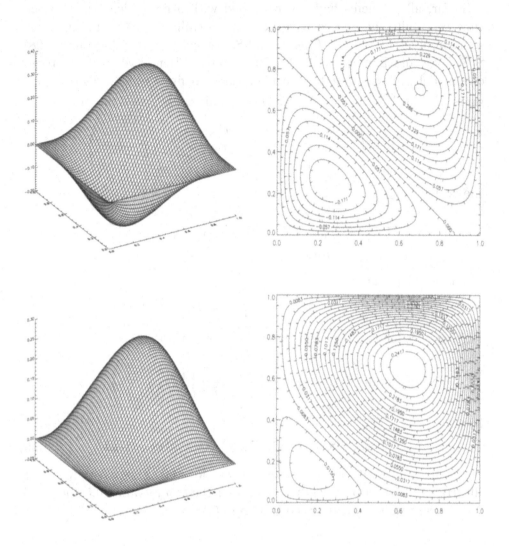

Figure 5.19. Example (5.9), finite differences, regular sparse grid spaces $V_n^{(1)}$: solution $u(\mathbf{x})$ and its contour lines for $x_3 = 0.5$ (top) and $x_3 = 0.875$ (bottom).

5.2. Flow problems and the Navier–Stokes equations

We now consider the application of an adaptive sparse grid method to the solution of incompressible flow problems. The governing relations are given by the Navier–Stokes equations

$$\partial_t \mathbf{u} + \nabla \cdot (\mathbf{u} \otimes \mathbf{u}) = \mathbf{f} - \nabla p + \nu \Delta \mathbf{u}, \quad (\mathbf{x}, t) \in \Omega \times [0, T], \quad (5.10)$$
$$\nabla \cdot \mathbf{u} = 0,$$
$$\mathbf{u}(\mathbf{x}, 0) = u^0(\mathbf{x}),$$

here in primitive variables, with velocity $\mathbf{u} = (u, v, w)^T$ and pressure p. The parameter ν denotes the kinematic viscosity; $\mathbf{f}$ is a prescribed volume force. The system has to be completed by proper initial and boundary conditions.

We apply the the pressure correction scheme (Chorin 1968, Temam 1969, Bell, Colella and Glaz 1989). To this end, the discretization of the Navier–Stokes equations is split into two subproblems.

Transport step. Calculate the auxiliary velocity $\hat{\mathbf{u}}^{n+1}$ by

$$\frac{\hat{\mathbf{u}}^{n+1} - \mathbf{u}^n}{k} + \mathbf{C}(\mathbf{u}^n, \mathbf{u}^{n-1}, \ldots) = \mathbf{f}^{n+1} - \nabla p^n + \nu \Delta \hat{\mathbf{u}}^{n+1}. \quad (5.11)$$

Here, $\mathbf{C}(\mathbf{u}^n, \mathbf{u}^{n-1}, \ldots)$ resembles a stabilized, space-adaptive discretization of $\nabla \cdot (\mathbf{u} \otimes \mathbf{u})$ by means of a Petrov–Galerkin/collocation approach using sparse grid interpolets. Adaptive refinement or coarsening takes place in every time-step steered by the size of the actual coefficients of the multiscale representation of the current solution approximation; see Griebel (1998), Koster (2002) and Griebel and Koster (2000) for further details.

Projection step. Calculate $\mathbf{u}^{n+1}$, p^{n+1} as the solution of

$$\frac{\mathbf{u}^{n+1} - \hat{\mathbf{u}}^{n+1}}{k} = -\nabla(p^{n+1} - p^n), \quad (5.12)$$
$$\nabla \cdot \mathbf{u}^{n+1} = 0.$$

Instead of the simple Euler scheme, higher-order time discretization techniques can be applied analogously. Of course, both subproblems have to be augmented by boundary conditions – not only for $\mathbf{u}$ and $\hat{\mathbf{u}}$, but also for the pressure p. The (in general non-physical) pressure boundary conditions especially are the subject of controversy (cf. Gresho and Sani (1987), Karniadakis and Sherwin (1999), Prohl (1997)), which is beyond the scope of this paper. In the following, we will assume periodic boundary conditions for $\mathbf{u}$ and p. The saddle point problem (5.12) is treated by solving the Schur complement equation

$$\nabla \cdot \nabla (p^{n+1} - p^n) = \frac{1}{k} \nabla \cdot \hat{\mathbf{u}}^{n+1}, \quad (5.13)$$

followed by the correction of the velocity

$$\mathbf{u}^{n+1} = \hat{\mathbf{u}}^{n+1} - k\nabla(p^{n+1} - p^n). \tag{5.14}$$

To this end, a weakly divergence-free adaptive discretization of $\nabla \cdot \nabla$ is applied together with sparse grid interpolets. The solver involves a transform to the lifting interpolets of Section 4.4, a diagonal scaling, and a backtransform; see Koster (2002) and Griebel and Koster (2000) for further details.

Merging of modons

Now, we apply the adaptive version (Griebel and Koster 2000, Koster 2002, Griebel and Koster 2003) of this sparse grid interpolet solver to the model problem of the interaction of three vortices in a 2D flow. Here we use the interpolets of Section 4.3 with $N = 6$.

The initial velocity is induced by three vortices, each with a Gaussian vorticity profile

$$\omega(\mathbf{x}, 0) = \omega_0 + \sum_{i=1}^{3} \omega_i \exp\left(\frac{-\|\mathbf{x} - \mathbf{x}_i\|^2}{\sigma_i^2}\right), \quad \mathbf{x} \in [0, 1]^2.$$

The first two vortices have the same positive sign $\omega_1, \omega_2 > 0$, and the third has a negative sign; see Figure 5.20.

The different parameters ω_0, ω_i, and σ_i are chosen such that the mean value of $\omega(.,0)$ vanishes and that $\omega(.,0)|_{\partial[0,1]^2}$ is almost ω_0 to allow for periodic boundary conditions.

Owing to the induced velocity field, the three vortices start to rotate around each other. In a later stage, the two same-sign vortices merge, which leads to a configuration of two counter-rotating vortices. This process is the basic mechanism in 2D turbulence, and it takes place, *e.g.*, in the shear layer problem of the next subsection or during the convergence of ω to the solution of the Joyce–Montgomery equation (Chorin 1998) for random initial vorticity fields.

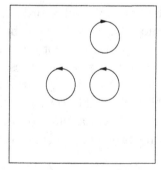

Figure 5.20. Initial configuration for the three vortices' interaction.

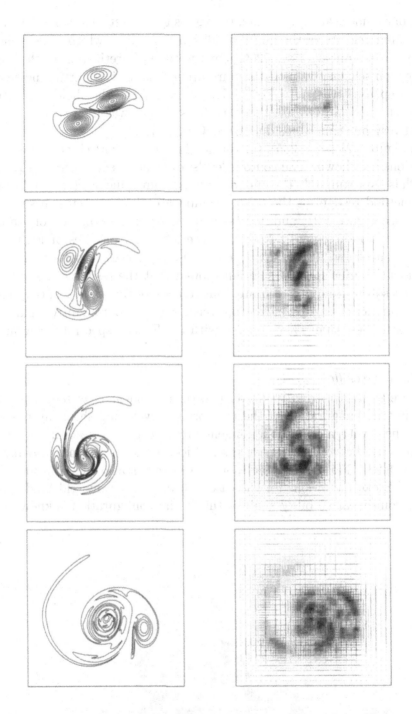

Figure 5.21. Isolines of ω at $t = 5$, 10, 15, 35 (from top to bottom). Left: adaptive wavelet solver ($\epsilon = 4 \cdot 10^{-4}$). Right: adaptive sparse grid.

In our numerical experiments, the viscosity was set to $\nu = 3.8 \cdot 10^{-6}$. The maximum velocity of the initial condition is $7 \cdot 10^{-2}$, which corresponds to a Reynolds number of ≈ 55200. For the time discretization in the Chorin projection scheme, we used the third-order Adams–Bashforth scheme. The time-step used was $dt = 10^{-2}$. The values of the threshold parameter ϵ were set to $\{8, 4, 1\} \cdot 10^{-4}$, and the finest level l was limited to $(10, 10)$ to avoid very fine time-steps due to the CFL-condition.

In Figure 5.21 (page 235), the contour lines of the vorticity of the adaptive solutions are shown. The contour levels are equally spaced from -1.5 to 3, which is approximately the minimum/maximum value of $\omega(.,0)$. Obviously, our method recognizes the arising complicated flow patterns in space and time (see Figure 5.21) and devotes many more degrees of freedom to these than to smooth regions. Note that the left-hand pictures of Figure 5.21 are enhanced by 20%. Further comparisons of these results with a conventional Fourier spectral solution showed that the results of the adaptive wavelet solver are quite accurate. But to achieve the same accuracy, only a small number of degrees of freedom was needed by our adaptive sparse grid approach – less than 1% compared with the Fourier spectral technique.

2D shear layer flow

Now we apply the adaptive wavelet solver to a shear layer model problem. The initial configuration of the temporally developing shear layer is a velocity field with a hyperbolic-tangent profile $u(y) = U \tanh(2y/\delta)$, $v = 0$, where δ is the vorticity shear layer thickness $\delta = 2U/\max(du/dy)$; see Figure 5.22. This initial condition is an exact stationary solution of the Navier–Stokes equations for the case $\nu = 0$, i.e., for inviscid fluids. However, from linear stability analysis this initial configuration is known to be

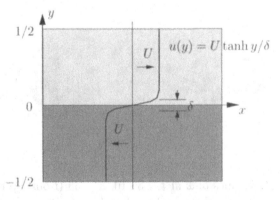

Figure 5.22. Velocity profile of the initial condition for the shear flow model problem. The thickness of the boundary layer is defined as $\delta = 2U/\max_y |\partial_y u(y)|$.

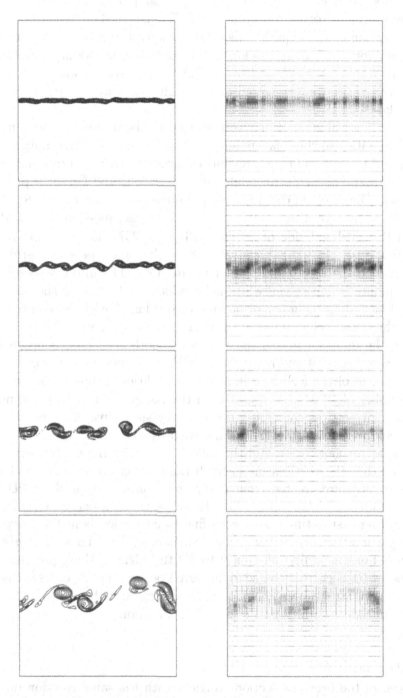

Figure 5.23. Isolines of the rotation for the values {0.1, 0.2, 0.3, 0.5, 0.75, 1.0, ..., 2.25, 2.5} at time $t = 4$, 8, 16, 36. Left: isolines of computed rotation. Right: adaptive sparse grid.

inviscidly unstable. Slight perturbations are amplified by Kelvin–Helmholtz instabilities and vortex roll-up occurs (Michalke 1964).

In our numerical experiments, the vorticity thickness δ was chosen so that ten vortices should develop in the computational domain $[0,1]^2$. The maximum velocity was $U = 1.67 \cdot 10^{-2}$, and the viscosity was $\nu = 3.8 \cdot 10^{-6}$. The instabilities were triggered by a superimposed weak white noise in the shear layer.

The numerical simulations were applied to the periodized version of the problem with two shear layers on $[0,1] \times [0,2]$. We used the same adaptive sparse grid solver as in the previous experiment, based on interpolets with $N = 6$, a stabilized Petrov–Galerkin collocation, and Chorin's projection method. The values of the threshold parameter ϵ were set to $\{12, 8, 4\} \cdot 10^{-4}$ and the finest level l was limited to $(10, 10)$. The time-step was $5 \cdot 10^{-3}$.

On the left-hand side of Figure 5.23 (page 237), the resulting rotation $\omega = \nabla \times \mathbf{u}$ for $\epsilon = 8 \cdot 10^{-4}$ is given. The evolution of the sparse grid points is shown on the right-hand side of Figure 5.23. The initial velocity is not very smooth owing to the white noise added for triggering the instability. Therefore, a large number of degrees of freedom (DOF) is required to resolve the initial velocity in the shear layer sufficiently well. Then, in a first phase, the diffusion makes the velocity smooth very fast, which leads to the strong decay of the number of DOF. This process is stopped by the development of the Kelvin–Helmholtz instabilities leading to an increase of the number of DOF ($4 < t < 10$). In the last phase ($t > 10$), the number of coherent vortices constantly decreases by successive merging. The final state comprises two counter-rotating vortices which dissipate.

Besides the generation and the roll-up of Kelvin–Helmholtz vortices, we also see that the vortices merge with time. This process, in which a few large vortices are created from many small ones, is typical for 2D flows; *cf.* Chapter 4.6 of Chorin (1998). It comes with an energy transfer from fine to coarse structures, *i.e.*, from fine to coarse levels in the sense of an isotropic multiscale representation (Novikov 1976). Thus 2D turbulence behaves fundamentally differently to 3D turbulence. Here, predominantly an energy transfer from coarse to fine levels occurs, that is, coarse structures decay to fine structures. The maximal level was limited to 10 to avoid overly small time-steps resulting from the CFL condition.

3D shear layer flow

As seen in the previous section, vortices with the same rotation direction merge successively in 2D flows. In 3D flows, however, this effect is no longer present. Here the additional third dimension allows for vortex tubes. They can be bent, folded, and stretched over time until they break apart. Their diameter reduces during stretching, and therefore smaller-scale structures

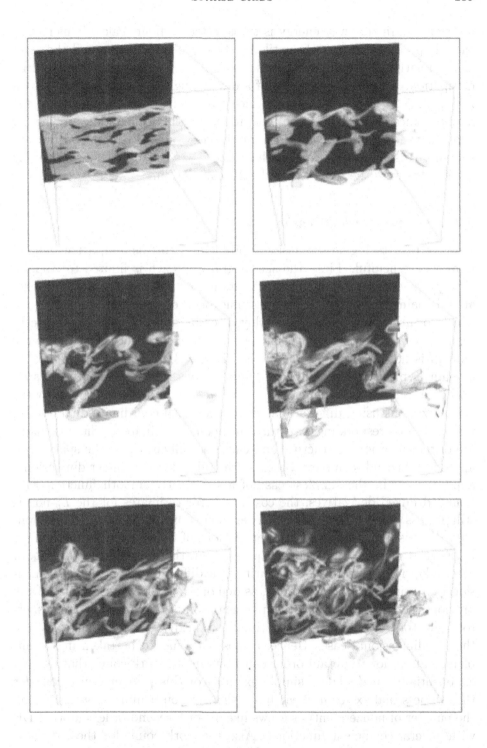

Figure 5.24. Iso-surfaces of the rotation for the 3D turbulent shear layer problem at time $0, 15, 30, 60, 90, 150$.

are created. In this way energy is transported to finer scales. This energy cascade proceeds recursively until the structures are so tiny and the associated relative Reynolds number is so small that energy is annihilated by dissipation. The different phases of a developing 3D shear layer are shown in Figure 5.24 (page 239), from which this increase of complexity in the flow becomes clear. The initial conditions for this calculation are the 3D analogues of the initial conditions for the previous 2D example. However, viscosity was increased, *i.e.*, we have set $\nu = 5.0410^{-6}$. The remaining parameters were the same as for the 2D calculation.

5.3. Problems of high dimensionality

We now turn to higher-dimensional problems. To this end, a few general remarks are helpful. First, the sparse grid approach is limited to (topologically) quadrilateral domains due to its tensor product nature. Complicated domains must be built up by gluing together quadrilateral patches on which locally sparse grids are employed, or by appropriate mapping techniques. Now, in higher dimensionalities, this question of the shape of the domain is not as important as in the 2D and 3D case, since complicated domains typically do not appear in applications. Conceptually, besides $\mathbb{R}^d$ itself, we use mainly hypercubes $[-a, a]^d$, $a > 0$, and their straightforward generalizations using different values of a for each coordinate direction, as well as the corresponding structures in polar coordinates. These domains are of tensor product structure and cause no difficulties to the sparse grid approach. Second, complexity issues are still crucial in higher-dimensional applications. In the simplest case of a sufficiently smooth function with bounded mixed derivatives, the cost and error estimates for the L_p-norms still possess $(\log N)^{d-1}$-terms which depend exponentially on the dimensionality d. This limits the practical applications of the sparse grid method for PDEs to at most 18 dimensions at present. Furthermore, in the nonsmooth case, adaptivity towards a singularity is a difficult task in higher dimensions. In principle, it can be done, as demonstrated in the examples of the previous sections. However, an implementation of refinement strategies, error indicators, and solvers with optimal complexity is more difficult than in the low-dimensional case. Hence, utmost care has to be taken in the implementation not to obtain order constants in the work count that depend exponentially on d. For a simple example of this problem, just consider the stiffness matrix for d-linear finite elements on a uniform grid. There, the number of nonzero entries grows like 3^d for a second-order elliptic PDE with general coefficient functions. Also the work count for the computation of each entry by a conventional quadrature scheme involves terms that grow exponentially with d. Here, at least in certain simple cases like the Laplacian, an algorithmic scheme can be applied, which is based on the so-called *unidirectional principle* (Bungartz 1998). It allows us to treat the single dimensions separately and results in linear complexity in both the degrees of freedom *and d*, for the matrix–vector product in a Galerkin sparse grid approach. However, note that, especially in the case of more general nonseparable operators, this is the challenging part of a sparse grid finite element implementation.

The next question concerns the sources of high-dimensional problems and PDEs. Here, besides pure integration problems stemming from physics and finance, typically models from the stochastic and data analysis world show up. For example, high-dimensional Laplace/diffusion problems and high-dimensional convection–diffusion problems result from diffusion approximation techniques and the Fokker–Planck equation. Examples are the description of queueing networks (Mitzlaff 1997, Shen, Chen, Dai and Dai 2002), reaction mechanisms in molecular biology (Sjöberg 2002, Elf, Lötstedt and Sjöberg 2001), the viscoelasticity in polymer fluids (Rouse 1953, Prakash and Öttinger 1999, Prakash 2000), or various models for the pricing of financial derivatives (Reisinger 2003). Furthermore, homogenization with multiple scales (Allaire 1992, Cioranescu, Damlamian and Griso 2002, Matache 2001, Matache and Schwab 2001, Hoang and Schwab 2003) as well as stochastic elliptic equations (Schwab and Todor 2002, 2003) result in high-dimensional PDEs. Next, we find quite high-dimensional problems in quantum mechanics. Here, the dimensionality of the Schrödinger equation grows with the number of considered atoms. Sparse grids have been used in this context by Garcke (1998), Garcke and Griebel (2000), Yserentant (2004) and Hackbusch (2001).

In the following, we illustrate the potential of sparse grids for problems of a higher dimensionality in the context of numerical quadrature and data mining.

Classification and regression in data mining
Data mining is the process to find hidden patterns, relations, and trends in large data sets. It plays an increasing role in commerce and science. Typical scientific applications are the post-processing of data in medicine (*e.g.*, CT data), the evaluation of data in astrophysics (*e.g.*, telescope and observatory data), and the grouping of seismic data, or the evaluation of satellite pictures (*e.g.*, NASA earth observing system). Financial and commercial applications are perhaps of greater importance. With the development of the internet and e-commerce, there are huge data sets collected, more or less automatically, which can be used for business decisions and further strategic planning. Here, applications range from contract management to risk assessment, from the segmentation of customers for marketing to fraud detection, stock analysis and turnover prediction.

Usually, the process of data mining (or knowledge discovery) can be separated into the planning step, the preparation phase, the mining phase (*i.e.*, machine learning) and the evaluation. To this end, association-analysis classification, clustering, and prognostics are to be performed. For a thorough overview of the various tasks arising in the data mining process, see Berry and Linoff (2000) and Cios, Pedrycz and Swiniarski (1998).

In the following, we consider the classification problem in detail. Here, a set of data points in d-dimensional feature space is given together with a class label in $\{-1, 1\}$, for example. From these data, a classifier must be constructed which allows us to predict the class of any newly given data point for future decision making. Widely used approaches are nearest neighbour methods, decision tree induction, rule learning, and memory-based reasoning. There are also classification algorithms based on adaptive multivariate regression splines, neural networks, support vector machines, and regularization networks. Interestingly, these latter techniques can be interpreted in the framework of regularization networks (Girosi, Jones and Poggio 1995). This approach allows a direct description of the most important neural networks, and it also allows for an equivalent description of support vector machines and n-term approximation schemes (Girosi 1998).

We follow Garcke, Griebel and Thess (2001) and consider a given set of already classified data (the training set):

$$S = \{(\boldsymbol{x}_i, y_i) \in \mathbb{R}^d \times \mathbb{R}\}_{i=1}^{M}.$$

We now assume that these data have been obtained by sampling of an unknown function f which belongs to some function space V defined over $\mathbb{R}^d$. The sampling process was disturbed by noise. The aim is now to recover the function f from the given data as well as possible. This is clearly an ill-posed problem since infinitely many solutions are possible. To get a well-posed, uniquely solvable problem, we have to restrict f. To this end, regularization theory (Tikhonov and Arsenin 1977, Wahba 1990) imposes an additional smoothness constraint on the solution of the approximation problem, and the regularization network approach considers the variational problem

$$\min_{f \in V} R(f)$$

with

$$R(f) = \frac{1}{M} \sum_{i=1}^{m} C(f(\boldsymbol{x}_i), y_i) + \lambda \Phi(f). \tag{5.15}$$

Here, $C(.,.)$ denotes an error cost function which measures the interpolation error and $\Phi(f)$ is a smoothness functional which must be well defined for $f \in V$. The first term enforces closeness of f to the data, the second term enforces smoothness of f, and the regularization parameter λ balances these two terms. We consider the case

$$C(x,y) = (x-y)^2 \quad \text{and} \quad \Phi(f) = \|Pf\|_2^2 \quad \text{with} \quad Pf = \nabla f.$$

The value of λ can be chosen according to cross-validation techniques (Allen 1972, Golub, Heath and Wahba 1979, Utreras 1979, Wahba 1985) or to some other principle, such as structural risk minimization (Vapnik 1982). We find exactly this type of formulation in the case $d = 2,3$ in many scattered data approximation methods (see Arge, Dæhlen and Tveito (1995) and Hoschek and Lasser (1992)), where the regularization term is usually physically motivated.

We now restrict the problem to a finite-dimensional subspace $V_N \subset V$. The function f is then replaced by

$$f_N = \sum_{j=1}^{N} \alpha_j \varphi_j(\boldsymbol{x}). \tag{5.16}$$

Here $\{\psi_j\}_{j=1}^{N}$ should span V_N and preferably should form a basis for V_N. The coefficients $\{\alpha_j\}_{j=1}^{N}$ denote the degrees of freedom. Note that the restriction to a suitably chosen finite-dimensional subspace involves some additional regularization (regularization by discretization) which depends on the choice of V_N. In this way we obtain from the minimization problem a feasible linear system. We thus have to minimize

$$R(f_N) = \frac{1}{M} \sum_{i=1}^{M} \left(f_N(\boldsymbol{x}_i) - y_i\right)^2 + \lambda \|Pf_N\|_{L_2}^2, \quad f_N \in V_N \tag{5.17}$$

in the finite-dimensional space V_N. We plug (5.16) into (5.17) and obtain

$$R(f_N) = \frac{1}{M} \sum_{i=1}^{M} \left(\sum_{j=1}^{N} \alpha_j \varphi_j(\boldsymbol{x}_i) - y_i\right)^2 + \lambda \left\|P \sum_{j=1}^{N} \alpha_j \varphi_j\right\|_{L_2}^2 \tag{5.18}$$

$$= \frac{1}{M} \sum_{i=1}^{M} \left(\sum_{j=1}^{N} \alpha_j \varphi_j(\boldsymbol{x}_i) - y_i\right)^2 + \lambda \sum_{i=1}^{N} \sum_{j=1}^{N} \alpha_i \alpha_j (P\varphi_i, P\varphi_j)_{L_2}. \tag{5.19}$$

Differentiation with respect to α_k, $k = 1, \ldots, N$, gives

$$0 = \frac{\partial R(f_N)}{\partial \alpha_k} = \frac{2}{M} \sum_{i=1}^{M} \left(\sum_{j=1}^{N} \alpha_j \varphi_j(\boldsymbol{x}_i) - y_i\right) \cdot \varphi_k(\boldsymbol{x}_i) + 2\lambda \sum_{j=1}^{N} \alpha_j (P\varphi_j, P\varphi_k)_{L_2}. \tag{5.20}$$

This is equivalent to $(k = 1, \ldots, N)$

$$\lambda \sum_{j=1}^{N} \alpha_j (P\varphi_j, P\varphi_k)_{L_2} + \frac{1}{M} \sum_{j=1}^{N} \alpha_j \sum_{i=1}^{M} \varphi_j(\boldsymbol{x}_i) \cdot \varphi_k(\boldsymbol{x}_i) = \frac{1}{M} \sum_{i=1}^{M} y_i \varphi_k(\boldsymbol{x}_i) \qquad (5.21)$$

and we obtain $(k = 1, \ldots, N)$

$$\sum_{j=1}^{N} \alpha_j \left[M\lambda(P\varphi_j, P\varphi_k)_{L_2} + \sum_{i=1}^{M} \varphi_j(\boldsymbol{x}_i) \cdot \varphi_k(\boldsymbol{x}_i) \right] = \sum_{i=1}^{M} y_i \varphi_k(\boldsymbol{x}_i). \qquad (5.22)$$

In matrix notation, we end up with the linear system

$$(\lambda C + B \cdot B^T)\alpha = By. \qquad (5.23)$$

Here, C is a square $N \times N$ matrix with entries $C_{j,k} = M \cdot (P\varphi_j, P\varphi_k)_{L_2}$, $j, k = 1, \ldots, N$, and B is a rectangular $N \times M$ matrix with entries $B_{j,i} = \varphi_j(\boldsymbol{x}_i)$, $i = 1, \ldots, M, j = 1, \ldots, N$. The vector y contains the data y_i and has length M. The unknown vector α contains the degrees of freedom α_j and has length N. With the gradient $P = \nabla$ in the regularization expression in (5.15), we obtain a Poisson problem with an additional term that resembles the interpolation problem. The natural boundary conditions for such a differential equation in $\Omega = [0, 1]^d$, for instance, are Neumann conditions. The discretization (5.16) gives us then the linear system (5.23) where C corresponds to a discrete Laplacian. To obtain the classifier f_N, we now have to solve this system.

Again, the curse of dimensionality prohibits us from using for V_N conventional finite element spaces living on a uniform grid. The complexity would grow exponentially with d. Instead, we used a sparse grid approach, namely the combination method, as already described for the 2D and 3D case in (4.41), (4.42), and Figure 4.14. These formulae can easily be generalized to the d-dimensional case. Here we consider the minimization problem on a sequence of grids, solve the resulting linear systems (5.23), and combine the solution accordingly. The complexity of the method is with respect to the number of levels n, as usual for regular sparse grids of the order $O(n^{d-1}2^n)$. With respect to the number M of training data, the cost complexity scales linearly in M for a clever implementation. This is a substantial advantage in comparison to most neural networks and support vector machines, which generally scale at least quadratically in M and are therefore not suited to problems with very large data sets.

We now apply our approach to different test data sets. Here we use both synthetic data generated by DatGen (Melli 2003) and real data from practical data mining applications. All the data sets are rescaled to $[0, 1]^d$.

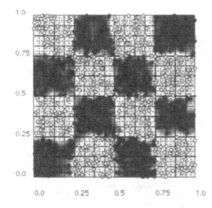 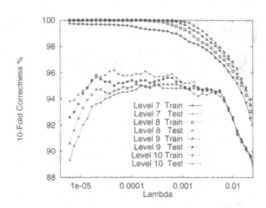

Figure 5.25. Left: chessboard data set, combination technique with level 10, $\lambda = 4.53999 \cdot 10^{-5}$. Right: plot of the dependence on λ (in logscale) and level.

To evaluate our method we give the correctness rates on testing data sets if available and the 10-fold cross-validation results on training and testing data sets. For the 10-fold cross-validation we proceed as follows. We divide the training data into 10 equally sized disjoint subsets. For $i = 1$ to 10, we pick the ith of these subsets as a further testing set and build the sparse grid combination classifier with the data from the remaining nine subsets. We then evaluate the correctness rates of the current training and testing set. In this way we obtain ten different training and testing correctness rates. The 10-fold cross-validation result is just the average of these ten correctness rates. For further details, see Stone (1974). For a critical discussion on the evaluation of the quality of classifier algorithms, see Salzberg (1997).

We first consider 2D problems with small sets of data that correspond to certain structures. Then we treat problems with huge sets of synthetic data with up to 5 million points.

The first example is taken from Ho and Kleinberg (1996) and Kaufman (1999). Here, 1000 training data points were given which are more or less uniformly distributed in $\Omega = [0,1]^2$. The associated data values are plus one or minus one depending on their location in Ω such that a 4×4 chessboard structure appears: see Figure 5.25 (left). We computed the 10-fold cross-validated training and testing correctness with the sparse grid combination method for different values of the regularization parameter λ and different levels n. The results are shown in Figure 5.25 (right).

We see that the 10-fold testing correctness is well around 95% for values of λ between $3 \cdot 10^{-5}$ and $5 \cdot 10^{-3}$. Our best 10-fold testing correctness was 96.20% on level 10 with $\lambda = 4.54 \cdot 10^{-5}$. The chessboard structure is thus reconstructed with less than 4% error.

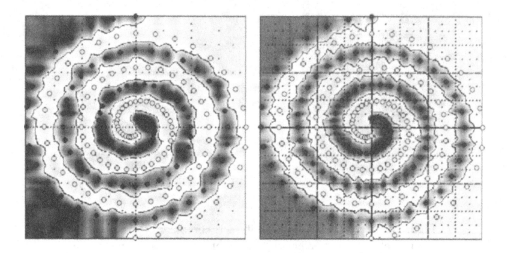

Figure 5.26. Spiral data set, sparse grid with level 6 (left) and 8 (right), $\lambda = 0.001$.

Another 2D example with structure is the spiral data set, first proposed by Wieland; see also Fahlmann and Lebiere (1990). Here, 194 data points describe two intertwined spirals: see Figure 5.26. This is surely an artificial problem, which does not appear in practical applications. However it serves as a hard test case for new data mining algorithms. It is known that neural networks can have severe problems with this data set, and some neural networks can not separate the two spirals at all. In Figure 5.26 we give the results obtained with our sparse grid combination method with $\lambda = 0.001$ for $n=6$ and $n=8$. Already for level 6, the two spirals are clearly detected and resolved. Note that here only 577 grid points are contained in the sparse grid. For level 8 (2817 sparse grid points), the shape of the two reconstructed spirals gets smoother and the reconstruction gets more precise.

The BUPA Liver Disorders data set from the Irvine Machine Learning Database Repository (Blake and Merz 1998) consists of 345 data points with six features plus a selector field used to split the data into two sets with 145 instances and 200 instances, respectively. Here we only have training data and can therefore only report our 10-fold cross-validation results. No comparison with unused test data is possible.

We compare with the two best results from Lee and Mangasarian (2001), the smoothed support vector machine (SSVM) introduced therein, and the feature selection concave minimization (FSV) algorithm due to Bradley and Mangasarian (1998). Table 5.1 gives the results for the 10-fold correctness.

Our sparse grid combination approach performs on level 3 with $\lambda = 0.0625$ at 69.23% 10-fold testing correctness. But our other results were also in this range. Our method performs only slightly worse here than the SSVM but

Table 5.1. Results for the BUPA liver disorders data set.

	SSVM	FSV	sparse grid combination method			
			level 1 $\lambda = 0.0001$	level 2 $\lambda = 0.1$	level 3 $\lambda = 0.0625$	level 4 $\lambda = 0.625$
train. %	70.37	68.18	83.28	79.54	90.20	88.66
test. %	70.33	65.20	66.89	66.38	69.23	68.74

Table 5.2. Results for a 6D synthetic massive data set, $\lambda = 0.01$.

	# of points	training corr.	testing corr.	total time (sec)	data matrix time (sec)	# of iterat.
level 1	50 000	87.9	88.1	158	152	41
	500 000	88.0	88.1	1570	1528	44
	5 million	88.0	88.0	15933	15514	46
level 2	50 000	89.2	89.3	1155	1126	438
	500 000	89.4	89.2	11219	11022	466
	5 million	89.3	89.2	112656	110772	490

clearly better than FSV. Note that the results for the robust linear program (RLP) algorithm (Bennett and Mangasarian 1992), the support vector machine using the 1-norm approach ($SVM_{\|\cdot\|_1}$), and the classical support vector machine ($SVM_{\|\cdot\|_2^2}$) (Bradley and Mangasarian 1998, Cherkassky and Mulier 1998, Vapnik 1995) were reported to be somewhat worse in Lee and Mangasarian (2001).

Next, we produced a 6D data set with 5 million training points and 20 000 points with DatGen (Melli 2003) for testing. We used the call

```
datgen -r1 -X0/100,R,O :0/100,R,O:0/100,R,O:0/100,R,O:0/200,
R,O:0/200,R,O -R2 -C2/4 -D2/5 -T10/60 -O502 0000 -p -e0.15.
```

The results are given in Table 5.2. On level one, a testing correctness of 88% was achieved already, which is quite satisfying for this data. We see that really huge data sets of 5 million points could be handled. We also give the CPU time which is needed for the computation of the matrices $G_1 = B_1 \cdot B_1^T$. Here, more than 96% of the computing time is spent on the matrix assembly. Again, the execution times scale linearly with the number of data points.

Analogously, we can at present deal with up to 18-dimensional problems before storage limitations on our available computers stop us. This moderate number of dimensions, however, is still sufficient for many practical applications. In very high-dimensional data, there exist mostly strong correlations and dependencies between the dimensions. Then, in a preprocessing step, the effective dimensions can be determined, for instance, by means of principal component analysis, and the dimension can be reduced in many cases to 18 or less.

To reach 18 dimensions, a generalization of the sparse grid combination technique to simplicial basis functions (Garcke and Griebel 2001a, 2002) is needed. Note that the sparse grid combination technique can be parallelized in a straightforward way (Garcke and Griebel 2001b, Garcke, Hegland and Nielsen 2003). Finally, the sparse grid technique can be used in a dimension-adaptive fashion (Hegland 2002, 2003), which further enhances the method's capabilities. This approach will be discussed for the example of numerical integration now.

Integration

The computation of high-dimensional integrals is a central part of computer simulations in many application areas such as statistical mechanics, financial mathematics, and computational physics. Here, the arising integrals usually cannot be solved analytically, and thus numerical approaches are required. Furthermore, often a high-accuracy solution is needed, and thus such problems can be computationally quite challenging even for parallel supercomputers. Conventional algorithms for the numerical computation of such integrals are usually limited by the curse of dimensionality. However, for special function classes, such as spaces of functions which have bounded mixed derivatives, Smolyak's construction (Smolyak 1963) (see (4.1)) can overcome this curse to a certain extent. In this approach, multivariate quadrature formulas are constructed using combinations of tensor products of appropriate 1D formulas. In this way, the number of function evaluations and the numerical accuracy become independent of the dimension of the problem up to logarithmic factors. Smolyak's construction is simply our sparse grid approach. It has been applied to numerical integration by several authors, using the midpoint rule (Baszenski and Delvos 1993), the rectangle rule (Paskov 1993), the trapezoidal rule (Bonk 1994a), the Clenshaw–Curtis rule (Cools and Maerten 1997, Novak and Ritter 1998), the Gauss rules (Novak and Ritter 1997), and the Gauss–Patterson rules (Gerstner and Griebel 1998, Petras 2000) as the 1D basis integration procedure. The latter approach, in particular, achieves the highest polynomial exactness of all nested quadrature formulas and shows very good results for sufficiently smooth multivariate integrands. Further studies have been made concerning extrapolation methods (Bonk 1994a), discrepancy measures (Frank and Heinrich 1996), and complexity questions (Wasilkovski and Woźniakowski 1999).

There is also a large variety of other methods for the numerical integration of multivariate functions such as Monte Carlo and quasi-Monte Carlo methods (Niederreiter 1992), lattice rules (Sloan and Joe 1994), adaptive subdivision methods (Genz and Malik 1980, Dooren and Ridder 1976), and approximation methods based on neural networks (Barron 1994, Mhaskar 1996). Each of these methods is particularly suitable for functions from a certain function class and has a complexity which is then also independent or nearly independent of the problem's dimensionality.

Despite the large improvements of the quasi-Monte Carlo and sparse grid methods over the Monte Carlo method, their convergence rates will suffer more and more with rising dimension owing to their respective dependence on the dimension in the logarithmic terms. Therefore, one aim of recent numerical approaches has been to reduce the dimension of the integration problem without affecting the accuracy unduly.

In some applications, the different dimensions of the integration problem are not equally important. For example, in path integrals the number of dimensions corresponds to the number of time-steps in the time discretization. Typically, the first steps in the discretization are more important than the last steps since they determine the outcome more substantially. In other applications, although the dimensions seem to be of the same importance at first sight, the problem can be transformed into an equivalent one where the dimensions are not. Examples are the Brownian bridge discretization or the Karhunen–Loeve decomposition of stochastic processes.

Intuitively, problems where the different dimensions are not of equal importance might be easier to solve: numerical methods could concentrate on the more important dimensions. Interestingly, complexity theory also reveals that integration problems with weighted dimensions can become tractable even if the unweighted problem is not (Wasilkovski and Woźniakowski 1999). Unfortunately, classical adaptive numerical integration algorithms (Genz and Malik 1980, Dooren and Ridder 1976) cannot be applied to high-dimensional problems, since the work overhead in order to find and adaptively refine in important dimensions would be too large.

To this end, a variety of algorithms have been developed that try to find and quantify important dimensions. Often, the starting point of these algorithms is Kolmogorov's superposition theorem: see Kolmogorov (1956, 1957). Here, a high-dimensional function is approximated by sums of lower-dimensional functions. A survey of this approach from the perspective of approximation theory is given in Khavinson (1997). Further results can be found in Rassias and Simsa (1995) and Simsa (1992). Analogous ideas are followed in statistics for regression problems and density estimation.

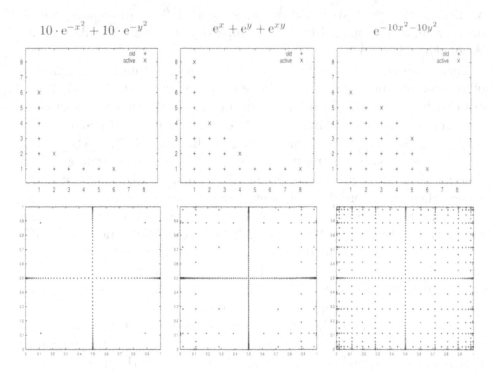

Figure 5.27. The resulting index sets and corresponding sparse grids for TOL = 10^{-15} for some isotropic test functions.

Here, examples are so-called additive models (Hastie and Tibshirani 1990), multivariate adaptive regression splines (MARS) (Friedman 1991), and the ANOVA decomposition (Wahba 1990, Yue and Hickernell 2002); see also Hegland and Pestov (1999). Other interesting techniques for dimension reduction are presented in He (2001). If the importance of the dimensions is known *a priori*, techniques such as importance sampling can be applied in Monte Carlo methods (Kalos and Whitlock 1986). For the quasi-Monte Carlo method, a sorting of the dimensions according to their importance leads to a better convergence rate (yielding a reduction of the effective dimension). The reason for this is the better distributional behaviour of low-discrepancy sequences in lower dimensions than in higher ones (Caflisch, Morokoff and Owen 1997). The sparse grid method, however, *a priori* treats all dimensions equally and thus gains no immediate advantage for problems where dimensions are of different importance.

In Gerstner and Griebel (2003), we developed a generalization of the conventional sparse grid approach which is able to adaptively assess the dimensions according to their importance, and thus reduces the dependence of the computational complexity on the dimension. This is quite in the spirit of Hegland (2002, 2003). The dimension-adaptive algorithm tries to find

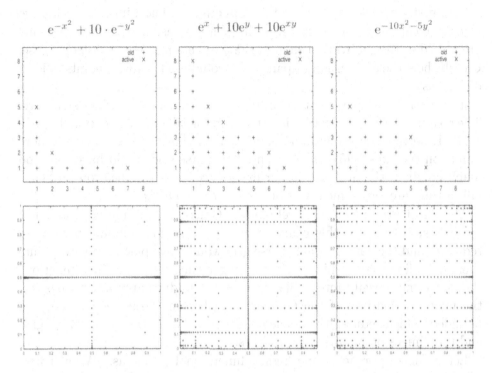

Figure 5.28. The resulting index sets and corresponding sparse grids for TOL = 10^{-15} for some anisotropic test functions.

important dimensions automatically and adapts (places more integration points) in those dimensions. To achieve this efficiently, a data structure for a fast bookkeeping and searching of generalized sparse grid index sets is necessary.

We will now show the performance of the dimension-adaptive algorithm in numerical examples. First, we consider some 2D test functions, which allows us to show the resulting grids and level index sets. In these cases, the exact value of the integral is known (or can be computed quickly). The second example is a path integral of 32 dimensions in which the integral value is also known beforehand. The third example is a 256-dimensional application problem from finance where the exact value is unknown. We use the well-known Patterson formulas (Patterson 1986) for univariate quadrature in all examples. These were shown to be a good choice for the sparse grid construction by Gerstner and Griebel (1998).

Let us first consider simple combinations of exponential functions defined over $[0,1]^2$. In Figures 5.27 and 5.28 we depict the level index sets used by the algorithm as well as the resulting dimension-adapted sparse grids for some isotropic and some anisotropic functions, respectively. In these examples, the selected error threshold is TOL = 10^{-15}.

The first example is a sum of 1D functions. The dimension-adaptive algorithm correctly selects no indices in joint dimensions. Also, more points are placed in the x-direction than in the y-direction in the anisotropic case. Clearly, here the conventional sparse grid would use too many points in joint directions.

The second example is not separable, nor does it have product structure. The resulting level index set is almost triangular, like the conventional sparse grid. However, the dimension-adaptive algorithm chooses to select more points on the axes, while the conventional sparse grid would have used too many points in the interior. In our experience, many application problems fall in this category, which we would call *nearly additive*.

The third example is the well-known Gaussian hat function, and has product structure. In this example, many points in joint dimensions are required, and the conventional sparse grid would have placed too few points there. At first sight, this is a surprising result, since product functions should be more easily integrable by a tensor product approach. However, the mixed derivatives of the Gaussian can become large even if they are bounded, which reduces the efficiency of both the conventional sparse grid and the dimension-adaptive approaches.

Let us now approach some higher-dimensional problems. We will first consider an initial value problem given by the linear PDE

$$\frac{\partial u}{\partial t} = \frac{1}{2} \cdot \frac{\partial^2 u}{\partial x^2}(x,t) + v(x,t) \cdot u(x,t),$$

with initial condition $u(x,0) = f(x)$. The solution of this problem can be obtained with the Feynman–Kac formula as

$$u(x,t) = E_{x,0}\big(f(\xi(t)) \cdot e^{\int_0^t v(\xi(r), t-r)\, dr}\big),$$

where ξ represents a Wiener path starting at $\xi(0) = x$. The expectation $E_{x,0}$ can be approximated by a discretization of time using a finite number of time-steps $t_i = i \cdot \Delta t$ with $\Delta t = t/d$. The integral in the exponent is approximated by a 1D quadrature formula such as a sufficiently accurate trapezoidal rule.

The most natural way to discretize the Wiener path is by a random walk, i.e., by the recursive formula

$$\xi_k = \xi_{k-1} + \sqrt{\Delta t}\, z_k,$$

where $\xi_0 = x$ and z_k are normally distributed random variables with mean zero and variance one. The dimensions in the random walk discretization are all of the same importance since all the variances are identical to Δt.

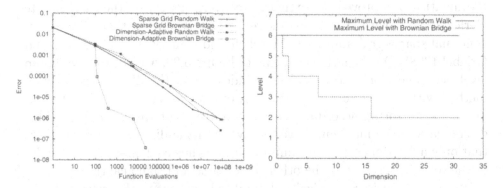

Figure 5.29. Computational results for the path integral ($d = 32$): integration error vs. number of function evaluations (left) and maximum level over all dimensions (sorted) for the dimension-adaptive algorithm with Brownian bridge discretization (right).

In the Brownian bridge discretization (Caflisch et al. 1997), however, the path is discretized using a future and a past value

$$\xi_k = \frac{1}{2}(\xi_{k-h} + \xi_{k+h}) + \sqrt{\frac{h \cdot \Delta t}{2}} \cdot z_k.$$

Starting with $\xi_0 := x$ and $\xi_d := x + \sqrt{t}\, z_d$, the subsequent values to be computed are $\xi_{d/2}, \xi_{d/4}, \xi_{3d/4}, \xi_{d/8}, \xi_{3d/8}, \xi_{5d/8}, \xi_{7d/8}, \xi_{1d/16}, \xi_{3d/16}, \ldots$ with corresponding $h = 1/2, 1/4, 1/4, 1/8, 1/8, 1/8, 1/8, 1/16, 1/16, \ldots$. The Brownian bridge leads to a concentration of the total variance in the first few steps of the discretization and thus to a weighting of the dimensions.

Let us now consider the concrete example (Morokoff and Caflisch 1995)

$$v(x, t) = \left(\frac{1}{t+1} + \frac{1}{x^2 + 1} - \frac{4x^2}{(x^2 + 1)^2}\right),$$

with initial condition $u(x, 0) = \frac{1}{x^2+1}$. The exact solution is then

$$u(x, t) = \frac{t+1}{x^2 + 1}.$$

The results for $d = 32$, $t = 0.02$ and $x = 0$ are shown on the left-hand side of Figure 5.29. We see the integration error plotted against the number of function evaluations in a log-log scale. Here, the conventional sparse grid method is compared with the dimension-adaptive algorithm for the random walk and Brownian bridge discretizations. In this example, the conventional sparse grid is for the random walk discretization obviously close to the optimum, since the dimension-adaptive method cannot improve on the performance. The conventional sparse grid gains no advantage from the Brownian bridge discretization, but the convergence rate of the dimension-adaptive algorithm is dramatically improved. Note that the convergence rate of the quasi-Monte Carlo method (with Brownian bridge) is comparable to that of the conventional sparse grid approach (Morokoff and Caflisch 1995, Gerstner and Griebel 1998). On the right-hand side of Figure 5.29, we plot the maximum level per dimension of the final level index set of the dimension-adaptive method with and without the Brownian bridge discretization. Here the dimensions are sorted according to this quantity. For the Brownian bridge discretization, the maximum level decays with the dimension. This shows that only a few dimensions are important and thus contribute substantially to the total integral while the other dimensions add significantly less.

Let us now consider a typical collateralized mortgage obligation problem, which involves several tranches, which in turn derive their cash flows from an underlying pool of mortgages (Caflisch et al. 1997, Paskov and Traub 1995). The problem is to estimate the expected value of the sum of present values of future cash flows for each tranche. Let us assume that the pool of mortgages has a $21\frac{1}{3}$ year maturity and cash flows are obtained monthly. Then the expected value requires the evaluation of an integral of dimension $d = 256$ for each tranche,

$$\int_{\mathbb{R}^d} v(\xi_1, \ldots, \xi_d) \cdot g(\xi_1) \cdot \ldots \cdot g(\xi_d) \, \mathrm{d}\xi_1 \cdots \mathrm{d}\xi_d,$$

with Gaussian weights $g(\xi_i) = (2\pi\sigma^2)^{-1/2} e^{-\xi_i^2/2\sigma^2}$. The sum of the future cash flows v is basically a function of the interest rates i_k (for month k),

$$i_k := K_0 e^{\xi_1 + \cdots + \xi_k} i_0$$

with a certain normalizing constant K_0 and an initial interest rate i_0 (for details see the first example in Caflisch et al. (1997) and compare Gerstner and Griebel (1998) and Paskov and Traub (1995)). Again the interest rates can be discretized using either a random walk or the Brownian bridge construction. For the numerical computation, the integral over $\mathbb{R}^d$ is transformed into an unweighted integral on $[0,1]^d$ with the help of the inverse normal distribution.

In Figure 5.30, we again compare the conventional sparse grid method with the dimension-adaptive method for the random walk and the Brownian bridge discretization. The error is computed against an independent quasi-Monte Carlo calculation. Note also that in this example the convergence rate of the conventional sparse grid approach is comparable to the quasi-Monte Carlo method (Gerstner and Griebel 1998).

We see again that a weighting of the dimensions does not influence the convergence of the conventional sparse grid method. But for the dimension-adaptive method the amount of work is again substantially reduced (by several orders of magnitude) for the same accuracy when the Brownian bridge discretization is used, and thus higher accuracies can be obtained.

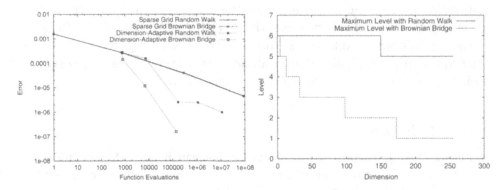

Figure 5.30. Computational results for the CMO problem ($d = 256$): integration error vs. number of function evaluations (left) and maximum level over all dimensions (sorted) for the dimension-adaptive algorithm with and without Brownian bridge discretization (right).

In this example the dimension-adaptive method also gives better results than the conventional sparse grid method for the random walk discretization. This implies that the conventional sparse grid uses too many points in mixed dimensions for this problem. The problem seems to be intrinsically lower-dimensional and nearly additive (Caflisch et al. 1997).

At present, we are working to carry this dimension-adaptive approach over to PDEs.

6. Concluding remarks

In this contribution we have given an overview of the basic principles and properties of sparse grids as well as a report on the state of the art concerning sparse grid applications. Starting from the dominant motivation – breaking the curse of dimensionality – we discussed the underlying tensor product approach, based upon different 1D multiscale bases such as the classical piecewise linear hierarchical basis, general hierarchical polynomial bases, interpolets, or wavelets. We then presented various resulting sparse grid constructions and discussed their properties with respect to computational complexity, discretization error, and smoothness requirements. The approach can be extended to nonsmooth solutions by adaptive refinement methods. We demonstrated the effectiveness of sparse grids in a series of applications. The presented numerical results include 2D and 3D PDE model problems, flow problems, and even two non-PDE applications in higher dimensions, namely numerical quadrature and data mining.

Since their introduction slightly more than a decade ago, sparse grids have seen a very successful development and a variety of different applications.

Especially for higher-dimensional scenarios, we are convinced that sparse grids, together with dimension adaptivity, will also have a thriving future.

For readers who want to stay up-to-date on sparse grid research, we refer to the sparse grid bibliography at www.ins.uni-bonn.de/info/sgbib, which gives roughly 300 articles from the past 40 years.

REFERENCES

S. Achatz (2003a), Adaptive finite Dünngitter-Elemente höherer Ordnung für elliptische partielle Differentialgleichungen mit variablen Koeffizienten, Dissertation, Institut für Informatik, TU München.

S. Achatz (2003b), 'Higher order sparse grids methods for elliptic partial differential equations with variable coefficients', *Computing* **71**, 1–15.

G. Allaire (1992), 'Homogenization and two-scale convergence', *SIAM J. Math. Anal.* **21**, 1482–1516.

A. Allen (1972), *Regression and the Moore–Penrose Pseudoinverse*, Academic Press, New York.

E. Arge, M. Dæhlen and A. Tveito (1995), 'Approximation of scattered data using smooth grid functions', *J. Comput. Appl. Math.* **59**, 191–205.

K. Babenko (1960), 'Approximation by trigonometric polynomials in a certain class of periodic functions of several variables', *Soviet Math. Dokl.* **1**, 672–675. Russian original in *Dokl. Akad. Nauk SSSR* **132** (1960), 982–985.

R. Balder (1994), Adaptive Verfahren für elliptische und parabolische Differentialgleichungen auf dünnen Gittern, Dissertation, Institut für Informatik, TU München.

R. Balder and C. Zenger (1996), 'The solution of multidimensional real Helmholtz equations on sparse grids', *SIAM J. Sci. Comp.* **17**, 631–646.

R. Balder, U. Rüde, S. Schneider and C. Zenger (1994), Sparse grid and extrapolation methods for parabolic problems, in *Proc. International Conference on Computational Methods in Water Resources, Heidelberg 1994* (A. Peters et al., eds), Kluwer Academic, Dordrecht, pp. 1383–1392.

A. Barron (1993), 'Universal approximation bounds for superpositions of a sigmoidal function', *IEEE Trans. Inform. Theory* **39**, 930–945.

A. Barron (1994), 'Approximation and estimation bounds for artificial neural networks', *Machine Learning* **14**, 115–133.

G. Baszenski and F. Delvos (1993), Multivariate Boolean midpoint rules, in *Numerical Integration IV* (H. Brass and G. Hämmerlin, eds), Vol. 112 of *International Series of Numerical Mathematics*, Birkhäuser, Basel, pp. 1–11.

G. Baszenski, F. Delvos and S. Jester (1992), Blending approximations with sine functions, in *Numerical Methods of Approximation Theory 9* (D. Braess and L. Schumaker, eds), Vol. 105 of *International Series of Numerical Mathematics*, Birkhäuser, Basel, pp. 1–19.

B. J. C. Baxter and A. Iserles (2003), 'On the foundations of computational mathematics', in *Handbook of Numerical Analysis*, Vol. 11 (F. Cucker, ed.), Elsevier, pp. 3–35.

R. Becker and R. Rannacher (1996), 'A feed-back approach to error control in finite element methods: Basic analysis and examples', *East–West J. Numer. Math.* **4**, 237–264.

J. Bell, P. Colella and H. Glaz (1989), 'A second order projection method for the incompressible Navier–Stokes equations', *J. Comput. Phys.* **85**, 257–283.

R. Bellmann (1961), *Adaptive Control Processes: A Guided Tour*, Princeton University Press.

K. Bennett and O. Mangasarian (1992), 'Robust linear programming discrimination of two linearly inseparable sets', *Optimiz. Methods and Software* **1**, 23–34.

M. Berry and G. Linoff (2000), *Mastering Data Mining*, Wiley.

C. Blake and C. Merz (1998), 'UCI repository of machine learning databases'. www.ics.uci.edu/~mlearn/MLRepository.html

T. Bonk (1994a), Ein rekursiver Algorithmus zur adaptiven numerischen Quadratur mehrdimensionaler Funktionen, Dissertation, Institut für Informatik, TU München.

T. Bonk (1994b), A new algorithm for multi-dimensional adaptive numerical quadrature, in *Adaptive Methods: Algorithms, Theory, and Applications* (W. Hackbusch and G. Wittum, eds), Vol. 46 of *Notes on Numerical Fluid Mechanics*, Vieweg, Braunschweig/Wiesbaden, pp. 54–68.

P. Bradley and O. Mangasarian (1998), Feature selection via concave minimization and support vector machines, in *Machine Learning: Proc. 15th International Conference; ICML '98* (J. Shavlik, ed.), Morgan Kaufmann, pp. 82–90.

H.-J. Bungartz (1992a), An adaptive Poisson solver using hierarchical bases and sparse grids, in *Iterative Methods in Linear Algebra* (P. de Groen and R. Beauwens, eds), Elsevier, Amsterdam, pp. 293–310.

H.-J. Bungartz (1992b), Dünne Gitter und deren Anwendung bei der adaptiven Lösung der dreidimensionalen Poisson-Gleichung, Dissertation, Institut für Informatik, TU München.

H.-J. Bungartz (1996), Concepts for higher order finite elements on sparse grids, in *Proc. International Conference on Spectral and High Order Methods, Houston 1995* (A. Ilin and L. Scott, eds), Houston J. Math., pp. 159–170.

H.-J. Bungartz (1997), 'A multigrid algorithm for higher order finite elements on sparse grids', *ETNA* **6**, 63–77.

H.-J. Bungartz (1998), Finite elements of higher order on sparse grids, Habilitationsschrift, Institut für Informatik, TU München and Shaker Verlag, Aachen.

H.-J. Bungartz and S. Dirnstorfer (2003), 'Multivariate quadrature on adaptive sparse grids', *Computing* **71**, 89–114.

H.-J. Bungartz and T. Dornseifer (1998), Sparse grids: Recent developments for elliptic partial differential equations, in *Multigrid Methods V* (W. Hackbusch and G. Wittum, eds), Vol. 3 of *Lecture Notes in Computational Science and Engineering*, Springer, Berlin/Heidelberg.

H.-J. Bungartz and M. Griebel (1999), 'A note on the complexity of solving Poisson's equation for spaces of bounded mixed derivatives', *J. Complexity* **15**, 167–199.

H.-J. Bungartz and W. Huber (1995), First experiments with turbulence simulation on workstation networks using sparse grid methods, in *Computational*

Fluid Dynamics on Parallel Systems (S. Wagner, ed.), Vol. 50 of *Notes on Numerical Fluid Mechanics*, Vieweg, Braunschweig/Wiesbaden.

H.-J. Bungartz, M. Griebel and U. Rüde (1994*a*), 'Extrapolation, combination, and sparse grid techniques for elliptic boundary value problems', *Comput. Meth. Appl. Mech. Eng.* **116**, 243–252.

H.-J. Bungartz, M. Griebel, D. Röschke and C. Zenger (1994*b*), 'Pointwise convergence of the combination technique for the Laplace equation', *East–West J. Numer. Math.* **2**, 21–45.

H.-J. Bungartz, M. Griebel, D. Röschke and C. Zenger (1994*c*), Two proofs of convergence for the combination technique for the efficient solution of sparse grid problems, in *Domain Decomposition Methods in Scientific and Engineering Computing* (D. Keyes and J. Xu, eds), Vol. 180 of *Contemporary Mathematics*, AMS, Providence, RI, pp. 15–20.

H.-J. Bungartz, M. Griebel, D. Röschke and C. Zenger (1996), 'A proof of convergence for the combination technique using tools of symbolic computation', *Math. Comp. Simulation* **42**, 595–605.

R. Caflisch, W. Morokoff and A. Owen (1997), 'Valuation of mortgage backed securities using Brownian bridges to reduce effective dimension', *J. Comput. Finance* **1**, 27–46.

J. Carnicer, W. Dahmen and J. Pena (1996), 'Local decomposition of refinable spaces', *Appl. Comp. Harm. Anal.* **3**, 127–153.

V. Cherkassky and F. Mulier (1998), *Learning from Data: Concepts, Theory and Methods*, Wiley.

A. Chorin (1968), 'Numerical solution of the Navier–Stokes equations', *Math. Comp.* **22**, 745–762.

A. Chorin (1998), *Vorticity and Turbulence*, Springer.

C. Chui (1992), *An Introduction to Wavelets*, Academic Press, Boston.

C. Chui and Y. Wang (1992), 'A general framework for compactly supported splines and wavelets', *J. Approx. Theory* **71**, 263–304.

D. Cioranescu, A. Damlamian and G. Griso (2002), 'Periodic unfolding and homogenization', *CR Acad. Sci. Paris, Ser. I* **335**, 99–104.

K. Cios, W. Pedrycz and R. Swiniarski (1998), *Data Mining Methods for Knowledge Discovery*, Kluwer.

A. Cohen (2003), *Numerical Analysis of Wavelet Methods*, Vol. 32 of *Studies in Mathematics and its Applications*, North-Holland.

A. Cohen and I. Daubechies (1996), 'A new technique to estimate the regularity of refinable functions', *Rev. Mat. Iberoamer.* **12**, 527–591.

A. Cohen, W. Dahmen and R. DeVore (2001), 'Adaptive wavelet methods for elliptic operator equations', *Math. Comp.* **70**, 27–75.

A. Cohen, I. Daubechies and J. Feauveau (1992), 'Biorthogonal bases of compactly supported wavelets', *Comm. Pure Appl. Math.* **45**, 485–560.

R. Cools and B. Maerten (1997), Experiments with Smolyak's algorithm for integration over a hypercube, Technical Report, Department of Computer Science, Katholieke Universiteit Leuven.

W. Dahmen and A. Kunoth (1992), 'Multilevel preconditioning', *Numer. Math.* **63**, 315–344.

I. Daubechies (1988), 'Orthogonal bases of compactly supported wavelets', *Comm. Pure Appl. Math.* **41**, 909–996.

I. Daubechies (1992), *Ten Lectures on Wavelets*, Vol. 61 of *CBMS–NSF Regional Conf. Series in Appl. Math.*, SIAM.

I. Daubechies (1993), 'Orthonormal bases of compactly supported wavelets II', *SIAM J. Math. Anal.* **24**, 499–519.

I. Daubechies and W. Sweldens (1998), 'Factoring wavelet transforms into lifting steps', *J. Fourier Anal. Appl.* **4**, 245–267.

F. Delvos (1982), 'd-variate Boolean interpolation', *J. Approx. Theory* **34**, 99–114.

F. Delvos (1990), 'Boolean methods for double integration', *Math. Comp.* **55**, 683–692.

F. Delvos and W. Schempp (1989), *Boolean Methods in Interpolation and Approximation*, Vol. 230 of *Pitman Research Notes in Mathematics*, Longman Scientific and Technical, Harlow.

G. Deslauriers and S. Dubuc (1989), 'Symmetric iterative interpolation processes', *Constr. Approx.* **5**, 49–68.

R. DeVore (1998), Nonlinear approximation, in *Acta Numerica*, Vol. 7, Cambridge University Press, pp. 51–150.

R. DeVore, S. Konyagin and V. Temlyakov (1998), 'Hyperbolic wavelet approximation', *Constr. Approx.* **14**, 1–26.

J. Dick, I. Sloan, X. Wang and H. Woźniakowski (2003), Liberating the weights, Technical Report AMR03/04, University of New South Wales.

D. Donoho (2000), 'High-dimensional data analysis: The curses and blessings of dimensionality'. Aide-Memoire.

D. Donoho and P. Yu (1999), Deslauriers–Dubuc: Ten years after, in Vol. 18 of *CRM Proceedings and Lecture Notes* (G. Deslauriers and S. Dubuc, eds).

P. V. Dooren and L. D. Ridder (1976), 'An adaptive algorithm for numerical integration over an n-dimensional cube', *J. Comp. Appl. Math.* **2**, 207–217.

T. Dornseifer (1997), Diskretisierung allgemeiner elliptischer Differentialgleichungen in krummlinigen Koordinatensystemen auf dünnen Gittern, Dissertation, Institut für Informatik, TU München.

T. Dornseifer and C. Pflaum (1996), 'Elliptic differential equations on curvilinear bounded domains with sparse grids', *Computing* **56**, 607–615.

J. Elf, P. Lötstedt and P. Sjöberg (2001), Problems of high dimension in molecular biology, in *17th Gamm Seminar, Leipzig 2001* (W. Hackbusch, ed.), pp. 1–10.

K. Eriksson, D. Estep, P. Hansbo and C. Johnson (1996), *Adaptive Finite Elements*, Springer, Berlin/Heidelberg.

G. Faber (1909), 'Über stetige Funktionen', *Mathematische Annalen* **66**, 81–94.

S. Fahlmann and C. Lebiere (1990), The cascade-correlation learning architecture, in *Advances in Neural Information Processing Systems*, Vol. 2 (Touretzky, ed.), Morgan-Kaufmann.

A. Frank (1995), Hierarchische Polynombasen zum Einsatz in der Datenkompression mit Anwendung auf Audiodaten, Diplomarbeit, Institut für Informatik, TU München.

K. Frank and S. Heinrich (1996), 'Computing discrepancies of Smolyak quadrature rules', *J. Complexity* **12**, 287–314.

K. Frank, S. Heinrich and S. Pereverzev (1996), 'Information complexity of multivariate Fredholm integral equations in Sobolev classes', *J. Complexity* **12**, 17–34.

J. Friedman (1991), 'Multivariate adaptive regression splines', *Ann. Statist.* **19**, 1–141. With discussion and a rejoinder by the author.

J. Garcke (1998), Berechnung von Eigenwerten der stationären Schrödingergleichung mit der Kombinationstechnik, Diplomarbeit, Institut für Angewandte Mathematik, Universität Bonn.

J. Garcke and M. Griebel (2000), 'On the computation of the eigenproblems of hydrogen and helium in strong magnetic and electric fields with the sparse grid combination technique', *J. Comput. Phys.* **165**, 694–716.

J. Garcke and M. Griebel (2001a), Data mining with sparse grids using simplicial basis functions, in *Proc. 7th ACM SIGKDD International Conference on Knowledge Discovery and Data Mining, San Francisco, USA* (F. Provost and R. Srikant, eds), pp. 87–96.

J. Garcke and M. Griebel (2001b), On the parallelization of the sparse grid approach for data mining, in *Large-Scale Scientific Computations, Third International Conference, LSSC 2001, Sozopol, Bulgaria* (S. Margenov, J. Wasniewski and P. Yalamov, eds), Vol. 2179 of *Lecture Notes in Computer Science*, pp. 22–32.

J. Garcke and M. Griebel (2002), 'Classification with sparse grids using simplicial basis functions', *Intelligent Data Analysis* **6**, 483–502.

J. Garcke, M. Griebel and M. Thess (2001), 'Data mining with sparse grids', *Computing* **67**, 225–253.

J. Garcke, M. Hegland and O. Nielsen (2003), Parallelisation of sparse grids for large scale data analysis, in *Proc. International Conference on Computational Science 2003 (ICCS 2003), Melbourne, Australia* (P. Sloot, D. Abramson, A. Bogdanov, J. Dongarra, A. Zomaya and Y. Gorbachev, eds), Vol. 2659 of *Lecture Notes in Computer Science*, Springer, pp. 683–692.

A. Genz and A. Malik (1980), 'An adaptive algorithm for numerical integration over an n-dimensional rectangular region', *J. Comp. Appl. Math.* **6**, 295–302.

T. Gerstner (1995), Ein adaptives hierarchisches Verfahren zur Approximation und effizienten Visualisierung von Funktionen und seine Anwendung auf digitale 3D Höhenmodelle, Diplomarbeit, Institut für Informatik, TU München.

T. Gerstner (1999), Adaptive hierarchical methods for landscape representation and analysis, in *Process Modelling and Landform Evolution*, Vol. 78 of *Lecture Notes in Earth Sciences*, Springer.

T. Gerstner and M. Griebel (1998), 'Numerical integration using sparse grids', *Numer. Alg.* **18**, 209–232.

T. Gerstner and M. Griebel (2003), 'Dimension-adaptive tensor-product quadrature', *Computing* **71**, 65–87.

F. Girosi (1998), 'An equivalence between sparse approximation and support vector machines', *Neural Computation* **10**, 1455–1480.

F. Girosi, M. Jones and T. Poggio (1995), 'Regularization theory and neural networks architectures', *Neural Computation* **7**, 219–265.

G. Golub, M. Heath and G. Wahba (1979), 'Generalized cross validation as a method for choosing a good ridge parameter', *Technometrics* **21**, 215–224.

W. Gordon (1969), Distributive lattices and the approximation of multivariate functions, in *Approximation with Special Emphasis on Spline Functions* (I. Schoenberg, ed.), Academic Press, New York, pp. 223–277.

W. Gordon (1971), 'Blending function methods of bivariate and multivariate interpolation and approximation', *SIAM J. Numer. Anal.* **8**, 158–177.

W. Gordon and C. Hall (1973), 'Transfinite element methods: Blending-function interpolation over arbitrary curved element domains', *Numer. Math.* **21**, 109–129.

R. Graham, D. Knuth and O. Patashnik (1994), *Concrete Mathematics*, Addison-Wesley, Reading.

P. Gresho and R. Sani (1987), 'On pressure boundary conditions for the incompressible Navier–Stokes equations', *Int. J. Numer. Meth. Fluids* **7**, 371–394.

M. Griebel (1991a), Parallel multigrid methods on sparse grids, in *Multigrid Methods III* (W. Hackbusch and U. Trottenberg, eds), Vol. 98 of *International Series of Numerical Mathematics*, Birkhäuser, Basel, pp. 211–221.

M. Griebel (1991b), A parallelizable and vectorizable multi-level algorithm on sparse grids, in *Parallel Algorithms for Partial Differential Equations* (W. Hackbusch, ed.), Vol. 31 of *Notes on Numerical Fluid Mechanics*, Vieweg, Braunschweig/Wiesbaden, pp. 94–100.

M. Griebel (1992), 'The combination technique for the sparse grid solution of PDEs on multiprocessor machines', *Parallel Processing Letters* **2**, 61–70.

M. Griebel (1993), Sparse grid multilevel methods, their parallelization and their application to CFD, in *Proc. Parallel Computational Fluid Dynamics 1992* (J. Häser, ed.), Elsevier, Amsterdam, pp. 161–174.

M. Griebel (1994a), 'Multilevel algorithms considered as iterative methods on semi-definite systems', *SIAM J. Sci. Statist. Comput.* **15**, 547–565.

M. Griebel (1994b), *Multilevelmethoden als Iterationsverfahren über Erzeugendensystemen*, Teubner Skripten zur Numerik, Teubner, Stuttgart.

M. Griebel (1998), 'Adaptive sparse grid multilevel methods for elliptic PDEs based on finite differences', *Computing* **61**, 151–179.

M. Griebel and W. Huber (1995), Turbulence simulation on sparse grids using the combination method, in *Parallel Computational Fluid Dynamics, New Algorithms and Applications* (N. Satofuka, J. Périaux and A. Ecer, eds), Elsevier, Amsterdam, pp. 75–84.

M. Griebel and S. Knapek (2000), 'Optimized tensor-product approximation spaces', *Constr. Approx.* **16**, 525–540.

M. Griebel and F. Koster (2000), Adaptive wavelet solvers for the unsteady incompressible Navier–Stokes equations, in *Advances in Mathematical Fluid Mechanics* (J. Malek, J. Necas and M. Rokyta, eds), Springer, pp. 67–118.

M. Griebel and F. Koster (2003), Multiscale methods for the simulation of turbulent flows, in *Numerical Flow Simulation III* (E. Hirschel, ed.), Vol. 82 of *Notes on Numerical Fluid Mechanics and Multidisciplinary Design*, Springer, pp. 203–214.

M. Griebel and P. Oswald (1994), 'On additive Schwarz preconditioners for sparse grid discretizations', *Numer. Math.* **66**, 449–463.

M. Griebel and P. Oswald (1995a), 'On the abstract theory of additive and multiplicative Schwarz algorithms', *Numer. Math.* **70**, 161–180.

M. Griebel and P. Oswald (1995b), 'Tensor product type subspace splittings and multilevel iterative methods for anisotropic problems', *Adv. Comput. Math.* **4**, 171–206.

M. Griebel and T. Schiekofer (1999), An adaptive sparse grid Navier–Stokes solver in 3D based on the finite difference method, in *Proc. ENUMATH97* (H. Bock, G. Kanschat, R. Rannacher, F. Brezzi, R. Glowinski, Y. Kuznetsov and J. Periaux, eds), World Scientific, Heidelberg.

M. Griebel and M. Schweitzer (2002), 'A particle-partition of unity method, part II: Efficient cover construction and reliable integration', *SIAM J. Sci. Comp.* **23**, 1655–1682.

M. Griebel and V. Thurner (1993), 'Solving CFD-problems efficiently by the combination method', *CFD-News* **3**, 19–31.

M. Griebel and V. Thurner (1995), 'The efficient solution of fluid dynamics problems by the combination technique', *Int. J. Numer. Meth. Heat Fluid Flow* **5**, 251–269.

M. Griebel, W. Huber and C. Zenger (1993a), A fast Poisson solver for turbulence simulation on parallel computers using sparse grids, in *Flow Simulation on High-Performance Computers I* (E. Hirschel, ed.), Vol. 38 of *Notes on Numerical Fluid Mechanics*, Vieweg, Braunschweig/Wiesbaden.

M. Griebel, W. Huber and C. Zenger (1996), Numerical turbulence simulation on a parallel computer using the combination method, in *Flow Simulation on High-Performance Computers II* (E. Hirschel, ed.), Vol. 52 of *Notes on Numerical Fluid Mechanics*, Vieweg, Braunschweig/Wiesbaden, pp. 34–47.

M. Griebel, P. Oswald and T. Schiekofer (1999), 'Sparse grids for boundary integral equations', *Numer. Math.* **83**, 279–312.

M. Griebel, M. Schneider and C. Zenger (1992), A combination technique for the solution of sparse grid problems, in *Iterative Methods in Linear Algebra* (P. de Groen and R. Beauwens, eds), Elsevier, Amsterdam, pp. 263–281.

M. Griebel, C. Zenger and S. Zimmer (1993b), 'Multilevel Gauss–Seidel-algorithms for full and sparse grid problems', *Computing* **50**, 127–148.

M. Gromov (1999), *Metric Structures for Riemannian and Non-Riemannian Spaces*, Vol. 152 of *Progress in Mathematics*, Birkhäuser.

W. Hackbusch (1985), *Multigrid Methods and Applications*, Springer, Berlin/Heidelberg.

W. Hackbusch (1986), *Theorie und Numerik elliptischer Differentialgleichungen*, Teubner, Stuttgart.

W. Hackbusch (2001), 'The efficient computation of certain determinants arising in the treatment of Schrödinger's equation', *Computing* **67**, 35–56.

W. Hahn (1990), Parallelisierung eines adaptiven hierarchischen Dünngitterverfahrens, Diplomarbeit, Institut für Informatik, TU München.

K. Hallatschek (1992), 'Fouriertransformation auf dünnen Gittern mit hierarchischen Basen', *Numer. Math.* **63**, 83–97.

T. Hastie and R. Tibshirani (1990), *Generalized Additive Models*, Chapman and Hall.

T. He (2001), *Dimensionality Reducing Expansion of Multivariate Integration*, Birkhäuser.

M. Hegland (2002), Additive sparse grid fitting, in *Proc. 5th International Conference on Curves and Surfaces, Saint-Malo, France 2002*. Submitted.

M. Hegland (2003), Adaptive sparse grids, in *Proc. 10th Computational Techniques and Applications Conference, CTAC-2001* (K. Burrage and R. Sidje, eds), Vol. 44 of *ANZIAM J.*, pp. C335–C353.

M. Hegland and V. Pestov (1999), Additive models in high dimensions, Technical Report 99-33, MCS-VUW research report.

P. Hemker (1995), 'Sparse-grid finite-volume multigrid for 3D problems', *Adv. Comput. Math.* **4**, 83–110.

P. Hemker and P. de Zeeuw (1996), BASIS3: A data structure for 3-dimensional sparse grids, in *Euler and Navier–Stokes Solvers Using Multi-Dimensional Upwind Schemes and Multigrid Acceleration* (H. Deconinck and B. Koren, eds), Vol. 56 of *Notes on Numerical Fluid Mechanics*, Vieweg, Braunschweig/Wiesbaden, pp. 443–484.

P. Hemker and C. Pflaum (1997), 'Approximation on partially ordered sets of regular grids', *Appl. Numer. Math.* **25**, 55–87.

P. Hemker, B. Koren and J. Noordmans (1998), 3D multigrid on partially ordered sets of grids. *Multigrid Methods V*, Vol. 3 of *Lecture Notes in Computational Science and Engineering*, Springer, Berlin/Heidelberg, pp. 105–124.

J. Hennart and E. Mund (1988), 'On the h- and p-versions of the extrapolated Gordon's projector with applications to elliptic equations', *SIAM J. Sci. Statist. Comput.* **9**, 773–791.

J. Heroth (1997), Are sparse grids suitable for the tabulation of reduced chemical systems?, Technical Report, TR 97-2, Konrad-Zuse-Zentrum für Informationstechnik Berlin.

F. Hickernell, I. Sloan and G. Wasilkowski (2003), On tractability of weighted integration for certain Banach spaces of functions, Technical Report AMR03/08, University of New South Wales.

R. Hiptmair and V. Gradinaru (2003), 'Multigrid for discrete differential forms on sparse grids', *Computing* **71**, 17–42.

E. Hlawka (1961), 'Funktionen von beschränkter Variation in der Theorie der Gleichverteilung', *Ann. Math. Pure Appl.* **54**, 325–333.

T. Ho and E. Kleinberg (1996), 'Checkerboard dataset'.
www.cs.wisc.edu/math-prog/mpml.html

V. Hoang and C. Schwab (2003), High-dimensional finite elements for elliptic problems with multiple scales, Technical Report 2003-14, Seminar für Angewandte Mathematik, ETH Zürich.

R. Hochmuth (1999), Wavelet bases in numerical analysis and restricted nonlinear approximation, Habilitationsschrift, Freie Universität Berlin.

R. Hochmuth, S. Knapek and G. Zumbusch (2000), Tensor products of Sobolev spaces and applications, Technical Report 685, SFB 256, Universität Bonn.

J. Hoschek and D. Lasser (1992), *Grundlagen der Geometrischen Datenverarbeitung*, Teubner, chapter 9.

W. Huber (1996a), Numerical turbulence simulation on different parallel computers using the sparse grid combination method, in *Proc. EuroPar '96, Lyon 1996* (L. Bougé, P. Fraigniaud, A. Mignotte and Y. Robert, eds), Vol. 1124 of *Lecture Notes in Computer Science*, Springer, Berlin/Heidelberg, pp. 62–65.

W. Huber (1996b), Turbulenzsimulation mit der Kombinationsmethode auf Workstation-Netzen und Parallelrechnern, Dissertation, Institut für Informatik, TU München.

M. Kalos and P. Whitlock (1986), *Monte Carlo Methods*, Wiley.

G. Karniadakis and S. Sherwin (1999), *Spectral/hp Element Methods for CFD*, Oxford University Press.

L. Kaufman (1999), Solving the quadratic programming problem arising in support vector classification, in *Advances in Kernel Methods: Support Vector Learning* (B. Schölkopf, C. Burges and A. Smola, eds), MIT Press, pp. 146–167.

S. Khavinson (1997), *Best Approximation by Linear Superposition (Approximate Nomography)*, Vol. 159 of *AMS Translations of Mathematical Monographs*, AMS, Providence, RI.

S. Knapek (2000a), Approximation and Kompression mit Tensorprodukt-Multiskalen-Approximationsräumen, Dissertation, Institut für Angewandte Mathematik, Universität Bonn.

S. Knapek (2000b), Hyperbolic cross approximation of integral operators with smooth kernel, Technical Report 665, SFB 256, Universität Bonn.

S. Knapek and F. Koster (2002), 'Integral operators on sparse grids', *SIAM J. Numer. Anal.* **39**, 1794–1809.

A. Kolmogorov (1956), 'On the representation of continuous functions of several variables by superpositions of continuous functions of fewer variables', *Dokl. Akad. Nauk SSSR* **108**, 179–182. In Russian; English Translation in *Amer. Math. Soc. Transl.* (2) **17** (1961), 369–373.

A. Kolmogorov (1957), 'On the representation of continuous functions of several variables by superpositions of continuous functions of one variable and addition', *Dokl. Akad. Nauk SSSR* **114**, 953–956. In Russian; English Translation in *Amer. Math. Soc. Transl.* (2) **28** (1963), 55–59.

B. Koren, P. Hemker and P. de Zeeuw (1996), Semi-coarsening in three directions for Euler-flow computations in three dimensions, in *Euler and Navier–Stokes Solvers Using Multi-Dimensional Upwind Schemes and Multigrid Acceleration* (H. Deconinck and B. Koren, eds), Vol. 56 of *Notes on Numerical Fluid Mechanics*, Vieweg, Braunschweig/Wiesbaden, pp. 547–567.

F. Koster (2002), Multiskalen-basierte Finite Differenzen Verfahren auf adaptiven dünnen Gittern, Dissertation, Institut für Angewandte Mathematik, Universität Bonn.

A. Krommer and C. Ueberhuber (1994), *Numerical Integration on Advanced Computer Systems*, Vol. 848 of *Lecture Notes in Computer Science*, Springer, Berlin/Heidelberg.

F. Kupka (1997), Sparse Grid Spectral Methods for the Numerical Solution of Partial Differential Equations with Periodic Boundary Conditions, Dissertation, Institut für Mathematik, Universität Wien.

Y. Lee and O. Mangasarian (2001), 'SSVM: A smooth support vector machine for classification', *Comput. Optimiz. Appl.* **20**, 5–22.

P. Lemarié (1988), 'Ondelettes à localisation exponentielle', *J. Math. Pures Appl.* **67**, 222–236.

C. Liem, T. Lu and T. Shih (1995), *The Splitting Extrapolation Method*, World Scientific, Singapore.

R. Lorentz and P. Oswald (1998), Multilevel finite element Riesz bases in Sobolev spaces, in *Proc. 9th International Conference on Domain Decomposition* (P. Bjoerstad et al., eds), Domain Decomposition Press, Bergen, pp. 178–187.

S. Martello and P. Toth (1990), *Knapsack Problems: Algorithms and Computer Implementations*, Wiley, Chichester.

A. Matache (2001), Sparse two-scale FEM for homogenization problems, Technical Report 2001-09, Seminar für Angewandte Mathematik, ETH Zürich.

A. Matache and C. Schwab (2001), Two-scale FEM for homogenization problems, Technical Report 2001-06, Seminar für Angewandte Mathematik, ETH Zürich.

G. Melli (2003), 'DatGen: A program that creates structured data', web site: www.datasetgenerator.com

Y. Meyer (1992), *Wavelets and Operators*, Cambridge University Press.

H. Mhaskar (1996), 'Neural networks and approximation theory', *Neural Networks* **9**, 711–722.

A. Michalke (1964), 'On the inviscid instability of the hyperbolic tangent velocity profile', *J. Fluid Mech.* **19**, 543–556.

V. Milman (1988), 'The heritage of P. Levy in geometrical functional analysis', *Asterisque* **157–158**, 273–301.

V. Milman and G. Schechtman (1986), *Asymptotic Theory of Finite-Dimensional Normed Spaces*, Vol. 1200 of *Lecture Notes in Mathematics*, Springer.

U. Mitzlaff (1997), Diffusionsapproximation von Warteschlangensystemen, Dissertation, Institut für Mathematik, Technische Universität Clausthal.

W. Morokoff and R. Caflisch (1995), 'Quasi-Monte Carlo integration', *J. Comput. Phys.* **122**, 218–230.

W. Mulder (1989), 'A new multigrid approach to convection problems', *J. Comput. Phys.* **83**, 303–323.

N. Naik and J. van Rosendale (1993), 'The improved robustness of multigrid elliptic solvers based on multiple semicoarsened grids', *SIAM J. Numer. Anal.* **30**, 215–229.

H. Niederreiter (1992), *Random Number Generation and Quasi-Monte-Carlo Methods*, SIAM, Philadelphia.

P. Niyogi and F. Girosi (1998), 'Generalization bounds for function approximation from scattered noisy data', *Adv. Comput. Math.* **10**, 51–80.

E. Novak and K. Ritter (1996), 'High dimensional integration of smooth functions over cubes', *Numer. Math.* **75**, 79–98.

E. Novak and K. Ritter (1997), Simple cubature formulas for d-dimensional integrals with high polynomial exactness and small error, Technical Report, Institut für Mathematik, Universität Erlangen–Nürnberg.

E. Novak and K. Ritter (1998), The curse of dimension and a universal method for numerical integration, in *Multivariate Approximation and Splines* (G. Nürnberger, J. Schmidt and G. Walz, eds), International Series in Numerical Mathematics, Birkhäuser, Basel, pp. 177–188.

E. Novikov (1976), 'Dynamics and statistics of a system of vortices', *Sov. Phys. JETP* **41,5**, 937–943.

P. Oswald (1994), *Multilevel Finite Element Approximation*, Teubner Skripten zur Numerik, Teubner, Stuttgart.

S. Paskov (1993), 'Average case complexity of multivariate integration for smooth functions', *J. Complexity* **9**, 291–312.

S. Paskov and J. Traub (1995), 'Faster valuation of financial derivatives', *J. Portfolio Management* **22**, 113–120.

T. Patterson (1986), 'The optimum addition of points to quadrature formulae', *Math. Comp.* **22**, 847–856.

A. Paul (1995), Kompression von Bildfolgen mit hierarchischen Basen, Diplomarbeit, Institut für Informatik, TU München.

A. Peano (1976), 'Hierarchies of conforming finite elements for plane elasticity and plate bending', *Comp. Math. Appl.* **2**, 211–224.

K. Petras (2000), 'On the Smolyak cubature error for analytic functions', *Adv. Comput. Math.* **12**, 71–93.

A. Pfaffinger (1997), Funktionale Beschreibung und Parallelisierung von Algorithmen auf dünnen Gittern, Dissertation, Institut für Informatik, TU München.

C. Pflaum (1992), Anwendung von Mehrgitterverfahren auf dünnen Gittern, Diplomarbeit, Institut für Informatik, TU München.

C. Pflaum (1996), Diskretisierung elliptischer Differentialgleichungen mit dünnen Gittern, Dissertation, Institut für Informatik, TU München.

C. Pflaum (1998), 'A multilevel algorithm for the solution of second order elliptic differential equations on sparse grids', *Numer. Math.* **79**, 141–155.

C. Pflaum and A. Zhou (1999), 'Error analysis of the combination technique', *Numer. Math.* **84**, 327–350.

G. Pöplau and F. Sprengel (1997), Some error estimates for periodic interpolation on full and sparse grids, in *Curves and Surfaces with Applications in CAGD* (A. Le Méhauté, C. Rabut and L. Schumaker, eds), Vanderbilt University Press, Nashville, Tennessee, pp. 355–362.

J. Prakash (2000), Rouse chains with excluded volume interactions: Linear viscoelasticity, Technical Report 221, Berichte der Arbeitsgruppe Technomathematik, Universität Kaiserslautern.

J. Prakash and H. Öttinger (1999), 'Viscometric functions for a dilute solution of polymers in a good solvent', *Macromolecules* **32**, 2028–2043.

A. Prohl (1997), *Projection and Quasi-Compressibility Methods for Solving the Incompressible Navier–Stokes Equations*, Advances in Numerical Mathematics, B. G. Teubner.

T. Rassias and J. Simsa (1995), *Finite Sums Decompositions in Mathematical Analysis*, Wiley.

C. Reisinger (2003), Numerische Methoden für hochdimensionale parabolische Gleichungen am Beispiel von Optionspreisaufgaben, Dissertation, Universität Heidelberg.

P. Rouse (1953), 'A theory of the linear viscoelastic properties of dilute solutions of coiling polymers', *J. Chem. Phys.* **21**, 1272–1280.

S. Salzberg (1997), 'On comparing classifiers: Pitfalls to avoid and a recommended approach', *Data Mining and Knowledge Discovery* **1**, 317–327.

T. Schiekofer (1998), Die Methode der finiten Differenzen auf dünnen Gittern zur Lösung elliptischer und parabolischer partieller Differentialgleichungen, Dissertation, Institut für Angewandte Mathematik, Universität Bonn.

T. Schiekofer and G. Zumbusch (1998), Software concepts of a sparse grid finite difference code, in *Proc. 14th GAMM-Seminar on Concepts of Numerical Software* (W. Hackbusch and G. Wittum, eds), Notes on Numerical Fluid Mechanics, Vieweg, Braunschweig/Wiesbaden.

S. Schneider (2000), Adaptive Solution of Elliptic PDE by Hierarchical Tensor Product Finite Elements, Dissertation, Institut für Informatik, TU München.

C. Schwab and R. Todor (2002), Sparse finite elements for stochastic elliptic problems, Technical Report 2002-05, Seminar für Angewandte Mathematik, ETH Zürich.

C. Schwab and R. Todor (2003), 'Sparse finite elements for stochastic elliptic problems: Higher order moments', *Computing* **71**, 43–63.

M. Schweitzer (2003), *A Parallel Multilevel Partition of Unity Method for Elliptic Partial Differential Equations*, Vol. 29 of Lecture Notes in Computational Science and Engineering, Springer.

R. Sedgewick and P. Flajolet (1996), *Analysis of Algorithms*, Addison-Wesley, Reading.

X. Shen, H. Chen, J. Dai and W. Dai (2002), 'The finite element method for computing the stationary distribution on an SRBM in a hypercube with applications to finite buffer queueing networks', *Queuing Systems* **42**, 33–62.

J. Simsa (1992), 'The best L^2-approximation by finite sums of functions with separable variables', *Aequationes Mathematicae* **43**, 284–263.

P. Sjöberg (2002), Numerical solution of the master equation in molecular biology, Master's Thesis, Department of Scientific Computing, Uppsala Universität.

I. Sloan (2001), QMC integration: Beating intractability by weighting the coordinate directions, Technical Report AMR01/12, University of New South Wales.

I. Sloan and S. Joe (1994), *Lattice Methods for Multiple Integration*, Oxford University Press.

I. Sloan and H. Woźniakowski (1998), 'When are quasi-Monte Carlo algorithms efficient for high dimensional integrals ?', *J. Complexity* **14**, 1–33.

S. Smolyak (1963), 'Quadrature and interpolation formulas for tensor products of certain classes of functions', *Soviet Math. Dokl.* **4**, 240–243. Russian original in *Dokl. Akad. Nauk SSSR* **148** (1963), 1042–1045.

F. Sprengel (1997a), Interpolation and Wavelet Decomposition of Multivariate Periodic Functions, Dissertation, FB Mathematik, Universität Rostock.

F. Sprengel (1997b), 'A unified approach to error estimates for interpolation on full and sparse Gauss–Chebyshev grids', *Rostocker Math. Kolloq.* **51**, 51–64.

R. Stevenson (1996), Piecewise linear (pre)-wavelets on non-uniform meshes, in *Multigrid Methods IV* (W. Hackbusch and G. Wittum, eds), Vol. 3 of Lecture Notes in Computational Science and Engineering, Springer.

M. Stone (1974), 'Cross-validatory choice and assessment of statistical predictions', *J. Royal Statist. Soc.* **36**, 111–147.

T. Störtkuhl (1995), Ein numerisches adaptives Verfahren zur Lösung der biharmonischen Gleichung auf dünnen Gittern, Dissertation, Institut für Informatik, TU München.

W. Sweldens (1997), 'The lifting scheme: A construction of second generation wavelets', *SIAM J. Math. Anal.* **29**, 511–546.

M. Talagrand (1995), 'Concentration of measure and isoperimetric inequalities in product spaces', *Publ. Math. IHES* **81**, 73–205.

R. Temam (1969), 'Sur l'approximation de la solution des équations de Navier–Stokes par la méthode des fractionaires (II)', *Arch. Rat. Mech. Anal.* **33**, 377–385.

V. Temlyakov (1989), *Approximation of Functions with Bounded Mixed Derivative*, Vol. 178 of *Proc. Steklov Inst. of Math.*, AMS, Providence, RI.

V. Temlyakov (1993), 'On approximate recovery of functions with bounded mixed derivative', *J. Complexity* **9**, 41–59.

V. Temlyakov (1994), *Approximation of Periodic Functions*, Nova Science, Commack, New York.

A. Tikhonov and V. Arsenin (1977), *Solutions of Ill-Posed Problems*, W. H. Winston, Washington DC.

J. Traub and H. Woźniakowski (1980), *A General Theory of Optimal Algorithms*, Academic Press, New York.

J. Traub, G. Wasilkowski and H. Woźniakowski (1983), *Information, Uncertainty, Complexity*, Addison-Wesley, Reading.

J. Traub, G. Wasilkowski and H. Woźniakowski (1988), *Information-Based Complexity*, Academic Press, New York.

H. Triebel (1992), *Theory of Function Spaces II*, Birkhäuser.

F. Utreras (1979), Cross-validation techniques for smoothing spline functions in one or two dimensions, in *Smoothing Techniques for Curve Estimation* (T. Gasser and M. Rosenblatt, eds), Springer, Heidelberg, pp. 196–231.

V. Vapnik (1982), *Estimation of Dependences Based on Empirical Data*, Springer, Berlin.

V. Vapnik (1995), *The Nature of Statistical Learning Theory*, Springer.

G. Wahba (1985), 'A comparison of GCV and GML for choosing the smoothing parameter in the generalized splines smoothing problem', *Ann. Statist.* **13**, 1378–1402.

G. Wahba (1990), *Spline Models for Observational Data*, Vol. 59 of *Series in Applied Mathematics*, SIAM, Philadelphia.

G. Wasilkovski and H. Woźniakowski (1995), 'Explicit cost bounds of algorithms for multivariate tensor product problems', *J. Complexity* **11**, 1–56.

G. Wasilkovski and H. Woźniakowski (1999), 'Weighted tensor product algorithms for linear multivariate problems', *J. Complexity* **15**, 402–447.

A. Werschulz (1995), The complexity of the Poisson problem for spaces of bounded mixed derivatives, Technical Report CUCS-016-95, Columbia University.

H. Woźniakowski (1985), 'A survey of information-based complexity', *J. Complexity* **1**, 11–44.

H. Yserentant (1986), 'On the multi-level splitting of finite element spaces', *Numer. Math.* **49**, 379–412.

H. Yserentant (1990), 'Two preconditioners based on the multi-level splitting of finite element spaces', *Numer. Math.* **58**, 163–184.

H. Yserentant (1992), Hierarchical bases, in *Proc. ICIAM '91, Washington 1991* (R. O'Malley et al., eds), SIAM, Philadelphia.

H. Yserentant (2004), On the regularity of the electronic Schrödinger equation in Hilbert spaces of mixed derivatives. *Numer. Math.*, in press.

R. Yue and F. Hickernell (2002), 'Robust designs for smoothing spline ANOVA models', *Metrika* **55**, 161–176.

C. Zenger (1991), Sparse grids, in *Parallel Algorithms for Partial Differential Equations* (W. Hackbusch, ed.), Vol. 31 of *Notes on Numerical Fluid Mechanics*, Vieweg, Braunschweig/Wiesbaden.

O. Zienkiewicz, D. Kelly, J. Gago and I. Babuška (1982), Hierarchical finite element approaches, error estimates, and adaptive refinement, in *The Mathematics of Finite Elements and Applications IV* (J. Whiteman, ed.), Academic Press, London.

G. Zumbusch (1996), *Simultaneous hp Adaptation in Multilevel Finite Elements*, Shaker, Aachen.

Complete search in continuous global optimization and constraint satisfaction

Arnold Neumaier

Institut für Mathematik, Universität Wien,

Nordbergstraße 15, A-1090 Wien, Austria

E-mail: `Arnold.Neumaier@univie.ac.at`

`www.mat.univie.ac.at/~neum/`

This survey covers the state of the art of techniques for solving general-purpose constrained global optimization problems and continuous constraint satisfaction problems, with emphasis on complete techniques that provably find all solutions (if there are finitely many). The core of the material is presented in sufficient detail that the survey may serve as a text for teaching constrained global optimization.

After giving motivations for and important examples of applications of global optimization, a precise problem definition is given, and a general form of the traditional first-order necessary conditions for a solution. Then more than a dozen software packages for complete global search are described.

A quick review of incomplete methods for bound-constrained problems and recipes for their use in the constrained case follows; an explicit example is discussed, introducing the main techniques used within branch and bound techniques. Sections on interval arithmetic, constrained propagation and local optimization are followed by a discussion of how to avoid the cluster problem. Then a discussion of important problem transformations follows, in particular of linear, convex, and semilinear (= mixed integer linear) relaxations that are important for handling larger problems.

Next, reliability issues – centring on rounding error handling and testing methodologies – are discussed, and the COCONUT framework for the integration of the different techniques is introduced. A list of challenges facing the field in the near future concludes the survey.

CONTENTS

1	Introduction	272
2	Why global optimization?	277
3	Basic ideas	279
4	Problem formulation	282
5	First-order optimality conditions	284
6	Software for complete global optimization	287
7	Incomplete methods for simple constraints	291
8	Bound-constrained approximation	295
9	Pure branching methods	298
10	Box reduction: an example	300
11	Interval arithmetic	302
12	The branch and bound principle	307
13	The role of local optimization	310
14	Constraint propagation	313
15	The cluster problem and second-order information	318
16	Linear and convex relaxations	323
17	Semilinear constraints and MILP	329
18	Semilinear relaxations	335
19	Other problem transformations	339
20	Rigorous verification and certificates	341
21	Test problems and testing	344
22	The COCONUT environment	346
23	Challenges for the near future	348
	References	351

1. Introduction

Consider everything. Keep what is good. Avoid evil whenever you recognize it.
St. Paul, ca. 50 A.D. (*The Bible*, 1 Thess. 5:21–22)

1.1. Early history

As the above quote shows, continuous global optimization or constraint satisfaction and the associated global search methods have been part of the art of successful living since antiquity. In the mathematical literature published before 1975, there are occasional references related to the topic, few and important enough to mention them individually. (Please inform me about other significant work on continuous global optimization published before 1975 not mentioned here!) Several independent strands of work (probably done in ignorance of each other) are discernible.

(1) Markowitz and Manne (1957) and Dantzig *et al.* (Dantzig, Johnson and White 1958, Dantzig 1960) used piecewise linear approximations for the approximate global minimization of separable nonconvex programs, formulating them as mixed integer linear programs. Land and Doig (1960) and Little, Murty, Sweeney and Karel (1963) introduced the branch and bound technique for discrete optimization, applicable to mixed integer linear programs. Motzkin and Strauss (1965) showed that solving the (discrete) maximum clique problem is equivalent to finding the global minimum (or maximum) of a special nonconvex quadratic program. Falk and Soland (1969) gave the first piecewise linear relaxations of nonconvex problems, thus making them available for obtaining bounds in a branch and bound scheme. Beale and Tomlin (1970) and Tomlin (1970) introduced special ordered sets, defining the way piecewise linear functions are still handled today in mixed integer linear programming solvers. McCormick (1972) introduced the now frequently used linear relaxations for products and quotients, which made the solution of general factorable global optimization problems accessible to the branch and bound technique.

(2) Moore (1962) showed in Part 4 of his PhD thesis, which introduced interval arithmetic to numerical analysis (following an unpublished technical report by Moore and Yang (1959)), that by repeated subdivision and simple interval evaluation, the range – hence in particular the global minimum – of a rational function over a box can be determined in principle to arbitrary accuracy. Skelboe (1974) improved this basic but excessively slow method by embedding it into (what would today be called) a branch and bound scheme for continuous variables, giving (what is now called) the Moore–Skelboe algorithm. Moore's thesis (Moore 1962) also showed that interval methods can be used to prove the nonexistence of solutions of nonlinear systems in a box (which nowadays is used to discard boxes in a branch and bound scheme) and to reduce the region where a solution can possibly lie (which is now used to avoid excessive splitting). Kahan (1968) discovered that interval techniques can also be used to prove the existence of solutions of nonlinear systems in a box (and hence to verify feasibility). Krawczyk (1969) simplified Moore's methods for systems of equations; the Krawczyk operator based on his paper is used in several state-of-the-art global solvers. (See also the historical remarks in Hansen (2001) and Madsen and Skelboe (2002).) Piyavskii (1972) introduced complete global optimization methods based on Lipschitz constants, which are similar in flavour to interval methods.

(3) Tsuda and Kiono (1964) introduced the Monte Carlo method for finding solutions of systems of equations; and the thesis by Mockus (1966) applied it to global optimization. Becker and Lago (1970) were the first to use clustering methods, and Törn (1972, 1974) suggested combining these with local optimization, a combination which currently defines the most efficient class of stochastic global optimization algorithms (Janka 2000). Holland (1973) introduced genetic algorithms, still a popular (although usually slow) stochastic heuristics for global optimization.

As a recognizable mathematical discipline with diverse solution methods for precisely formulated problems involving continuous variables, the field essentially dates back to 1975 when the first book containing exclusively global optimization papers appeared: the volume *Towards Global Optimization*, edited by Dixon and Szegő (1975). In the almost 30 years since the publication of this landmark volume, tremendous progress has been made, and many signs indicate that the field is now ripe for manifold applications in science and engineering.

1.2. Scope

Global optimization is the task of finding the absolutely best set of admissible conditions to achieve an objective under given constraints, assuming that both are formulated in mathematical terms. It is much more difficult than convex programming or finding local minimizers of nonlinear programs, since the gap between the necessary KKT (Karush–Kuhn–Tucker) conditions for optimality and known sufficient conditions for global optimality is tremendous.

Many famous hard optimization problems, such as the travelling salesman problem or the protein folding problem, are global optimization problems. The truth of the famous unresolved conjecture $P \neq NP$ (Garey and Johnson 1979) would imply (Murty and Kabadi 1987, Pardalos and Schnitger 1988) that there are no general algorithms that solve a given global optimization problem in time that is polynomial in the problem description length. However, some large-scale global optimization problems have been solved by current methods, and a number of software packages are available that reliably solve most global optimization problems in small (and sometimes larger) dimensions. The author maintains a web site on Global (and Local) Optimization (1995) that contains many links to online information about the subject.

The different algorithms can be classified according to the degree of rigour with which they approach the goal.

- An *incomplete* method uses clever intuitive heuristics for searching but has no safeguards if the search gets stuck in a local minimum.
- An *asymptotically complete* method reaches a global minimum with certainty, or at least with probability one, if allowed to run indefinitely long, but has no means of knowing when a global minimizer has been found.

- A *complete* method reaches a global minimum with certainty, assuming exact computations and indefinitely long run-time, and knows after a finite time that an approximate global minimizer has been found (to within prescribed tolerances).
- A *rigorous* method reaches a global minimum with certainty and within given tolerances even in the presence of rounding errors, except in near-degenerate cases, where the tolerances may be exceeded.

(Often, the label *deterministic* is used to characterize the last two categories of algorithms; however, this label is slightly confusing since many incomplete and asymptotically complete methods are deterministic, too.)

1.3. Complete search

Complete methods (and *a fortiori* rigorous ones) are (in exact arithmetic) guaranteed to find the global minimizer (within some tolerances) with a predictable amount of work. Here predictable only means relative to known problem characteristics such as Lipschitz constants or other global information (needed for the convergence proof, but usually not for the algorithm itself). The bound on the amount of work is usually very pessimistic – exponential in the problem characteristics. It is only a weak guarantee that does not ensure that the algorithm is efficient in any sense, but it guarantees the absence of systematic deficiencies that prevent finding (ultimately) a global minimizer.

The simplest complete method for bound-constrained problems is *grid search*, where all points on finer and finer grids are tested, and the best point on each grid is used as a starting point for a local optimization. Since the number of points on a grid grows exponentially with the dimension, grid search is efficient only in one and two dimensions. More efficient complete methods generally combine branching techniques with one or several techniques from local optimization, convex analysis, interval analysis and constraint programming.

Generally, complete methods (including approximation methods that reduce the problem to one treated by complete methods) are more reliable than incomplete methods since, to the extent they work (which depends on the difficulty of the problem), they have built-in guarantees.

Complete methods with finite termination require more or less detailed access to global information about the problem. In most complete codes, this is obtained using interval arithmetic (which provides global control of nonlinearities) in an automatic differentiation-like manner (*cf.* Section 16), traversing a computational graph either explicitly, or implicitly by operator overloading. If only black box function (and sometimes gradient) evaluation routines are available, complete methods will find the global minimizer with certainty after a finite time, but will know when this is the case only after an exponentially expensive dense search (*cf.* Theorem 9.1 below). Thus, for complete black box algorithms, stopping must be based on heuristic recipes.

Good heuristics and probabilistic choices (similar to but usually simpler than those for incomplete methods) also play a role in complete methods, mainly to provide good feasible points cheaply that benefit the complete search.

1.4. About the contents

In this survey, the reader will be introduced to theory and techniques that form the backbone of the packages implementing complete or even rigorous algorithms. The core of the material is presented in sufficient detail that the survey may serve as a text for teaching constrained global optimization.

We deliberately exclude methods specific to special problem classes (such as distance geometry or protein folding (Neumaier 1997)), and methods specific to combinatorial optimization (Nemhauser and Wolsey 1988, 1989, Wolsey 1998). Moreover, the discussion of incomplete methods is limited to a short overview, and to techniques that remain useful for complete methods.

No attempt has been made to be objective in selection and evaluation of the material; even for the topics I discuss, there is often much more in the references quoted. Instead I have tried to give personal value judgements whenever I found it appropriate. At the present state of the art, where so many methods compete and reliable comparative information is only just beginning to become available, this seems justified. Thus I discuss the methods that I find most interesting, most useful, and most promising. I hope that my selection bias will be justified by the future. Also, while I try to give accurate references, I do not always refer to the first paper discussing a concept or method but rather quote convenient books or articles summarizing the relevant information, where available.

As one can see from the list of current codes for complete global optimization given in Section 6, none of these codes makes use of all available state-of-the-art techniques. Indeed, in the past, many research groups on global optimization worked with little knowledge of or consideration for related areas. It is hoped that this survey will help to change this lack of communication across the borders of the various traditions in global optimization.

Reviews from other perspectives, with less emphasis on the complete search aspect, are given in Gray *et al.* (1997), Pintér (1996*b*), Törn (2000), and Törn, Ali and Viitanen (1999). For recent books and other basic references, see Section 3.

2. Why global optimization?

Superficially, global optimization is just a stronger version of local optimization, whose great usefulness in practice is undisputed. Instead of searching for a locally unimprovable feasible point we want the globally best point in the feasible region. In many practical applications, finding the globally best point is desirable but not essential, since any sufficiently good feasible point is useful and usually an improvement over what is available without optimization. For such problems, there is little harm in doing an incomplete search; and indeed, this is all that can be achieved for many large-scale problems or for problems where function values (and perhaps derivatives) are available only through a black box routine that does not provide global information.

However, there are many problem classes where it is indispensable to do a complete search, in particular the following.

- **Hard feasibility problems** (*e.g.*, robot arm design, *cf.* Lee and Mavroidis (2002) and Lee, Mavroidis and Merlet (2002)), where local methods do not return useful information since they generally get stuck in local minimizers of the merit function, not providing feasible points (though continuation methods are applicable for polynomial systems in low dimensions).
- **Computer-assisted proofs** (*e.g.*, the proof of the Kepler conjecture by Hales (1998)), where inequalities must be established with mathematical guarantees.
- **Safety verification problems**, where treating nonglobal extrema as worst cases may severely underestimate the true risk (emphasized in the context of robust control by Balakrishnan and Boyd (1992)).
- Many problems in **chemistry** (*cf.* below), where often only the global minimizer (of the free energy) corresponds to the situation matching reality.
- **Semi-infinite programming**, where the optimal configurations usually involve global minimizers of auxiliary problems.

These problems, as well as the fact that algorithms doing a complete search are significantly more reliable and give rise to more interesting mathematics, justify our focus on complete solution techniques.

To show the relevance of global optimization for both pure and applied mathematics, we sketch here a number of typical applications. Of course, this is only the tip of an iceberg ....

(i) Graph structure. Many problems in graph theory are global optimization problems. For example, the *maximum clique problem* asks for the maximal number of mutually adjacent vertices in a given graph. By a well-known theorem of Motzkin and Strauss (1965), an equivalent formulation is the indefinite quadratic program

$$\begin{aligned}\max\quad & x^T A x \\ \text{s.t.}\quad & e^T x = 1, \quad x \geq 0,\end{aligned}$$

where A is the adjacency matrix of the graph and e is the vector of all ones. Since the maximum clique problem is *NP*-hard, the same holds for all classes of global optimization problems that contain indefinite quadratic programming.

(ii) Packing problems. The problem is to place a number of k-dimensional ($k \leq 4$) objects of known shape within a number of larger regions of k-space of known shape in such a way that there is no overlap and a measure of waste is minimized. The simplest packing problem is the *knapsack problem* where a maximal number of objects of given weights is to be placed into a container with given maximum weight capacity. Many packing problems arise in industry; but there are also a number of famous packing problems in geometry, of which the 300 year-old *Kepler problem* of finding the densest packing of equal spheres in Euclidean 3-space was only solved recently by Hales (1998), reducing the problem to several thousand linear programs and some interval calculations to ensure rigorous handling of rounding errors. (The proof is still disputed because of the difficulty to check it for correctness; *cf.* Lagarias (2002). A proof based on rigorous global optimization algorithms would probably be more transparent.)

(iii) Scheduling problems. The problem is to match tasks (or people) and slots (time intervals, machines, rooms, airplanes, *etc.*) such that every task is handled in exactly one slot and additional constraints are satisfied. If there are several feasible matchings, one that minimizes some cost or dissatisfaction measure is wanted. Simple scheduling problems such as the *linear assignment problem* can be formulated as linear programs and are solved very efficiently, but already the related *quadratic assignment problem* is one of the hardest global optimization problems, where already most instances with about 30 variables are at the present limit of tractability: *cf.* Anstreicher (2003).

(iv) Nonlinear least squares problems. In many applications, we need to fit data to functional expressions. This leads to optimization problems with an objective function of a form such as

$$f(\theta) = \sum_l \|y_l - F(x_l, \theta)\|^2,$$

where x_l, y_l are given data vectors and θ is a parameter vector. Under certain assumptions, the most likely value of θ is the global minimizer; it generally must have a small objective function value at noise level if the model is to be deemed adequate. If the F_l are nonlinear in θ, a nonconvex optimization problem results that frequently has spurious local minima far above the noise level. A particularly obnoxious case is obtained for data fitting problems in *training neural networks*.

(v) **Protein folding.** The protein folding problem (Neumaier 1997) consists in finding the equilibrium configuration of the N atoms in a protein molecule with given amino acid sequence, assuming the forces between the atoms are known. These forces are given by the gradient of the $3N$-dimensional potential energy function $V(x_1, \ldots, x_N)$, where x_i denotes the coordinate vector of the ith atom, and the equilibrium configuration is given by the global minimizer of V. Because short-range repulsive forces act like packing constraints, there are numerous local minima.

(vi) **Chemical equilibrium problems.** (Floudas 1997, McDonald and Floudas 1995.) The task here is to find the number and composition of the phases of a mixture of chemical substances allowed to relax to equilibrium. Local optimization of the associated Gibbs free energy is notorious for giving wrong (nonglobal) solutions. The need to solve such problems was one of the main driving forces for the development of constrained global optimization packages in the chemical engineering community, which is still among the leaders in the field.

(vii) **Robotics.** For applications in robotics, see Neumaier (2003b).

(viii) **Other applications.** Many more applications can be found in the books by Pintér (1996a), Floudas and Pardalos (1990), Floudas *et al.* (1999) and Jaulin *et al.* (2001).

3. Basic ideas

In the following, we discuss complete methods for finding the global minimizer(s) of an objective function subject to constraints. Such problems are typically much more difficult than local optimization problems, since it is often hard to decide whether a local minimizer found is global, and since nonlocal space-covering techniques are needed to avoid being trapped in a region with only nonglobal local minimizers.

Basic to almost all complete global optimization algorithms is the *branching principle* (Section 9). This technique consists in splitting (branching) the original problem recursively into subproblems which are sooner or later easy to solve. In pure branching methods, the more prospective branches are split more frequently, while in *branch and bound methods* we compute bounds on the objective function for each subproblem, in the hope of being able to eliminate many subproblems at an early stage.

The very useful technique of *constraint propagation*, discussed in Section 14, allows us to reduce the feasible region in many cases by exploiting properties of *separable constraints* of the form

$$\sum_{k \in K} q_k(x_k) \in \mathbf{b}$$

with simple, often linear or quadratic functions q_k of a single variable only. This technique may save a lot of branching steps and thus speeds up the branch and bound procedure. This is a reason why special care should be taken in presenting (or transforming) the problem in a form which has as much separability as possible, and we introduce the notion of a *semiseparable program* adapted to this feature. Section 18 addresses ways to transform general problems into semiseparable form by introducing appropriate extra variables. Semiseparable programs are also amenable to approximation by a *mixed integer linear program* (MILP), the only class of global optimization problems that has long been successfully solvable, even for large problems. We shall not discuss techniques for solving MILPs (see, *e.g.*, Bixby et al. (2000), Nemhauser and Wolsey (1988, 1989), Wolsey (1998)), but we show how to approximate (and indeed rigorously relax) general global optimization problems by MILPs in Sections 17 and 18.

In order to be able to eliminate subproblems quickly, it is important that one can easily locate good feasible points. This is usually done by local optimization (often in a somewhat rudimentary form); see Section 13. However, especially for problems with many local extrema, it is important to use heuristics which (hopefully) prevent a local method being trapped in a high-lying local minimum. A suitable *tunnelling technique* is discussed in Section 13.

Another basic principle, discussed in Section 16, is that of *outer approximation* of the feasible domain and *underestimation* of the objective function, in order to obtain *relaxed problems* which are convex and hence solvable by local methods. Indeed, this is the traditional way to obtain the bounds on the subproblem. In particular, we consider the use of *cutting planes* and more general cutting surfaces. *Nonconvex relaxations* are also of interest if they can be solved efficiently.

A useful tool for the automatic construction of tight bound constraints, outer approximations and underestimating functions in nonlinear problems is *interval arithmetic*. Though little known in the optimization community, interval arithmetic is an elegant way of calculating with bound constraints, intervals, and simple higher-dimensional geometric shapes like boxes and parallelepipeds. Its most prominent feature is that it allows strict estimates of the approximation error in linear and quadratic approximations of nonlinear functions over a box, thereby providing non-local information even in large boxes. In Section 11, we shall give a very short introduction to this subject (just sufficient for writing programs); a more leisurely introduction embedded into a standard numerical analysis course can be found in Neumaier (2001b), and a much more extensive treatment in Neumaier (1990). Interval arithmetic can also be used to certify rigorously the validity of calculations with finite precision arithmetic, and some such applications to optimization are briefly treated in Section 20. The state of the art in 1996 of certified global optimization with interval methods may be found in Kearfott (1996b).

3.1. Basic references

A basic reference on most aspects of global optimization is the *Handbook of Global Optimization* by Horst and Pardalos (1995). It contains chapters written by the experts in the respective subfields, on global optimality conditions, complexity issues, concave minimization, dc methods, indefinite quadratic programming, complementarity problems, minimax problems, multiplicative programming, Lipschitz optimization, fractional programming, network problems, continuation methods, interval methods, and stochastic methods (including simulated annealing).

Some more recent books present the state of the art in deterministic global optimization from different perspectives: The interval point of view is in Kearfott's book *Rigorous Global Search* (Kearfott 1996b). The constraint propagation perspective is in the book *Numerica* by Van Hentenryck, Michel and Deville (1997b); see also the tutorial by Lustig and Puget (2001). The convex analysis perspective is in the books *Deterministic Global Optimization* by Floudas (1999) and *Convexification and Global Optimization in Continuous and Mixed-Integer Nonlinear Programming* by Tawarmalani and Sahinidis (2002b). An attempt to give a synthetic view of the field (but mostly restricted to discrete optimization) is in the book *Logic-Based Methods for Optimization* by Hooker (2000); see also his survey (Hooker 2002).

A comprehensive background on local optimization (needed as part of most good global optimization algorithms) can be found in the book *Numerical Optimization* by Nocedal and Wright (1999). For interior point methods, this should be complemented by Wright (1997).

Other books on global optimization methods include those of Forgó (1988), Hansen (1992a), Horst, Pardalos and Thoai (1995), Horst and Tuy (1990), Mockus (1989), Pardalos and Rosen (1987), Pintér (1996a), Ratschek and Rokne (1984, 1988), Törn and Žilinskas (1989), Van Laarhoven and Aarts (1987), Zhigljavsky (1991), and proceedings of conferences on global optimization include Bliek, Jermann and Neumaier (2003), Bomze, Csendes, Horst and Pardalos (1996), Dixon and Szegő (1975), Floudas and Pardalos (1992, 1996, 2000) and Grossmann (1996). The *Journal of Global Optimization* is devoted exclusively to papers on global optimization and its applications.

4. Problem formulation

In the present context, a global optimization problem is specified in the form

$$\min \ f(x)$$
$$\text{s.t.} \quad x \in \mathbf{x}, \ F(x) \in \mathbf{F}, \ x_I \text{ integral.} \tag{4.1}$$

Here

$$\mathbf{x} = [\underline{x}, \overline{x}] = \{x \in \mathbb{R}^n \mid \underline{x} \le x \le \overline{x}\}$$

is a bounded or unbounded *box* in $\mathbb{R}^n$ (with $\underline{x} \in (\mathbb{R} \cup \{-\infty\})^n, \overline{x} \in (\mathbb{R} \cup \{\infty\})^n, \underline{x} \le \overline{x}$), and x_I denotes the subvector $(x_{i_1}, \ldots, x_{i_l})^T$ of x when $I = (i_1, \ldots, i_l)$ is a list of indices. Inequalities between vectors are interpreted componentwise. Here $f : \mathbf{x} \to \mathbb{R}$ is a continuous objective function, $F : \mathbf{x} \to \mathbb{R}^m$ is a vector of m continuous constraint functions $F_1(x), \ldots, F_m(x)$, and $\mathbf{F}$ is a box in $\mathbb{R}^m$ defining the constraints on $F(x)$. The *feasible domain* is defined by

$$C = \{x \in \mathbf{x} \mid x_I \text{ integral}, F(x) \in \mathbf{F}\}. \tag{4.2}$$

Points in C are called *feasible*, and a *solution* of (4.1) is a feasible point $\hat{x} \in C$ such that

$$f(\hat{x}) = \min_{x \in C} f(x). \tag{4.3}$$

A *local minimizer* only satisfies $f(\hat{x}) \le f(x)$ for all $x \in C$ in some neighbourhood of $\hat{x}$, and the solutions are precisely the *global minimizers*, i.e., the local minimizers with smallest objective function value. A local (global) *solver* is an algorithm or programming package designed for finding a local (global) minimizer. (We avoid the ambiguous term *optimizer* which may denote either a minimizer or a solver.)

The difficulties in global optimization stem mainly from the fact that there are generally many local minimizers but only one of them is the global minimizer (or just a few), and that the feasible region may be disconnected. (Consider, e.g., the set of positions in the Alps above a certain altitude.) Already a linear objective function has one minimizer in each connected component of the feasible domain, and local descent methods usually fail if they start in the wrong component.

Even the *constraint satisfaction problem*, i.e., the problem of deciding whether the feasible set is nonempty (and finding a feasible point) is frequently highly nontrivial, and may be essentially as difficult as the optimization problem itself (cf. Section 13). The usual device of minimizing a suitable measure of infeasibility does not work when the constraints are sufficiently nonlinear, since this measure has itself local minima in which descent methods often get stuck.

Usually, it is possible to reformulate a global optimization problem such that f and F are *smooth*, i.e., twice continuously differentiable. Note that (4.1) is sufficiently flexible to take care of:

- free variables x_i: take $\underline{x}_i = -\infty$, $\overline{x}_i = \infty$;
- nonnegative variables x_i: take $\underline{x}_i = 0$, $\overline{x}_i = \infty$;
- binary variables x_i: take $\underline{x}_i = 0$, $\overline{x}_i = 1$, $i \in I$;
- equality constraints $F_i(x) = 0$: take $\underline{F}_i = \overline{F}_i = 0$;
- inequality constraints $F_i(x) \leq 0$: take $\underline{F}_i = -\infty$, $\overline{F}_i = 0$.

If I is not empty then, if f and F are linear then (4.1) is called a *mixed integer linear program* (MILP); and if f and F are convex, and $\underline{F}_i = -\infty$ for all nonlinear F_i, (4.1) is called a *mixed integer nonlinear program* (MINLP). Strictly speaking, this term should apply for all problems (4.1); however, the current techniques for MINLP use the convexity in an essential way, so that it is appropriate to reserve the term for the convex case. Nonconvex mixed integer global optimization problems have received little attention, but see, e.g., Floudas (1995), Grossmann (2002), Sahinidis (1996), and Tawarmalani and Sahinidis (2002b).

The only class of global optimization problems that can be reliably solved for many large problem instances (say, $\approx 10^5$ variables and $|I| \approx 10^3$) is the class of MILPs. This is due to the fact that after fixing the integer variables one is left with a linear program, which can be solved efficiently. Instead of trying all integer combinations separately, branching techniques (branch and bound, branch and cut), combined with preprocessing the resulting linear programs, drastically cut down the number of cases to be looked at. MINLP shares with MILP the feature that fixing all integer variables leads to a tractable problem, in this case a convex nonlinear program, for which every local minimizer is a solution; however, the dimensions are here more limited since nonlinear programming codes are significantly slower than their linear counterparts.

Most constrained global optimization is nowadays best viewed as an adaptation of mixed integer programming technology to nonlinear problems. Historically, however, many of the techniques were devised independently by groups working in integer programming, combinatorial optimization, unconstrained optimization, interval analysis, and constraint logic programming.

Other important classes of global optimization problems are as follows:

- simply constrained: if $\dim F = 0$,
- continuous: if $I = \emptyset$,
- bound-constrained: if simply constrained and continuous,
- separable: if $f(x) = \sum_{k=1}^{n} f_k(x_k)$ and $F(x) = \sum_{k=1}^{n} F_k(x_k)$,

- factorable: if f and F are obtained by applying a finite sequence of arithmetic operations and unary elementary functions to constants and the x_k,
- reverse convex: if f, F are concave, and $\underline{F}_i = -\infty$ for all nonlinear F_i,
- DC: if f, F are differences of convex functions.

5. First-order optimality conditions

Nonlinear programming provides the following necessary (*Karush–John*) *optimality conditions* for local minimizers. We assume that f, F are continuously differentiable, and denote by $f'(x)$ and $F'(x)$ the derivatives at x. Note that $f'(x)$ is a row vector and $F'(x)$ a matrix, the *Jacobian*.

Theorem 5.1. (Karush (1939), John (1948)) For every local minimizer $\hat{x}$ of (4.1) (which defines the notation) there are a number $\kappa \geq 0$ and a vector y, not both zero, such that the row vector

$$g^T = \kappa f'(\hat{x}) + y^T F'(\hat{x}) \tag{5.1}$$

satisfies

$$g_i \begin{cases} \geq 0 & \text{if } \underline{x}_i = \hat{x}_i < \overline{x}_i, \quad i \notin I, \\ \leq 0 & \text{if } \underline{x}_i < \hat{x}_i = \overline{x}_i, \quad i \notin I, \\ = 0 & \text{if } \underline{x}_i < \hat{x}_i < \overline{x}_i, \quad i \notin I, \end{cases} \tag{5.2}$$

$$y_i \begin{cases} \geq 0 & \text{if } \underline{F}_i < F_i(\hat{x}) = \overline{F}_i, \\ \leq 0 & \text{if } \underline{F}_i = F_i(\hat{x}) < \overline{F}_i, \\ = 0 & \text{if } \underline{F}_i < F_i(\hat{x}) < \overline{F}_i. \end{cases} \tag{5.3}$$

Note that there is no restriction on g_i if $i \in I$ or $\underline{x}_i = \overline{x}_i$, and no restriction on y_i if $\underline{F}_i = F_i(\hat{x}) = \overline{F}_i$. Buried implicitly in results of Mangasarian (1969), and spelled out explicitly in Neumaier and Schichl (2003), is the observation that one may in fact assume that either κ or the subvector y_J of y is nonzero, where J is the set of indices i such that either $F_i(x)$ is nonconvex and $F_i(\hat{x}) = \underline{F}_i$ or $F_i(x)$ is nonconcave and $F_i(\hat{x}) = \overline{F}_i$. (In particular, J does not contain any index i such that $F_i(x)$ is linear.) In view of the homogeneity of the statement of the theorem, one can therefore scale the multipliers such that

$$\kappa + y_J^T D y_J = 1, \tag{5.4}$$

where D is an arbitrary diagonal matrix with positive entries. This condition is relevant for the application of exclusion box techniques (*cf.* Section 15).

We say that $\hat{x}$ satisfies a *constraint qualification* (CQ) if (5.1)–(5.3) hold for some $\kappa > 0$. In this case, one can scale g, κ, y to enforce $\kappa = 1$ and obtains the more frequently used *Kuhn–Tucker conditions* (Kuhn and Tucker 1951, Kuhn 1991). A sufficient condition for the constraint qualification is that the rows of $F'(\hat{x})$ are linearly independent; various weaker conditions guaranteeing CQ are known.

If $\kappa = 1$ then y is called an optimal *Lagrange multiplier* corresponding to $\hat{x}$ (it need not be unique). In this case, g is the gradient of the associated *Lagrangian* (Lagrange 1797)

$$L(x, y) = f(x) + y^T F(x)$$

at $x = \hat{x}$.

Note that minimizers with huge Lagrange multipliers are best considered as points nearly violating the constraint qualification, so that (5.1) holds with $y = O(1)$ and tiny κ.

If there are only nonnegativity constraints and equality constraints,

$$C = \{x \geq 0 \mid F(x) = b\},$$

corresponding to $\underline{x}_i = 0$, $\overline{x}_i = \infty$, $\underline{F}_i = \overline{F}_i = b_i$ then the conditions (5.3) are vacuous, and (5.2) reduces to the traditional *complementarity condition*

$$\min(g_i, x_i) = 0 \quad \text{for all } i.$$

Example 5.2. We consider the problem

$$\begin{aligned}\min \quad & f(x) = -x_1 - 2x_2 \\ \text{s.t.} \quad & F(x) = (x_1 - 1)^2 + (x_2 - 1)^2 = 1, \quad x_1, x_2 \in [-1, 1].\end{aligned} \tag{5.5}$$

The feasible region is a quarter circle, and the contour lines of the objective function are linear, decreasing in the direction indicated in Figure 5.1. This implies that there is a unique maximizer at P, a local minimizer at Q and a global minimizer at R. The solution is therefore $\hat{x} = (0, 1)$. Since there are only two variables, we analysed the problem graphically, but we could as well have proceeded symbolically as follows.

Assuming for simplicity the validity of the CQ, we find for the gradient of the Lagrangian

$$g = \begin{pmatrix} -1 \\ -2 \end{pmatrix} + y \begin{pmatrix} 2x_1 - 2 \\ 2x_2 - 2 \end{pmatrix}.$$

The Kuhn–Tucker conditions require that $g_i \geq 0$ if $\hat{x}_i = -1$, $g_i \leq 0$ if $\hat{x}_i = 1$, and $g_i = 0$ otherwise. This leaves three cases for each component, and a total of $3 \cdot 3 = 9$ cases. If we assume $|\hat{x}_1|, |\hat{x}_2| < 1$ we must have $g = 0$, hence $\hat{x}_1 = 1 + 1/2y$, $\hat{x}_2 = 1 + 1/y$. Since $\hat{x}$ must be feasible, $y < 0$, and since $F(\hat{x}) = 1$, $y = -\frac{1}{2}\sqrt{5}$, $\hat{x} = (1 - 1/\sqrt{5}, 1 - 2/\sqrt{5})^T$, which is the local maximizer P. If we assume $\hat{x}_1 = -1$ or $\hat{x}_2 = -1$, or $\hat{x}_1 = \hat{x}_2 = 1$,

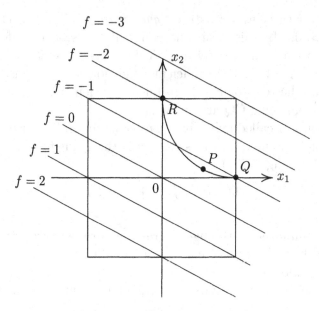

Figure 5.1. Feasible region and contour lines for Example 5.2.

we find a contradiction with $F(\hat{x}) = 1$. (These are 6 cases!) If we assume $|\hat{x}_2| < 1 = \hat{x}_1$ we find Q, and for $|\hat{x}_1| < 1 = \hat{x}_2$ we find R. Thus we have three feasible points satisfying the Kuhn–Tucker conditions, and a comparison of their function values shows that R is the global minimizer.

In general, we have three cases for each two-sided inequality and two for each one-sided inequality; since the number of independent choices must be multiplied, the total number of cases grows exponentially with the number of inequalities in the problem formulation. Hence this symbolic approach is limited to problems with few inequality constraints. Even then it only works if the resulting nonlinear equations are symbolically solvable and have few solutions only. Thus, in general, we need to resort to numerical methods.

We draw several conclusions from the example. First, there is a combinatorial aspect to the continuous global optimization problem, so that it resembles a mixed integer problem. Second, several cases can often be excluded by a single argument, which is the basis for the branch and bound approach to global optimization. Third, the Karush–John or Kuhn–Tucker conditions do not distinguish between maxima and minima (and other 'stationary' points); all these would have to be enumerated in a naive approach. Since there may be an exponential number of Kuhn–Tucker points, additional techniques are needed to reduce the search space. Lagrange multiplier techniques involving second-order conditions will address this last point; *cf.* Theorem 15.1.

6. Software for complete global optimization

Here we list some of the better complete global optimization codes available on the Web, with short comments on scope and method. Several of the codes (LGO, BARON, SBB, DICOPT) can be called from the GAMS modelling system (GAMS World 2003), allowing for very convenient input. Input from the AMPL modelling system (Fourer, Gay and Kernighan 1993) will be possible through an AMPL to GAMS translator available within the COCONUT environment; *cf.* Section 22. Only two of the codes (GlobSol and Numerica) are rigorous solvers.

6.1. Some branching codes using function values only

The codes listed use black box function evaluation routines, and have heuristic stopping rules, so that the actual implementation yields an incomplete search only.

(i) DIRECT: Divide Rectangles (in Fortran)
Gablonsky and Kelley (2001)
ftp://ftp.math.ncsu.edu/FTP/kelley/iffco/DIRECTv204.tar.gz

Direct.m, a MATLAB implementation of DIRECT
www4.ncsu.edu/~definkel/research/

Implementations of a simple and efficient global optimization method by Jones, Perttunen and Stuckman (1993) for bound-constrained problems. DIRECT is based on branching and a Pareto principle for box selection.

(ii) MCS: Multilevel Coordinate Search
Huyer and Neumaier (1999)
www.mat.univie.ac.at/~neum/software/mcs/

A MATLAB program for bound-constrained global optimization using function values only. MCS is based on branching and sequential quadratic programming.

(iii) LGO: Lipschitz Global Optimization (commercial)
Pintér (1996a, 1999)
is.dal.ca/~jdpinter/l_s_d.htm

An integrated development environment for global optimization problems with Lipschitz-continuous objective and constraints. LGO is based on branching and stochastic estimation of Lipschitz constants; constraints other than simple bounds are handled by L_1 penalty terms, but interior convex constraints by projection penalties. (LGO also has options for incomplete search methods; in many cases, these give better results than the branching option.)

6.2. Some branch and bound codes

The codes listed use global information (generally from required symbolic problem input). They have finite termination with guarantee that the global minimizer is found within certain tolerances; in difficult cases storage or time limits may be exceeded, however, leading to appropriate error messages. All codes use at least basic constraint propagation, but differ considerably in the other techniques implemented.

Not listed are the many MILP codes available (see the Global Optimization web page mentioned in the Introduction).

(i) BARON: Branch-And-Reduce Optimization Navigator (commercial)
Sahinidis *et al.* (Ryoo and Sahinidis 1996, Sahinidis 1996, 2000, 2003, Tawarmalani and Sahinidis 2002*b*, 2004)
archimedes.scs.uiuc.edu/baron/baron.html

A general-purpose solver for optimization problems with nonlinear constraints and/or integer variables. Fast specialized solvers for many linearly constrained problems. BARON is based on branching and box reduction using convex and polyhedral relaxation and Lagrange multiplier techniques.

(ii) GlobSol: Global Solver (in Fortran 90)
Kearfott (1996*b*, 2003)
www.mscs.mu.edu/~globsol/

Branch and bound code for global optimization with general factorable constraints, with rigorously guaranteed results (even round-off is accounted for correctly). GlobSol is based on branching and box reduction using interval analysis to verify that a global minimizer cannot be lost.

(iii) LINGO (commercial)
Gau and Schrage (2004)
www.lindo.com/cgi/frameset.cgi?leftlingo.html;lingof.html

Branch and bound code for global optimization with general factorable constraints, including nondifferentiable expressions. LINGO is based on linear relaxations and mixed integer reformulations. C and Excel interfaces are available.

(iv) Frontline Interval Global Solver (commercial)
Nenov and Fylstra (2003)
www.solver.com/technology5.htm

This solver is based on interval methods and linear relaxations. Visual Basic and Excel interfaces are available.

(v) ALIAS (in C)
Merlet (ALIAS-C++ 2003, Merlet 2001)
www-sop.inria.fr/coprin/logiciels/ALIAS/ALIAS-C++/ALIAS-C++.html

Branch and bound environment for solving constraint satisfaction problems (and rudimentary global optimization). A toolkit of interval analysis and constraint programming techniques with a Maple interface for symbolic preprocessing.

(vi) Numerica
Van Hentenryck et al. (1997b)
Branch and bound code for constrained optimization (with mathematically rigorous results). This code (no longer available) was based on branching and box reduction using interval analysis and deeper constraint propagation techniques. The box reduction and interval analysis algorithms of Numerica are now available in *ILOG Solver* (commercial) at
www.ilog.com/products/solver/

(vii) αBB
Adjiman, Floudas and others (Adjiman et al. 1996, 1998a, 1998b, Androulakis et al. 1995)
titan.princeton.edu/soft.html#abb

Branch and bound code for nonlinear programs. The site currently has the description only, no code. αBB is based on branch and bound by convex underestimation, using interval analysis to write nonlinearities in DC (difference of convex functions) form.

(viii) LaGO
Nowak, Alperin and Vigerske (2003)
www-iam.mathematik.hu-berlin.de/~eopt/#Software

Branch and bound code for mixed integer nonconvex nonlinear programming, using block-separable structure and convex underestimation. The site currently has the description only, no code.

(ix) GloptiPoly (in MATLAB)
Henrion and Lasserre (2003a, 2003b, 2003c)
www.laas.fr/~henrion/software/gloptipoly/

Global optimization of polynomial nonlinear programs, using semidefinite relaxations. Currently limited to problems with fewer than 20 variables.

(x) SOSTOOLS (in MATLAB)
Prajna, Papachristodoulou and Parrilo (2002)
control.ee.ethz.ch/~parrilo/sostools/index.html

MATLAB toolbox for solving sums of squares (SOS) optimization programs. Allows the solution of polynomial global optimization problems.

(xi) cGOP
Visweswaran and Floudas (1996a, 1996b)
titan.princeton.edu/soft.html

Branch and bound code for linearly constrained global optimization problems with an objective containing linear, bilinear, and convex terms, using convex relaxations.

(xii) MINLPBB (commercial)
Fletcher and Leyffer (1994, 1998)
www-unix.mcs.anl.gov/~leyffer/solvers.html

Branch and bound code for mixed integer nonlinear programming; finding the global optimum is guaranteed only if all constraints become convex when all integer variables are fixed. Problems with AMPL input can be solved online via NEOS at
www-neos.mcs.anl.gov/neos/solvers/MINCO:MINLP-AMPL/

MINLP uses standard mixed integer programming techniques and filter methods for the local subproblems.

(xiii) DICOPT (commercial)
Duran and Grossmann (1986), Kocis and Grossmann (1989)
www.gams.com/solvers/dicopt/main.htm

Solver for mixed integer monlinear programming (MINLP) problems. Finding the global optimum is guaranteed only if all constraints become convex when all integer variables are fixed.

(xiv) SBB (commercial)
Bussiek and Drud (2001)
www.gams.com/solvers/solvers.htm#SBB

Branch and bound code for mixed integer nonlinear programming; finding the global optimum is guaranteed only if all constraints become convex when all integer variables are fixed. Problems with GAMS input can be solved online via NEOS at
www-neos.mcs.anl.gov/neos/solvers/MINCO:SBB-GAMS/

SBB uses standard mixed integer programming techniques and sequential quadratic programming methods for the local subproblems.

MINLPBB, DICOPT and SBB are borderline cases in our list since they do not use truly global techniques for the continuous variables, and will not be discussed further.

7. Incomplete methods for simple constraints

As mentioned in the Introduction, numerical methods for global optimization can be classified into four categories according to the available guarantees. We shall be mainly concerned with complete methods; however, since incomplete and asymptotically complete methods are frequently successful and, for many difficult problems, are the only feasible choice, we give an overview of the main possibilities. In these categories are several deterministic and most stochastic methods. For some of the latter, it is possible to prove convergence with probability arbitrarily close to 1 (if running arbitrarily long), but this does not yet guarantee convergence. (Moreover, the assumptions underlying the convergence proofs are frequently not verifiable for particular examples.)

The simplest incomplete method is *multiple random start*, consisting of picking random starting points and performing local optimizations from these points, in the hope that one of them is in the basin of attraction of the global minimizer. Most stochastic techniques can be regarded as devices for speeding up this basic method, by picking the points more carefully and by doing only rudimentary local optimization, or optimizing only selectively.

Most of the research on incomplete search has been concentrated on global optimization methods for simply constrained problems only. Many different incomplete methods are known for simply constrained problems, and we sort them into four categories: *local descent techniques*, including among others multiple random start, clustering (Boender, Rinnooy Kan, Timmer and Stougie 1982), tunnelling (Levy and Montalvo 1985), and smoothing methods (Guddat, Guerra Vasquez and Jongen 1990, Kostrowicki and Scheraga 1996, Moré and Wu 1997, Stillinger 1985); *response surface techniques*, including Bayesian stochastic techniques Mockus (1989, 1994) and related techniques (Björkman and Holmström 2000, Jones 2001, Jones, Schonlau and Welch 1998); *nonmonotonic search techniques*, including among others tabu search (Glover 1989/90, Glover and Laguna 1997), simulated annealing (Kirkpatrick, Geddat, Jr. and Vecchi 1983, Ingber 1993, Van Laarhoven and Aarts 1987), and deterministic variants such as threshold accepting (Dueck and Scheuer 1990); *ensemble methods*, including genetic algorithms (Holland 1973, Forrest 1993, Michalewicz 1996) and variants such as ant colony minimization (Dorigo, Maniezzo and Colorni 1991).

No attempt is made to be representative or complete on referencing or describing the large literature on incomplete techniques; we only mention the 1975 book by Dixon and Szegő (1975), which marks the start of a tradition of comparing different global optimization methods, an excellent exposition of stochastic global optimization methods for bound-constrained problems on the Web by Törn (2000), and another web survey of (mainly incomplete) methods by Gray et al. (1997). For incomplete search in combinatorial optimization (where the underlying ideas are also called *metaheuristics*), see, *e.g.*, Aarts and Lenstra (1997), Yagiura and Ibaraki (2001).

Instead of describing technical details of the various methods (these vary from author to author and even from paper to paper), we give an informal view of the ideas, strengths and weaknesses of one method from every category, each based on analogies to natural processes where more or less global optima are reached. While these techniques are motivated by nature it is important to remember that processes in nature need not be the most efficient ones; at best they can be assumed to be efficient *given the conditions under which they have to operate* (namely an uncertain and changing environment that is potentially hazardous to those operating in it). Indeed, much of our present technology has vastly surpassed natural efficiency by unnatural means, and it would be surprising if it were different in global optimization. Even assuming that nature solves truly *global* optimization problems (a disputable assumption), simple lower estimates for the number of elementary steps – roughly corresponding to function evaluations – available to natural processes to converge are (in chemistry and in biology) in the range of 10^{15} or even more. This many function evaluations are unacceptable for present day computers, and will be so in the near future.

With a limited number of function evaluations, the quality of incomplete methods depends a lot on details of the implementation; comparisons on relative efficiency are virtually missing. Indeed, the techniques must generally be tuned to special classes of applications in order to be fast and competitive, which makes general-purpose comparisons difficult and inconclusive.

Smoothing (= homotopy = continuation) methods are based on the intuition that, in nature, macroscopic features are usually an average effect of microscopic details; averaging smooths out the details in such a way as to reveal the global picture. A huge valley seen from far away has a well-defined and simple shape; only by looking more closely, the many local minima are visible, more and more at smaller and smaller scales. The hope is that by smoothing a rugged objective function surface, most or all local minima disappear, and the remaining major features of the surface only show a single minimizer. By adding more and more details, the approximations made by the smoothing are undone, and finally one ends up at the global minimizer of the original surface.

In mathematical terms, one has to define a homotopy by introducing an additional parameter t into the problem in such a way that $t = 0$ gives the original problem, while $t = 1$ gives either a related convex problem or a related problem with a unique and known global minimizer. (There are various ways of doing this; homotopies whose parameter has a natural interpretation in the context of the original problem usually perform better.) Then a sequence of local problems is solved for $t = t_1, t_2 \ldots, t_N$, where the t_i

form a decreasing sequence starting at 1 and ending at 0. Each time, the solution of the previous problem is taken as the starting point for the current problem. The quality of the final local minimizer depends on the homotopy, and is frequently the global or at least a good local minimizer.

There is no theoretical work on conditions that would ensure convergence to the global minimum. In particular, it is quite possible for such a method to miss the global minimum. However, for properly chosen homotopies, smoothing methods at least give good local minima with a small number of function evaluations. (For more theory on homotopy methods, see, *e.g.*, Guddat *et al.* (1990, Section 6.3).)

Response surface techniques are designed specifically for the global optimization of functions that are very expensive to evaluate. They construct in each iteration an interpolation or approximation *surrogate function* of known analytic form. The surrogate function is then subjected to global optimization, *e.g.*, by some form of multiple random start (started at a selection of the current points). The resulting optimizers (or some points where the feasible region has only been sparsely explored) are taken as new evaluation points. Since this is sequential global optimization, each step is much more expensive than the others, but the reduction of the number of function values needed gives (for sufficiently expensive function evaluations) a net gain in speed.

In principle, these methods may have convergence guarantees if the point selection strategy is well chosen; but this is irrelevant in view of the fact that for expensive functions, only few (perhaps up to 1000) function evaluations are admissible.

Simulated annealing takes its intuition from the fact that the heating (annealing) and slow cooling of a metal brings it into a more uniformly crystalline state that is believed to be the state where the free energy of bulk matter takes its global minimum. (Incidentally, even for the simplest potential energy functions, it is still an unsolved problem whether this is indeed true with mathematical rigour. Apart from that, even very pure crystals still have defects; *i.e.*, the global minimum is not quite achieved in nature.) The role of temperature is to allow the configurations to reach higher energy states with a probability given by Boltzmann's exponential law, so that they can overcome energy barriers that would otherwise force them into local minima. This is quite unlike line search methods and trust region methods on which good local optimization programs are based.

In its original form, the simulated annealing method is provably convergent in a probabilistic sense but exceedingly slow; various *ad hoc* enhancements make it much faster. In particular, except for simple problems, success depends very much on the implementation used.

Genetic algorithms make use of analogies to biological evolution by allowing mutations and crossing over between candidates for good local optima in the hope of deriving even better ones. At each stage, a whole population of configurations is stored. Mutations have a similar effect to random steps in simulated annealing, and the equivalent of lowering of the temperature is a rule for more stringent selection of surviving or mating individuals.

The ability to leave regions of attraction to local minimizers is, however, drastically enhanced by crossing over. This is an advantage if, with high probability, the crossing rules produce offspring of similar or even better fitness (objective function value); if not, it is a severe disadvantage. Therefore the efficiency of a genetic algorithm (compared with simulated annealing-type methods) depends in a crucial way on the proper selection of crossing rules. The effect of interchanging coordinates is beneficial mainly when these coordinates have a nearly independent influence on the fitness, whereas if their influence is highly correlated (such as for functions with deep and narrow valleys not parallel to the coordinate axes), genetic algorithms have much more difficulties. Thus, unlike simulated annealing, successful tuning of genetic algorithms requires a considerable amount of insight into the nature of the problem at hand.

Both simulated annealing methods and genetic algorithms are, in their simpler forms, easy to understand and easy to implement, features that invite potential users of optimization methods to experiment with their own versions. The methods often work, if only slowly, and may be useful tools for applications where function values are not very expensive and the primary interest is to find (near-)solutions *now*, even when the reliability is uncertain and only subglobal optima are reached.

To make simulated annealing methods and genetic algorithms efficient, clever enhancements exploiting expert knowledge about the problem class at hand are essential. Theoretical work on explaining the effectiveness of useful enhancements is completely lacking. Also, I have not seen careful comparisons of the various options available and their comparative evaluation on standard collections of test problems.

In general, incomplete methods tend to fail systematically to find the global optimum on the more difficult problems in higher dimensions, but they frequently give relatively good points with a reasonable amount of effort. Beyond a certain number of function evaluations (that depends on the problem), progress slows down drastically if the global optimum has not yet been located already. This is unlikely to change in the future, although new heuristics and variations of old ones are discovered almost every year.

For general-purpose global optimization, the most promising incomplete methods appear to be clustering methods (see the recent comparison by Janka (2000)), being fairly robust and fast. In particular, the multilevel clustering algorithm by Boender et al. (1982), as implemented by Csendes (1988), can be recommended. Among incomplete algorithms adapted to problem structure, I would favour smoothing methods (if a natural homotopy is available) and tabu-search-like strategies (since these have a kind of memory).

8. Bound-constrained approximation

For general constraints, incomplete techniques are much less developed. Only the smoothing techniques extend without difficulties to general constraints. To use the other incomplete techniques, it is customary to rewrite problems with general constraints in an equivalent or approximately equivalent form with either simple constraints only, for which the methods of the previous section apply, or into a mixed integer linear problem (MILP), since highly efficient software is available for solving the latter (Bixby et al. 2000). Both transformations are of great practical importance and widely used. Solving the transformed (equivalent or approximate) problem yields an approximate solution for the original problem, and local optimization from this approximate solution gives the global minimizer of the original problem if the approximation was good enough, and usually a good local minimizer otherwise.

In this section we treat the approximation of general constrained problems by bound-constrained problems using penalty and barrier functions. The approximation of nonlinear problems by mixed integer linear programs is treated in Section 18.

8.1. Penalty and barrier formulations

Traditionally (see Fiacco and McCormick (1990)), constraints that cannot be handled explicitly are accounted for in the objective function, using simple l_1 or l_2 penalty terms for constraint violations, or logarithmic barrier terms penalizing the approach to the boundary. In both cases, the reformulation changes the solution, so that this is an instance of an approximation method, and the result should be used as a starting point for a subsequent local optimization of the original problem. There are also so-called exact penalty functions whose optimization gives the exact solution (see, for example, Nocedal and Wright (1999)); however, this only holds if the penalty parameter is large enough, and what is large enough cannot be assessed without having global information.

The use of more general transformations gives rise to more precisely quantifiable approximation results. In particular, if it is known in advance that all constraints apart from the simple constraints are soft constraints only (so that some violation is tolerated), one may pick a transformation that incorporates prescribed tolerances into the reformulated simply constrained problem, using the following variation of a similar, but less flexible result of Dallwig, Neumaier and Schichl (1997), given in Huyer and Neumaier (2003).

Theorem 8.1. (Soft optimality theorem) Given $\Delta, \underline{\sigma}_i, \overline{\sigma}_i > 0, f_0 \in \mathbb{R}$, let

$$q(x) = \frac{f(x) - f_0}{\Delta + |f(x) - f_0|},$$

$$\delta_i(x) = \begin{cases} (F_i(x) - \underline{F}_i)/\underline{\sigma}_i, & \text{if } F_i(x) \leq \underline{F}_i, \\ (F_i(x) - \overline{F}_i)/\overline{\sigma}_i, & \text{if } F_i(x) \geq \overline{F}_i, \\ 0, & \text{otherwise,} \end{cases}$$

$$r(x) = \frac{2\sum \delta_i^2(x)}{1 + \sum \delta_i^2(x)}.$$

Then the merit function

$$f_{\text{merit}}(x) = q(x) + r(x)$$

has its range bounded by $]-1, 3[$, and the global minimizer $\hat{x}$ of f_{merit} in $\mathbf{x}$ either satisfies

$$F_i(\hat{x}) \in [\underline{F}_i - \underline{\sigma}_i, \overline{F}_i + \overline{\sigma}_i] \quad \text{for all } i, \tag{8.1}$$

$$f(\hat{x}) \leq \min\{f(x) \mid F(x) \in \mathbf{F}, x \in \mathbf{x}\}, \tag{8.2}$$

or one of the following two conditions holds:

$$\{x \in \mathbf{x} \mid F(x) \in \mathbf{F}\} = \emptyset, \tag{8.3}$$

$$f_0 < \min\{f(x) \mid F(x) \in \mathbf{F}, x \in \mathbf{x}\}. \tag{8.4}$$

Here (8.1) says that a soft version of the constraints is satisfied. The numbers $\underline{\sigma}_i$ and $\overline{\sigma}_i$ measure the degree to which the lower and upper bounds in the constraint $F_i(x) \in \mathbf{F}_i$ may be softened; suitable values are in many practical applications available from the meaning of the constraints.

Inequality (8.2) says that f_{merit} has a lower global minimum value (attained at a point satisfying the soft constraints) than the global minimum value of the original problem (on the hard version of the constraints). Thus little is lost from a practical point of view.

The degenerate cases (8.3)–(8.4) account for the possibility of an empty feasible set (8.3), and for a choice of f_0 that was too small. If a feasible point is already known we may choose f_0 as the function value of the best feasible point known (at the time of posing the problem), thus eliminating the possibility (8.3). If none is known, f_0 should be chosen as a fairly large value to avoid (8.4); it can be reset (and the optimization restarted) when a feasible point becomes available during the search.

In spite of the absolute value in the definition of $q(x)$, f_{merit} is continuously differentiable if f and F have this property. A suitable value for Δ is the median of the values $|f(x) - f_0|$ for an initial set of trial points x (in the context of global optimization often determined by a space-filling design (McKay, Beckman and Conover 1979, Owen 1992, Owen 1994, Sacks, Welch, Mitchell and Wynn 1989, Tang 1993)).

8.2. Projection penalties

A little-known result by Pintér (1996a) may be used to get in certain cases (in particular, for linear and convex quadratic constraints) an exact reformulation as a non-smooth but Lipschitz-continuous simply constrained problem. The idea is to project infeasible points to the feasible domain.

To accommodate linear constraints (or convex quadratic ones), Pintér assumes that x_0 is a known interior point. For arbitrary $\gamma > 0$ we now define the modified objective function

$$\overline{f}(x) := f(\overline{x}) + \gamma \|\overline{x} - x\|^2, \tag{8.5}$$

where

$$\overline{x} = \lambda x_0 + (1 - \lambda)x \tag{8.6}$$

and $\lambda = \lambda_x \geq 0$ is minimal such that $\overline{x}$ satisfies the linear constraints. This is well-defined, since $\lambda = 1$ always works by the choice of x_0. Each constraint contributes a lower bound $\in [0, 1]$ for λ, and the largest of these bounds is the desired value. In particular, a linear constraint $a^T x \leq \alpha$ contributes a nonzero lower bound

$$\lambda \geq (a^T x - \alpha)/(a^T x - a^T x_0)$$

if both numerator and denominator of the right-hand side are positive. A convex quadratic constraint similarly yields a quadratic inequality that can easily be solved for λ. (Convexity can be weakened to star-shapedness with respect to x_0.)

The modified objective function (8.5) is Lipschitz-continuous, but non-smooth at all points where the ray (8.6) hits a lower-dimensional face of the feasible domain. Note that to evaluate (8.5), function values are needed only at points satisfying the linear (or convex quadratic) constraints.

An interior point can be found by solving a linear program or convex second-order cone program. If no interior point exists since the feasible set is in a lower-dimensional subspace, each feasible point has the form $x = x_0 + Cz$ with $z \in \mathbf{z}$, where x_0 is in the relative interior of the feasible domain, and $\mathbf{z}$ a box with $0 \in \text{int } \mathbf{z}$. Both x_0 and C can be found by techniques from convex analysis for finding a maximal independent set of points in the affine subspace spanned by the feasible set. Reposing the optimization problem in terms of z reduces the dimension and yields a problem in which 0 is an interior point.

9. Pure branching methods

We begin our analysis of complete methods for global optimization by looking at the options for methods that can access no global information about a problem. The information is made available via black box routines that provide *local information* only, *i.e.*, function values and possibly gradients or Hessians at single points. A necessary and sufficient condition for complete methods based on local information only is given by the following important *density theorem* due to Törn and Žilinskas (1989). It formalizes the simple observation that after finitely many local evaluations there are still many 'holes', *i.e.*, balls not containing an already evaluated point, and there are many functions (Neumaier 2003a) that have the known function values, gradients and Hessians at the evaluation points but an arbitrarily low function value at the centre of such a ball.

Theorem 9.1. Any method based on local information only that converges for every continuous f to a global minimizer of f in a feasible domain C must produce a sequence of points $x^1, x^2, \ldots$ that is dense in C.

Conversely, for any such method,

$$\liminf_{l \to \infty} f(x^l) = \min\{f(x) \mid x \in C\}.$$

A global optimization method based on local information only is called *convergent* if it satisfies the hypothesis of this density theorem. (Actual implementations of a convergent global optimization method usually are not truly convergent since they must have built-in termination criteria that are necessarily heuristic.)

Convergence is a *minimal* requirement and does not make an algorithm good! For example, exhaustive grid search is convergent but far too slow in dimensions > 2. (Compare with local optimization with line searches along the steepest descent direction, which is globally convergent but frequently very slow.) In a sense, the density theorem says that any convergent method must be ultimately exhaustive, though it may delay the detailed exploration of unpromising regions. Since, in practice, only a limited number of points can be explored, the behaviour of a pure branching method is governed by its ability to find a good ordering of the points to be evaluated for which premature termination has no severe effect.

Three good asymptotically complete general-purpose global optimization algorithms based on local information only are currently available: DIRECT (Jones *et al.* 1993), MCS (Huyer and Neumaier 1999) and LGO (Pintér

1996a). All work for bound-constrained problems only and need the approximation techniques of Section 8 for more general problems. (Some of these are built into LGO, but must be coded by the user for DIRECT and MCS.) All three algorithms enforce convergence by employing a branching scheme. They differ in how and when to split, and what is done within each box.

A *branching scheme* generates a sequence of rooted trees of boxes whose leaves cover the feasible set. At least one point in each box is evaluated. The first tree just has the original box as root and only leaf. Each other tree is obtained from the previous one by splitting one or several leaves. If the diameters of all boxes at all leaves converge to zero, convergence of the algorithm is straightforward.

The convergence to zero of the diameters is ensured by appropriate *splitting rules* that define when and how a box is split. For example, convergence is guaranteed when in each of a sequence of rounds,

- we always split the oldest box along the oldest side, and possibly split finitely many other boxes, or
- we always split the longest box along the longest side, and possibly split finitely many other boxes (where length = sum of length of sides),

provided that each split of the oldest (or longest) box produces boxes whose volume is at most a fixed fraction < 1 of the unsplit box. The possibility of 'and finitely many other boxes' (but not many if the code is to be robust!) can be used with considerable flexibility without destroying the convergence property.

Apart from the convergence requirement, the key to efficiency is a proper balance of global and local search. This is achieved in DIRECT by splitting in each round all boxes for which the pair (v, f) (where v is the volume and f the midpoint function value) is not dominated by another such pair. Here (v, f) is dominated by (v', f') if both $v' < v$ and $f' > f$. In particular, the box of largest volume and the box with the best function value are never dominated and hence always split. MCS instead uses domination of pairs (l, f), where l is a suitably assigned level, and in addition employs local optimization steps (using line searches and sequential bound-constrained quadratic programs) from appropriate candidate points. LGO uses lower bounds

$$L \geq \max_{k,l} \|f(x_k) - f(x_l)\|/\|x_k - x_l\|$$

on Lipschitz constants L obtained from the previous function evaluations to decide on the promising boxes to split first. (Upper bounds on L, and hence bounds on function values, cannot be obtained from local information only.)

The combination of a suitable branching strategy with the heuristic methods discussed earlier would make the latter complete, and appears to be a fruitful research direction.

To improve on the density theorem we must find ways to throw away irrelevant parts of the feasible domain that are guaranteed not to contain a global minimizer. To be able to do this reliably, some kind of global information is necessary. This is utilized by box reduction techniques, discussed in Section 10 using a simple example, and afterwards in more depth.

10. Box reduction: an example

Box reduction techniques are based on a more or less sophisticated interplay of several components: logical constraint propagation, interval analysis, convex relaxations and duality arguments involving Lagrange multipliers. Before giving a more formal treatment, we illustrate simple arguments of each of these components by reconsidering Example 5.2.

Suppose that a local solver has already produced the local minimizer $\hat{x} = \binom{0}{1}$ for the problem (5.5) discussed in Example 5.2, perhaps as the best local minimizer found by minimizing from a few random starting points. We use box reduction to check whether there is possibly a better feasible point. In fact, we know already that this is not the case, but we obtained this knowledge in a way that works only for very simple problems. Thus we want to do it again, using only techniques of wide applicability.

The idea of box reduction is to use various arguments that allow us to shrink the box without losing any feasible point that is at least as good as the best point found already. Since $\hat{x}$ is feasible with objective function value -2, any such point satisfies

$$f(x) = -x_1 - 2x_2 \leq -2, \tag{10.1}$$
$$F(x) = (x_1 - 1)^2 + (x_2 - 1)^2 = 1, \tag{10.2}$$
$$x_1 \in [-1, 1], \quad x_2 \in [-1, 1]. \tag{10.3}$$

Constraint propagation (see Section 14) is a very cheap and easily formalizable process that gives important initial range reductions in many otherwise difficult problems. It consists in deducing better bounds for a variable by using the other bounds and one of the constraints. In particular, (10.1) implies $x_2 \geq 1 - x_1/2 \geq 0.5$ since $x_1 \leq 1$, and $x_1 \geq 2 - 2x_2 \geq 0$ since $x_2 \leq 1$, reducing the bounds to

$$x_1 \in [0, 1], \quad x_2 \in [0.5, 1].$$

Similarly, (10.2) implies $(x_1 - 1)^2 = 1 - (x_2 - 1)^2 \geq 1 - 0.25 = 0.75$, hence $x_1 \leq 1 - \sqrt{0.75} < 0.14$, giving the improved bound

$$x_1 \in [0, 0.14].$$

This bound could be used to improve again x_2 using (10.1); and by alternating use of (10.1) and (10.2) one would obtain a sequence of boxes shrinking towards $\hat{x}$. This is a special feature of this simple example. In most cases, this simple substitution process gives no or only very little improvements after the first few good reductions. (Look at a problem with the constraints $x_1 + x_2 = 0$, $x_1 - x_2 = 0$, $x_1, x_2 \in [-1, 1]$ to see why.)

Interval analysis (see Section 11) can be applied in a number of different ways. Here we use it to produce linear relaxations of the nonlinear constraint. The Jacobian of $F(x)$ at $x \in \mathbf{x} = ([0, 0.14], [0.5, 1])^T$ is

$$F'(x) = (2x_1 - 2, 2x_2 - 2) \in ([-2, -1.72], [-1, 0]) = F'(\mathbf{x}).$$

The mean value theorem implies that, for any $\tilde{x} \in \mathbf{x}$,

$$F(x) \in F(\tilde{x}) + F'(\mathbf{x})(x - \tilde{x}) \quad \text{if } x \in \mathbf{x}.$$

Using $\tilde{x} = \hat{x}$ we find

$$1 \in 1 + [-2, -1.72]x_1 + [-1, 0](x_2 - 1) = [1 - 2x_1, 2 - 1.72x_1 - x_2];$$

the interval evaluation needs no case distinction since x_1 and $x_2 - 1$ happen to have constant sign. The lower bound gives no new information, but the upper bound leads to the new constraint

$$1.72x_1 + x_2 \leq 1.$$

By its derivation, this constraint is weaker than (10.2).

But since it is linear, the constraint is quite useful for *relaxation techniques* (see Section 16). It allows us to create a convex relaxation of the problem. Indeed, we may look at the relaxed linear program

$$\begin{aligned} \min \quad & -x_1 - 2x_2 \\ \text{s.t.} \quad & 1.72x_1 + x_2 \leq 1, \quad 0 \leq x_1 \leq 0.14, \quad 0.5 \leq x_2 \leq 1. \end{aligned} \quad (10.4)$$

By construction, every feasible point better than the best point is feasible for (10.4), hence the minimum of (10.4) will be a *lower bound* on the best-possible objective function value of the original problem. Solving (10.4) gives the solution $\hat{x} = \binom{0}{1}$ with function value -2. Since this lower bound equals the best function value found so far for the original problem, the original problem has global minimum -2. This is a happy accident due to special circumstances: our problem already had a linear objective function, and the global minimizer was at a corner of the feasible set. (But as we shall see, we can adapt the technique to work much more generally if the box is narrow enough.)

Multiplier techniques can be used in a variety of ways. Here we use them to check whether there is a second, undiscovered global minimizer. We use the Lagrange multiplier $\hat{\lambda} = 2$ associated with the linear constraint of (10.4) at the solution. The associated linear combination $-x_1 - 2x_2 + 2(1.72x_1 + x_2 - 1)$ is bounded by the best-known function value -2 of the original problem, giving $2.44x_1 - 2 \leq -2$, hence $x_1 \leq 0$. Thus we must have $x_1 = 0$, and constraint propagation using (10.1) implies $x_2 = 1$. Thus the box has been reduced to $\hat{x}$, showing that it is the only global minimizer.

What generalizes? The problem discussed was deliberately kept simple so that the complete solution process could be demonstrated explicitly. In general, constraint propagation only gives limited reduction. Similarly, relaxed linear or convex programs usually only give a lower bound on the smallest-possible objective function value, but the linear combination derived from the Lagrange multipliers frequently contains useful information that can be exploited by constraint propagation to get a further significant box reduction.

If the reduction process stalls or becomes slow, the box is split into two or more smaller boxes. On the smaller boxes, the same techniques may prove effective, and we alternate box reduction and box splitting until all box sizes are below some termination threshold. Usually, only very few boxes remain if sufficiently good reduction techniques are used (pathological exceptions include min $x - x$ s.t. $x \in [0, 1]$). If no box remains, the problem is guaranteed to have no feasible point.

The total number of boxes processed is a measure of the difficulty of a problem for the particular algorithm used. Simple problems (like the example discussed above) only need a single box; in the worst case, an exponential number of boxes may be needed. In the latter case, time and storage limitations may force a premature termination; in this case the best point found is not verified to be a global minimizer.

11. Interval arithmetic

Interval analysis, the study of theory and algorithms for computing with intervals, is a large subject; see Moore (1979) (introductory), Neumaier (2001b) (embedded in a numerical analysis context) and Neumaier (1990) (advanced). Its importance for global optimization stems from several, interrelated facts.

- Interval analysis gives easily computable (though sometimes only very crude) bounds on the range expressions.
- Interval analysis allows one to control nonlinearities in a simple way (via centred forms).
- Interval analysis extends classical analysis in its ability to provide *semi-local* existence and optimality conditions, valid within a *pre-specified* local region around some point, while classical analysis generally only asserts the existence of such neighbourhoods without providing a simple way to find them.

We give here a short introduction to the basics and mention the main techniques useful for global optimization. General references on interval methods in global optimization include Adjiman et al. (1998a), Benhamou, McAllister and Van Hentenryck (1994), Benhamou and Older (1997), Berner (1996), Dallwig et al. (1997), Epperly and Pistikopoulos (1997), Epperly and Swaney (1996), Jansson (1994), Jansson and Knüppel (1992, 1995), Hansen (1980, 1992a), Kearfott (1996b), Kearfott, Novoa and Chenyi Hu (1991), Neumaier (1996), Ratschek and Rokne (1984, 1988), and Van Hentenryck et al. (1997b).

If $\mathbf{a}$ and $\mathbf{b}$ are two intervals we define for $\circ \in \{+,-,*,/,\hat{\ }\}$ the binary operation

$$\mathbf{a} \circ \mathbf{b} := \Box \{\tilde{a} \circ \tilde{b} \mid \tilde{a} \in \mathbf{a}, \tilde{b} \in \mathbf{b}\}, \tag{11.1}$$

provided the right-hand side is defined. Here

$$\Box S = [\inf S, \sup S]$$

denotes the *interval hull* of a set of real numbers, *i.e.*, the tightest interval containing S. A monotonicity argument gives for addition and subtraction

$$\mathbf{a} + \mathbf{b} = [\underline{a} + \underline{b}, \overline{a} + \overline{b}], \tag{11.2}$$
$$\mathbf{a} - \mathbf{b} = [\underline{a} - \overline{b}, \overline{a} - \underline{b}], \tag{11.3}$$

and for multiplication and division

$$\mathbf{a} * \mathbf{b} = \Box\{\underline{a}\underline{b}, \underline{a}\overline{b}, \overline{a}\underline{b}, \overline{a}\overline{b}\}, \tag{11.4}$$
$$\mathbf{a}/\mathbf{b} = \Box\{\underline{a}/\underline{b}, \underline{a}/\overline{b}, \overline{a}/\underline{b}, \overline{a}/\overline{b}\} \quad \text{if } 0 \notin \mathbf{b}; \tag{11.5}$$

in most cases only two of these products or quotients need to be computed. We also define elementary functions $\varphi \in \{\text{sqr}, \text{sqrt}, \exp, \log, \sin, \cos, \text{abs}, \ldots\}$ of an interval $\mathbf{a}$ (and similarly $-\mathbf{a}, \mathbf{a}_+$, *etc.*) by

$$\varphi(\mathbf{a}) := \Box\{\varphi(\tilde{a}) \mid \tilde{a} \in \mathbf{a}\} \tag{11.6}$$

whenever the right-hand side is defined. Again $\varphi(\mathbf{a})$ can be computed from the value of φ at the end-points of $\mathbf{a}$ and the interior extremal values, depending on the monotonicity properties of φ. Note, however, that $|\mathbf{a}|$ is defined as $\sup \text{abs}(\mathbf{a})$, since this expression figures prominently in estimates involving interval techniques.

For interval vectors (=boxes) $\mathbf{x}$, analogous definitions apply. We also need the *interior*

$$\text{int } \mathbf{x} = \{\tilde{x} \in \mathbb{R}^n \mid \underline{x} < \tilde{x} < \overline{x}\}$$

of a box $\mathbf{x} \subseteq \mathbb{R}^n$.

For details and a systematic study of interval operations see Neumaier (1990); we only remark here that some rules familiar from real arithmetic fail, and in particular the interval evaluation of different expressions equivalent in real arithmetic may give different results. For example (with $-\mathbf{a} := 0 - \mathbf{a} = [-\overline{a}, -\underline{a}]$),

$$\mathbf{a} + (-\mathbf{a}) = \mathbf{a} - \mathbf{a} \neq 0 \quad \text{except when } \underline{a} = \overline{a}.$$

Therefore, we also use the converse *inner operations*

$$\mathbf{a} \oplus \mathbf{b} := [\underline{a} + \overline{b}, \overline{a} + \underline{b}], \tag{11.7}$$

$$\mathbf{a} \ominus \mathbf{b} := [\underline{a} - \underline{b}, \overline{a} - \overline{b}]. \tag{11.8}$$

Here, expressions of the form $\pm\infty \mp \infty$ in (11.7) or (11.8) must be interpreted as $-\infty$ for the lower bounds and as $+\infty$ for the upper bounds. Note that the result of an inner operation is not necessarily an interval since it may happen that the lower bound is larger than the upper bound; giving an empty 'interval'.

All these operations are very simple to program. Note that many implementations of interval arithmetic are rather slow since they take care to guarantee correct (and often optimal) outward rounding, needed when interval arithmetic is used for mathematically rigorous certification (see Section 20). For global optimization without certification, *unsafe* interval arithmetic, which uses the standard rounding for floating point operations, and hence is significantly faster but may lose containment of points that lie too close to the boundary, usually suffices if certain safety measures are taken. But it is significantly harder to ensure robust behaviour with unsafe interval arithmetic since occasionally the solution is lost, too.

Important. When using unsafe interval arithmetic, proper safeguards must be taken at places (such as inner operations and intersections) where intervals might become (spuriously) empty owing to accumulation of round-off errors. In place of an empty result, a thin interval formed from the arithmetic mean of the two intersecting bounds should be returned in a robust implementation.

As already mentioned, an interval evaluation $f(\mathbf{x})$ of some expression f often overestimates the desired *range*

$$\text{Range}(f, \mathbf{x}) = \{f(x) \mid x \in \mathbf{x}\}$$

of a function. However, under very mild conditions (Neumaier 1990, Section 1.4), the evaluation over small boxes satisfies

$$f(\mathbf{x}) \subseteq \text{Range}(f, \mathbf{x}) + O(\varepsilon) \quad \text{if } \overline{x} - \underline{x} = O(\varepsilon);$$

we refer to this as the *linear approximation property* of simple interval evaluation.

Better enclosures, especially for small ε, can be obtained by *centred forms*; the simplest of these (but not the most efficient: see Chapter 2 of Neumaier

(1990) for better methods based on slopes) is the *mean value form*. Owing to the mean value theorem, we have

$$f(x) \in f(z) + f'(\mathbf{x})(x - z) \quad \text{if } x, z \in \mathbf{x}. \tag{11.9}$$

In particular, Range$(f, \mathbf{x})$ is contained in $f(z) + f'(\mathbf{x})(\mathbf{x} - z)$, and it can be shown that, under mild conditions,

$$f(z) + f'(\mathbf{x})(\mathbf{x} - z) \subseteq \text{Range}(f, \mathbf{x}) + O(\varepsilon^2) \quad \text{if } \overline{x} - \underline{x} = O(\varepsilon);$$

we say that the mean value form (as other centred forms) has the *quadratic approximation property*. Recently, centred forms based on higher-order Taylor expansions have found considerable attention; these are able to give significantly sharper bounds in cases where simple interval evaluation suffers from severe dependence. See the survey Neumaier (2002) and the numerical comparisons in Makino and Berz (2003); *cf.* also Carrizosa, Hansen and Messine (2004).

Apart from interval evaluation and centred forms, we need *interval Newton methods* for verifying solutions of nonlinear systems of equations. The prototype (but again not the most efficient method; see Neumaier (1990, Chapter 5) for better methods based on slopes and Gauss–Seidel iteration) is the Krawczyk (1969) method. To check for solutions of $F(x) = 0$ with $x \in \mathbf{x}$, Krawczyk multiplies the vector version of (11.9) by a matrix C and subtracts it from x to find

$$x \in K(\mathbf{x}, z) := z - CF(z) + (I - CF'(\mathbf{x}))(\mathbf{x} - z).$$

For $z \in \mathbf{x}$, the resulting *Krawczyk operator* $K(\mathbf{x}, z)$ (*cf.* Krawczyk (1969), Kahan (1968)) has the following properties, typical for interval Newton operators:

(i) any zero $x \in \mathbf{x}$ of F lies in $\mathbf{x} \cap K(\mathbf{x}, z)$,

(ii) if $\mathbf{x} \cap K(\mathbf{x}, z) = \emptyset$ then $\mathbf{x}$ contains no zero of F,

(iii) if $K(\mathbf{x}, z) \subseteq \text{int } \mathbf{x}$ then $\mathbf{x}$ contains a unique zero of F.

Properties (i) and (ii) follow directly from the above derivation, while (iii) is a simple consequence of Banach's fixed point theorem.

The most important part is (iii), since, applied to the Karush–John conditions, it allows the elimination of large regions around a local minimizer; *cf.* Section 15. However, (i) and (ii) are also useful as ways of reducing a box or eliminating it, if it contains no zero. This is implemented in GlobSol (Kearfott 1996*b*) and Numerica (Van Hentenryck *et al.* 1997*b*).

Another useful interval Newton operator with analogous properties is

$$x \in N(\mathbf{x}, z) := z - (CF[\mathbf{x}, z])^H (CF(z)),$$

where C is an approximate inverse of the interval slope $F[\mathbf{x}, z]$ and $\mathbf{A}^H \mathbf{b}$ is an enclosure for the set of solutions of $Ax = b, A \in \mathbf{A}, b \in \mathbf{b}$ computed, *e.g.*, by the Hansen–Bliek method (Bliek 1992, Hansen 1992*b*, Neumaier 1999).

11.1. Convexity check

Interval analysis can be used to check the convexity of a function $f : \mathbf{x} \to \mathbb{R}$ in some box $\mathbf{x}$. Let $\mathbf{G}$ be a matrix of intervals (usually simply called an *interval matrix*), calculated as an enclosure of $f''(x)$ for $x \in \mathbf{x}$, then, with $r = \max\{\overline{x}_k - \underline{x}_k \mid k = 1, \ldots, n\}$, the linear approximation property implies that $|\overline{G} - \underline{G}| = O(r)$. Such a statement implies that $|\mathbf{G} - \tilde{G}| = O(r)$ for all individual matrices $\tilde{G} \in \mathbf{G}$, with absolute values taken component-wise. In particular, if $\hat{G}$ is positive definite then, provided the underlying box is not too wide, all matrices in $\mathbf{G}$ are definite, too; and if this is the case, f is convex in $\mathbf{x}$. The following constructive criterion for simultaneously checking the definiteness of all members of an interval matrix was given in Neumaier (1996).

Theorem 11.1. (Sufficient conditions for convexity) Let $f : \mathbf{x} \to \mathbb{R}$ be twice continuously differentiable on the compact box $\mathbf{x}$, and suppose that $\mathbf{G}$ is a symmetric interval matrix such that

$$f''(x) \in \mathbf{G} \quad \text{for all } x \in \mathbf{x}. \tag{11.10}$$

(i) If some symmetric matrix $G_0 \in \mathbf{G}$ is positive definite and all symmetric matrices in $\mathbf{G}$ are nonsingular then they are all positive definite, and f is uniformly convex in $\mathbf{x}$.

(ii) In particular, this holds if the midpoint matrix

$$\check{G} = (\sup \mathbf{G} + \inf \mathbf{G})/2$$

is positive definite with inverse C, and the preconditioned radius matrix

$$\Delta = |C|\operatorname{rad} \mathbf{G},$$

where $|C|$ is the componentwise absolute value of C and

$$\operatorname{rad} \mathbf{G} = (\sup \mathbf{G} - \inf \mathbf{G})/2,$$

satisfies the condition

$$\|\Delta\| < 1 \tag{11.11}$$

(in an arbitrary norm).

Proof. (i) Since the eigenvalues are continuous functions of the matrix entries and the product of the eigenvalues (the determinant) cannot vanish, no eigenvalue changes sign. Hence the eigenvalues of all matrices in $\mathbf{G}$ are positive, since this is the case for the positive definite member. Thus all symmetric matrices in **G** are positive definite. By well-known results, uniform convexity of f now follows from (11.10).

(ii) $G_0 = \check{G}$ belongs to **G**, and condition (11.11) implies strong regularity of the interval matrix **G** (Neumaier 1990, Section 4.1) and hence nonsingularity of all matrices in **G**. Thus (i) applies. □

In many cases, the Hessian of the augmented Lagrangian can be shown to have the form
$$f''(x) = \sum u_i A_i, \quad u_i \in \mathbf{u}_i,$$
with constructively available real matrices A_i and intervals $\mathbf{u}_i = [\check{u}_i - r_i, \check{u}_i + r_i]$. In this case, the above result can be strengthened (with virtually the same proof) by replacing $\check{G}$ and Δ with
$$\check{G} = \sum \check{u}_i A_i$$
and
$$\Delta' = \sum r_i |CA_i|,$$
respectively. Indeed, it is not difficult to see that for $\mathbf{G} = \sum \mathbf{u}_i A_i$, we always have $0 \leq \Delta' \leq \Delta$, so that the refined test is easier to satisfy.

Other sufficient conditions for convexity based on scaled Gerschgorin theorems and semidefinite programming, form the basis of the αBB method (Adjiman et al. 1996, Androulakis et al. 1995) and are given in Adjiman et al. (1998a) and Adjiman et al. (1998b).

12. The branch and bound principle

The branch and bound principle is a general label (invented in Land and Doig (1960) and Little et al. (1963)) to denote methods to split a problem recursively into subproblems which are sooner or later eliminated by showing that the subproblem cannot lead to a point better than (or as least as good as) the best point found so far. The latter is often checked by computing lower bounds on the objective function, and the splitting produces new branches in the tree of all subproblems tried, according to so-called *branching rules*; hence the name 'branch and bound'. But in practice, the subproblems are best treated in a more flexible fashion, allowing us also to eliminate subproblems only partially.

General references for branch and bound in global optimization include Beale (1979, 1988), Berner (1996), Epperly and Pistikopoulos (1997), Epperly and Swaney (1996), Floudas (1995), Grossmann (1990), Horst et al. (1995), Jansson and Knüppel (1992), Kearfott (1996b), Pintér (1996a) and Van Hentenryck et al. (1997b). A thorough discussion of branch and bound in discrete optimization, with many algorithmic choices that are of potential interest in general global optimization, is given in Parker and Rardin (1988).

For a global optimization problem

$$\begin{aligned}\min\ & f(x) \\ \text{s.t.}\ & x \in \mathbf{x}^{\text{init}}, \quad F(x) \in \mathbf{F}, \quad x_I \text{ integral},\end{aligned} \tag{12.1}$$

a natural way to define subproblems is to choose boxes $\mathbf{x} \subseteq \mathbf{x}^{\text{init}}$ of the initial box $\mathbf{x}^{\text{init}}$, and to consider the subproblems

$$\begin{aligned}\min\ & f(x) \\ \text{s.t.}\ & x \in \mathbf{x}, \quad F(x) \in \mathbf{F}, \quad x_I \text{ integral},\end{aligned} \tag{12.2}$$

that is, each subproblem is characterized by (and stored as) the box over which the problem is solved. The branching process then consists in splitting a box $\mathbf{x}$ into two or several smaller boxes whose union is $\mathbf{x}$. The most typical branching rule is to select a *bisection coordinate* j and to split the jth component of the box at a *bisection point* ξ. Thus, the current box $\mathbf{x}$ is replaced by two sub-boxes $\mathbf{x}^{\text{low}}, \mathbf{x}^{\text{upp}}$ with

$$\begin{aligned}\mathbf{x}_k^{\text{low}} &= \mathbf{x}_k^{\text{upp}} = \mathbf{x}_k \quad \text{if } k \neq j, \\ \mathbf{x}_j^{\text{low}} &= [\underline{x}_j, \xi], \quad \mathbf{x}_j^{\text{upp}} = [\xi, \overline{x}_j].\end{aligned} \tag{12.3}$$

This branching rule is termed *bisection*. The bisection point ξ is often taken as the *midpoint* $\xi = (\overline{x}_j + \underline{x}_j)/2$ of the interval $\mathbf{x}_j$; but this fails when there are infinite bounds and is inefficient when the interval ranges over several orders of magnitude. In this case, a more useful bisection point is a *safeguarded geometric mean*, defined by

$$\xi = \operatorname{sign} \underline{x}_j \sqrt{\underline{x}_j \overline{x}_j} \quad \text{if } 0 < \underline{x}_j \overline{x}_j < \infty,$$

and otherwise

$$\begin{aligned}\xi &= 0 & \text{if } \underline{x}_j < 0 < \overline{x}_j, \\ \xi &= \min(\mu, q\overline{x}_j) & \text{if } \underline{x}_j = 0, \\ \xi &= \max(-\mu, q\underline{x}_j) & \text{if } \overline{x}_j = 0, \\ \xi &= q^{-1}\underline{x}_j & \text{if } \underline{x}_j > 0, \\ \xi &= q^{-1}\overline{x}_j & \text{if } \overline{x}_j < 0,\end{aligned}$$

where $q \in\ ]0,1[$ is a fixed constant (such as $q = 0.01$) and variables whose initial interval contains 0 are assumed to be most likely of magnitude μ.

The branching coordinate is more difficult to choose, but the speed of a branch and bound algorithm may be heavily affected by this choice. For a good algorithm, the choice should be scaling-invariant, but the details depend on how the algorithm treats the individual subproblems.

Sometimes, a *trisection* branching rule is used which splits some component of a box into three intervals. Also, *multisection* branching rules may be employed; only one natural choice is described here. Suppose we know that a sub-box $\mathbf{x}^0$ of $\mathbf{x}$ cannot contain a solution of (12.1). (In practice, $\mathbf{x}^0$ would be the intersection of an exclusion box with $\mathbf{x}$; see Section 11.) Then we can cover $\mathbf{x} \setminus \mathbf{x}^0$ by (at most) $2n$ sub-boxes, namely, for $j = 1, \ldots, n$,

$$\begin{aligned}
\mathbf{x}_k^{2j-1} &= \mathbf{x}_k^{2j} = \mathbf{x}_k^0 \quad \text{if } k < j, \\
\mathbf{x}_j^{2j-1} &= [\underline{x}_j, \underline{x}_j^0], \quad \mathbf{x}_j^{2j} = [\overline{x}_j^0, \overline{x}_j], \\
\mathbf{x}_k^{2j-1} &= \mathbf{x}_k^{2j} = \mathbf{x}_k \quad \text{if } k > j.
\end{aligned} \quad (12.4)$$

However, this may yield long and thin slices and is then rather inefficient.

For a comparison of some branching rules for bound-constrained problems see Csendes and Ratz (1997), Ratz (1996), Ratz and Csendes (1995).

The *bounding rule* in its classical variant requires the solution of a *convex relaxation*, i.e., a convex (and often linear) optimization problem whose feasible set contains the feasible set of the subproblem (*outer approximation*) and whose objective function is at no feasible point larger than the original objective function (*underestimation*). If the convex problem is infeasible, the subproblem is infeasible, too, and can be discarded. If the convex problem is feasible, its solution provides a lower bound on $f(x)$, and when this lower bound is larger than the value of f^{best} for some feasible point x^{best} known (stored in a list of *best feasible points found so far*) we conclude that the subproblem no longer contributes to the solution of the global optimization problem and hence can be discarded.

Clearly, this procedure is equivalent to adding the constraint $f(x) \leq f^{\text{best}}$ to the definition of the subproblem and checking infeasibility of the resulting reduced subproblem. This suggests a more general approach to defining subproblems by adding other *cuts*, i.e., derived inequalities that have to be satisfied at a global minimizer. If these inequalities are linear, the cuts define hyperplanes and are referred to as *cutting planes*; cf. Section 16. Branch and bound methods using cuts are frequently labelled *branch and cut*.

Another important approach to handling subproblems uses constraint propagation and related techniques that define *reduction* (also called *tightening, narrowing, filtering* or *pruning*) rules which serve to reduce (as much as easily possible) the box defining a subproblem without changing its feasible set. If reduction results in an empty box, the subproblem is eliminated; if not, the subproblem may still have been reduced so much that many branching steps are saved. Fast and simple reduction rules use constraint propagation, discussed in Section 14; more expensive rules are discussed in Section 15. The balancing of work done in reduction versus work saved through less branching is a delicate matter, which at present more or less depends on *ad hoc* recipes.

Note that reduction techniques may be applied not only to the original constraints but to all constraints that must be satisfied at the global minimizer. This includes cutting planes (see Section 16) and the equations and inequalities derived from the Karush–John optimality conditions (see Section 5). In particular, software based on interval techniques (GlobSol (Kearfott 1996b), Numerica (Van Hentenryck *et al.* 1997b)) make essential use of the latter.

13. The role of local optimization

Local optimization routines are an important part of most global solvers. They are used for two different purposes.

(i) To find feasible points if the feasible domain has a complicated definition, and to find better local minimizers when (after successful tunnelling) a feasible point better than the previously best local minimizer has been found.

(ii) To solve auxiliary optimization problems such as relaxations of the original problem (for generating improved bounds) or bound-constrained approximations (for tunnelling).

13.1. Relaxation

The auxiliary local optimization problems that need to be solved are simpler in structure since they 'relax' the problem in some way. A *relaxation* is a modification of the original problem whose solution is tractable and gives some information about the possible location of the global minimizer. In the past, mainly linear and convex relaxation have been used, since for these, local optimization provides global solutions, which usually implies useful global information about the original problem. We shall discuss various ways of obtaining and using linear and convex relaxations in Section 16. Nonconvex relaxations may be useful, too, if they are reliably solvable to global optimality. We therefore discuss semilinear relaxations – which can be solved by MILP techniques – in Section 18.

13.2. Tunnelling

One may consider solving a global optimization problem as a *sequential nonlinear programming method* (SNLP), where local optimization (NLP) steps that improve a feasible point to local optimality alternate with tunnelling steps that produce better (nearly) feasible points by some *tunnelling procedure*. For complete methods based on branching, the 'tunnelling' is done by finding nearly feasible points during inspection of the subproblems.

The success of the tunnelling step depends on the details of looking for such points. One strategy (Dallwig et al. 1997) proceeds by solving on selected sub-boxes nonlinear least squares problems that minimize the sum of squares of the constraint violations, and (if a best feasible point with function value f^{best} is already available) the violation of $f(x) \leq f^{\text{best}} - \Delta$, where $\Delta \geq 0$ is some measure of minimal gain in function value. Alternatively, one may use the soft optimality theorem (Theorem 8.1) in place of least squares. (See also Guddat et al. (1990) for tunnelling by continuation.) Thus, in a sense, the global optimization of (4.1) consists in solving a sequence of harder and harder feasibility problems

$$\begin{aligned}&\text{find} \quad x\\ &\text{s.t.} \quad x \in \mathbf{x}, \quad F(x) \in \mathbf{F}, \quad x_I \text{ integral}, \quad\quad (13.1)\\ &\quad\quad f(x) \leq f^{\text{best}} - \Delta.\end{aligned}$$

Typical global optimization methods spend perhaps 5% of their time on finding a global minimizer, and the remaining 95% on the verification that there is no significantly better feasible point, *i.e.*, showing that the feasibility problem (13.1) has no solution. Also, hard problems need a significant amount of time to find the first feasible point. Thus the initial and final (dominant) stages of a global optimization solution process are essentially identical to that for a feasibility problem.

In particular, general feasibility problems, also called *constraint satisfaction problems*, can be as hard as general global optimization problems, and the techniques needed for solving constraint satisfaction problems are essentially the same as those for solving global optimization problems.

13.3. General considerations

Considerations of superlinear convergence of local optimization algorithms imply that one generally uses *sequential quadratic programming* (SQP) techniques, which solve a sequence of related quadratic programs whose solution converges (under certain conditions, *cf.* below) to a local minimizer of the original problem; if the starting point is feasible (which, initially, need not be the case), the function value of the local minimizer is at or below that of the starting point.

To give the reader a rough idea of times and difficulties, here are some (completely unreliable but catchy) rules of thumb. If the time needed to solve a linear program of a certain size is LP then solving a problem of comparable size and sparsity structure may take perhaps the time

$$QP = 5 * LP$$

for a convex quadratic program,

$$QP' = 10 * LP$$

for a local minimizer of a nonconvex quadratic program,
$$SQP = 30 * QP$$
for a convex nonlinear program,
$$SQP' \geq 200 * QP$$
for a local minimizer of a nonconvex nonlinear program,
$$GLP_f \geq 100 * SQP$$
for *finding* a global minimizer of a nonconvex nonlinear program, and
$$GLP_v \geq 1000 * SQP$$
for *verifying* that it is a global minimizer.

We now comment on the properties of local optimization software that are important for their use in global optimization. Usually, it is more important that the local solver is fast than that it is very robust (*i.e.*, guaranteed to succeed), since lack of robustness in some of the local optimizations is made up for by the structure of the global solution process. To help control the amount of work done in the local part, it should be possible to force a premature return with a less than optimal point when some limit (of time or number of function values) is exceeded. Nevertheless, the local solver should be good to ensure that solving a problem with a unique minimizer (which is automatically global) by the global solver does not take much longer than a good local solver would need.

Modern nonlinear programming codes are usually 'globally convergent' in some sense. The global convergence proofs (to a *local* minimizer only!) usually make more or less stringent assumptions that imply the absence of difficulties in finding feasible points. Formally, we may say that a local optimization algorithm is *globally convergent* if there is a continuous function $d_{\text{feas}} : \mathbb{R}^n \to \mathbb{R}$ (defining a 'distance to feasibility') such that

$$d_{\text{feas}}(x) \geq 0, \quad \text{with equality if and only if } x \text{ is feasible}$$

and the algorithm produces for arbitrary continuous problems and arbitrary starting points a sequence of $x^l \in \mathbb{R}^n$ satisfying one of the following conditions:

(i) x^l converges to the set of points satisfying the Karush–John conditions (and, possibly, second-order necessary conditions);
(ii) $d_{\text{feas}}(x^l) \to 0$ and $f(x^l) \to -\infty$;
(iii) x_l converges to the set of points where the objective or some constraint function is not continuously differentiable;
(iv) $d_{\text{feas}}(x^l) \to 0$, $\|x^l\| \to \infty$;
(v) $d_{\text{feas}}(x^l)$ converges to a nonzero local minimum of d_{feas}.

Conditions (i) and (ii) characterize the achievement of the optimization goal, while conditions (iii)–(v) characterize various modes of unavoidable failure. Failures of type (iii) or (iv) are usually attributed to bad modelling or bad choice of the optimization methods. Some methods such as bundle methods can cope with lack of differentiability, hence do not lead to case (iii).

A failure of type (v) is unavoidable if there is no feasible point. However, failures of type (v) may happen for problems with nonconvex constraints even though feasible points exist. One could say that from a local point of view, an optimization problem is *easy* (for an algorithm) if (v) cannot occur whenever a feasible point exists. A local algorithm may be considered good if among its easy problems are all problems with convex constraints only, and all problems satisfying certain strong versions (Burke 1991) of the Mangasarian–Fromovitz constraint qualification (Mangasarian and Fromovitz 1967). Ideally, a good local algorithm would provide in these cases a certificate of infeasibility whenever it detects case (v).

14. Constraint propagation

In many cases, general constraints can be used to reduce the size of a box in the branching scheme. The general technique is called *constraint propagation* and was pioneered in constraint logic (Cleary 1987, Older and Vellino 1993) and interval analysis (Neumaier 1988), but has also forerunners in presolve techniques in mathematical programming (Mangasarian and McLinden 1985, Lodwick 1989, Anderson and Anderson 1995). See Babichev *et al.* (1993), Benhamou *et al.* (1994), Benhamou and Older (1997), Chen and van Emden (1995), Hager (1993), Hyvönen and De Pascale (1996), Kearfott (1991), Van Hentenryck (1989), Van Hentenryck, Michel and Benhamou (1997a), Van Hentenryck *et al.* (1997b) for further developments, and the COCONUT report (Bliek *et al.* 2001) for an extensive recent survey.

We follow here the set-up by Dallwig *et al.* (1997), which handles linear constraints (and more generally block-separable constraints) without the need to decompose the constraints into primitive pieces defined by single operations. (In the following, if J is a list of indices, x_J denotes the subvector of x formed by the components with index in J.)

Proposition 14.1. Let the q_k be real-valued functions defined on $\mathbf{x}_{J_k}$.

(i) If (for suitable $\overline{q}_k, \overline{s}$)

$$\overline{q}_k \geq \sup\{q_k(x_{J_k}) \mid x_{J_k} \in \mathbf{x}_{J_k}\}, \quad \overline{s} \geq \sum_k \overline{q}_k, \qquad (14.1)$$

then, for arbitrary $\underline{a}$,

$$x \in \mathbf{x}, \quad \underline{a} \leq \sum_k q_k(x_{J_k}) \implies q_k(x_{J_k}) \geq \underline{a} - \overline{s} + \overline{q}_k \quad \text{for all } k. \quad (14.2)$$

(ii) If
$$\underline{q}_k \leq \inf\{q_k(x_{J_k}) \mid x_{J_k} \in \mathbf{x}_{J_k}\}, \quad \underline{s} \leq \sum_k \underline{q}_k, \tag{14.3}$$
then, for arbitrary $\underline{a}$,
$$x \in \mathbf{x}, \quad \sum_k q_k(x_{J_k}) \leq \bar{a} \implies q_k(x_{J_k}) \leq \bar{a} - \underline{s} + \underline{q}_k \text{ for all } k. \tag{14.4}$$

Proof. The assumptions of part (i) imply
$$q_k(x_{J_k}) \geq \underline{a} - \sum_{l \neq k} q_l(x_{J_l}) \geq \underline{a} - \sum_{l \neq k} \bar{q}_l \geq \underline{a} + \bar{q}_k - \bar{s},$$
hence the conclusion in (14.2) holds. Part (ii) is proved in the same way. □

The proposition is applied as follows to reduce the size of boxes by tightening bound constraints. Suppose that $x \in \mathbf{x}$. For any constraint of the form
$$\underline{a} \leq \sum_k q_k(x_{J_k}) \tag{14.5}$$
we form the quantities (14.1). (This is straightforward if the q_k depend on a single variable x_k only, $J_k = \{k\}$; in the most important cases, q_k is linear or quadratic in x_k, and the supremum is very easy to calculate; in more complicated cases, upper (resp. lower) bounds can be calculated with interval arithmetic.) Then one checks the condition
$$\underline{a} \leq \bar{s}. \tag{14.6}$$
If it is violated then (14.5) is clearly inconsistent with $x \in \mathbf{x}$ (and in the branch and bound application, the corresponding subproblem can be discarded). If (14.6) holds, one can exploit the conclusion in (14.2), provided that one can compute the set of $x_{J_k} \in \mathbf{x}_{J_k}$ (or a superset) such that
$$q_k(x_{J_k}) \geq \bar{q}_k + \underline{a} - \bar{s}. \tag{14.7}$$
If $\underline{a}$ is sufficiently close to $\bar{s}$ then x_{J_k} will be forced to be close to the global maximum of q_k over the interval $\mathbf{x}_{J_k}$, thus reducing the component $\mathbf{x}_{J_k}$ and hence the box $\mathbf{x}$. This procedure can be applied for each k in turn to get an optimally reduced box. One can similarly proceed for block-separable constraints of the form $\sum q_k(x_{J_k}) \leq \bar{a}$. (The reader might wish to reconsider the example in Section 10 in the light of the above result.)

In the separable case ($J_k = \{k\}$), computing the set of x_k with (14.7) is easy, especially for linear or quadratic q_k. If q_k is nonmonotonic, it may happen that the resulting set is disconnected; then one has to make a choice between taking its convex hull – which is an interval – or of considering splitting the box into sub-boxes corresponding to the connected components.

In the case of two-sided constraints $\sum q_k(x_{J_k}) \in \mathbf{a}$, which includes the equality constraint $\sum q_k(x_{J_k}) = q_0$ for $\mathbf{a} = q_0$, one can combine (14.2) and (14.4) using interval arithmetic as follows. (See (11.7) for the inner addition $\oplus$.)

Proposition 14.2. Suppose that

$$\mathbf{q}_k \supseteq \Box\{q_k(x_{J_k}) \mid x_{J_k} \in \mathbf{x}_{J_k}\}, \quad \mathbf{r} \supseteq \mathbf{a} - \sum_k \mathbf{q}_k. \tag{14.8}$$

(i) If $0 \notin \mathbf{r}$ then the conditions

$$x \in \mathbf{x}, \quad \sum_k q_k(x_{J_k}) \in \mathbf{a} \tag{14.9}$$

are inconsistent.

(ii) Any x satisfying (14.9) also satisfies

$$q_k(x_{J_k}) \in \mathbf{r} \oplus \mathbf{q}_k \quad \text{for all } k. \tag{14.10}$$

Proof. (14.9) implies $0 \in \mathbf{a} - \sum q_k(x_{J_k}) \subseteq \mathbf{a} - \sum \mathbf{q}_k$, hence $0 \in \mathbf{r}$. Now suppose that $0 \in \mathbf{r}$. In the notation of the previous proposition we have

$$q_k(x_{J_k}) \in [\underline{a} - \overline{s} + \overline{q}_k, \overline{a} - \underline{s} + \underline{q}_k] = [\underline{a} - \overline{s}, \overline{a} - \underline{s}] \oplus \mathbf{q}_k,$$

and since $\mathbf{r} = \mathbf{a} - [\underline{s}, \overline{s}] = [\underline{a} - \overline{s}, \overline{a} - \underline{s}]$, this implies (14.10). $\square$

Again, condition (14.10) can be used to reduce $\mathbf{x}_{J_k}$ whenever

$$\mathbf{q}_k \not\subseteq \mathbf{r} \oplus \mathbf{q}_k. \tag{14.11}$$

We give details for the most important case of quadratic (and including linear) functions, dropping indices for a moment.

Proposition 14.3. Let $\mathbf{c}$ be an interval, $a, b \in \mathbb{R}$, and put

$$\mathbf{d} := (b^2 + 4a\mathbf{c})_+, \quad \mathbf{w} := \sqrt{\mathbf{d}} \text{ (if } \mathbf{d} \neq \emptyset).$$

Then

$$\{x \in \mathbb{R} \mid ax^2 + bx \in \mathbf{c}\} = \begin{cases} \emptyset & \text{if } \mathbf{d} = \emptyset, \\ \emptyset & \text{if } a = b = 0 \notin \mathbf{c}, \\ \mathbb{R} & \text{if } a = b = 0 \in \mathbf{c}, \\ \frac{\mathbf{c}}{b} & \text{if } a = 0, \\ \frac{-b-\mathbf{w}}{2a} \cup \frac{-b+\mathbf{w}}{2a} & \text{otherwise.} \end{cases}$$

Proof. $ax^2 + bx = \tilde{c} \in \mathbf{c}$ is equivalent to $x = \tilde{c}/b$ when $a = 0$, and to $x = (-b \pm \sqrt{b^2 + 4a\tilde{c}})/2a$ otherwise; in the latter case, the expression under the square root must be nonnegative and hence lies in $\mathbf{d}$. Since the varying $\tilde{c}$ occurs in these formulas only once, the range over $\tilde{c} \in \mathbf{c}$ is given by $\mathbf{c}/b$ if $a = 0$ and by $(-b \pm \sqrt{\mathbf{d}})/2a$ otherwise (use monotonicity!). $\square$

Note that the differences in Proposition 14.1 and the numerators in Proposition 14.3 may suffer from severe cancellation of leading digits, which requires attention in an actual implementation.

In the application to reducing boxes, one must of course intersect these formulae with the original interval. If the empty set results, the subproblem corresponding to the box **x** can be eliminated. (But remember to be cautious when using unsafe interval arithmetic!) If a disjoint union of two intervals results we either split the box into two boxes corresponding to the two intervals or leave $\mathbf{x}_k$ unchanged; the first alternative is advisable only when the gap in the interval is quite large.

All reduction techniques may be used together with the technique of *shaving*, which may be seen as an adaptation of the probing technique in mixed integer programming. The idea is to try to remove a fraction of the range $[\underline{x}_i, \overline{x}_i]$ of some variable x_i by restricting the range to a small subrange $[\underline{x}_i, \xi]$ or $[\xi, \overline{x}_i]$ at one of the two end-points of that variable, and testing whether reducing the small slab obtained in this way results in an empty intersection. If this is the case, the range of x_i can be restricted to the complementary interval $[\xi, \overline{x}_i]$ and $[\underline{x}_i, \xi]$, respectively. While more expensive, it reduces the overestimation in the processing of constraints which contain a variable several times. In practice, one would perhaps try to shave away 10% of the length of an interval.

14.1. Consistency concepts

In constraint logic programming (see the book by Van Hentenryck *et al.* (1997b) and the references at the beginning of this section), there are many consistency concepts that describe the strength of various reduction techniques. Essentially, a box is *consistent* with respect to a set of reduction procedures if their application does not reduce the box. A simple recursive argument invoking the finiteness of machine-representable boxes shows that every box can be reduced to a consistent box with finitely many applications of the reduction procedures in an arbitrary order. (Depending on the rules used, the resulting reduced box – called a *fixed point* of the reduction procedures – may or may not depend on the order of applying the rules.)

From a practical point of view, it is not advisable to apply the available rules until the fixed point is reached. The reason is that frequently the first few reductions are substantial, and later ones only reduce the box by tiny fractions; the convergence speed may be arbitrarily slow. For example, for the pair of constraints $x_1 + x_2 = 0$, $x_1 - qx_2 = 0$ with $q \in]0, 1[$, the unique fixed point (with respect to the simple reduction described above) reduces the volume in each step by a factor of q. For q close to one this is very inefficient compared to, say, a linear programming relaxation (which gives the result immediately).

Thus one has to be selective in practice, using suitable strategic rules for when to use which reduction strategy. The choice is usually done by various *ad hoc* recipes that balance the likely gain and the amount of work needed. Moreover, fine-grained interaction between different computations to avoid some unnecessary computation, such as that described in Granvilliers (2001) may be decisive in getting optimal performance.

14.2. Semiseparable constraints

With some more work, the above techniques can also be utilized for semiseparable constraints. We need the following result (essentially a form of the Cauchy–Schwarz inequality).

Lemma 14.4. *If A is a rectangular matrix such that $A^T A$ is nonsingular then*
$$|u_k| \leq \sqrt{((A^T A)^{-1})_{kk}} \|Au\|_2 \quad \text{for all } u.$$

Proof. Let $A = QR$ be an orthogonal factorization of A with $Q^T Q = I$ and R square nonsingular. Then $A^T A = R^T R$ and $\|Au\|_2 = \|Ru\|_2$. Since
$$|u_k| = |(R^{-T} e^{(k)})^T Ru| \leq \|R^{-T} e^{(k)}\|_2 \|Ru\|_2,$$
$$\|R^{-T} e^{(k)}\|_2^2 = (e^{(k)})^T R^{-1} R^{-T} e^{(k)} = ((R^T R)^{-1})_{kk},$$
the assertion follows. $\square$

Now suppose that we have a semiseparable inequality of the form
$$\sum_k q_k(x_k) + (x - x^0)^T H(x - x^0) \leq \bar{a}, \tag{14.12}$$

with possibly nonsymmetric H. Using a modified Cholesky factorization (Gill and Murray 1974, Schnabel and Eskow 1990)
$$H + H^T = R^T R - D$$

with a (nonnegative) diagonal matrix D, we can rewrite (14.12) as
$$0 \leq \frac{1}{2} \|R(x - x^0)\|_2^2 \leq \bar{a} - \sum_k q_k(x_k) + \frac{1}{2}(x - x^0)^T D(x - x^0). \tag{14.13}$$

The right-hand side of (14.13) is a separable quadratic form, hence can be written as $\bar{a} - \sum \tilde{q}_k(x_k)$ with $\tilde{q}_k(x_k) = q_k(x_k) - \frac{1}{2} D_{kk}(x_k - x_k^0)^2$. Therefore, Proposition 14.1(ii) applies. Moreover, one gets the extra inequality
$$\|R(x - x^0)\|_2^2 \leq 2(\bar{a} - \underline{s}),$$

which together with the lemma gives the further inequalities

$$|x_k - x_k^0| \leq \sqrt{2(\bar{a} - \underline{s})((R^T R)^{-1})_{kk}}, \qquad (14.14)$$

which may help to reduce $\mathbf{x}_k$.

14.3. Block-separable constraints

For only block-separable constraints ($|J_k| > 1$), the q_k are multivariate, and we need to resort to suboptimal interval techniques.

How to exploit the enclosures from Proposition 14.1 and 14.2 to reduce the box depends on the special form of the q_k. In many cases, we can in turn solve the conditions directly for each variable involved, substituting an enclosing interval for all other variables.

If this is not possible directly we can use the mean value form (or another centred form) to rewrite a constraint $F_i(x) \in \mathbf{F}_i$ as

$$F_i(\xi) + F'(\mathbf{x})(x - \xi) \cap \mathbf{F}_i \neq \emptyset;$$

this is now a separable expression with interval coefficients that can be processed as above to reduce the box. This way of proceeding, dating back to Neumaier (1988), is called *conditioning* in Van Hentenryck *et al.* (1997b), and is used in the Numerica package. Similarly, by using a Taylor expansion to second order with an interval Hessian, we get a semiseparable expression with interval coefficients that can in principle be processed as above. (However, the interval coefficients cause here additional complications.) In the context of Taylor models of arbitrary order, a variation of this (with thin coefficients and an interval remainder term) has been used in the linear dominated bounder of Makino (1998); *cf.* the discussion in Makino and Berz (2003) and Neumaier (2002).

15. The cluster problem and second-order information

When programming a simple branch and bound algorithm for global optimization, one quickly notices that it is fairly easy to eliminate boxes far away from the global minimizer, while, especially in higher dimensions, there remains a large cluster of tiny boxes in a neighbourhood of the global minimizer that is difficult to eliminate. The occurrence of this situation is called the *cluster problem*. Often, algorithms try to avoid the cluster problem by providing only a Δ-optimal solution; *i.e.*, the program stops when it has shown that there is no feasible point with an objective function value of $f^{\text{best}} - \Delta$, where f^{best} is the function value of the best feasible point found so far. However, when Δ is small (as we want) then the cluster problem is still present, although to a less pronounced degree.

Kearfott and Du (1994) studied the cluster problem for unconstrained global optimization, and discovered that the source of the problem was the limited accuracy with which the function values were bounded. In particular, they showed that the cluster problem disappears if, for x in a box of diameter $O(\varepsilon)$, one can bound the overestimation of $f(x^{\text{best}}) - f(x)$ by $O(\varepsilon^3)$. Here we give a simplified version of their result.

Let $\widehat{x} \in \mathbb{R}^n$ be a global minimizer of $f(x)$, and let $\widehat{G}$ be the Hessian at $\widehat{x}$. Near the global minimizer, we have

$$f(x) = f(\widehat{x}) + \frac{1}{2}(x - \widehat{x})^T \widehat{G}(x - \widehat{x}) + O(\|x - \widehat{x}\|^3)$$

since the gradient vanishes at $\widehat{x}$. Suppose we can bound the objective function value over a box of diameter ε with an accuracy of $\Delta = K\varepsilon^{s+1}$, $s \leq 2$. Then no box of diameter ε containing a point x with $\frac{1}{2}(x - \widehat{x})^T \widehat{G}(x - \widehat{x}) + O(\|x - \widehat{x}\|^3) \leq \Delta$ can be eliminated. For sufficiently small Δ, this describes a nearly ellipsoidal region with volume proportional to $\sqrt{(2\Delta)^n / \det \widehat{G}}$, and any covering by boxes of diameter ε contains at least $\text{const} \sqrt{(2\Delta)^n / (\varepsilon^n \det \widehat{G})}$ boxes. The number of uneliminated boxes is therefore proportional to at least

$$(\text{const}/\varepsilon)^{n/2} / \sqrt{\det \widehat{G}} \quad \text{if } s = 0,$$

$$\sqrt{\text{const}^n / \det \widehat{G}} \quad \text{if } s = 1,$$

$$\sqrt{(\text{const} \cdot \varepsilon)^n / \det \widehat{G}} \quad \text{if } s = 2.$$

We see that for $s = 0$, the number grows immensely as ε gets small. For $s = 1$, the number of boxes needed – while (for small ε) essentially independent of ε – may still grow exponentially with the dimension, and it is especially large for problems where the Hessian at the solution is ill-conditioned. However, the number is guaranteed to be small (for small ε) when $s = 2$.

For pure constraint satisfaction problems, a similar cluster effect is present (Schichl and Neumaier 2004), but with order reduced by one; thus the quadratic approximation property available from methods exploiting first-order information only (such as centred forms) already avoids the cluster effect, except in degenerate cases. However, especially near poorly conditioned solutions, the size of boxes that can be eliminated is significantly larger if second-order information is used (Schichl and Neumaier 2004). In the case of nonisolated solution sets, some clustering seems unavoidable, but Lyapunov–Schmidt reduction techniques (Neumaier 2001a) might prove useful. The problem of covering nonisolated solution sets efficiently with a small number of boxes is discussed in considerable algorithmic detail by

Xuan-Ha Vu, Sam-Haroud and Silaghi (2002, 2003) for the case of pure constraint satisfaction problems; see also Chapter 7 of the COCONUT report (Bliek et al. 2001).

For constrained global optimization, similar arguments as for the unconstrained case apply in a reduced manifold with the result that, in the formulas, n must be replaced by $n - a$, where a is the maximal number of constraints active at the solution, with linearly independent constraint gradients.

Clearly, to bound the overestimation over a box of diameter $O(\varepsilon)$ by $O(\varepsilon^3)$ requires that we know the Hessian up to $O(\varepsilon)$, and that we are able to bound the deviation from a quadratic model. (Actually, the above argument shows that $o(\varepsilon^2)$ is sufficient, but this still requires the knowledge of the Hessian up to $o(1)$.) Thus it is necessary to have access to second-order information. Unfortunately, in higher dimensions, no cheap method is known that bounds function values over an arbitrary narrow box of diameter $O(\varepsilon)$ close to a minimizer by $O(\varepsilon^3)$. In a single dimension, cheap methods are known; see Cornelius and Lohner (1984) and Neumaier (1990, Section 2.4). In dimension > 1, peeling methods together with Taylor expansions work with an effort that grows like $O(n^3 \cdot 3^n)$; see the discussion in Neumaier (2002, Section 5).

Fortunately, however, it turns out that by using interval Hessian matrices (which, for 3-times differentiable functions have the required $O(\varepsilon)$ accuracy: see Neumaier (1990, Section 1.4)), there are several ways to avoid the cluster problem, at least when the global minimizer is nondegenerate, *i.e.*, satisfies the second-order sufficient conditions for a local minimizer.

15.1. Using Hessian information

Explicit global Hessian information can be used, as in GlobSol (Kearfott 1996b) and Numerica (Van Hentenryck et al. 1997b), by interval Newton methods (see Section 11) applied to the Karush–John conditions discussed in Section 5. These may verify the existence of a unique solution of the Karush–John conditions (Theorem 5.1 and equation (5.4)) in some box around the best point found, and hence allow us to shrink that box to a single point.

Alternatively, we may use global Hessian information to verify the second-order sufficient conditions for a global minimizer given in Neumaier (1996). They apply to smooth nonlinear programs of the form

$$\min \ f(x) \\ \text{s.t.} \quad x \in \mathbf{x}, \ F(x) = 0. \tag{15.1}$$

Thus it is necessary to introduce slack variables to rewrite general inequality constraints as equality constraints. The sufficient condition is as follows.

Theorem 15.1. Let $\widehat{x}$ be a Kuhn–Tucker point for the nonlinear program (15.1), with associated multiplier z, and let

$$y := f'(\widehat{x})^T - F'(\widehat{x})^T z, \tag{15.2}$$

$$D = \mathrm{Diag}\left(\sqrt{\frac{2|y_1|}{\overline{x}_1 - \underline{x}_1}}, \ldots, \sqrt{\frac{2|y_n|}{\overline{x}_n - \underline{x}_n}}\right). \tag{15.3}$$

If, for some continuously differentiable function $\varphi : \mathbb{R}^m \to \mathbb{R}$ with

$$\varphi(0) = 0, \quad \varphi'(0) = z^T, \tag{15.4}$$

the *generalized augmented Lagrangian*

$$\widehat{L}(x) := f(x) - \varphi(F(x)) + \frac{1}{2}\|D(x - \widehat{x})\|_2^2 \tag{15.5}$$

is convex in $[u, v]$, then $\widehat{x}$ is a global solution of (15.1). Moreover, if $\widehat{L}(x)$ is strictly convex in $[u, v]$, this solution is unique.

A choice for φ that works in some neighbourhood of a strong global minimizer (*i.e.*, one in which sufficient second-order conditions for local optimality hold) is given in Neumaier (1996), together with further implementation hints. The convexity can be checked by means of interval arithmetic; see Section 11. If these conditions hold in some box, we can shrink this box to a single point.

We can use any of these techniques to construct boxes **y** that are guaranteed to contain no global minimizer unless already detected, resulting in exclusion constraints. An *exclusion constraint* is a constraint of the form

$$x \notin \mathbf{y}.$$

It can be used to reduce an arbitrary box **x** by intersecting it with **y** and taking the interval hull, which may result in a smaller box. If there were no reduction but the intersection were strictly contained in **x**, then we might also want to resort to multisection: *cf.* (12.4). Interesting exclusion boxes are those that are constructed around local minimizers, since this helps fighting the cluster problem.

It is possible (though probably not most efficient) to base global optimization algorithms on exclusion methods alone; see the work of Georg *et al.* (Allgower, Erdmann and Georg 2002, Georg 2001, Georg 2002), who also give associated complexity results.

15.2. Backboxing

Whenever we have a tentative approximate global minimizer $\tilde{x}$, we try to find simultaneously a large box **x** and a tiny box **z** such that any global minimizer $\widehat{x} \in \mathbf{x}$ satisfies $\widehat{x} \in \mathbf{z}$. This allows us to use **x** as an exclusion region while **z** is stored in an output list as a box containing a putative minimizer: this is termed *backboxing*. (After terminating the branching process, these boxes need to be checked again for possible elimination.)

Since we expect that $\tilde{x}$ has a function value optimal within $O(\varepsilon)$, but knowing that this only enforces that $\tilde{x}$ has an accuracy of $O(\sqrt{\varepsilon})$ (possibly less in the case of singular Hessians), we start with a box

$$\mathbf{x} = [\tilde{x} - \sqrt{\varepsilon}u, \tilde{x} + \sqrt{\varepsilon}u]$$

for some vector u reflecting the scaling of the variables, and apply the available reduction techniques until no significant improvement results. Call the resulting box **z**. If second-order techniques are used to do the box reduction, then **z** is usually a tiny box or empty.

If **z** is empty, $\tilde{x}$ was not a good approximation but we know that **x** contains no solution. If **z** is nonempty, it is likely that **z** contains a solution. Indeed this is always the case if *only* interval Newton-like reduction techniques are used and $\mathbf{z} \subseteq \text{int } \mathbf{x}$. (This requires some qualifying conditions that ensure that we can verify sufficient existence conditions such as those in Neumaier (1990, Chapter 5.3–4).) Thus we may store **z** in a list of output boxes together with a flag whether existence (and possibly uniqueness) was verified.

If **z** is still a box of significant size, we must have been close to a degeneracy; splitting would probably not improve this and lead to an exponential number of boxes; thus it is preferable to put this box also in the list of output boxes to indicate that a low resolution candidate for a solution has been found. (This way of handling degeneracies is due to Kearfott (1996a).)

No matter what case we have been in, we always know that **x** cannot contain a solution not yet in the output list. Therefore, we may add the exclusion constraint $x \notin \mathbf{x}$ to the problem description. However, one can often make **x** even bigger. So we try recursively

$$\mathbf{x}^0 = \mathbf{x}, \ \mathbf{z}^0 = \mathbf{z}, \quad \text{but } \mathbf{z}^0 = \text{mid}(\mathbf{x}) \text{ if } \mathbf{z} = \emptyset,$$
$$\mathbf{x}^l = 2\mathbf{x}^{l-1} \ominus \mathbf{z}^{l-1}, \quad \mathbf{z}^l = \text{reduce}(\mathbf{x}^l);$$

using the available ways of reducing $\mathbf{x}^l$, stopping when $\mathbf{z}^l \subseteq \mathbf{x}^{l-1}$ or $\mathbf{z}^l = \mathbf{x}^l$. (For the inner subtraction $\ominus$, see (11.8).) Then we have the generally stronger new exclusion constraint $x \notin \mathbf{x}^l$. (This way of generating exclusion constraints, using interval Newton methods, is due to Van Iwaarden (1986), who calls the technique *backboxing*, and is part of GlobSol (Kearfott 1996b).) Recent methods by Schichl and Neumaier (2004) for constructing large exclusion boxes can be combined with this iterative approach.

15.3. Finite termination

Closely related to the cluster problem is the question of finite termination, *i.e.*, whether branch and bound algorithms find (assuming exact arithmetic) a global optimizer with a finite amount of branching only. This is not easy to achieve, and in practice, most algorithms are content with working towards ε-optimality, *i.e.*, finding a (nearly) feasible point within ε of the true but unknown optimal function value.

Theoretical finite termination guarantees are available only for problems where the optimum is attained at extreme points (Al-Khayyal and Sherali 2000, Shectman and Sahinidis 1998). However, in practice, algorithms based on the explicit use of second-order interval information (either via interval Newton operators or via second-order sufficient conditions) have finite termination behaviour on problems with a nondegenerate global minimizer, and it is likely that this can be proved theoretically.

In the case of degeneracies, behaviour of branch and bound methods can become arbitrarily poor. However, the situation may improve in cases where the degeneracy can be removed by identifying and eliminating redundant constraints causing the degeneracy. To do this rigorously requires care; see Huyer and Neumaier (2004) for first results in this direction.

16. Linear and convex relaxations

One of the highly developed sides of global optimization is the use of *linear and convex relaxations* to find a lower bound for the value of the objective function, which makes it possible to discard boxes where this lower bound is larger than the function value f^{best} of the best feasible point found so far. The details are well covered in several books (Floudas 1995, 1999, Tawarmalani and Sahinidis 2002*b*), so we are brief here and only describe the basic issues and recent extensions that are not widely known.

16.1. Reformulation-linearization

McCormick (1976) introduced the notion of a factorable function (composed of finitely many unary or binary operations), and constructed non-smooth relations for such functions. Kearfott (1991), and perhaps others before him, noticed that by introducing intermediate variables, every factorable optimization problem can be rewritten in a form in which all constraints are unary, $z = \varphi(x)$, or binary, $z = x \circ y$. Independently, Ryoo and Sahinidis (1996) proposed using in place of these constraints implied linear constraints (so-called *linear relaxations*) to generate a set of linear inequalities defining a polyhedral outer approximation. Since the objective can be represented by a single variable and another constraint, this allows one to find a linear programming relaxation for arbitrary factorable optimization problems.

Linear relaxations for unary operations are easily found by a simple graphical analysis of the various elementary functions. In particular, for a convex function, the secant between the end-points of the graph is an overestimate, and any tangent is an underestimate; frequently, taking the two tangents at the end-points is already quite useful. For concave functions, the reverse situation holds, and in the general case one may also need to consider bitangents, and tangent secants. Since powers can be written in terms of exp, log and the product, the only binary operations that need to be analysed are products and quotients.

Assuming that bounds $x \in \mathbf{x}$, $y \in \mathbf{y}$ for the factors are available, McCormick proposed for the product $z = xy$ the relaxations

$$\underline{y}x + \underline{x}y - \underline{x}\,\underline{y} \leq z \leq \underline{y}x + \overline{x}y - \overline{x}\,\underline{y},$$
$$\overline{y}x + \overline{x}y - \overline{x}\,\overline{y} \leq z \leq \overline{y}x + \underline{x}y - \underline{x}\,\overline{y},$$

which follow immediately from $(x - \underline{x})(y - \underline{y}) \geq 0$ and three similar inequalities. Al-Khayyal and Falk (1983) showed later that these inequalities are indeed best-possible in the sense that any other generally valid linear inequality is a consequence of these and the bound constraints. (One says they form the *convex and concave envelope*.)

For the quotient $x = z/y$, exactly the same formulas are valid with $\mathbf{x} = \mathbf{z}/\mathbf{y}$, but remarkably, one does not get the envelope in this way. For example, the following inequality, due to Zamora and Grossmann (1999), is not implied.

Proposition 16.1. *Let $\mathbf{x}, \mathbf{y}, \mathbf{z}$ be nonnegative intervals. If $x \in \mathbf{x}$, $y \in \mathbf{y}$, and $z = xy \in \mathbf{z}$ then*

$$xy \geq \left(\frac{z + \sqrt{\overline{z}\,\underline{z}}}{\sqrt{\overline{z}} + \sqrt{\underline{z}}}\right)^2. \tag{16.1}$$

Inequality (16.1) describes a convex set in the nonnegative orthant of $\mathbb{R}^3$, although the inequality itself is not convex. However, it is the prototype of a convex conic constraint (see below) and can be exploited by solvers for second-order cone programs. Therefore, adding this constraint gives a relaxation that may be tighter than the McCormick relaxation. The general formulas for the convex envelope of a quotient, derived explicitly in Tawarmalani and Sahinidis (2002b), are quite complicated.

Crama (1993) showed that the following bounds define the optimal convex relaxation of a product of factors bounded in $[0,1]$.

Proposition 16.2. *If $x_i \in [0,1]$ $(i = 1:n)$ and $z = x_1 \cdots x_n$ then*

$$1 - n + \sum_{k=1}^{n} x_k \leq z, \quad 0 \leq z \leq x_i \leq 1 \ (i = 1:n). \tag{16.2}$$

More generally, arbitrary multilinear functions have an optimal convex relaxation (the envelope) defined by finitely many linear inequalities. For this result, and for other methods for getting linear relaxations of nonconvex programs based on the *reformulation-linearization* technique: see Rikun (1997), Sherali (1997), Sherali and Tuncbilec (1995, 1997a, 1997b), Al-Khayyal, Larsen and van Voorhis (1995), and Audet, Hansen, Jaumard and Savard (2000). Also related is the *lift-and-project* technique in mixed integer linear programming; see, *e.g.*, Balas, Ceria and Cornuejols (1993).

Note that this approach using the factorable or a nearly factorable form generally results in problems with many more variables than in the original problem formulation. Nevertheless, since the resulting linear or convex programs are extremely sparse, the technique can be very useful, especially for larger problems. In particular, this is the main workhorse of the global optimization packages BARON (Tawarmalani and Sahinidis 2002b) and LINGO (Gau and Schrage 2004). Because linear programming solvers are currently much more reliable and faster than general convex solvers, the convex envelopes used in BARON are in fact approximated by a number of linear constraints computed adaptively with a variant of the sandwich algorithm of Rote et al. (Burkard, Hamacher and Rote 1992, Rote 1992).

Independently of the way a linear relaxation is produced (see below for alternatives that work without additional variables), the information in the linear relaxation can be exploited not only to get lower bounds on the objective or to eliminate a subproblem, but also to reduce the box. Cheap marginals-based range reduction techniques using Lagrangian multipliers are described in Tawarmalani and Sahinidis (2002b) and are implemented in BARON. Recent results of Lebbah et al. (2004) and Lebbah, Rueher and Michel (2002) show that the more expensive approach of minimizing and maximizing each variable with respect to a linear relaxation (which BARON 5.0 did only at the root node of the branch tree) may give a significant speed-up on difficult constraint satisfaction problems, and are now part of the default strategy in BARON 6.0.

16.2. Semidefinite relaxations

Starting quite recently, a large number of papers appeared that propose the use of *semidefinite relaxations*, or *convex conic relaxations*, to solve polynomial constraint satisfaction and global optimization problems: see, *e.g.*, Chesi and Garulli (2001), Jibetean and De Klerk (2003), Fujie and Kojima (1997), Kim and Kojima (2001, 2003), Kim, Kojima and Waki (2003), Kojima, Kim and Waki (2003), Kojima and Tuncel (2000), Lasserre (2001), Meziat (2003), Gatermann and Parrilo (2004), Parrilo (2003), Parrilo and Lall (2003), Parrilo and Sturmfels (2003). These techniques are implemented in two software packages, GloptiPoly (Henrion and Lasserre 2003a, 2003b, 2003c) and SOSTOOLS (Prajna et al. 2002). Being developed completely independently from the mainstream in global optimization, these packages do not incorporate any of the other global techniques, and hence are currently restricted to problems with few variables (say, below 20). But since they are able to solve many of these problems to global optimality without doing any branching, their combination with the other techniques, in particular with branch and bound, appears to be highly promising.

The background of these methods is that constraints of the form

$$\sum_k x_k A_k \text{ is positive semidefinite,} \qquad (16.3)$$

where the A_k are symmetric (or complex Hermitian) matrices, so-called *semidefinite constraints*, define convex sets, and that constraints of the form

$$\|Ax - b\|_2 \leq a^T x + \alpha$$

or

$$\|Ax - b\|_2^2 \leq x_i x_j, \quad x_i, x_j \in \mathbb{R}_+,$$

so-called *second-order cone constraints*, describe convex conic sections. Problems with a linear or convex quadratic objective and an arbitrary number of such constraints (in addition to linear constraints) can be efficiently solved using interior point methods. Therefore, convex conic and semidefinite relaxations of nonlinear constraints can be efficiently exploited. Books and surveys emphasizing the nonlinear case include Alizadeh (2003), de Klerk (2002), Lobo, Vandenberghe, Boyd and Lebret (1998), Todd (2001), Vandenberghe and Boyd (1996) and Wolkowicz, Saigal and Vandenberghe (2000); for software, see the semidefinite programming homepage (Helmberg 2003) and the package SeDuMi (see SeDuMi (2001) and Sturm (1999)), on which both GloptiPoly and SOSTOOLS are based.

The basic idea behind semidefinite relaxations is the observation that, given any set of basis functions $\varphi_i(x)$ and any nonnegative weight function $w(x)$, the matrix M with components

$$M_{ik} = w(x)\varphi_i(x)\varphi_k(x)$$

is always symmetric and positive semidefinite. If the φ_i and w are polynomials then the entries of M are also polynomials, and by introducing auxiliary variables z_j for the elements of a basis of polynomials of sufficiently high degree, one can write both the entries of M and any polynomial objective or constraint as a linear combination of the z_j. The condition that M is positive semidefinite therefore gives rise to a semidefinite constraint. Possible choices for $w(x)$ can easily be made up from the constraints. Moreover, given an equality constraint, any multiple by a polynomial is another equality constraint, and given two inequality constraints $u(x) \geq 0$ and $v(x) \geq 0$, their product is again such a constraint. Thus lots of additional polynomial constraints can be generated and used. Results from algebraic geometry can then be invoked to show that infeasibility and ε-optimality can always be achieved by using sufficiently high degrees, without the need for any problem splitting. Apparently, in many cases, relatively low degrees suffice, which is fortunate since the number of intermediate variables would otherwise become excessively large. Moreover, problem symmetry can be exploited by using basis sets with corresponding symmetry properties (Gatermann and Parrilo 2004).

The conic and semidefinite relaxations produced in this way also result in problems with many more variables than in the original problem formulation, but since semidefinite relaxations are often much stronger than linear relaxations, the effort required to solve these large problems may be well spent if a subproblem is solved without the need to split it into many smaller pieces. Since problems with semidefinite constraints involving larger matrices are more expensive to solve than those with convex conic constraints, the latter are in principle preferable, but conclusive results on the best way of using or combining the various possible relaxations are not yet available.

For semidefinite relaxations of certain fractional functions see Tawarmalani and Sahinidis (2001, 2002a, 2002b).

16.3. Relaxations without extra variables

In place of introducing additional variables for nonlinear intermediate expressions, it is also possible to relax the original constraints directly. Apart from McCormick's non-smooth convex relations (McCormick 1976), which are difficult to use, this can be done in two different ways.

The first possibility is to write the constraints as a difference of convex functions (DC representation). The package αBB (see Adjiman et al. (1996, 1998a, 1998b), Adjiman and Floudas (1996), Androulakis et al. (1995)) uses DC-techniques, by separating in each inequality constraint $h(x) \leq 0$ a recognizable linear, convex or concave parts from a 'general' remainder. Linear and convex parts are kept, concave parts are overestimated by secant-type constructions, and general terms are made convex by adding a nonpositive separable quadratic function. This ensures that a convex underestimating inequality results. More specifically, if $f(x)$ is twice continuously differentiable at all x in a neighbourhood of a box **x** and D is a diagonal matrix with nonnegative entries, then

$$f_{\rm rel}(x) := f(x) + \frac{1}{2}(x - \underline{x})^T D(x - \overline{x})$$

is an underestimator of $f(x)$ on the box, and the amount of underestimation is bounded by

$$|f_{\rm rel}(x) - f(x)| \leq \frac{1}{8} \operatorname{rad} \mathbf{x}^T D \operatorname{rad} \mathbf{x}, \qquad (16.4)$$

attained at the midpoint. (At the vertices there is no overestimation.) If the Hessian $G(x)$ lies in the interval matrix $\mathbf{G}$ for all $x \in \mathbf{x}$ (such a $\mathbf{G}$ can be found by interval evaluation of the Hessian, e.g., using automatic differentiation) and all symmetric matrices in $\mathbf{G} + D$ are positive semidefinite then f_{rel} is convex. The latter condition can be checked as in Theorem 11.1; the difficulty is to choose D in such a way that this condition holds and the underestimation bound in (16.4) is kept small but the work in getting D remains reasonable (Adjiman et al. 1998a, 1998b). Recent, more advanced convexity-enforcing corrections are discussed in Akrotirianakis and Floudas (2004).

More general DC-techniques are treated extensively from a mostly theoretical point of view in the book by Horst and Tuy (1990); see also the overview in Tuy (1995). Apart from what is used in αBB (and described above), these techniques have not materialized in available codes; however, see, e.g., Le Thi Hoai An and Pham Dinh Tao (1998) for some recent numerical results.

General techniques for recognizing convexity automatically are discussed in forthcoming work of Fourer (2003) and Maheshwari, Neumaier and Schichl (2003). Other questions related to the semidefiniteness of an interval matrix are discussed in Jaulin and Henrion (2004).

The second possibility is to use centred forms. The Frontline Interval Global Solver constructs linear enclosures based on a centred form (in fact a first-order Taylor form),

$$f(x) \in \mathbf{f} + c^T(x - z), \tag{16.5}$$

using forward propagation in an automatic differentiation-like manner, described in Kolev and Nenov (2001). Since the coefficients of the linear term in (16.5) are real numbers, this directly gives two parallel linear functions which underestimate and overestimate $f(x)$:

$$\underline{f} + c^T(x - z) \le f(x) \le \overline{f} + c^T(x - z).$$

The COCONUT environment constructs a centred form using slopes and automatic differentiation-like backward propagation, according to formulas given in Schichl and Neumaier (2003), from which linear enclosures are constructed. Indeed, given a centred form

$$f(x) = \tilde{f} + \tilde{s}^T(x - z), \quad \tilde{f} \in \mathbf{f}, \ \tilde{s} \in \mathbf{s},$$

we have the linear underestimator

$$f(x) \ge \gamma + c^T(x - \underline{x}),$$

where

$$\gamma = \underline{f} + \overline{s}^T(\underline{x} - z), \quad c_i = \frac{\underline{s}_i(\overline{x}_i - z_i) - \overline{s}_i(\underline{x}_i - z_i)}{\overline{x}_i - \underline{x}_i}.$$

A similar formula provides a linear overestimator. Geometrically, the formulas amount to enclosing the double cone defined by the centred form by a pair of hyperplanes; since linear functions are separable, the formulas derived from an analysis of the univariate case can be applied componentwise.

17. Semilinear constraints and MILP

Let us call a constraint *semilinear* if, for arguments x in a *bounded* box **x**, it is equivalent to a finite list of linear constraints and integer constraints; usually the latter involve additional auxiliary variables. The objective function $f(x)$ is called *semilinear* if the inequality $f(x) \le x_0$, where x_0 is an additional variable, is semilinear. A *semilinear program* is an optimization problem with a semilinear objective function and a bounded feasible domain defined by semilinear constraints only. Since we can rewrite an arbitrary global optimization problem

$$\min \quad f(x)$$
$$\text{s.t.} \quad x \in C$$

in the form

$$\min \quad x_0$$
$$\text{s.t.} \quad x \in C, \ f(x) \le x_0,$$

it is clear from the definition that any semilinear program can be rewritten as a mixed integer linear program by the introduction of additional variables.

The remarkable fact that every factorable optimization problem can be arbitrarily closely approximated by semilinear programs (see Section 18) implies that we can use MILP software to obtain arbitrarily good approximate solutions of factorable optimization problems. To make this observation computationally useful we need to handle two tasks.

(i) Find interesting classes of semilinear constraints and constructive procedures for translating such constraints into linear and integer constraints.

(ii) Show how to approximate factorable constraints by semilinear constraints: see Section 18.

In this section we look at task (i). This is in principle well known, but usually considered to be part of the modelling process. For good overviews of the modelling related issues see, *e.g.*, Floudas (1995, Section 7.4), Williams (1999) and (in German) Kallrath (2002). Here we simply give the underlying mathematical substance.

All linear constraints and integer constraints are trivially semilinear. A *binary constraint*

$$z \in \{0, 1\}$$

is semilinear, since it can be written in the equivalent form

$$z \in [0, 1], \quad z \in \mathbb{Z}.$$

We call a list x_K of variables constrained by

$$\sum_{k \in K} x_k = 1, \quad x_k \in \{0, 1\} \; (k \in K), \tag{17.1}$$

where K is some index set, a *binary special ordered set* (BSOS) (*cf.* Dantzig et al. (1958), Beale and Tomlin (1970)). Clearly, the constraint

$$x_K \text{ is a BSOS} \tag{17.2}$$

is also semilinear. Because (17.1) can hold only if all but one of the x_k ($k \in K$) vanish, (17.2) is equivalent to requiring that

$$x_K = e^{(k)} \quad \text{for some } k, \tag{17.3}$$

where $e^{(k)}$ is the unit vector with a one in position k and zeros elsewhere. (A binary special ordered set of size two is just a pair of complementary binary variables, and one of its variables is redundant.) Note that a BSOS is a special ordered set of type 1, and can be handled efficiently by most MILP codes. Since special ordered sets, defined more generally as sets of variables such that at most one – type 1 – or two (which then must be adjacent) – type 2 – are nonzero, are ubiquitous in MILP formulations, any MILP solver has special facilities to make efficient use of special ordered sets.

Many techniques for translating semilinear constraints are consequences of the following basic result.

Theorem 17.1. Let $F_k : C_0 \to \mathbb{R}^{m_k} \; (k = 1, \ldots, d)$ be scalar- or vector-valued functions such that

$$F_k(x) \geq \underline{F}_k \quad \text{for all } x \in C_0 \tag{17.4}$$

with $\underline{F}_k \in \mathbb{R}^{m_k}$. Then there is a point $x \in C_0$ such that

$$F_1(x) \geq 0 \; \vee \; \cdots \; \vee \; F_d(x) \geq 0 \tag{17.5}$$

if and only if there are $z \in \mathbb{R}^d$ and $x \in C_0$ such that

$$z \text{ is a BSOS}, \tag{17.6}$$

$$F_k(x) \geq \underline{F}_k(1 - z_k) \quad \text{for all } k = 1, \ldots, d. \tag{17.7}$$

(The symbol $\vee$ denotes the logical operation **or**. The operation **and** is implicitly given by the comma, and we follow the convention that **and** is binding more strongly than **or**.)

Proof. If (17.5) holds then $F_k(x) \geq 0$ for some k, and $z = e^{(k)}$ satisfies (17.6) and (17.7). Conversely, if (17.6) holds then $z = e^{(k)}$ for some k, and (17.7) implies $F_k(x) \geq 0$; the other constraints in (17.7) are automatically satisfied because of (17.4). □

Note that (17.4) can always be satisfied if C_0 is bounded and the F_k are continuous.

A constraint of the form (17.5) is called a *disjunctive constraint*. The theorem implies that *linear* disjunctive constraints, where all $F_k(x)$ are affine functions of x, are semilinear if the F_k have *known*, finite lower bounds on the feasible domain (*bound qualification*), since then (17.7) consists of linear constraints. In the following, *we shall always silently assume the bound qualification*. (In practice, this is usually enforced where necessary by ad hoc 'big M' domain restrictions. In rigorous solvers, this is of course forbidden.)

More generally, linear disjunctive constraints of the form

$$A_1 x \in \mathbf{b}_1 \quad \vee \quad \cdots \quad \vee \quad A_d x \in \mathbf{b}_d \tag{17.8}$$

are semilinear, since we can rewrite each $A_k x \in \mathbf{b}_k$ in the form

$$\begin{pmatrix} A_k x - \underline{b}_k \\ \overline{b}_k - A_k x \end{pmatrix} \geq 0.$$

Note that we can rewrite (17.8) in the equivalent form

$$A_k x \in \mathbf{b}_k \quad \text{for some} \quad k \in \{1 : d\}. \tag{17.9}$$

Since many practically relevant constraints can be cast in the form (17.8), this makes the theorem a very useful tool for recognizing semilinear constraints and translating them into a MILP formulation. (There is also an extended literature on disjunctive programming not based on transformations to MILP; for pointers see Balas (1979), Jeroslow (1977), Sherali and Shetty (1980).)

For example, *semicontinuous (semi-integer) variables* are variables x_k constrained by

$$x_k = 0 \quad \vee \quad x_k \in \mathbf{a} \tag{17.10}$$

and

$$x_k = 0 \quad \vee \quad x_k \in \mathbf{a}, \quad x_k \in \mathbb{Z}, \tag{17.11}$$

respectively, which are semilinear constraints.

A *numerical special ordered set* (NSOS) is a vector $\lambda \in \mathbb{R}^d$ such that

$$\lambda \geq 0, \quad \sum_{k=1}^{d} \lambda_k = 1,$$

at most two λ_k are nonzero, and nonzero λ_k must have adjacent indices.

Since the latter condition can be formulated as

$$\lambda_k + \lambda_{k+1} = 1 \quad \text{for some } k,$$

it is disjunctive; hence the constraint

$$x_K \text{ is an NSOS} \tag{17.12}$$

is semilinear. Note that an NSOS is a special ordered set of type 2, and can be handled efficiently by most MILP codes.

An *exclusion constraint* of the form

$$x \notin \text{int } \mathbf{x}, \tag{17.13}$$

where $\mathbf{x}$ is a box, is semilinear since it is a disjunction of the constraints

$$x_k \leq \underline{x}_k \quad \vee \quad x_k \geq \overline{x}_k.$$

Propositional constraints are semilinear: if x_k denotes a binary variable which has the value 1 if and only if a corresponding logical proposition P_k holds, then

$$P_1 \vee \cdots \vee P_K \quad \text{iff} \quad x_1 + \cdots + x_K \geq 1,$$
$$P_1 \wedge \cdots \wedge P_K \quad \text{iff} \quad x_k = 1 \quad \text{for } k = 1:K,$$
$$P_1 \Leftrightarrow P_2 \quad \text{iff} \quad x_1 = x_2,$$
$$P_1 \Rightarrow P_2 \quad \text{iff} \quad x_1 \leq x_2,$$
$$P_1 \vee \cdots \vee P_K \Rightarrow P_{K+1} \vee \cdots \vee P_L \quad \text{iff} \quad x_k \leq x_{K+1} + \cdots + x_L \quad \text{for } k = 1:K.$$

Conditional linear constraints of the form

$$Ax \in \mathbf{a} \quad \text{if } Bx < b \tag{17.14}$$

are semilinear since (17.14) is equivalent to

$$Ax \in \mathbf{a} \quad \vee \quad (Bx)_1 \geq b_1 \quad \vee \quad \cdots \quad \vee \quad (Bx)_d \geq b_d,$$

where d is the number of rows of B and b. (Conditional linear constraints with $=$ or $\leq$ in place of $<$ in (17.14) are apparently not semilinear in general since their disjunctive form contains strict inequalities, which – according to our definition – are not regarded as linear constraints. However, conditional linear constraints where the condition involves only integer variables and rational coefficients are semilinear since the condition can be replaced by an equivalent strict inequality condition.)

Certain *minimum and maximum constraints* are also semilinear. A constraint of the form

$$a^T x \leq \min_{i=1:d}(Ax - b)_i \tag{17.15}$$

is equivalent to the linear constraints

$$a^T x \leq (Ax - b)_i \quad \text{for } i = 1:d.$$

The reverse constraint
$$a^T x \geq \min_{i=1:d}(Ax-b)_i \quad (17.16)$$
is equivalent to the linear disjunctive constraint
$$a^T x \geq (Ax-b)_1 \vee \cdots \vee a^T x \geq (Ax-b)_d.$$
Similarly, a constraint of the form
$$a^T x \geq \max_{i=1:d}(Ax-b)_i \quad (17.17)$$
is equivalent to the linear constraints
$$a^T x \geq (Ax-b)_i \quad \text{for} \quad i=1:d,$$
and the reverse constraint
$$a^T x \leq \max_{i=1:d}(Ax-b)_i \quad (17.18)$$
is equivalent to the linear disjunctive constraint
$$a^T x \leq (Ax-b)_1 \vee \cdots \vee a^T x \leq (Ax-b)_d.$$
The constraints
$$a^T x = \min_{i=1:d}(Ax-b)_i, \quad (17.19)$$
$$a^T x = \max_{i=1:d}(Ax-b)_i \quad (17.20)$$
are also semilinear, since they are equivalent to (17.15), (17.16) and (17.17), (17.18), respectively. In particular, *linear complementarity constraints* (Billups and Murty 2000), defined by
$$\min(a^T x - \alpha, b^T x - \beta) = 0 \quad (17.21)$$
are semilinear. Their MILP reformulation needs a single binary variable only since the associated BSOS has size two.

Linear complementarity constraints arise in bilevel programming (see, e.g., Ben-Ayed (1993), Luo, Pang and Ralph (1996), Outrata, Kočvara and Zowe (1998), Shimizu, Ishizuka and Bard (1997), Vicente and Calamai (1994)), in which the inner optimization problem is a linear program. See, e.g., Grossmann and Floudas (1987) for solving bilevel programs as mixed integer problems.

Constraints of the form
$$a^T x - \alpha \leq |b^T x - \beta|, \quad (17.22)$$
$$a^T x - \alpha = |b^T x - \beta|, \quad (17.23)$$
$$a^T x - \alpha \geq |b^T x - \beta| \quad (17.24)$$

are semilinear since we can write the absolute value in the form $|b^T x - \beta| = \max(\beta - b^T x, b^T x - \beta)$; again a single binary variable suffices for the MILP formulation. In particular, a constraint

$$\alpha \leq |x| \leq \beta \tag{17.25}$$

can be modelled as

$$(\alpha + \beta)z - \beta \leq x \leq (\alpha + \beta)z - \alpha, \quad z \in \{0, 1\}.$$

Certain other *piecewise linear constraints* are also semilinear. Of particular interest are those of the form

$$a^T x \leq \varphi(x_i), \tag{17.26}$$

where φ is a continuous, piecewise linear function of a *single* variable with a finite number of derivative discontinuities. Let $\xi_0 < \xi_1 < \cdots < \xi_d$ be a list of *nodes* such that $x_i \in [\xi_0, \xi_d]$ and φ is linear in each interval $[\xi_{k-1}, \xi_k]$. Then

$$\varphi(\xi) = \varphi_k + \varphi'_k(\xi - \xi_k) \quad \text{for } \xi \in [\xi_{k-1}, \xi_k], \tag{17.27}$$

where

$$\varphi_k = \varphi(\xi_k), \quad \varphi'_k = \frac{\varphi(\xi_k) - \varphi(\xi_{k-1})}{\xi_k - \xi_{k-1}}.$$

Therefore, (17.26) can be rewritten as a disjunction of the d constraints

$$x_i \in [\xi_{k-1}, \xi_k], \quad a^T x \leq \varphi_k + \varphi'_k(x_i - \xi_k)$$

for $k = 1, \ldots, d$. Since these are linear constraints, (17.26) is equivalent to a linear disjunctive constraint. The constraints

$$a^T x \geq \varphi(x_i), \tag{17.28}$$
$$a^T x = \varphi(x_i), \tag{17.29}$$

are semilinear by the same argument, with $\geq$ or $=$ in place of $\leq$.

Piecewise linear constraints may also be modelled by NSOS (*cf.* (17.12)); see Beale (1988, Section 10.3) and Beale and Forrest (1976), Beale and Tomlin (1970), Burkard et al. (1992), Dantzig et al. (1958) and Tomlin (1970). Indeed, if $\varphi(x)$ is piecewise linear with nodes $\xi_{1:d}$ and corresponding function values $\varphi_k = \varphi(\xi_k)$ then we may write an arbitrary argument x as

$$x = \sum \xi_k \lambda_k, \quad \lambda \text{ is an NSOS}, \tag{17.30}$$

and find

$$\varphi(x) = \sum \varphi_k \lambda_k.$$

Therefore, if we add the semilinear constraints (17.30), we may replace each occurrence of $\varphi(x)$ by $\sum \varphi_k \lambda_k$. This even works for unbounded variables and for general separable constraints $\sum \varphi_l(x_l) \in \mathbf{a}$ with piecewise linear φ_l. Many modern MILP programs have special features that allow them to handle piecewise linear constraints using special ordered sets of type 2.

Many combinatorial constraints are semilinear. For example, *all-different constraints* of the form

$$\text{the components of } x_K \text{ are distinct integers} \tag{17.31}$$

are semilinear, since we can rewrite them as

$$x_k \in \mathbb{Z} \quad \text{for } k \in K; \quad |x_j - x_k| \geq 1 \quad \text{for } j, k \in K, \; j \neq k.$$

A *cardinality constraint*

$$\text{the number of nonzero } x_k \; (k \in K) \text{ is in } \mathbf{s}$$

is semilinear if we know that x_K is integral and nonnegative. Indeed, an equivalent condition is the existence of binary numbers z_k such that

$$z_k = 1 \quad \text{if } x_k > 0,$$
$$z_k = 0 \quad \text{if } x_k < 1,$$
$$\sum_{k \in K} z_k \in \mathbf{s},$$

and these are semilinear constraints.

Cardinality rules (Yan and Hooker 1999), *i.e.*, constraints of the form

$$\geq j \text{ components of } x_J \text{ equal } 1 \quad \Rightarrow \quad \geq k \text{ components of } x_K \text{ equal } 1$$

for binary $x_{J \cup K}$, can clearly be written in terms of cardinality constraints and hence are semilinear, too.

18. Semilinear relaxations

The preceding results are of importance for general global optimization since every factorable global optimization problem can be approximated arbitrarily well by semilinear programs. Furthermore, these approximations can be made in a way to provide rigorous relaxations, so that solving the resulting semilinear programs after a MILP reformulation can be used to obtain lower bounds in a branch and bound scheme.

The ideas go back to Markowitz and Manne (1957) and Dantzig (1960) for approximate separable nonconvex programming using piecewise linear constraints. (A Lagrangian method by Falk and Soland (1969) gives piecewise linear relaxations, but in general, these do not yield arbitrarily good approximations.) With a trick due to Pardalos and Rosen (1987) (for the special case of indefinite quadratic programs, but not in the context of approximations) that allows one to transform multivariate quadratic expressions into separable form, everything extends easily to the semiseparable case; see (18.3) below. For indefinite quadratic programs, this is discussed in detail in Horst et al. (1995).

With a suitable reformulation, arbitrary factorable optimization problems (and many nonfactorable ones) can be rewritten in such a way that the objective function is linear and all constraints are either semilinear, or of the form (17.26), (17.28), (17.29) with continuous functions of a single variable. To see this, we introduce an auxiliary variable for every intermediate result; then the objective function is just a variable, hence linear, and the constraints are simple constraints or equations involving a single operation only,

$$x_k = \varphi(x_i), \tag{18.1}$$
$$x_k = x_i \circ x_j \quad (\circ \in \{+, -, *, /, \hat{\ }\}). \tag{18.2}$$

The problem formulation in terms of constraints of the form (18.1) and (18.2) together with a simple objective min $\pm x_i$ and simple bounds (and possibly integrality constraints) is called the *ternary form* of a global optimization problem.

To find a semilinear relaxation, we note that the equations (18.1) have the form (17.29) and hence can be handled as in the previous section. The equations (18.2) are linear if $\circ \in \{+, -\}$. For $\circ = /$, we get equivalent constraints $x_i = x_k x_j$, and for $\circ = \hat{\ }$ (the power), we can rewrite the constraint $x_k = x_i^{x_j}$ as

$$y_k = x_j y_i, \quad y_k = \log x_k, \quad y_i = \log x_i.$$

(Powers with constant exponents are treated as a case of (18.1).) It remains to consider products. But $x_k = x_i x_j$ is equivalent to

$$\alpha x_i + \beta x_j = u, \quad \alpha x_i - \beta x_j = v,$$
$$w = v^2, \quad w + 4\alpha\beta x_k = u^2,$$

for arbitrary $\alpha, \beta \neq 0$. The first two are linear constraints in x_i, x_j, u, v, and the others are of the form (17.29). This proves that the reformulation can always be done.

However, it is clear that in most cases many fewer intermediate variables need to be introduced since affine expressions $a^T x + \alpha$ can be left intact, as can all expressions depending only on a single variable. Moreover, as we shall see in a moment, quadratic and bilinear expressions can be handled more efficiently.

Therefore, it is advisable to do in a first step only those substitutions needed to transform the problem such that the new objective function

$f(x) =: F_0(x)$ and the components $F_i(x)$ $(i = 1 : m)$ of the new constraint function vector $F(x)$ are *semiseparable*, i.e., of the form

$$F_i(x) = \sum_{(j,k)\in K_i} \varphi_j(x_k) + x^T H_i x + c_i^T x + \gamma_i \quad (i = 0 : m) \tag{18.3}$$

with nonlinear univariate functions φ_j and (in general extremely sparse) matrices H_i. Note that linear terms may be absorbed into the sum, and quadratic and bilinear terms into $x^T H_i x$.

In a second step, the quadratic terms are rewritten as a weighted sum of squares,

$$x^T H_i x = \frac{1}{2} \sum_{j \in J_i} d_j (r_j^T x)^2. \tag{18.4}$$

This is always possible, usually in many ways; for example, by a spectral factorization

$$H_i + H_i^T = QDQ^T, \quad D \text{ diagonal,}$$

which gives

$$2x^T H_i x = (Q^T x)^T D(Q^T x) = \sum D_{kk}(Q^T x)_k^2.$$

(For numerical stability we need to take care of scaling issues, to ensure that no unavoidable cancellation of significant digits takes place.) Using (18.4) and substituting new variables for the $r_j^T x$, we see that we can achieve in this second step the *separable form*

$$F_i(x) = \sum_{(j,k)\in K_i} \varphi_j(x_k) + c_i^T x + \gamma_i \quad (i = 0 : m) \tag{18.5}$$

with increased K_i. Constraints of the form $F_i(x) \leq \overline{F}_i$ are now replaced by

$$\sum_{(j,k)\in K_i} y_j + c_i^T x + \gamma_i \leq \overline{F}_i,$$

$$y_j \geq \varphi_j(x_k) \quad \text{for } (j,k) \in K_i,$$

and similarly for the objective function. Constraints of the form $F_i(x) \geq \underline{F}_i$ are replaced by

$$\sum_{(j,k)\in K_i} y_j + c_i^T x + \gamma_i \geq \underline{F}_i,$$

$$y_j \leq \varphi_j(x_k) \quad \text{for } (j,k) \in K_i.$$

Finally two-sided constraints $F_i(x) \in \mathbf{F}_i$ with finite $\mathbf{F}_i$ are replaced by

$$\sum_{(j,k)\in K_i} y_j + c_i^T x + \gamma_i \in \mathbf{F}_i,$$

$$y_j = \varphi_j(x_k) \quad \text{for} \quad (j,k) \in K_i.$$

Thus, in this third step, the required form has been achieved, and generally much more parsimoniously. (A few more variables could be saved by leaving in each nonlinear $F_i(x)$ one of the nonlinear terms unsubstituted.)

So far, no approximation has occurred; the reformulated problem is equivalent to the original one. In a final *approximation step*, constraints of the form (17.26), (17.28), (17.29) are replaced by piecewise linear constraints. If we only need an approximation (as is traditionally done (Beale and Forrest 1976)), then we use a piecewise linear interpolant to φ.

However, with only a little more work, *outer approximations* can be constructed if we have two piecewise linear approximations $\underline{\varphi}, \overline{\varphi}$ with the same nodes $\xi_0 < \cdots < \xi_d$, satisfying

$$\underline{\varphi}(\xi) \leq \varphi(\xi) \leq \overline{\varphi}(\xi) \quad \text{for} \quad \xi \in [\xi_0, \xi_d]. \tag{18.6}$$

To get $\xi_0 = \underline{x}_i$ and $\xi_d = \overline{x}_i$ we need good bounds $\mathbf{x}_i$ on x_i, which can usually be calculated by constraint propagation (see Section 14). The bounding functions $\underline{\varphi}$ and $\overline{\varphi}$ can be found by exploiting convexity properties of φ, which are well known for elementary functions and can be determined with interval analysis for factorable univariate functions. Given (18.6), the constraint (17.26) *implies* (and not only approximates) the semilinear constraints

$$x_i \in [\xi_{k-1}, \xi_k], \quad a^T x \leq \overline{\varphi}_k + \overline{\varphi}'_k(x_i - \xi_k) \quad \text{for some } k,$$

the constraint (17.28) implies the semilinear constraints

$$x_i \in [\xi_{k-1}, \xi_k], \quad \underline{\varphi}_k + \underline{\varphi}'_k(x_i - \xi_k) \leq a^T x \quad \text{for some } k,$$

and the constraint (17.29) implies the semilinear constraints

$$x_i \in [\xi_{k-1}, \xi_k], \quad \underline{\varphi}_k + \underline{\varphi}'_k(x_i - \xi_k) \leq a^T x \leq \overline{\varphi}_k + \overline{\varphi}'_k(x_i - \xi_k) \quad \text{for some } k.$$

Moreover, by adaptively adding additional nodes we can make the gap between the bounds in (18.6) arbitrarily small, and the approximation by these semilinear constraints becomes arbitrarily good (at the cost of higher complexity, of course).

As one can see, the complexity of the resulting MILP formulation depends on the number of nonlinear operations (but in a problem-dependent fashion because of the quadratic bilinear terms), and grows linearly with the number of nodes used in the piecewise linear approximation. Hence it is an efficient technique only if the number of nonlinear operations is not too large, and the approximation not too close.

19. Other problem transformations

Linear or convex relaxations of a global optimization problem may be viewed as transformations of the problem or of its nonlinear (resp. nonconvex) constraints. There is a number of other useful transformations of constraints or groups of constraints.

19.1. General cuts

A *redundant constraint*, or simply a *cut*, is an inequality (or sometimes an equation) not in the original problem formulation that must hold for any global minimizer; if the inequality is linear, it is called a *cutting plane* (Gomory 1960). A lot is known about cutting planes in mixed integer linear programming (see, *e.g.*, Nemhauser and Wolsey (1988), Nemhauser and Wolsey (1989), Wolsey (1998)); we are here particularly interested in techniques for the smooth case.

We have already met several kinds of derived constraints that cut off part of the feasible region.

- The constraint $f(x) \leq f^{\text{best}}$ cuts off points worse than the best feasible point found so far (with function value f^{best}).

- Exclusion constraints, discussed in Section 15, cut off a region around local minimizers that do not contain any other, better minimizer.

- The linear, convex and semilinear relaxations of constraints, discussed in Sections 16 and 18, are of course special cases of cuts.

For bound-constrained indefinite quadratic programs, Vandenbussche (2003) generalized techniques for mixed integer linear programming to find cuts which lead to excellent results on this problem class.

Surely there is much more to be explored here.

19.2. Symbolic transformations

The quality of all techniques considered so far may depend strongly on the form in which a problem is posed. Symbolic techniques may be employed to change the given form into another, perhaps more advantageous form. Unfortunately, it is not clear which transformations are most valuable, and the best transformations must usually be found on a problem-specific *ad hoc* basis. A recent example of a very difficult constraint satisfaction problem in robotics that only yielded to such an approach is described by Lee et al. (2002). We simply list here a few of the techniques that may be of interest.

Frequently, problems involving *trigonometric variables* can be replaced by equivalent problems with only polynomial equations, using (as recommended in ALIAS-C++ (2003)) the additional constraint

$$s^2 + c^2 = 1$$

together with the substitution

$$s = \sin\varphi, \quad c = \cos\varphi, \quad s/c = \tan\varphi, \quad c/s = \cot\varphi,$$

and similar rules for trigonometric functions of half or double angles.

Techniques from *algebraic geometry* can be applied to 'solve' polynomial equations symbolically or to bring them into a special form that may be useful. In particular, *Gröbner basis* methods (see, e.g., Buchberger and Winkler (1998), Faugere, Gianni, Lazard, and Mora (1993), Stetter (1997)) provide normal forms that have a triangular structure and thus allow a complete enumeration of solutions for small polynomial problems. The work grows exponentially with the number of variables. Elimination theory (see, e.g., Cox, Little and O'Shea (1998), Emiris and Canny (1995), Emiris and Mourrain (1999), Jónsson and Vavasis (2001), Moller and Stetter (1995), Bondyfalat, Mourrain and Pan (2000), Mourrain and Pan (1997), Mourrain, Pan and Ruatta (2003)) provides different, often less expensive techniques for potential simplifications by the elimination of variables. The results are often expressed in terms of determinants, and their exploitation by global solution techniques is not well explored. While the matrices arising in elimination theory appear to be related to those in semidefinite relaxations of polynomial systems, the connection apparently received little attention (Datta 2001, Hanzon and Jibetean 2003).

Unfortunately, the equations resulting from completely automatic algebraic techniques are often numerically unstable. If this is the case, function evaluations need either rational arithmetic or higher precision. In such cases, interval evaluation including their refinements suffer from excessive cancellation and provide only very weak global information. It would be very desirable to have flexible tools that do only partial elimination but provide stable reduced equations of some sort.

Hanzon and Jibetean (2003) apply these techniques to find the global minimizer of multivariate polynomials, giving attention also to the case where the minimum is achieved at infinity.

Automatic differentiation (Berz, Bischof, Corliss and Griewank 1996, Coleman and Verma 1998, Griewank 1989, 1991, 1992, Griewank and Corliss 1991) is a now classical technique for obtaining high-quality derivatives analytically and cheaply by transforming a program for function evaluation into a program for the evaluation of derivatives. This technique can be applied directly to create the Karush–John optimality conditions (Theorem 5.1 and equation (5.4)) as additional constraints for constraint propagation or for verifying optimality, as is done in the COCONUT environment (*cf.* Section 22) and (with a weaker form of (5.4)) in Numerica (Van Hentenryck *et al.* 1997*b*) and GlobSol (Kearfott 1996*b*).

On the other hand, the automatic differentiation techniques can also be adapted to provide evaluations of interval derivatives, slopes (Bliek (1992, 1997), with improvements in Schichl and Neumaier (2003)), linear enclosures (Nenov and Fylstra 2003), and second-order slopes (Kolev 1997).

20. Rigorous verification and certificates

The reliability of claimed results is the most poorly documented aspect of current global optimization software. Indeed, as was shown by Neumaier and Shcherbina (2004), even famous state-of-the-art solvers like CPLEX8.0 (and many other commercial MILP codes) may lose an integral global solution of an innocent-looking mixed integer linear program. In our testing of global solvers within the COCONUT project we noticed many other cases where global solutions were lost or feasible problems were declared infeasible, probably because of ill-conditioned intermediate calculations that lead to rounding errors not covered by the built-in tolerances.

For the solution of precise mathematical problems (such as the Kepler problem (Hales 1998)), but also for safety-critical optimization problems, it is necessary to have a complete mathematical guarantee that the global minimizer has been found. This requires special attention since numerical computations are affected by rounding errors. Fortunately, interval arithmetic, if performed with directed (outward) rounding, is able to give mathematical guarantees even in the presence of rounding errors.

20.1. Rounding in the problem definition

Many problems contain floating-point constants in their formulation. Therefore, frequently, the translation of the problems into an internal format involves floating-point computations which introduce rounding errors. Unfortunately, none of the currently available modelling systems allows one to control these rounding errors or any rounding errors made in a presolve phase used to simplify the problem formulation. The rigorous solvers available (GlobSol and Numerica) have special input modes for constants, but cannot be fed with problems generated from AMPL or GAMS input (the format for most test problems in the collections available: see Section 21). It is hoped that future releases of modelling systems provide options that allow for the passing of either symbolic constants or interval-valued coefficients computed from the input, so that the exact problem, or at least nearly exact but rigorous relaxations of the exact problem, can be recovered.

20.2. Rounding in the solution process

Most current solvers simply implement algorithms valid in exact arithmetic, and do not consider rounding errors, except by allowing for certain nonrigorous *ad hoc* tolerances in testing feasibility.

On the other hand, certain solvers (in the above list of solvers, GlobSol and Numerica) do only rigorous computations – by enclosing all numbers in intervals accounting for the rounding errors. However, they do not make use of convexity arguments leading to linear or convex programs that can be solved by local techniques, and hence have a competitive disadvantage for numerous problems. The main reason seems to be that, until recently, making linear (or convex) programming rigorous was very expensive compared with the traditional approximate approach, and time-consuming to implement.

Neumaier and Shcherbina (2004) showed that it is possible to certify the results of linear optimization problems with finite bounds by simple pre- and post-processing, without having to modify the solvers. Jansson (2003, 2004b) extended this to the case of unbounded variables (where only a little more work is needed unless a large number of unbounded variables is present), and Jansson (2004a) extended the approach further to the case of convex programs.

The availability of these cheap, easy-to-use methods for certifying the results of linear and convex optimization programs is likely to change this in the near future. First results in this direction are presented by Lebbah *et al.* (2002, 2004), who report rigorous results for a combination of constraint propagation, interval Newton and linear programming methods that significantly outperform other rigorous solvers (and also the general-purpose solver BARON) on a number of difficult constraint satisfaction problems.

20.3. Certification of upper bounds

Apart from controlling rounding errors in the computation of bounds, care must also be taken in using objective function values as upper bounds on the objective function. This is permitted only if the argument is feasible. However, especially in the presence of equality constraints, the arguments are often not exactly feasible but satisfy the constraints only within certain tolerances. In these cases, a rigorous upper bound on the objective can be obtained only if the existence of a feasible point in a small box around the approximate point can be proved rigorously, and the objective function is then evaluated at this box. This requires the use of interval Newton techniques (*cf.* Section 11). However, since there are frequently fewer equality constraints than variables, the standard existence tests must be modified to take account of this, and also to handle inequalities correctly. For a description of the main techniques currently available to certify the existence of feasible points, see, e.g., Kearfott (1996a, 1996b).

20.4. Certificates of infeasibility

If an optimization problem (or a subproblem in a box generated by branch and bound) has no feasible point, a *certificate of infeasibility* can often be given allowing an easy check that this is the case. For linear constraints, the following result applies (Neumaier and Shcherbina 2004), which only uses basic interval arithmetic.

Theorem 20.1. The set of points satisfying

$$x \in \mathbf{x}, \quad Ax \in \mathbf{b} \tag{20.1}$$

is empty if and only if there is a multiplier vector y such that

$$(y^T A)\mathbf{x} \cap y^T \mathbf{b} = \emptyset. \tag{20.2}$$

Proof. If x satisfies (20.1) then the left-hand side of (20.2) contains $y^T A x$ and hence is nonempty. Thus (20.2) implies that (20.1) cannot be satisfied. The converse is a simple consequence of the Lemma of Farkas and the fact (Neumaier 1990, Section 3.1) that $a^T \mathbf{x} = \{a^T x \mid x \in \mathbf{x}\}$. □

Thus a certificate of infeasibility consists in a multiplier vector y satisfying (20.2), and is, e.g., a byproduct of phase 1 of a simplex algorithm.

If there are nonlinear constraints, there are simple certificates for infeasibility of

$$x \in \mathbf{x}, \quad F(x) \in \mathbf{F},$$

such as a multiplier vector y with

$$y^T F(\mathbf{x}) \cap y^T \mathbf{F} = \emptyset, \tag{20.3}$$

where $F(\mathbf{x})$ is an interval evaluation of F at the box $\mathbf{x}$, or

$$y^T F(\xi) + (y^T F'(\mathbf{x}))(\mathbf{x} - \xi) \cap y^T \mathbf{F} = \emptyset, \tag{20.4}$$

where $F'(\mathbf{x})$ is an interval evaluation of the Jacobian F' at the box $\mathbf{x}$. Similarly, if a feasible point with objective function value f^{best} is known, then a multiplier vector y with

$$\min_{x \in \mathbf{x}}(f(x) + y^T F(x)) > f^{\text{best}} + \sup y^T \mathbf{F} \tag{20.5}$$

is a certificate that the box $\mathbf{x}$ cannot contain a global minimizer. The left-hand side can be bounded from below by interval evaluation or a centred form, giving a verifiable sufficient condition. In the linear case, (20.5) reduces to half of Theorem 20.1.

It is not difficult to show that, for convex constraints, a certificate of infeasibility can be constructed in complete analogy to the linear case. But in the nonconvex case, there is no guarantee that such a certificate exists (or can be found easily if it exists). Moreover, local solvers may fail because they are not able to find a feasible point, even if one exists. Indeed, finding a feasible point is in the latter case already a global problem that cannot be handled by local methods.

Good certificates of infeasibility of the form (20.4) are, however, available for small boxes not too close to the feasible domain. This follows from the quadratic approximation property of centred forms. A suitable multiplier vector y can be obtained in this case from the linearized problem.

Thus, in combination with branching, we can certify the nonexistence of a solution in a covering of almost all of the initial box.

20.5. Certification of global minimizers

One may also be interested in providing a minimal number of mathematically rigorous certificates that constitute a proof that some point in a narrow computed box is in fact a global minimizer. These certificates are mathematically valid only if the corresponding conditions have been evaluated in exact arithmetic; and additional safeguards are needed to ensure their validity in finite precision arithmetic. Virtually nothing has been done so far with regard to this problem.

21. Test problems and testing

An important part of the development of global optimization software is the careful testing of proposed methods.

For useful test problem collections, see, *e.g.*, The COCONUT Benchmark (2002), Floudas *et al.* (1999), GLOBAL Library (2002), Huyer and Neumaier (1999), Janka (2000), Jansson and Knüppel (1995) and Walster, Hansen and Sengupta (1985). In particular, Huyer and Neumaier (1999) contains a test suite containing the traditional global optimization test set of low-dimensional problems by Dixon and Szegő (1975), together with test results for DIRECT, MCS, and many incomplete global optimization methods. Janka (2000) (see also Khompatraporn, Pintér and Zabinsky (2004), Mongeau, Karsenty, Rouz and Hiriart-Urruty (2000)) contains a comparison of stochastic global optimization routines on a large number of low-dimensional test problems from different sources, and Jansson and Knüppel (1995) and Walster *et al.* (1985) contain test results for some interval methods on a large number of low-dimensional test problems. Testing in higher dimensions has been much more limited, although this is about to change.

The documentation and availability of test problems has been considerably simplified by coding them in one of the widely used modelling languages. AMPL (Fourer *et al.* 1993) and GAMS (GAMS World 2003) are two flexible and convenient algebraic modelling languages enabling rapid prototyping and model development. They are of widespread use in the optimization community, as attested by the large number of existing interfaces with state-of-the-art optimization solvers.

The recent *Handbook of Test Problems in Local and Global Optimization* (Floudas *et al.* 1999) contains a large collection of test problems for local and global optimization problems, both academic and from real applications. (Unfortunately, the book contains a significant number of inaccuracies; see Shcherbina (2002).) The algebraic test problems of this collection are available in the GAMS modelling language, and the differential-algebraic problems are supplied in the MINOPT modelling language. All test problems can be downloaded from the *Handbook*'s web site. A recent web site by GAMS World (GLOBAL Library 2002) started collecting a library, GlobalLib, of real-life global optimization problems with industrial relevance, coded in GAMS, but currently most problems on this site are without computational results. 131 algebraic test problems from the *Handbook* are all included and constitute about a third of the 397 test problems currently available at GlobalLib.

Test problems for local optimization should also pass global optimization solvers; the traditional test set for low-dimensional unconstrained problems is that by Moré, Garbow and Hillstrom (1981), with optional bounds from Gay (1984). A number of these problems have in fact several local minimizers and are therefore global optimization problems. Bob Vanderbei maintains a large collection of AMPL files for constrained nonlinear optimization problems (Vanderbei 2000) from practical applications; also included are the major part of the CUTE collection and the more academic but useful low-dimensional problem collection of Hock and Schittkowski (1981).

The COCONUT Benchmark (2002) (*cf.* Shcherbina *et al.* (2003)) is a collection of nearly 1300 AMPL models, containing the CUTE part of the Vanderbei test collection, AMPL versions of the problems from GlobalLib (collection from summer 2002), and a large collection of pure constraint satisfaction problems from various places. All problems are annotated with best-known function values (or even solutions) and some statistical information such as the number of variables and constraints.

The COCONUT project has extensive test results on its benchmark for a number of solvers, made public on the COCONUT web site. Extensive benchmarking results for *local* optimization by Mittelmann (2002) are also available online. See also Dolan and Moré (2000, 2001).

Bussiek, Drud, Meeraus and Pruessner (2003) report on obtaining reliable and repeatable comparisons in global optimization. The ACM Transaction of Mathematical Software (TOMS 2002) recommends considering advice in Johnson (2002) for performing computational experiments. The Mathematical Programming Society has guidelines (Jackson, Boggs, Nash and Powell 1990/91) for reporting results of computational experiments based on Crowder, Dembo and Mulvey (1979), Greenberg (1990) and Ratliff and Pierskalla (1981). See also Barr *et al.* (1995).

22. The COCONUT environment

This survey is part of an attempt to integrate various existing complete approaches to global optimization into a uniform whole. This is the goal of the COCONUT project, sponsored by the European Union. The COCONUT consortium provides on its homepage (COCONUT 2001) a modular solver environment for nonlinear global optimization problems with an open-source kernel, which can be expanded by commercial and open-source solver components (inference engines). The following information is taken from Schichl (2004).

The application programming interface (API) of the COCONUT environment is designed to make the development of the various module types independent of each other and independent of the internal model representation. It is a collection of open-source C++ classes protected by the LGPL license model (GNU Lesser General Public License 1999), so that it can be used as part of commercial software. It uses the FILIB++ (Lerch *et al.* 2001) library for interval computations and the matrix template library MTL (Siek and Lumsdaine 1999) for the internal representation of various matrix classes. Support for dynamic linking relieves the user from recompilation when modules are added or removed. In addition, it is designed for distributed computing, and will probably be developed further (in the years after the end of the COCONUT project) to support parallel computing as well.

The solution algorithm is an advanced branch and bound scheme which proceeds by working on a set of search nodes, each representing a subproblem of the optimization problem to be solved. A complete optimization problem is always represented by a *single* directed acyclic graph (DAG). The vertices of the graph represent operators similar to computational trees. Constants and variables are sources, objective and constraints are sinks of the DAG. This DAG is optimally small in the sense that it contains every subexpression of objective function and constraints only once.

For expression DAGs, special forward and backward evaluators are provided to allow function evaluation and automatic differentiation-like tasks. Currently implemented are real function values, function ranges, gradients (real, interval), and slopes. In the near future, evaluators for Hessians (real, interval) and second-order slopes (see, *e.g.*, Schichl and Neumaier (2004)) will be provided as well.

A strategy engine is the main part of the algorithm. It makes decisions, directs the search, and invokes the various modules. The strategy engine consists of the logic core ('search') which is essentially the main solution loop, special decision makers for determining the next action at every point in the algorithm. It calls management modules, report modules, and inference engines in a sequence defined by programmable search strategies.

The strategy engine can be programmed using a simple strategy language based on the language Python (in which, for instance, most of the interface of the web search engine Google is written). Since it is interpreted, (semi-)interactive and automatic solution processes are possible, and even debugging and single-stepping of strategies is supported. The language is object-oriented, provides dynamically typed objects, and is garbage-collecting. These features make the system easily extendable. Furthermore, the strategy engine manages the search graph and the search database. The strategy engine uses a component framework to communicate with the inference engines. This makes it possible to launch inference engines dynamically (on need) to avoid memory overload. Since the strategy engine is itself a component, even multilevel strategies are possible.

Corresponding to every type of problem change, a class of inference engines is designed: model analysis (*e.g.*, find convex part), model reduction (*e.g.*, pruning, fathoming), model relaxation (*e.g.*, linear relaxation), model splitting (*e.g.*, bisection), model glueing (*e.g.*, undo excessive splitting), computing of local information (*e.g.*, probing, local optimization). Several state-of-the-art techniques are already provided, including interfaces to local nonlinear solvers and linear programming systems, rigorous point verifiers, exclusion box generators, constraint propagation, linear relaxations, a splitter, and a box-covering module.

Changes suggested by an inference engine and approved of by the strategy engine are performed by appropriate management modules. Report modules produce human-readable or machine-readable files for checkpointing or external use.

The open design of the solver architecture, and its extensibility to include both open source modules and commercial programs, was chosen in the hope that the system will be a unique platform for global optimization in the future, serving the major part of the community, bringing their members closer together.

Several researchers and companies from outside the COCONUT project have already agreed to complement our efforts in integrating the known techniques by contributing to the COCONUT environment.

23. Challenges for the near future

We end the survey by listing a number of challenges that researchers in global optimization, and those working on software systems and support, may wish to face to improve the state of the art.

(1) Ensuring reliability is perhaps the most pressing issue. While in theory essentially all techniques discussed here can be made fully rigorous, many of them with little computational overhead (see Section 20), only very few solvers do this. As a result, even otherwise excellent solvers (such as CPLEX 8.0 for linear mixed integer problems) occasionally lose the solution and give completely misleading results without warning, and global solvers based on these inherit the problems unless properly safeguarded. Safe bounds can guard against all errors due to finite precision arithmetic. Programming bugs are another possible source of loss of solutions, and can be discovered only through extensive testing on benchmarking suites with known solutions.

Also under the reliability heading are improvements relevant for computer-assisted proofs, especially the documentation of certificates that give a short and complete proof (that can be checked independently) that the solution is indeed correct.

(2) Better compiler (or even hardware) support for automatic differentiation, outward rounded interval arithmetic, and related techniques (Schichl and Neumaier 2003) based on computational graphs would significantly simplify its use in global optimization codes, and probably speed up the programs. (Code optimization would, however, need to provide an option that ensures that simplifications are only performed if they are mathematically safe even in finite precision arithmetic.) The SUN FORTE compiler (Walster 2000) already supports interval arithmetic. NAG (Cohen, Naumann and Riehme 2003, NAG 2003) is investigating the possible integration of automatic differentiation capabilities into its Fortran 95 compiler.

(3) Unbounded variables are perhaps the dominant reason for failure of current complete global optimization codes on problems with few variables. Unless the initial constraint propagation phase provides useful finite bounds, interval estimates are frequently meaningless since calculations with unbounded intervals rarely generate tight enclosures. Thus the bounding part of the search remains weak, and an excessive number of boxes is generated. Better techniques for handling problems with unbounded variables are therefore highly desirable. For unconstrained polynomial problems see Hanzon and Jibetean (2003).

(4) Unconstrained problems and bound-constrained problems in higher dimensions are harder for current solvers than highly constrained ones, since the lack of constraints gives little basis for attack with the known methods.

In particular, current complete solvers are quite slow on many nonzero residual least squares problems. Until better techniques become available, users should take advantage of available freedom in modelling by providing as many constraints as possible, *e.g.*, by adding for a least squares problem min $\|F(x)\|_2^2$ the additional constraints $F(x) \in \mathbf{F}$ for some reasonable box $\mathbf{F}$. While this may change the solution, it might be fully adequate for the application.

(5) Integrating techniques from mixed integer and semidefinite programming into the current solver frameworks appears to be a promising direction. Work in this direction has begun at various places, and it is already apparent that we can expect major improvements in speed.

(6) Problems with symmetries have many solutions that differ only in trivial rearrangements, sign changes, *etc.* However, it is not easy to avoid finding the solutions repeatedly or having to exclude repeatedly regions equivalent under symmetries. There are significant applications in cluster optimization (The Cambridge Cluster Database 1997), packing problems, and the optimal design of experiments (Sloane 2003). A recent paper by Margot (2003) handles the integer linear programming case, and Gatermann and Parrilo (2004) address the case of polynomial systems.

(7) The representation of nonisolated solution sets is another challenge where papers are slowly forthcoming (*e.g.*, Xuan-Ha Vu *et al.* (2002, 2003) for the case of continuous constraint satisfaction problems) and which has important applications in the modelling of devices with uncertain parameters or flexible parts (Neumaier 2003b). A related problem is that of parametric global optimization, where the same parametrized problem needs to be solved in dependence on a few parameters. (The result is then a function of these parameters instead of a single solution.) Apart from some discussion in Guddat *et al.* (1990), very little seems to have been done in this area.

(8) Problems with severe dependence among the variables have poor interval extensions and hence create difficulties for complete solvers. This applies in particular to problems containing nested functions $f(x)$ such as those arising from volume-preserving discrete dynamics, where $f(x) = f_n(x)$ with $f_1(x) = \varphi(x)$, $f_n(x) = \varphi(f_{n-1}(x))$, which suffer from a severe wrapping effect (Nedialkov and Jackson 2001, Neumaier 1993), and problems involving determinants of matrices of size > 3. Taylor methods (Makino and Berz 2003, Neumaier 2002) and reformulation techniques might help overcome these problems.

(9) Differential constraints are not of the factorable form that is the basis of all current global solvers. But they arise in optimal control problems, and it is well known that many of these (especially in space mission design and in chemical engineering) have multiple minima and hence would need a global optimization approach. Recently, some approximate methods (Esposito and Floudas 2000, Meyer, Floudas and Neumaier 2002) and a complete method (Papamichail and Adjiman 2002) (for the inverse monotone case) have become available, though these work at present only in very low dimensions.

(10) Constraints involving integrals are also not factorable; so far, they have received no attention in a global optimization context. They arise naturally in many stochastic optimization problems defined in terms of continuous random variables, since expectations or probabilities involving these are given by integrals. Examples are probabilistic safety factors in engineering and value at risk in finance.

(11) Large-scale problems are obviously hard owing to their size and the worst-case exponential behaviour of branch and bound algorithms. However, as in many combinatorial optimization problems, there may be many large-scale problems that are tractable if their problem structure is exploited. Extending the current methods to take advantage of such structure would make them much more widely applicable. Recent work of Boddy and Johnson (2003), who solved to completion large quadratic constraint satisfaction problems arising in oil refinery, including one with 13 711 variables, 17 892 constraints (of which 2 696 were nonlinear) gives rise to optimism.

All these problems show that much remains to be done and that we can expect further progress in the future.

Acknowledgements

I want to thank the Mathematics Department of the University of Colorado at Denver, and in particular Weldon Lodwick, for the opportunity to give there (in April/May 2001) a course with the same title, and the Mathematics and Computer Science Division of the Argonne National Laboratory, in particular Sven Leyffer, for the invitation to present my vision of the current techniques and challenges in global optimization in a lecture within the Global Optimization Theory Institute (2003).

I also want to thank Hermann Schichl for many discussions and for comments on earlier versions of this survey, and Christian Jansson and Nick Sahinidis for additional comments that improved the paper.

This survey is part of work done in the context of the COCONUT project (COCONUT 2001) sponsored by the European Union, with the goal of integrating various existing complete approaches to global optimization into a uniform whole (*cf.* Section 22). Funding by the European Union under the IST Project Reference Number IST-2000-26063 within the FET Open Scheme is gratefully acknowledged. A preliminary version of the survey was contained in Chapter 4 of the unpublished COCONUT report (Bliek *et al.* 2001).

REFERENCES

E. H. L. Aarts and J. K. Lenstra, eds (1997), *Local Search in Combinatorial Optimization*, Wiley, Chichester.

C. S. Adjiman, I. P. Androulakis, C. D. Maranas and C. A. Floudas (1996), 'A global optimization method αBB for process design', *Comput. Chem. Engin.* **20**, 419–424.

C. S. Adjiman, S. Dallwig, C. A. Floudas and A. Neumaier (1998a) 'A global optimization method, αBB, for general twice-differentiable constrained NLPs, I: Theoretical advances', *Comput. Chem. Engin.* **22**, 1137–1158.

C. S. Adjiman, I. P. Androulakis and C. A. Floudas (1998b), 'A global optimization method, αBB, for general twice-differentiable constrained NLPs, II: Implementation and computational results', *Comput. Chem. Engin.* **22**, 1159–1179.

C. S. Adjiman and C. A. Floudas (1996), 'Rigorous convex underestimators for general twice-differentiable problems', *J. Global Optim.* **9**, 23–40.

I. G. Akrotirianakis and C. A. Floudas (2004), 'A new class of improved convex underestimators for twice continuously differentiable constrained NLPs', *J. Global Optim.*, in press.

ALIAS-C++ (2003), *A C++ Algorithms Library of Interval Analysis for Equation Systems, Version 2.2*. User manual: `manual-alias-C++2.2.ps` at `www-sop.inria.fr/coprin/logiciels/ALIAS/ALIAS-C++/`

F. Alizadeh (2003), Semidefinite and second order cone programming foundations, algorithms and applications. Slides: `tutorialProsper.pdf` at `www.ima.umn.edu/talks/workshops/3-11.2003/alizadeh/`

F. A. Al-Khayyal and J. E. Falk (1983), 'Jointly constrained biconvex programming', *Math. Oper. Res.* **8**, 273–286.

F. A. Al-Khayyal and H. D. Sherali (2000), 'On finitely terminating branch-and-bound algorithms for some global optimization problems', *SIAM J. Optim.* **10**, 1049–1057.

F. A. Al-Khayyal, C. Larsen and T. van Voorhis (1995), 'A relaxation method for nonconvex quadratically constrained quadratic programs', *J. Global Optim.* **6**, 215–230.

E. Allgower, M. Erdmann and K. Georg (2002), 'On the complexity of exclusion algorithms for optimization', *J. Complexity* **18**, 573–588.

Le Thi Hoai An and Pham Dinh Tao (1998), 'A branch-and-bound method via DC optimization algorithm and ellipsoidal technique for box constrained nonconvex quadratic programming problems', *J. Global Optim.* **13**, 171–206.

E. D. Anderson and K. D. Anderson (1995), 'Presolving in linear programming', *Math. Program.* **71**, 221–245.

I. P. Androulakis, C. D. Maranas and C. A. Floudas (1995), 'αBB: A global optimization method for general constrained nonconvex problems', *J. Global Optim.* **7**, 337–363.

K. Anstreicher (2003), 'Recent advances in the solution of quadratic assignment problems', *Math. Program. B* **97**, 27–42.

C. Audet, P. Hansen, B. Jaumard and G. Savard (2000), 'A branch and cut algorithm for nonconvex quadratically constrained quadratic programming', *Math. Program.* **87**, 131–152.

A. B. Babichev, O. B. Kadyrova, T. P. Kashevarova, A. S. Leshchenko, and A. L. Semenov (1993), 'UniCalc: A novel approach to solving systems of algebraic equations', *Interval Computations* **3**, 29–47.

V. Balakrishnan and S. Boyd (1992), Global optimization in control system analysis and design, in *Control and Dynamic Systems: Advances in Theory and Applications*, Vol. 53 (C. T. Leondes, ed.), Academic Press, New York, pp. 1–56.

E. Balas (1979), 'Disjunctive programming', *Ann. Discrete Math.* **5**, 3–51.

E. Balas, S. Ceria and G. Cornuejols (1993), 'A lift-and-project cutting plane algorithm for mixed 0–1 programs', *Math. Program. A* **58**, 295–323.

R. S. Barr, B. L. Golden, J. P. Kelly, M. G. C. Resende, and W. R. Stewart (1995), 'Designing and reporting on computational experiments with heuristic methods', *J. Heuristics* **1**, 9–32.

E. M. L. Beale (1979), Branch and bound methods for mathematical programming systems, *Ann. Discrete Math.* **5** 201–219.

E. M. L. Beale (1988), *Introduction to Optimization*, Wiley, Chichester.

E. M. L. Beale and J. J. H. Forrest (1976), 'Global optimization using special ordered sets', *Math. Program.* **10**, 52–69.

E. M. L. Beale and J. A. Tomlin (1970), Special facilities in a general mathematical programming system for non-convex problems using ordered sets of variables, in *OR 69: Proc. Fifth Int. Conf. Oper. Res.* (J. Lawrence, ed.) Tavistock Publications, London, pp. 447–454.

R. W. Becker and G. V. Lago (1970), A global optimization algorithm, in *Proc. 8th Allerton Conf. Cicuits Systems Theory* (Monticello, Illinois, 1970), pp. 3–12.

O. Ben-Ayed (1993), 'Bilevel linear programming', *Comput. Oper. Res.* **20**, 485–501.

F. Benhamou and W. J. Older (1997), 'Applying interval arithmetic to real, integer, and boolean constraints', *J. Logic Program.* **32**, 1–24.

F. Benhamou, D. McAllister and P. Van Hentenryck (1994), CLP(intervals) revisited, in *Proc. International Symposium on Logic Programming*, MIT Press, Ithaca, NY, pp. 124–138.

S. Berner (1996), 'Parallel methods for verified global optimization: Practice and theory', *J. Global Optim.* **9**, 1–22.

M. Berz, C. Bischof, G. Corliss and A. Griewank, eds (1996), *Computational Differentiation: Techniques, Applications, and Tools*, SIAM, Philadelphia.

A. C. Billups and K. G. Murty (2000), 'Complementarity problems', *J. Comput. Appl. Math.* **124**, 303–318.

R. E. Bixby, M. Fenelon, Z. Gu, E. Rothberg and R. Wunderling (2000), MIP: Theory and practice: Closing the gap, in *System Modelling and Optimization: Methods, Theory and Applications* (M. J. D. Powell and S. Scholtes, eds), Kluwer, Dordrecht, pp. 19–49.

M. Björkman and K. Holmström (2000), 'Global optimization of costly nonconvex functions using radial basis functions', *Optim. Eng.* **1**, 373–397.

C. Bliek (1992), Computer methods for design automation, PhD thesis, Department of Ocean Engineering, MIT.

C. Bliek (1997), 'Fast evaluation of partial derivatives and interval slopes', *Reliable Comput.* **3**, 259–268.

C. Bliek, P. Spellucci, L. N. Vicente, A. Neumaier, L. Granvilliers, E. Monfroy, F. Benhamou, E. Huens, P. Van Hentenryck, D. Sam-Haroud and B. Faltings (2001), *Algorithms for Solving Nonlinear Constrained and Optimization Problems: The State of the Art.* A progress report of the COCONUT project: www.mat.univie.ac.at/~neum/glopt/coconut/StArt.html

C. Bliek, C. Jermann and A. Neumaier, eds (2003), *Global Optimization and Constraint Satisfaction*, Vol. 2861 of *Lecture Notes in Computer Science*, Springer, Berlin.

M. Boddy and D. Johnson (2003), A new method for the global solution of large systems of continuous constraints, in *Global Optimization and Constraint Satisfaction* (C. Bliek et al., eds), Springer, Berlin, pp. 142–156.

C. G. E. Boender, A. H. G. Rinnooy Kan, G. T. Timmer and L. Stougie (1982), 'A stochastic method for global optimization', *Math. Program.* **22**, 125–140.

I. M. Bomze, T. Csendes, R. Horst and P. M. Pardalos, eds (1996), *Developments in Global Optimization*, Kluwer, Dordrecht.

D. Bondyfalat, B. Mourrain and V. Y. Pan (2000), 'Solution of a polynomial system of equations via the eigenvector computation', *Linear Algebra Appl.* **319**, 193–209.

B. Buchberger and F. Winkler (1998), *Groebner Bases: Theory and Applications*, Cambridge University Press.

R. E. Burkard, H. Hamacher and G. Rote (1992), 'Sandwich approximation of univariate convex functions with an application to separable convex programming', *Naval Res. Logist.* **38**, 911–924.

J. V. Burke (1991), 'An exact penalization viewpoint of constrained optimization', *SIAM J. Control Optim.* **29**, 968–998.

M. R. Bussiek and A. S. Drud (2001), SBB: A new solver for mixed integer nonlinear programming. Slides: www.gams.com/presentations/or01/sbb.pdf
SBB user manual (2002): www.gams.com/solvers/sbb.pdf

M. R. Bussiek, A. S. Drud, A. Meeraus and A. Pruessner (2003), Quality assurance and global optimization, in *Global Optimization and Constraint Satisfaction* (C. Bliek et al., eds), Springer, Berlin, pp. 223–238.

The Cambridge Cluster Database (1997), web document: brian.ch.cam.ac.uk/CCD.html

E. Carrizosa, P. Hansen and F. Messine (2004), 'Improving interval analysis bounds by translations', *J. Global Optim.*, to appear.

H. M. Chen and M. H. van Emden (1995), Adding interval constraints to the Moore–Skelboe global optimization algorithm, in *Extended Abstracts of APIC'95, International Workshop on Applications of Interval Computations* (V. Kreinovich, ed.), *Reliable Comput.* (Supplement), pp. 54–57.

G. Chesi and A. Garulli (2001), On the characterization of the solution set of polynomial systems via LMI techniques, in *Proc. of the European Control Conference ECC 2001, Porto (Portugal), September 4–7, 2001*, pp. 2058–2063.

J. G. Cleary (1987), 'Logical arithmetic', *Future Computing Systems* **2**, 125–149.

COCONUT (2001), Continuous constraints: Updating the technology. Web site: www.mat.univie.ac.at/~neum/glopt/coconut.html

The COCONUT Benchmark (2002), A benchmark for global optimization and constraint satisfaction. Web document:
www.mat.univie.ac.at/~neum/glopt/coconut/benchmark.html

M. Cohen, U. Naumann and J. Riehme (2003), Differentiation-enabled Fortran 95 compiler technology. Manuscript:
www-unix.mcs.anl.gov/~naumann/nagfm.ps

T. F. Coleman and A. Verma (1998), 'The efficient computation of sparse Jacobian matrices using automatic differentiation', *SIAM J. Sci. Comput.* **19**, 1210–1233.

H. Cornelius and R. Lohner (1984), 'Computing the range of values of real functions with accuracy higher than second order', *Computing* **33**, 331–347.

D. Cox, J. Little and D. O'Shea (1998), *Using Algebraic Geometry*, Springer, New York.

Y. Crama (1993), 'Concave extensions for nonlinear 0–1 maximization problems', *Math. Program.* **61**, 53–60.

H. P. Crowder, R. S. Dembo and J. M. Mulvey (1979), 'On reporting computational experiments with mathematical software', *ACM Trans. Math. Software* **5**, 193–203.

T. Csendes (1988), Nonlinear parameter estimation by global optimization: Efficiency and reliability. *Acta Cybernetica* **8**, 361–370. Fortran and C code is at
ftp://ftp.jate.u-szeged.hu/pub/math/optimization/index.html
A Fortran 95 version by A. Miller is at
www.mat.univie.ac.at/~neum/glopt/contrib/global.f90

T. Csendes and D. Ratz (1997), 'Subdivision direction selection in interval methods for global optimization', *SIAM J. Numer. Anal.* **34**, 922–938.

S. Dallwig, A. Neumaier and H. Schichl (1997), GLOPT: A program for constrained global optimization, in *Developments in Global Optimization* (I. Bomze et al., eds), Kluwer, Dordrecht, pp. 19–36.

G. B. Dantzig (1960), 'On the significance of solving linear programming problems with some integer variables', *Econometrica* **28**, 30–44.

G. B. Dantzig, S. Johnson and W. White (1958), 'A linear programming approach to the chemical equilibrium problem', *Management Science* **5**, 38–43.

R. S. Datta (2001), Using semidefinite programming to minimize polynomials. Manuscript: www.math.berkeley.edu/~datta/ee227apaper.pdf

E. de Klerk (2002), *Aspects of Semidefinite Programming: Interior Point Algorithms and Selected Applications*, Kluwer, Dordrecht.

L. C. W. Dixon and G. P. Szegő (1975), *Towards Global Optimization*, Elsevier, New York.

E. D. Dolan and J. J. Moré (2000), Benchmarking optimization software with COPS. Technical report ANL/MCS-246, Argonne National Laboratory:
www-unix.mcs.anl.gov/~more/cops

E. D. Dolan and J. J. Moré (2001), Benchmarking optimization software with performance profiles. Technical report ANL/MCS-P861-1200, Argonne National Laboratory: www-unix.mcs.anl.gov/~more/cops

M. Dorigo, V. Maniezzo and A. Colorni (1991), 'The ant system: Optimization by a colony of cooperating agents', *IEEE Trans. Systems, Man, Cyber. Part B* **26**, 29–41.

G. Dueck and T. Scheuer (1990), 'Threshold accepting: A general purpose optimization algorithm appearing superior to simulated annealing', *J. Comput. Physics* **90**, 161–175.

M. A. Duran and I. E. Grossmann (1986), 'An outer approximation algorithm for a class of mixed integer nonlinear programs', *Math. Program.* **36**, 307–339.

I. Z. Emiris and J. F. Canny (1995), 'Efficient incremental algorithms for the sparse resultant and the mixed volume', *J. Symbolic Comput.* **20**, 117–149.

I. Z. Emiris and B. Mourrain (1999), 'Matrices in elimination theory', *J. Symbolic Comput.* **28**, 3–44.

T. G. W. Epperly and E. N. Pistikopoulos (1997), 'A reduced space branch and bound algorithm for global optimization', *J. Global Optim.* **11**, 287–311.

T. G. W. Epperly and R. E. Swaney (1996), Branch and bound for global NLP, Chapters 1–2 in *Global Optimization in Engineering Design* (I. E. Grossmann, ed.), Kluwer, Dordrecht.

W. R. Esposito and C. A. Floudas (2000), 'Deterministic global optimization in nonlinear optimal control problems', *J. Global Optim.* **17**, 97–126.

J. E. Falk and R. M. Soland (1969), 'An algorithm for separable nonconvex programming', *Management Sci.* **15**, 550–569.

J. C. Faugere, P. Gianni, D. Lazard, and T. Mora (1993), 'Efficient computation of zero-dimensional Groebner bases by change of ordering', *J. Symbolic Comput.* **16**, 329–344.

A. Fiacco and G. P. McCormick (1990), *Sequential Unconstrained Minimization Techniques*, Vol. 4 of *Classics in Applied Mathematics*, SIAM, Philadelphia.

R. Fletcher and S. Leyffer (1994), 'Solving mixed integer nonlinear programs by outer approximation', *Math. Program.* **66**, 327–349.

R. Fletcher and S. Leyffer (1998), 'Numerical experience with lower bounds for MIQP branch-and-bound', *SIAM J. Optim.* **8**, 604–616.

C. A. Floudas (1995), *Nonlinear and Mixed-Integer Optimization: Fundamentals and Applications*, Oxford University Press.

C. A. Floudas (1997), Deterministic global optimization in design, control, and computational chemistry, in *Large Scale Optimization with Applications, Part II: Optimal Design and Control* (L. T. Biegler *et al.*, eds), Springer, New York, pp. 129–184.
ftp://titan.princeton.edu/papers/floudas/ima.pdf

C. A. Floudas (1999), *Deterministic Global Optimization: Theory, Algorithms and Applications*, Kluwer, Dordrecht.

C. A. Floudas and P. M. Pardalos (1990), *A Collection of Test Problems for Constrained Global Optimization Algorithms*, Vol. 455 of *Lecture Notes in Computer Science*, Springer, Berlin.

C. A. Floudas and P. M. Pardalos, eds (1992), *Recent Advances in Global Optimization*, Princeton University Press.

C. A. Floudas and P. M. Pardalos, eds (1996), *State of the Art in Global Optimization*, Kluwer, Dordrecht.

C. A. Floudas and P. M. Pardalos, eds (2000), *Optimization in Computational Chemistry and Molecular Biology: Local and Global Approaches*, Kluwer, Dordrecht.

C. A. Floudas, P. M. Pardalos, C. S. Adjiman, W. R. Esposito, Z. H. Gümüs, S. T. Harding, J. L. Klepeis, C. A. Meyer and C. A. Schweiger (1999), *Handbook of Test Problems in Local and Global Optimization*, Kluwer, Dordrecht. titan.princeton.edu/TestProblems/

F. Forgó (1988), *Nonconvex Programming*, Akadémiai Kiadó, Budapest.

S. Forrest (1993), 'Genetic algorithms: Principles of natural selection applied to computation', *Science* **261**, 872–878.

R. Fourer (2003), Convexity recognition. In preparation.

R. Fourer, D. M. Gay and B. W. Kernighan (1993), *AMPL: A Modeling Language for Mathematical Programming*, Duxbury Press, Brooks/Cole Publishing Company. www.ampl.com/cm/cs/what/ampl/

T. Fujie and M. Kojima (1997), 'Semidefinite programming relaxation for nonconvex quadratic programs', *J. Global Optim.* **10**, 367–380.

J. M. Gablonsky and C. T. Kelley (2001), 'A locally-biased form of the DIRECT algorithm', *J. Global Optim.* **21**, 27–37.

GAMS World (2003), web document: www.gamsworld.org

M. R. Garey and D. S. Johnson (1979), *Computers and Intractability*, Freeman, San Francisco, CA.

K. Gatermann and P. Parrilo (2004), 'Symmetry groups, semidefinite programs, and sum of squares', math.AC/0211450, *J. Pure Appl. Algebra*, to appear. www.zib.de/gatermann/publi.html

C.-Y. Gau and L. Schrage (2004), Implementation and testing of a branch-and-bound based method for deterministic global optimization, in *Proceedings of the Conference Frontiers in Global Optimization, Santorini, Greece, June 2003* (C. A. Floudas and P. M. Pardalos, eds), Kluwer, Dordrecht.

D. M. Gay (1984), A trust-region approach to linearly constrained optimization, in *Numerical Analysis* (D. F. Griffiths, ed.), Vol. 1066 of *Lecture Notes in Mathematics*, Springer, Berlin, pp. 72–105.

K. Georg (2001), 'Improving the efficiency of exclusion algorithms', *Adv. Geom.* **1**, 193–210.

K. Georg (2002), 'A new exclusion test', *J. Comput. Appl. Math.* **152**, 147–160.

P. E. Gill and W. Murray (1974), 'Newton type methods for unconstrained and linearly constrained optimization', *Math. Program.* **7**, 311–350.

Global (and Local) Optimization (1995), web site:
www.mat.univie.ac.at/~neum/glopt.html

GLOBAL Library (2002), web document:
www.gamsworld.org/global/globallib.htm

Global Optimization Theory Institute (2003), Argonne National Laboratory, September 8–10: www-unix.mcs.anl.gov/~leyffer/gotit/

F. Glover (1989/90), 'Tabu Search'. Part 1: *ORSA J. Comput.* **1** (1989), 190–206. Part 2: *ORSA J. Comput.* **2** (1990), 4–32.

F. Glover and M. Laguna (1997), *Tabu Search*, Kluwer, Boston.

GNU Lesser General Public License (1999), web document:
www.gnu.org/copyleft/lesser.html

R. E. Gomory (1960), *An Algorithm for the Mixed Integer Problem*, RM-2597, The Rand Corporation.

L. Granvilliers (2001), 'Progress in the solving of a circuit design problem', *J. Global Optim.* **20**, 155–168.

P. Gray, W. Hart, L. Painton, C. Phillips, M. Trahan and J. Wagner (1997), A survey of global optimization methods. Web document:
www.cs.sandia.gov/opt/survey/

H. J. Greenberg (1990), 'Computational testing: Why, how, and how much', *ORSA J. Comput.* **2**, 94–97.

A. Griewank (1989), On automatic differentiation, in *Mathematical Programming* (M. Iri and K. Tanabe, eds), KTK Scientific Publishers, Tokyo, pp. 83–107.

A. Griewank (1991), Automatic evaluation of first and higher-derivative vectors, in *International Series of Numerical Mathematics*, Vol. 97, Birkhäuser, pp. 135–148.

A. Griewank (1992), 'Achieving logarithmic growth of temporal and spatial complexity in reverse automatic differentiation', *Optim. Methods Software* **1**, 35–54.

A. Griewank and G. F. Corliss (1991). *Automatic Differentiation of Algorithms*, SIAM, Philadelphia.

I. E. Grossmann (1990), 'Mixed-integer nonlinear programming techniques for the synthesis of engineering systems', *Research in Engineering Design* **1**, 205–228.

I. E. Grossmann, ed. (1996), *Global Optimization in Engineering Design*, Kluwer, Dordrecht.

I. E. Grossmann (2002), 'Review of nonlinear mixed-integer and disjunctive programming techniques', *Optim. Eng.* **3**, 227–252.
egon.cheme.cmu.edu/Group/Papers/MINLPOPTE.pdf

I. E. Grossmann and C. A. Floudas (1987), 'Active constraint strategy for flexibility analysis in chemical processes', *Comput. Chem. Eng.* **11**, 675–693.

J. Guddat, F. Guerra Vasquez and H. T. Jongen (1990), *Parametric Optimization: Singularities, Path Following and Jumps*, Wiley, Chichester.

G. D. Hager (1993), 'Solving large systems of nonlinear constraints with application to data modeling', *Interval Computations* **3**, 169–200.

T. C. Hales (1998), An overview of the Kepler conjecture. Manuscript:
math.MG/9811071–math.MG/9811078
citeseer.nj.nec.com/hales98overview.html

E. R. Hansen (2001), Publications related to early interval work of R. E. Moore. Web document:
interval.louisiana.edu/Moores_early_papers/bibliography.html

E. R. Hansen (1980), 'Global optimization using interval analysis: The multidimensional case', *Numer. Math.* **34**, 247–270.

E. R. Hansen (1992a), *Global Optimization Using Interval Analysis*, Dekker, New York.

E. R. Hansen (1992b), 'Bounding the solution of interval linear equations', *SIAM J. Numer. Anal.* **29**, 1493–1503.

B. Hanzon and D. Jibetean (2003), 'Global minimization of a multivariate polynomial using matrix methods', *J. Global Optim.* **27**, 1–23.
homepages.cwi.nl/~jibetean/polopt.ps

C. Helmberg (2003), Semidefinite programming. Web site:
www-user.tu-chemnitz.de/~helmberg/semidef.html

D. Henrion and J. B. Lasserre (2003a), 'GloptiPoly: Global optimization over polynomials with MATLAB and SeDuMi', *ACM Trans. Math. Software* **29**, 165–194.

D. Henrion and J. B. Lasserre (2003b), Solving global optimization problems over polynomials with GloptiPoly 2.1, in *Global Optimization and Constraint Satisfaction* (C. Bliek et al., eds), Springer, Berlin, pp. 43–58.

D. Henrion and J. B. Lasserre (2003c), Detecting global optimality and extracting solutions in GloptiPoly. Manuscript:
www.laas.fr/~henrion/papers/extract.pdf

W. Hock and K. Schittkowski (1981), *Test Examples for Nonlinear Programming Codes*, Vol. 187 of *Lecture Notes in Economics and Mathematical Systems*, Springer, Berlin.
ftp://plato.la.asu.edu/pub/donlp2/testenviron.tar.gz

J. Holland (1973), 'Genetic algorithms and the optimal allocation of trials', *SIAM J. Comput.* **2**, 88–105.

J. N. Hooker (2000), *Logic-Based Methods for Optimization: Combining Optimization and Constraint Satisfaction*, Wiley, New York.

J. N. Hooker (2002), 'Logic, optimization and constraint programming', *INFORMS J. Comput.* **14**, 295–321. ba.gsia.cmu.edu/jnh/joc2.ps

R. Horst and P. M. Pardalos, eds (1995), *Handbook of Global Optimization*, Kluwer, Dordrecht.

R. Horst and H. Tuy (1990), *Global Optimization: Deterministic Approaches*, 2nd edn, Springer, Berlin.

R. Horst, P. M. Pardalos and N. V. Thoai (1995), *Introduction to Global Optimization*, Kluwer, Dordrecht.

W. Huyer and A. Neumaier (1999), 'Global optimization by multilevel coordinate search', *J. Global Optim.* **14**, 331–355. See also further tests in www.mat.univie.ac.at/~neum/glopt/contrib/compbound.pdf

W. Huyer and A. Neumaier (2003), SNOBFIT: Stable Noisy Optimization by Branch and Fit. Manuscript:
www.mat.univie.ac.at/~neum/papers.html#snobfit

W. Huyer and A. Neumaier (2004), 'Integral approximation of rays and verification of feasibility', *Reliable Comput.* **10**, 195–207.
www.mat.univie.ac.at/~neum/papers.html#rays

E. Hyvönen and S. De Pascale (1996), Interval computations on the spreadsheet, in *Applications of Interval Computations* (R. B. Kearfott and V. Kreinovich, eds), Applied Optimization, Kluwer, Dordrecht, pp. 169–209.

L. Ingber (1993), 'Simulated annealing: Practice versus theory', *Math. Comput. Modelling* **18**, 29–57.

R. H. F. Jackson, P. T. Boggs, S. G. Nash and S. Powell (1990/91), 'Guidelines for reporting results of computational experiments: Report of the ad hoc committee', *Math. Program.* **49**, 413–426.

E. Janka (2000), A comparison of stochastic methods for global optimization. Web document: www.mat.univie.ac.at/~vpk/math/gopt_eng.html

C. Jansson (1994), On self-validating methods for optimization problems, in *Topics in Validated Computations* (J. Herzberger, ed.), Elsevier Science BV, pp. 381–438.

C. Jansson (2003), Rigorous error bounds for the optimal value of linear programming problems, in *Global Optimization and Constraint Satisfaction* (C. Bliek et al., eds), Springer, Berlin, pp. 59–70.

C. Jansson (2004a), 'A rigorous lower bound for the optimal value of convex optimization problems', *J. Global Optim.* **28**, 121–137.

C. Jansson (2004b), 'Rigorous lower and upper bounds in linear programming', *SIAM J. Optim.*, to appear.

C. Jansson and O. Knüppel (1992), A global minimization method: The multidimensional case. Technical report 92-1, TU Hamburg-Harburg, January.

C. Jansson and O. Knüppel (1995), 'Branch and bound algorithm for bound constrained optimization problems without derivatives', *J. Global Optim.* **7**, 297–333.

L. Jaulin and D. Henrion (2004), 'Linear matrix inequalities for interval constraint propagation', *Reliable Comput.*, to appear.
www.laas.fr/~henrion/Papers/jaulin_lmi.pdf

L. Jaulin, M. Kieffer, O. Didrit and E. Walter (2001), *Applied Interval Analysis*, Springer, London.

R. G. Jeroslow (1977), 'Cutting plane theory: Disjunctive methods', *Ann. Discrete Math.* **1**, 293–330.

D. Jibetean and E. De Klerk (2003), Global optimization of rational functions: A semidefinite programming approach. Manuscript:
www.optimization-online.org/DB_HTML/2003/05/654.html

F. John (1948), Extremum problems with inequalities as subsidiary conditions, in *Studies and Essays Presented to R. Courant on his 60th Birthday, January 8, 1948*, Interscience, New York, pp. 187–204. Reprinted as *Fritz John, Collected Papers*, Vol. 2 (J. Moser, ed.), Birkhäuser, Boston (1985), pp. 543–560.

D. S. Johnson (2002), A theoretician's guide to the experimental analysis of algorithms, in *Proc. 5th and 6th DIMACS Implementation Challenges* (M. Goldwasser, D. S. Johnson, and C. C. McGeoch, eds) AMS, Providence, RI, pp. 215–250. www.research.att.com/~dsj/papers/experguide.pdf

D. R. Jones (2001), 'A taxonomy of global optimization methods based on response surfaces', *J. Global Optim.* **21**, 345–383.

D. R. Jones, C. D. Perttunen and B. E. Stuckman (1993), 'Lipschitzian optimization without the Lipschitz constant', *J. Optim. Theory Appl.* **79**, 157–181.

D. R. Jones, M. Schonlau and W. J. Welch (1998), 'Efficient global optimization of expensive black-box functions', *J. Global Optim.* **13**, 455–492.

G. Jónsson and S. A. Vavasis (2001), Accurate solution of polynomial equations using Macaulay resultant matrices. Manuscript:
www.cs.cornell.edu/home/vavasis/vavasis.html

W. Kahan (1968), A more complete interval arithmetic: Lecture notes for an engineering summer course in numerical analysis, University of Michigan.

J. Kallrath (2002), *Gemischt-ganzzahlige Optimierung: Modellierung in der Praxis*, Vieweg.

W. Karush (1939), Minima of functions of several variables with inequalities as side constraints, MSc dissertation, Department of Mathematics, University of Chicago, IL.

R. B. Kearfott (1991), 'Decomposition of arithmetic expressions to improve the behavior of interval iteration for nonlinear systems', *Computing* **47**, 169–191.

R. B. Kearfott (1996a), A review of techniques in the verified solution of constrained global optimization problems, in *Applications of Interval Computations* (R. B. Kearfott and V. Kreinovich, eds), Kluwer, Dordrecht, pp. 23–60.

R. B. Kearfott (1996b), *Rigorous Global Search: Continuous Problems*, Kluwer, Dordrecht.

R. B. Kearfott (2003), GlobSol: History, composition, and advice on use, in *Global Optimization and Constraint Satisfaction* (C. Bliek et al., eds), Springer, Berlin, pp. 17–31.

R. B. Kearfott and K. Du (1994), 'The cluster problem in multivariate global optimization', *J. Global Optim.* **5**, 253–265.

R. B. Kearfott, M. Novoa and Chenyi Hu (1991), 'A review of preconditioners for the interval Gauss–Seidel method', *Interval Computations* **1**, 59–85.

C. Khompatraporn, J. Pintér and Z. B. Zabinsky (2004), 'Comparative assessment of algorithms and software for global optimization', *J. Global Optim.*, to appear.

S. Kim and M. Kojima (2001), 'Second order cone programming relaxation methods of nonconvex quadratic optimization problem', *Optim. Methods Software* **15**, 201–224.

S. Kim and M. Kojima (2003), 'Exact solutions of some nonconvex quadratic optimization problems via SDP and SOCP relaxations', *Comput. Optim. Appl.* **26**, 143–154.

S. Kim, M. Kojima and H. Waki (2003), Generalized Lagrangian duals and sums of squares relaxations of sparse polynomial optimization problems. Manuscript: math.ewha.ac.kr/~skim/Research/list.html

S. Kirkpatrick, C. D. Geddat, Jr., and M. P. Vecchi (1983), 'Optimization by simulated annealing', *Science* **220**, 671–680.

G. R. Kocis and I. E. Grossmann (1989), 'Computational experience with DICOPT solving MINLP problems in process systems engineering', *Comput. Chem. Eng.* **13**, 307–315.

M. Kojima and L. Tuncel (2000), 'Cones of matrices and successive convex relaxations of nonconvex sets', *SIAM J. Optim.* **10**, 750–778.

M. Kojima, S. Kim and H. Waki (2003), 'A general framework for convex relaxation of polynomial optimization problems over cones', *J. Oper. Res. Soc. Japan* **46**, 125–144.

L. V. Kolev (1997), 'Use of interval slopes for the irrational part of factorable functions', *Reliable Comput.* **3**, 83–93.

L. V. Kolev and I. P. Nenov (2001), 'Cheap and tight bounds on the solution set of perturbed systems of nonlinear equations', *Reliable Comput.* **7**, 399–408.

J. Kostrowicki and H. A. Scheraga (1996), Some approaches to the multiple-minima problem in protein folding, in *Global Minimization of Nonconvex Energy Functions: Molecular Conformation and Protein Folding* (P. M. Pardalos et al., eds), AMS, Providence, RI, pp. 123–132.

R. Krawczyk (1969), 'Newton-Algorithmen zur Bestimmung von Nullstellen mit Fehlerschranken', *Computing* **4**, 187–201.

H. W. Kuhn (1991), Nonlinear programming: A historical note, in *History of Mathematical Programming* (J. K. Lenstra et al., eds) North Holland, Amsterdam, pp. 82–96.

H. W. Kuhn and A. W. Tucker (1951), in *Nonlinear Programming, Proc. 2nd Berkeley Symp. Math. Stat. Prob.* (J. Neyman, ed.), University of California Press, Berkeley, CA, pp. 481–492.

J. C. Lagarias (2002), 'Bounds for local density of sphere packings and the Kepler conjecture', *Discrete Comput. Geom.* **27**, 165–193.

J. L. Lagrange (1797), *Théorie des Fonctions Analytiques*, Impr. de la République, Paris.

A. H. Land and A. G. Doig (1960), 'An automated method for solving discrete programming problems', *Econometrica* **28**, 497–520.

J. B. Lasserre (2001), 'Global optimization with polynomials and the problem of moments', *SIAM J. Optim.* **11**, 796–817.

Y. Lebbah, M. Rueher and C. Michel (2002), A global filtering algorithm for handling systems of quadratic equations and inequations, in *Principles and Practice of Constraint Programming, CP 2002* (P. van Hentenryck, ed.), Vol. 2470 of *Lecture Notes in Computer Science*, Springer, New York, pp. 109–123.

Y. Lebbah, C. Michel, M. Rueher, J.-P. Merlet and D. Daney (2004), Efficient and safe global constraints for handling numerical constraint systems, *SIAM J. Numer. Anal.*, to appear.

E. Lee and C. Mavroidis (2002), 'Solving the geometric design problem of spatial 3R robot manipulators using polynomial homotopy continuation', *J. Mech. Design, Trans. ASME* **124**, 652–661.

E. Lee, C. Mavroidis and J. P. Merlet (2002), Five precision points synthesis of spatial RRR manipulators using interval analysis, in *Proc. 2002 ASME Mechanisms and Robotics Conference, Montreal, September 29–October 2*, pp. 1–10.
robots.rutgers.edu/Publications.htm

M. Lerch, G. Tischler, J. Wolff von Gudenberg, W. Hofschuster and W. Krämer (2001), The Interval Library filib 2.0: Design, features and sample programs. Preprint 2001/4, Universität Wuppertal.
www.math.uni-wuppertal.de/wrswt/software/filib.html

A. V. Levy and A. Montalvo (1985), 'The tunneling algorithm for the global minimization of functions', *SIAM J. Sci. Statist. Comput.* **6**, 15–29.

J. D. Little, K. C. Murty, D. W. Sweeney and C. Karel (1963), 'An algorithm for the travelling salesman problem', *Oper. Res.* **11**, 972–989.

M. Lobo, L. Vandenberghe, S. Boyd, and H. Lebret (1998), 'Applications of second-order cone programming', *Linear Algebra Appl.* **284**, 193–228.
www.stanford.edu/~boyd/socp.html

W. A. Lodwick (1989), 'Constraint propagation, relational arithmetic in AI systems and mathematical programs', *Ann. Oper. Res.* **21**, 143–148.

Z.-Q. Luo, J.-S. Pang and D. Ralph (1996), *Mathematical Programs with Equilibrium Constraints*, Cambridge University Press, Cambridge.

I. J. Lustig and J.-F. Puget (2001), Program does not equal program: Constraint programming and its relationship to mathematical programming, *Interfaces* **31**, 29–53.

K. Madsen and S. Skelboe (2002), The early days of interval global optimization. interval.louisiana.edu/conferences/VC02/abstracts/MADS.pdf

C. Maheshwari, A. Neumaier and H. Schichl (2003), Convexity and concavity detection. In preparation.

K. Makino (1998), Rigorous analysis of nonlinear motion in particle accelerators, PhD thesis, Department of Physics and Astronomy, Michigan State University: bt.pa.msu.edu/makino/phd.html

K. Makino and M. Berz (2003), 'Taylor models and other validated functional inclusion methods', *Int. J. Pure Appl. Math.* **4**, 379–456.

O. L. Mangasarian (1969), *Nonlinear Programming*, McGraw-Hill, New York. Reprinted as *Classics in Applied Mathematics*, SIAM, Philadelphia (1994).

O. L. Mangasarian and S. Fromovitz (1967), 'The Fritz John necessary optimality conditions in the presence of equality and inequality constraints', *J. Math. Anal. Appl.* **17**, 37–47.

O. L. Mangasarian and L. McLinden (1985), 'Simple bounds for solutions of monotone complementarity problems and convex programs', *Math. Program.* **32**, 32–40.

F. Margot (2003), 'Exploiting orbits in symmetric ILP', *Math. Program.* **98**, 3–21

H. M. Markowitz and A. S. Manne (1957), 'On the solution of discrete programming problems', *Econometrica* **25**, 84-110.

G. P. McCormick (1972), Converting general nonlinear programming problems to separable nonlinear programming problems. Technical report T-267, George Washington University, Washington, DC.

G. P. McCormick (1976), 'Computability of global solutions to factorable nonconvex programs, Part I: Convex underestimating problems', *Math. Program.* **10**, 147–175.

C. M. McDonald and C. A. Floudas (1995), 'Global optimization for the phase and chemical equilibrium problem: Application to the NRTL equation', *Comput. Chem. Eng.* **19**, 1111–1139.

M. McKay, R. Beckman and W. Conover (1979), 'A comparison of three methods for selecting values of input variables in the analysis of output from a computer code', *Technometrics* **21**, 239–245.

J.-P. Merlet (2001), 'A parser for the interval evaluation of analytical functions and its applications to engineering problems', *J. Symbolic Comput.* **31**, 475–486.

C. A. Meyer, C. A. Floudas and A. Neumaier (2002), 'Global optimization with nonfactorable constraints', *Ind. Eng. Chem. Res.* **41**, 6413–6424.

R. J. Meziat (2003), Analysis of non convex polynomial programs by the method of moments. Manuscript:
www.optimization-online.org/ARCHIVE_DIGEST/2003-03.html

Z . Michalewicz (1996), *Genetic Algorithm + Data Structures = Evolution Programs*, 3rd edn, Springer, New York.

H. Mittelmann (2002), Benchmarks. Web site:
plato.la.asu.edu/topics/benchm.html

J. Mockus (1966), Multiextremal problems in design, PhD Thesis, Nauka.

J. Mockus (1989), *Bayesian Approach to Global Optimization*, Kluwer, Dordrecht.

J. Mockus (1994), 'Application of Bayesian approach to numerical methods of global and stochastic optimization', *J. Global Optim.* **4**, 347–356.

H. M. Moller and H. J. Stetter (1995), 'Multivariate polynomial equations with multiple zeros solved by matrix eigenproblems', *Numer. Math.* **70**, 311–329.

M. Mongeau, H. Karsenty, V. Rouz and J.-B. Hiriart-Urruty (2000), 'Comparison of public-domain software for black box global optimization', *Optim. Methods Software* **13**, 203–226.

R. E. Moore (1962), Interval arithmetic and automatic error analysis in digital computing. PhD thesis, Appl. Math. Statist. Lab. Rep. 25, Stanford University, Stanford, CA:
interval.louisiana.edu/Moores_early_papers/disert.pdf

R. E. Moore (1979), *Methods and Applications of Interval Analysis*, SIAM, Philadelphia.

R. E. Moore and C. T. Yang (1959), Interval analysis, I. Space Division Report LMSD285875, Lockheed Missiles and Space Co.
interval.louisiana.edu/Moores_early_papers/Moore_Yang.pdf

J. Moré and Z. Wu (1997), 'Global continuation for distance geometry problems', *SIAM J. Optim.* **7**, 814–836.

J. J. Moré, B. S. Garbow and K. E. Hillstrom (1981), 'Testing unconstrained optimization software', *ACM Trans. Math. Software* **7**, 17–41.

T. S. Motzkin and E. G. Strauss (1965), 'Maxima for graphs and a new proof of a theorem of Turan', *Canad. J. Math.* **17**, 533–540.

B. Mourrain and V. Y. Pan (1997), Solving special polynomial systems by using structured matrices and algebraic residues, in *Proc. Workshop on Foundations of Computational Mathematics* (F. Cucker and M. Shub, eds), Springer, Berlin, pp. 287–304.

B. Mourrain, Y. V. Pan and O. Ruatta (2003), 'Accelerated solution of multivariate polynomial systems of equations', *SIAM J. Comput.* **32**, 435–454.

K. G. Murty and S. N. Kabadi (1987), 'Some NP-complete problems in quadratic and nonlinear programming', *Math. Program.* **39**, 117–129.

NAG (2003), Differentiation enabled Fortran compiler technology. Web document:
www.nag.co.uk/nagware/research/ad_overview.asp

N. S. Nedialkov and K. R. Jackson (2001), A new perspective on the wrapping effect in interval methods for initial value problems for ordinary differential equations, in *Perspectives on Enclosure Methods* (U. Kulisch et al., eds), Springer, Berlin, pp. 219–264. www.cs.toronto.edu/NA/reports.html#ned.scan00

G. L. Nemhauser and L. A. Wolsey (1988), *Integer and Combinatorial Optimization*, Wiley, New York.

G. L. Nemhauser and L. A. Wolsey (1989), Integer programming, Chapter VI in *Optimization* (G. L. Nemhauser et al., eds), Vol. 1 of *Handbooks in Operations Research and Management Science*, North Holland, Amsterdam, pp. 447–527.

I. P. Nenov and D. H. Fylstra (2003), 'Interval methods for accelerated global search in the Microsoft Excel Solver', *Reliable Comput.* **9**, 143–159.

A. Neumaier (1988), The enclosure of solutions of parameter-dependent systems of equations, in *Reliability in Computing* (R. E. Moore, ed.), Academic Press, San Diego, pp. 269–286.
www.mat.univie.ac.at/~neum/publist.html#encl

A. Neumaier (1990), *Interval Methods for Systems of Equations*, Cambridge University Press.

A. Neumaier (1993), 'The wrapping effect, ellipsoid arithmetic, stability and confidence regions', *Computing Supplementum* **9**, 175–190.

A. Neumaier (1996), 'Second-order sufficient optimality conditions for local and global nonlinear programming', *J. Global Optim.* **9**, 141–151.

A. Neumaier (1997), 'Molecular modeling of proteins and mathematical prediction of protein structure', *SIAM Review* **39**, 407–460.

A. Neumaier (1999), 'A simple derivation of the Hansen–Bliek–Rohn–Ning–Kearfott enclosure for linear interval equations', *Reliable Comput.* **5**, 131–136. Erratum, *Reliable Comput.* **6** (2000), 227.

A. Neumaier (2001a), 'Generalized Lyapunov–Schmidt reduction for parametrized equations at near singular points', *Linear Algebra Appl.* **324**, 119–131.

A. Neumaier (2001b), *Introduction to Numerical Analysis*, Cambridge University Press, Cambridge.

A. Neumaier (2002), 'Taylor forms: Use and limits', *Reliable Comput.* **9**, 43–79.

A. Neumaier (2003a), 'Rational functions with prescribed global and local minimizers', *J. Global Optim.* **25**, 175–181.

A. Neumaier (2003b), Constraint satisfaction and global optimization in robotics. Manuscript: www.mat.univie.ac.at/~neum/papers.html#rob

A. Neumaier and H. Schichl (2003), Sharpening the Karush–John optimality conditions. Manuscript: www.mat.univie.ac.at/~neum/papers.html#kj

A. Neumaier and O. Shcherbina (2004), Safe bounds in linear and mixed-integer programming, *Math. Program. A* **99**, 283–296.
www.mat.univie.ac.at/~neum/papers.html#mip

J. Nocedal and S. J. Wright (1999), *Numerical Optimization*, Springer Series in Operations Research, Springer, Berlin.

I. Nowak, H. Alperin and S. Vigerske (2003), LaGO: An object oriented library for solving MINLPs, in *Global Optimization and Constraint Satisfaction* (C. Bliek et al., eds), Springer, Berlin, pp. 32–42.
www.math.hu-berlin.de/~alpe/papers/LaGO/

W. Older and A. Vellino (1993), Constraint arithmetic on real intervals, in *Constrained Logic Programming: Selected Research* (F. Benhameou and A. Colmerauer, eds), MIT Press.

J. Outrata, M. Kočvara and J. Zowe (1998), *Nonsmooth Approach to Optimization problems with Equilibrium Constraints*, Kluwer, Dordrecht.

A. B. Owen (1992), 'Orthogonal arrays for computer experiments, integration and visualization', *Statist. Sinica* **2**, 439–452.

A. B. Owen (1994), 'Lattice sampling revisited: Monte Carlo variance of means over randomized orthogonal arrays', *Ann. Statist.* **22**, 930–945.

I. Papamichail and C. S. Adjiman (2002), 'A rigorous global optimization algorithm for problems with ordinary differential equations', *J. Global Optim.* **24**, 1–33.

P. M Pardalos and J. B. Rosen (1987), *Constrained Global Optimization: Algorithms and Applications*, Vol. 268 of Lecture Notes in Computer Science, Springer, Berlin.

P. M. Pardalos and G. Schnitger (1988), 'Checking local optimality in constrained quadratic programming is NP-hard', *Oper. Res. Lett.* **7**, 33–35.

R. G. Parker and R. L. Rardin (1988), *Discrete Optimization*, Academic Press, San Diego, CA.

P. A. Parrilo (2003), 'Semidefinite programming relaxations for semialgebraic problems', *Math. Program. B* **96**, 293–320.

P. A. Parrilo and S. Lall (2003), 'Semidefinite programming relaxations and algebraic optimization in control', *Europ. J. Control* **9**, 307–321.

P. A. Parrilo and B. Sturmfels (2003), Minimizing polynomial functions, in *Algorithmic and quantitative real algebraic geometry*, Vol. 60 of *DIMACS Series in Discrete Mathematics and Theoretical Computer Science*, AMS, Providence, RI, pp. 83–99. www.arxiv.org/abs/math.OC/0103170

J. D. Pintér (1996a), *Global Optimization in Action*, Kluwer, Dordrecht.

J. D. Pintér (1996b), 'Continuous global optimization software: A brief review', *Optima* **52**, 1–8.

J. D. Pintér (1999), *LGO: A Model Development System for Continuous Global Optimization. User's Guide*, Pintér Consulting Services, Inc., Halifax, NS.

S. A. Piyavskii (1972), An algorithm for finding the absolute extremum of a function, *USSR Comput. Math. and Math. Phys.* **12** 57–67.

S. Prajna, A. Papachristodoulou and A. Parrilo (2002), SOSTOOLS: Sum of Squares Optimization Toolbox for MATLAB – User's guide. Manuscript: www.optimization-online.org/DB_HTML/2002/05/483.html

H. D. Ratliff and W. Pierskalla (1981), 'Reporting computational experience in operations research', *Oper. Res.* **29**, xi–xiv.

H. Ratschek and J. G. Rokne (1984), *New Computer Methods for the Range of Functions*, Ellis Horwood, Chichester.

H. Ratschek and J. G. Rokne (1988), *New Computer Methods for Global Optimization*, Wiley, New York.

D. Ratz (1996), On branching rules in second-order branch-and-bound methods for global optimization, in *Scientific Computation and Validation* (G. Alefeld et al., eds), Akademie-Verlag, Berlin.

D. Ratz and T. Csendes (1995), 'On the selection of subdivision directions in interval branch-and-bound methods for global optimization', *J. Global Optim.* **7**, 183–207.

A. D. Rikun (1997), 'A convex envelope formula for multilinear functions', *J. Global Optim.* **10**, 425–437.

G. Rote (1992), 'The convergence of the sandwich algorithm for approximating convex functions', *Computing* **48**, 337–361.

H. S. Ryoo and N. V. Sahinidis (1996), 'A branch-and-reduce approach to global optimization', *J. Global Optim.* **8**, 107–139.

J. Sacks, W. J. Welch, T. J. Mitchell and H. P. Wynn (1989), 'Design and analysis of computer experiments', *Statist. Sci.* **4**, 409–435.

N. V. Sahinidis (1996), 'BARON: A general purpose global optimization software package', *J. Global Optim.* **8**, 201–205.

N. V. Sahinidis (2000), BARON: Branch And Reduce Optimization Navigator – User's manual. Web document:
archimedes.scs.uiuc.edu/baron/baron.html

N. V. Sahinidis (2003), Global optimization and constraint satisfaction: The branch-and-reduce approach, in *Global Optimization and Constraint Satisfaction* (C. Bliek et al., eds), Springer, Berlin, pp. 1–16.

H. Schichl (2004), Global optimization in the COCONUT project, in *Numerical Software with Result Verification*, Vol. 2991 of *Lecture Notes in Computer Science*, Springer, Heidelberg, pp. 243–249.
www.mat.univie.ac.at/~herman/papers.html

H. Schichl and A. Neumaier (2003), Interval analysis on directed acyclic graphs for global optimization. Manuscript:
www.mat.univie.ac.at/~neum/papers.html#intdag

H. Schichl and A. Neumaier (2004), 'Exclusion regions for systems of equations', *SIAM J. Numer. Anal.* **42**, 383–408.
www.mat.univie.ac.at/~neum/papers.html#excl

R. B. Schnabel and E. Eskow (1990), 'A new modified Cholesky factorization', *SIAM J. Sci. Statist. Comput.* **11**, 1136–1158.

SeDuMi (2001), web site: fewcal.kub.nl/sturm/software/sedumi.html

O. Shcherbina (2002), Misprints and mistakes in Floudas et al. (1999). Web document:
www.mat.univie.ac.at/~neum/glopt/contrib/handbook_corr.html

O. Shcherbina, A. Neumaier, D. Sam-Haroud, Xuan-Ha Vu and Tuan-Viet Nguyen (2003), Benchmarking global optimization and constraint satisfaction codes, in *Global Optimization and Constraint Satisfaction* (C. Bliek *et al.*, eds), Springer, Berlin, pp. 211–222.
www.mat.univie.ac.at/~neum/papers.html#bench

J. P. Shectman and N. V. Sahinidis (1998), 'A finite algorithm for global minimization of separable concave programs', *J. Global Optim.* **12**, 1–36.

H. D. Sherali (1997), 'Convex envelopes of multilinear functions over a unit hypercube and over special discrete sets', *Acta Math. Vietnamica* **22**, 245–270.

H. D. Sherali and C. M. Shetty (1980), *Optimization with Disjunctive Constraints*, Springer, Berlin.

H. D. Sherali and C. H. Tuncbilec (1995), 'A reformulation-convexification approach for solving nonconvex quadratic programming problems', *J. Global Optim.* **7**, 1–31.

H. D. Sherali and C. H. Tuncbilec (1997a), 'New reformulation linearization/convexification relaxations for univariate and multivariate polynomial programming problems', *Oper. Res. Lett.* **21**, 1–9.

H. D. Sherali and C. H. Tuncbilec (1997b), 'Comparison of two reformulation-linearization technique based linear programming relaxations for polynomial programming problems', *J. Global Optim.* **10**, 381–390.

K. Shimizu, Y. Ishizuka and J. F. Bard (1997), *Nondifferentiable and Two-Level Programming*, Kluwer, Boston.

J. Siek and A. Lumsdaine (1999), 'The matrix template library: Generic components for high-performance scientific computing', *Computing in Science and Engineering* **18**, 70–78. www.osl.iu.edu/research/mtl/

S. Skelboe (1974), 'Computation of rational interval functions', *BIT* **14**, 87–95.
www.diku.dk/~stig/CompRatIntv.pdf

N. J. A. Sloane (2003), web document with tables of packings and designs:
www.research.att.com/~njas/

H. Stetter (1997), Stabilization of polynomial systems solving with Groebner bases, in *Proc. ACM Int. Symp. Symbolic Algebraic Computation*, pp. 117–124.

F. H. Stillinger (1985), 'Role of potential-energy scaling in the low-temperature relaxation behavior of amorphous materials', *Phys. Rev. B* **32**, 3134–3141.

J. F. Sturm (1999), 'Using SeDuMi 1.02, a MATLAB toolbox for optimization over symmetric cones', *Optim. Methods Software* **11–12**, 625–653.
fewcal.kub.nl/sturm/software/sedumi.html

B. Tang (1993), 'Orthogonal array-based Latin hypercubes', *J. Amer. Statist. Assoc.* **88**, 1392–1397.

M. Tawarmalani and N. V. Sahinidis (2001), Semidefinite relaxations of fractional programs via novel convexification techniques, *J. Global Optim.* **20** 137–158.

M. Tawarmalani and N. V. Sahinidis (2002*a*), Convex extensions and envelopes of lower semi-continuous functions, *Math. Program.* **93** 247–263.

M. Tawarmalani and N. V. Sahinidis (2002*b*), *Convexification and Global Optimization in Continuous and Mixed-Integer Nonlinear Programming: Theory, Algorithms, Software, and Applications*, Kluwer, Dordrecht.

M. Tawarmalani and N. V. Sahinidis (2004), 'Global optimization of mixed-integer nonlinear programs: A theoretical and computational study', *Math. Program.*, DOI 10.1007/s10107-003-0467-6.

M. Todd (2001), Semidefinite optimization, in *Acta Numerica*, Vol. 10, Cambridge University Press, pp. 515–560.

J. A. Tomlin (1970), Branch and bound methods for integer and non-convex programming, in *Integer and Nonlinear Programming* (J. Abadie, ed.), American Elsevier Publishing Company, New York, pp. 437–450.

TOMS (2002), Information for authors. Web document:
www.acm.org/toms/Authors.html#TypesofPapers

A. Törn (1972), Global optimization as a combination of local and global search, in *Proceedings of Computer Simulation Versus Analytical Solutions for Business and Economical Models, Gothenburg, August 1972* (W. Goldberg, ed.).

A. Törn (1974), Global optimization as a combination of global and local search, PhD Thesis, Abo Akademi University, HHAAA 13.

A. Törn (2000), Global optimization. Web document:
www.abo.fi/~atorn/Globopt.html

A. Törn and A. Žilinskas (1989), *Global Optimization*, Vol. 350 of *Lecture Notes in Computer Science*, Springer, Berlin.

A. Törn, M. Ali and S. Viitanen (1999), 'Stochastic global optimization: Problem classes and solution techniques', *J. Global Optim.* **14**, 437–447.

T. Tsuda and T. Kiono (1964), Application of the Monte Carlo method to systems of nonlinear algebraic equations, *Numerische Matematik* **6** 59–67.

H. Tuy (1995), DC optimization: Theory, methods and algorithms, in *Handbook of Global Optimization* (R. Horst and P. M. Pardalos eds), Kluwer, Dordrecht, pp. 149–216.

M. Ulbrich and S. Ulbrich (1996), Automatic differentiation: A structure-exploiting forward mode with almost optimal complexity for Kantorovich trees, in *Applied Mathematics and Parallel Computing* (H. Fischer *et al.*, eds), Physica-Verlag, Heidelberg, pp. 327–357.

L. Vandenberghe and S. Boyd (1996), 'Semidefinite programming', *SIAM Review* **38**, 49–95. www.stanford.edu/~boyd/reports/semidef_prog.ps

D. Vandenbussche (2003), Polyhedral approaches to solving nonconvex quadratic programs, PhD thesis, School of Industrial and Systems Engineering, Georgia Institute of Technology.

B. Vanderbei (2000), Nonlinear optimization models. Web site:
www.sor.princeton.edu/~rvdb/ampl/nlmodels/

P. Van Hentenryck (1989), *Constraint Satisfaction in Logic Programming*, MIT Press, Cambridge, MA.

P. Van Hentenryck, L. Michel and F. Benhamou (1997a), 'Newton: Constraint programming over non-linear constraints', *Sci. Program.* **30**, 83–118.

P. Van Hentenryck, L. Michel and Y. Deville (1997b), *Numerica: A Modeling Language for Global Optimization*, MIT Press, Cambridge, MA.

R. Van Iwaarden (1986), An improved unconstrained global optimization algorithm, PhD thesis, University of Colorado at Denver, CO.

P. J. M. Van Laarhoven and E. H. L. Aarts (1987), *Simulated Annealing: Theory and Applications*, Kluwer, Dordrecht.

L. N. Vicente and P. H. Calamai (1994), 'Bilevel and multilevel programming: A bibliography review', *J. Global Optim.* **5**, 291–306.

V. Visweswaran and C. A. Floudas (1996a), New formulations and branching strategies for the GOP algorithm, in *Global Optimization in Chemical Engineering* (I. E. Grossmann, ed.), Kluwer, Dordrecht, pp. 75–100.

V. Visweswaran and C. A. Floudas (1996b), Computational results for an efficient implementation of the GOP algorithm and its variants, in *Global Optimization in Chemical Engineering* (I. E. Grossmann, ed.), Kluwer, Dordrecht, pp. 111–153.

Xuan-Ha Vu, D. Sam-Haroud and M.-C. Silaghi (2002), Approximation techniques for non-linear problems with continuum of solutions, in *Proc. 5th Int. Symp. Abstraction, Reformulation Approximation, SARA'2002* (B. Y. Choueiry and T. Walsh, eds), Vol. 2371 of *Lecture Notes in Artificial Intelligence*, Springer, pp. 224–241.

Xuan-Ha Vu, D. Sam-Haroud and M.-C. Silaghi (2003), Numerical constraint satisfaction problems with non-isolated solutions, in *Global Optimization and Constraint Satisfaction* (C. Bliek et al., eds), Springer, Berlin, pp. 194–210.

W. Walster (2000), Interval arithmetic solves nonlinear problems while providing guaranteed results, FORTE TOOLS feature stories. Web manuscript:
wwws.sun.com/software/sundev/news/features/intervals.html

G. Walster, E. Hansen and S. Sengupta (1985), Test results for a global optimization algorithm, in *Numerical Optimization 1984* (P. T. Boggs et al., eds), SIAM, Philadelphia, pp. 272–287.

H. P. Williams (1999), *Model Solving in Mathematical Programming*, 4th edn, Wiley, Chichester.

H. Wolkowicz, R. Saigal and L. Vandenberghe, eds (2000), *Handbook on Semidefinite Programming*, Kluwer, Dordrecht.

L. A. Wolsey (1998), *Integer Programming*, Wiley.

S. J. Wright (1997), *Primal-Dual Interior-Point Methods*, SIAM, Philadelphia.

M. Yagiura and T. Ibaraki (2001), 'On metaheuristic algorithms for combinatorial optimization problems', *Systems and Computers in Japan* **32**, 33–55.

H. Yan and J. N. Hooker (1999), 'Tight representations of logical constraints as cardinality rules', *Math. Program.* **85**, 363–377.

J. M. Zamora and I. E. Grossmann (1999), 'A branch and contract algorithm for problems with concave univariate, bilinear and linear fractional terms', *J. Global Optim.* **14**, 217–249.

A. A. Zhigljavsky (1991), *Theory of Global Random Search*, Kluwer, Dordrecht.

Multiscale computational modelling of the heart

N. P. Smith, D. P. Nickerson, E. J. Crampin and P. J. Hunter*
*Bioengineering Institute,
The University of Auckland,
Private Bag 92019 Auckland,
New Zealand*
E-mail: p.hunter@auckland.ac.nz

A computational framework is presented for integrating the electrical, mechanical and biochemical functions of the heart. Finite element techniques are used to solve the large-deformation soft tissue mechanics using orthotropic constitutive laws based in the measured fibre-sheet structure of myocardial (heart muscle) tissue. The reaction–diffusion equations governing electrical current flow in the heart are solved on a grid of deforming material points which access systems of ODEs representing the cellular processes underlying the cardiac action potential. Navier–Stokes equations are solved for coronary blood flow in a system of branching blood vessels embedded in the deforming myocardium and the delivery of oxygen and metabolites is coupled to the energy-dependent cellular processes. The framework presented here for modelling coupled physical conservation laws at the tissue and organ levels is also appropriate for other organ systems in the body and we briefly discuss applications to the lungs and the musculo-skeletal system. The computational framework is also designed to reach down to subcellular processes, including signal transduction cascades and metabolic pathways as well as ion channel electrophysiology, and we discuss the development of ontologies and markup language standards that will help link the tissue and organ level models to the vast array of gene and protein data that are now available in web-accessible databases.

---

* Supported by grants from the NZ Royal Society (to the Centre for Molecular Biodiscovery and the New Zealand Institute of Mathematics and its Applications) and the NZ Foundation for Research Science and Technology.

CONTENTS

1	Introduction	372
2	Finite element modelling of cardiac tissue	374
3	Continuum modelling	384
4	Cellular modelling	399
5	Vascular blood flow	412
6	Future directions	421
7	Conclusion	425
	References	426

1. Introduction

The application of numerical and computational techniques to interpret, simulate and ultimately elucidate the physiological function of complex whole organ systems has developed rapidly within the last two decades. This has been due partly to the availability of new experimental data at the cell, tissue and organ levels and partly to the improvement in computational algorithms and the parallel increase in availability of high performance computing resources.

Understanding the integrative function of an organ requires the modelling of biological processes at multiple spatial and temporal scales. Efficient contraction of the myocardium (heart muscle) to pump blood from the ventricles depends crucially on the organization of structural proteins at the tissue level and the operation of contractile proteins at the subcellular level. At all levels these processes are nonlinear and time-dependent.

In this review we analyse the mechanical, electrical and biochemical function of the heart using models which are based on the anatomy and biophysics of the cells, tissues and organ. The framework developed here for the heart is, however, applicable to all organ systems of the body and we briefly illustrate its application to the lungs and musculo-skeletal system at the end of the article.

1.1. Anatomy and physiology of the heart

The heart is a four-chambered pump with two priming chambers, the left and right atria, and two primary pumping chambers, the left and right ventricles (LV and RV, respectively). Blood is pumped from the RV via the pulmonary artery to the lungs at a peak pressure of about 4 kPa and from the LV via the aorta to the rest of the body at a peak pressure of about 17 kPa. Blood is ejected from the ventricles by the contraction of the myocardium when triggered by changes in the electrical potential across the muscle cell (myocyte) outer membrane (sarcolemma). The electrical potential inside the myocytes is normally maintained at -85 mV relative to the extracellular fluid. Electric current flow into a cell (inward current) raises the potential (depolarizes the cell) and initiates wave propagation along the electrically conducting cells which are also electrically connected by gap junction proteins. These electrically connected cells act as a cable but with continual regeneration of the signal via inward current through voltage-sensitive ion channels (see later). The whole electrical activation process is initiated under normal physiological conditions by the heart's pacemaker cells, which reside in the right atrium and are called the sino-atrial (SA) node. The signal spontaneously generated by the SA node propagates through the atrial muscle and then to the ventricles via a specialized conducting region (the atrio-ventricular node or AV node). From here the electrical wave travels down specialized conducting fibres called the bundle of His to a network of Purkinje fibres, which spread out through the inner region of the ventricular walls (the endocardium) and make an electrical connection with the myocytes and begin the propagating wavefront that rapidly spreads throughout the myocardium. The entire myocardium is normally activated within 50 ms and the mechanical contraction lasts a further 200–300 ms.

Myocardial tissue has been shown in a number of microstructural studies to consist of layers of interconnected sheets of tissue separated by cleavage planes (Le Grice *et al.* 1995, Nielsen *et al.* 1991). The layers of cells within a sheet are bound tightly together (3 to 4 cells thick) by endomysial collagen (see Figure 1.1). The cells are roughly cylindrical, about 20 μm in diameter, and about 100 μm long. The long axis of the myocytes is called the fibre axis and this varies by up to 150° across the wall, as illustrated in Figure 1.1. The direction orthogonal to the fibres within a sheet is called the sheet axis and the direction orthogonal to the sheet is referred to as the sheet normal. The mechanical and electrical properties of myocardial tissue are orthotropic with these microstructurally based fibre, sheet and sheet-normal axes forming the local material axes to which the constitutive laws are referred (see later). Microstructural detail is critical to both mechanical and electrical function of biological organs and often varies spatially within an organ. It is necessary, therefore, to characterize anisotropic conductivity and mechanical properties in continuum models.

We begin our analysis of heart function by introducing finite element descriptions of organ geometry and microstructure in Section 2. The continuum-based equations and techniques for their solution are outlined in Section 3. Cellular models and their spatial–temporal coupling into the continuum framework are covered in Section 4. Coronary blood flow (blood supply to heart muscle) is considered in Section 5. Finally the challenges and scope for extending this framework and application to other organs are considered in Sections 6 and 7.

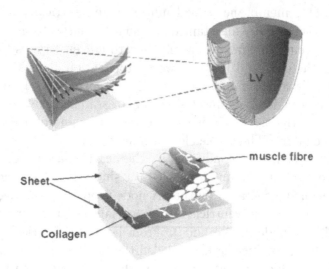

Figure 1.1. Microstructural organization of cardiac tissue. A transmural segment of tissue removed from the left ventricular wall in the upper figure contains muscle fibres that vary in orientation by about 150° (shown by the white lines). These fibres (muscle cells or 'myocytes') are bound tightly together in sheets 3 or 4 cells thick (see lower figure), that are loosely coupled to facilitate mechanical shearing. From Le Grice et al. (1995), with permission.

2. Finite element modelling of cardiac tissue

The development of anatomically based representations of organ geometry and tissue properties is the essential first step in the construction of an integrated model of whole organ function. In the following section a finite element method for representation of these fields is introduced. In Section 2.2 methods for determining nodal parameter values used in finite element interpolation functions are outlined.

2.1. Representation of fields

The finite element method is one of the most commonly used approaches to numerically represent spatially distributed fields. Traditionally, finite element models use low-order Lagrangian interpolation and, with very few exceptions, only one type of interpolation scheme in the solution domain. For a detailed introduction to the finite element method readers are referred to Zienkiewicz and Taylor (1994). In this review we present a different approach which uses high-order Hermitian interpolation to provide improved efficiency and convergence properties (Hunter 1975, Bradley, Pullan and Hunter 1997) for representing the nonlinear and C^1-continuous variation typical of many biological fields.

In the notation used below, ξ_i is the local finite element coordinate in the ith direction ($0 \leq \xi_i \leq 1$), u represents the field variable and u_n its value at local node n of a given element. $\Delta(n,e)$ maps the local node n of element e to its unique global number for a particular mesh. Continuity of the derivative of u with respect to ξ_i across element boundaries is achieved by defining additional nodal parameters which are the partial derivatives of the field variable u with respect to the local coordinate $\left(\frac{\partial u}{\partial \xi_i}\right)_n$. Illustrating this concept in one dimension, basis functions are chosen to ensure that nodal derivatives which contribute to the interpolation within an element are shared by two adjacent elements in order to maintain derivative continuity across different boundaries. Derivative continuity requires cubic interpolation and therefore four-element parameters. These parameters are specified as the nodal value and its derivative with respect to ξ for two nodes per element in one dimension. Thus:

$$u(\xi) = a + b\xi + c\xi^2 + d\xi^3, \quad \text{and therefore} \quad \frac{du}{d\xi} = b + 2c\xi + 3d\xi^2,$$

where

$$a = u_1, \quad b = u'_1, \quad c = 3u_2 - 3u_1 - 2u'_1 - u'_2, \quad d = u'_1 + u'_2 + 2u_1 - 2u_2,$$

or, rearranging,

$$u(\xi) = H_1^0(\xi)u_1 + H_1^1(\xi)u'_1 + H_2^0(\xi)u_2 + H_2^1(\xi)u'_2 \tag{2.1}$$

where $u'_n = \left(\frac{du}{d\xi}\right)_n$, and the four one-dimensional cubic-Hermite basis functions are illustrated in Figure 2.1.

To constrain continuity across element boundaries in the global coordinate system, rather than in the local ξ coordinate system, a further modification is required. The derivative $\left(\frac{du}{d\xi}\right)_n$ defined at node n is dependent upon the element ξ-coordinate in the two adjacent elements. It is more useful to define a global node derivative $\left(\frac{du}{ds}\right)_n$ where s is arclength, and then use

$$\left(\frac{du}{d\xi}\right)_n = \left(\frac{du}{ds}\right)_{\Delta(n,e)} \cdot \left(\frac{ds}{d\xi}\right)_n \tag{2.2}$$

where $\left(\frac{ds}{d\xi}\right)_n$ is an element 'scale factor' which scales the arclength derivative of global node Δ to the ξ-coordinate derivative of element node n.

Achieving derivative continuity in three dimensions for the element shown in Figure 2.2 requires eight values per node:

$$u, \quad \frac{\partial u}{\partial \xi_1}, \quad \frac{\partial u}{\partial \xi_2}, \quad \frac{\partial u}{\partial \xi_3}, \quad \text{and} \quad \frac{\partial^2 u}{\partial \xi_1 \partial \xi_2}, \quad \frac{\partial^2 u}{\partial \xi_1 \partial \xi_3}, \quad \frac{\partial^2 u}{\partial \xi_2 \partial \xi_3}, \quad \frac{\partial^3 u}{\partial \xi_1 \partial \xi_2 \partial \xi_3}.$$

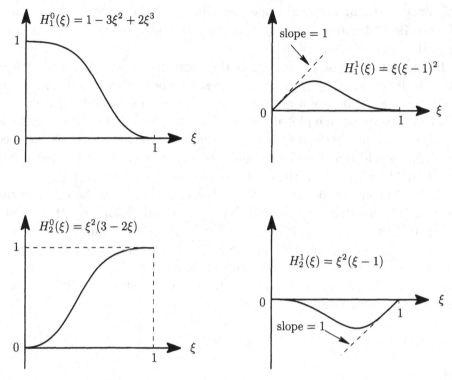

Figure 2.1. Cubic-Hermite basis functions.

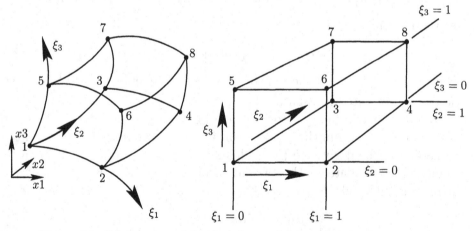

Figure 2.2. Schematic and local node numbering of a 3D element shown in physical space on the left and transformed into ξ space on the right.

The second-order cross-derivative terms arise from the need to maintain cubic interpolation along every element edge in each ξ direction. To illustrate, we consider a particular edge of a three-dimensional element where moving between nodes corresponds to only increasing in ξ_1. The cubic variation in u can be uniquely specified by u and $\frac{\partial u}{\partial \xi_1}$ at each node. However, $\frac{\partial u}{\partial \xi_2}$, $\frac{\partial u}{\partial \xi_3}$ and $\frac{\partial^2 u}{\partial \xi_2 \partial \xi_3}$ should also vary cubically and are completely independent of these parameters. Thus six additional parameters are specified for each node. Tricubic interpolation of these nodal parameters is given by

$$u(\xi_1, \xi_2, \xi_3) = \tag{2.3}$$

$$\sum_{k=1}^{2}\sum_{j=1}^{2}\sum_{i=1}^{2} \left[H_i^0(\xi_1) H_j^0(\xi_2) H_k^0(\xi_3) u_n + H_i^1(\xi_1) H_j^0(\xi_2) H_k^0(\xi_3) \left(\frac{\partial u}{\partial \xi_1}\right)_n + \right.$$

$$H_i^0(\xi_1) H_j^1(\xi_2) H_k^0(\xi_3) \left(\frac{\partial u}{\partial \xi_2}\right)_n + H_i^0(\xi_1) H_j^0(\xi_2) H_k^1(\xi_3) \left(\frac{\partial u}{\partial \xi_3}\right)_n +$$

$$H_i^1(\xi_1) H_j^1(\xi_2) H_k^0(\xi_3) \left(\frac{\partial^2 u}{\partial \xi_1 \partial \xi_2}\right)_n + H_i^1(\xi_1) H_j^0(\xi_2) H_k^1(\xi_3) \left(\frac{\partial^2 u}{\partial \xi_1 \partial \xi_3}\right)_n +$$

$$\left. H_i^0(\xi_1) H_j^1(\xi_2) H_k^1(\xi_3) \left(\frac{\partial^2 u}{\partial \xi_2 \partial \xi_3}\right)_n + H_i^1(\xi_1) H_j^1(\xi_2) H_k^1(\xi_3) \left(\frac{\partial^3 u}{\partial \xi_1 \partial \xi_2 \partial \xi_3}\right)_n \right]$$

where the nodal index $n = 4(k-1) + 2(j-1) + i$, and

$$H_1^0(\xi) = 1 - 3\xi^2 + 2\xi^3, \quad H_1^1(\xi) = \xi(\xi - 1)^2, \tag{2.4}$$

$$H_2^0(\xi) = \xi^2(3 - 2\xi) \quad \text{and} \quad H_2^1(\xi) = \xi^2(\xi - 1),$$

as shown in Figure 2.1. To simplify notation, let

$$u(\xi_1, \xi_2, \xi_3) = \sum_m \psi_m(\boldsymbol{\xi}) \bar{u}_m, \tag{2.5}$$

where $\bar{u}_m$ represents the vector of nodal values and their derivatives and ψ_m are formed from the products of the one-dimensional basis functions given in equation (2.4), and the summation over m is taken over all these quantities.

As in the one-dimensional case, to preserve derivative continuity in physical x-coordinate space rather than in ξ-coordinate space the global node derivatives need to be specified with respect to physical arclength. There are now three arclengths to consider: s_i, measuring arclength along the ξ_i-coordinate, for $i = 1, 2, 3$. Thus

$$\left(\frac{\partial u}{\partial \xi_1}\right)_n = \left(\frac{\partial u}{\partial s_1}\right)_{\Delta(n,e)} \cdot \left(\frac{ds_1}{d\xi_1}\right)_n, \tag{2.6}$$

$$\left(\frac{\partial u}{\partial \xi_2}\right)_n = \left(\frac{\partial u}{\partial s_2}\right)_{\Delta(n,e)} \cdot \left(\frac{ds_2}{d\xi_2}\right)_n,$$

$$\left(\frac{\partial u}{\partial \xi_3}\right)_n = \left(\frac{\partial u}{\partial s_3}\right)_{\Delta(n,e)} \cdot \left(\frac{ds_3}{d\xi_3}\right)_n,$$

$$\left(\frac{\partial^2 u}{\partial \xi_1 \partial \xi_2}\right)_n = \left(\frac{\partial^2 u}{\partial s_1 \partial s_2}\right)_{\Delta(n,e)} \cdot \left(\frac{ds_1}{d\xi_1}\right)_n \cdot \left(\frac{ds_2}{d\xi_2}\right)_n,$$

$$\left(\frac{\partial^2 u}{\partial \xi_1 \partial \xi_3}\right)_n = \left(\frac{\partial^2 u}{\partial s_1 \partial s_3}\right)_{\Delta(n,e)} \cdot \left(\frac{ds_1}{d\xi_1}\right)_n \cdot \left(\frac{ds_3}{d\xi_3}\right)_n,$$

$$\left(\frac{\partial^2 u}{\partial \xi_2 \partial \xi_3}\right)_n = \left(\frac{\partial^2 u}{\partial s_2 \partial s_3}\right)_{\Delta(n,e)} \cdot \left(\frac{ds_2}{d\xi_2}\right)_n \cdot \left(\frac{ds_3}{d\xi_3}\right)_n,$$

$$\left(\frac{\partial^3 u}{\partial \xi_1 \partial \xi_2 \partial \xi_3}\right)_n = \left(\frac{\partial^3 u}{\partial s_1 \partial s_2 \partial s_3}\right)_{\Delta(n,e)} \cdot \left(\frac{ds_1}{d\xi_1}\right)_n \cdot \left(\frac{ds_2}{d\xi_2}\right)_n \cdot \left(\frac{ds_3}{d\xi_3}\right)_n.$$

Equations (2.3)–(2.6) provide a means of interpolating spatially varying fields that are C^1-continuous.

We have used two alternative approaches to modelling the geometry of the heart. One uses the prolate spheroidal coordinate system shown in Figure 2.3 with bicubic-linear interpolation for the λ-coordinate (bicubic-Hermite in the (ξ_1, ξ_2) plane corresponding to the ventricular surfaces and linear in ξ_3 through the wall) and trilinear interpolation for the θ and μ coordinates. The other approach uses rectangular Cartesian coordinates and fully tricubic-Hermite interpolation.

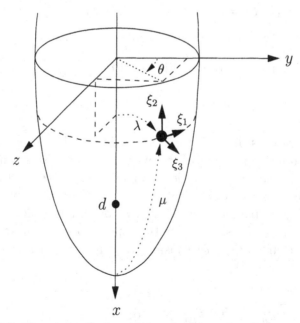

Figure 2.3. The prolate spheroidal and rectangular Cartesian coordinates for the ventricular models shown in Figure 2.4.

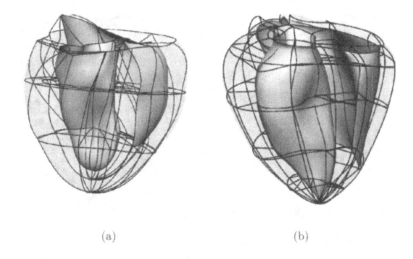

(a)   (b)

Figure 2.4. Anatomically accurate canine (a) and porcine (b) finite element models. (a) uses bilinear/cubic-Hermite interpolation based in prolate spheroidal coordinates to represent canine cardiac ventricles, while (b) uses tricubic-Hermite interpolation based in rectangular Cartesian coordinates to represent porcine cardiac ventricles.

As with geometry, cubic-Hermite interpolation can effectively be used to represent the variation in microstructurally based material directions as demonstrated in Figure 2.5, which shows the microstructural field embedded in the geometric models shown in Figure 2.4.

The next step in mesh construction is the calculation of values of variables u_n and their partial derivatives with respect to the local material directions ξ_i at nodes in the finite element mesh. This nonlinear fitting process is outlined in the following section.

2.2. Nonlinear fitting

The geometric fitting algorithm which is used to determine nodal field variables in the finite element tissue models from data is a variant of the well-known Iterative Closest Point (ICP) algorithm, where the surface fit is improved over several iterations. There are a number of implementations of this approach for fitting three-dimensional geometric models (Rusinkiewicz and Levoy 2001). The implementation described here includes a Sobolev smoothing constraint for sparse and scattered data and has been previously illustrated for fitting anatomically based bicubic-Hermite surface meshes (Young, Hunter and Smaill 1989, Bradley et al. 1997). Here, its implementation for fitting faces of three-dimensional volume meshes is given. The

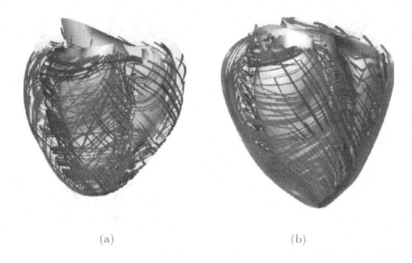

Figure 2.5. Anatomically accurate canine (a) and porcine (b) finite element models. Two layers of streamlines (one on the outer epicardial surface and one midway through the wall) are used to visualize the epicardial and midwall fibre directions which vary continuously through the wall.

nodal positions of this mesh are optimized to achieve the minimum distance between data points and the closest face of the mesh. This is done by minimizing an objective function based on an L_2 distance norm, calculated from projections of the data points on to the mesh faces.

The data for this fitting process are generated from a number of different sources, dependent on organ type and species. Data for human models are typically created by digitizing slices from the Visible Human (VH) database, which has become a common standard for researchers and provides a way of comparing and sharing geometries (Ackerman 1998). Non-invasive imaging techniques such as Magnetic Resonance Imaging (Young, Fayad and Axel 1996) and Computed Tomography (Kantor *et al.* 1999) provide additional and expanding sources of data. Animal models of organs such as the heart have been measured directly (Nielsen *et al.* 1991, Vetter and McCulloch 2000) and have also been measured from casts taken of the heart (Kassab, Rider, Tang and Fung 1993).

An initial mesh topology is extracted from the cloud of data points by selecting data points on the outside of the data cloud. These serve as nodes for the mesh which are then used to generate elements in a systematic manner. The nodes are chosen to construct a regular mesh with the minimal number of elements.

Data projection

The digitization process generates a Cartesian data set with geometric positions $z_d, d = 1, \ldots, N$. For each data point we can find the position on a given two-dimensional mesh face which has the smallest distance to the data point. This position is the orthogonal projection of the data point on to the mesh and has geometric position $u = x$. The point x is also given by the local element coordinate ξ_d as is shown in Figure 2.6. A least-squares distance function, D, between a data point and its projection may be expressed by

$$D(\xi_1, \xi_2) = \|x(\xi_1, \xi_2) - z_d\|^2, \qquad (2.7)$$

where z_d are the spatial coordinates of the data point and $x(\xi_1, \xi_2)$ is interpolated using a bicubic-Hermite basis function which is a reduced form of equation (2.3):

$$x(\xi_1, \xi_2) = \sum_{j=1}^{2}\sum_{i=1}^{2}\left[H_i^0(\xi_1)H_j^0(\xi_2)x_n + H_i^1(\xi_1)H_j^0(\xi_2)\left(\frac{\partial x}{\partial \xi_1}\right)_n \right. \qquad (2.8)$$

$$\left. + H_i^0(\xi_1)H_j^1(\xi_2)\left(\frac{\partial x}{\partial \xi_2}\right)_n + H_i^1(\xi_1)H_j^1(\xi_2)\left(\frac{\partial^2 x}{\partial \xi_1 \partial \xi_2}\right)_n\right]$$

with the derivatives with respect to ξ converted using

$$\left(\frac{\partial x}{\partial \xi_\alpha}\right) = \left(\frac{\partial x}{\partial s_\alpha}\right)\left(\frac{\partial s_\alpha}{\partial \xi_\alpha}\right),$$

and

$$\left(\frac{\partial^2 x}{\partial \xi_\alpha \partial \xi_\beta}\right) = \left(\frac{\partial^2 x}{\partial s_\alpha \partial s_\beta}\right)\left(\frac{\partial s_\alpha}{\partial \xi_\alpha}\right)\left(\frac{\partial s_\beta}{\partial \xi_\beta}\right),$$

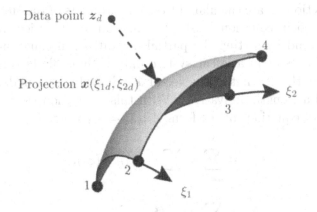

Figure 2.6. Projection of data point z_d on to an element face and the resulting closest point $x(\xi_{1d}, \xi_{2d})$.

for cross-derivatives ($\alpha, \beta = 1, 2$ where no summation is implied). Note $x(\xi_1, \xi_2)$ here refers to the continuum field evaluated at material points ξ_1, ξ_2, while x_n refers to the vector of nodal parameters used in the interpolation. Thus the local coordinates of the projection point (ξ_{1d}, ξ_{2d}), and hence its global coordinates, can be determined by solving the following nonlinear simultaneous equations using the Newton–Raphson procedure:

$$\frac{\partial D}{\partial \xi_1} = 0, \qquad \frac{\partial D}{\partial \xi_2} = 0. \qquad (2.9)$$

Objective function minimization

For a given projection of the data points on to the mesh (*i.e.*, ξ_d is held constant), the objective function to be minimized is the sum-of-squares of the individual errors

$$F(\bar{x}) = \sum_{d=1}^{N} w_d \| x(\xi_{1d}, \xi_{2d}) - z_d \|^2, \qquad (2.10)$$

where w_d is a weight for each data point and $\bar{x}$ is a vector of mesh parameters. Weights w_d are set to unity unless there is a clear difference in quality of the data (*e.g.*, when combining data sets acquired using different methods).

The fitting problem is to find the set of mesh parameters that minimizes this objective function. For simplicity the formulation of the linear system is illustrated below for a one-dimensional element, and the two- and three-dimensional formulations can be inferred.

Substituting equation (2.8) into equation (2.10) and differentiating, we obtain

$$\frac{\partial F}{\partial \bar{x}_m} = 2 \sum_{d=1}^{N} w_d \left(\sum_{r=1}^{4} \psi_r(\xi_d) \bar{x}_r - z_d \right) \psi_m(\xi_d), \qquad (2.11)$$

where the functions ψ are calculated from the products of the Hermite basis functions formed in equation (2.8). A minimum of the objective function can thus be found by setting the partial derivatives in equation (2.11) to zero. This will result in a linear system only if the scale factors are kept constant during the fit; that is, the vector $\bar{x}$ will contain the nodal positions and the nodal arclength derivatives. With this restriction we can obtain a linear system of equations of the form $A_{rm} \bar{x}_r = b_m$ where

$$A_{rm} = \sum_{d=1}^{N} w_d \sum_{r=1}^{4} \psi_r(\xi_d) \psi_m(\xi_d), \qquad (2.12)$$

$$b_m = \sum_{d=1}^{N} w_d \psi_m(\xi_d) z_d, \qquad (2.13)$$

and N is the total number of data points.

A linear system of equations governing the entire mesh can then be found by assembling a global stiffness matrix from all the element matrices, and solved to yield new nodal positions, derivatives and scale factors. The updated scale factors alter the shape of the mesh, since the scale factors are involved explicitly in the interpolation functions as described in equation (2.8). The steps described above are therefore repeated until convergence is achieved.

Sobolev smoothing

To deal with an insufficient number of data points, fitting noisy data or fitting data that has an uneven spread (see Figure 2.7), a smoothness constraint (Young et al. 1989) can be introduced by adding a second term to the objective function equation (2.10). This is known as Sobolev smoothing with a penalty function, in two dimensions, of the form

$$F_s(x) = \int_0^1 \int_0^1 \left\{ \alpha_1 \left\| \frac{\partial x}{\partial \xi_1} \right\|^2 + \alpha_2 \left\| \frac{\partial x}{\partial \xi_2} \right\|^2 + \alpha_3 \left\| \frac{\partial^2 x}{\partial \xi_1^2} \right\|^2 \right.$$
$$\left. + \alpha_4 \left\| \frac{\partial^2 x}{\partial \xi_2^2} \right\|^2 + \alpha_5 \left\| \frac{\partial^2 x}{\partial \xi_1 \partial \xi_2} \right\|^2 \right\} d\xi_1 \, d\xi_2, \qquad (2.14)$$

where α_i ($i = 1, \ldots, 5$) are the Sobolev weights (penalty parameters). Each term has a distinct effect on the final shape of the fitted object. The first two terms (α_1, α_2) control the arclength, while the third and fourth terms (α_3, α_4) control the arc curvature in the ξ_1 and ξ_2 directions, respectively. The last term (α_5) represents the face area. For instance, if the weight

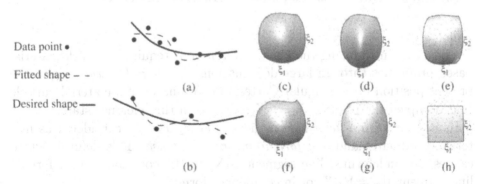

Figure 2.7. Unbiased fitting due to (a) scattered, and (b) sparse data. Effects of Sobolev weights in a 2D surface fit with (c) no smoothing (d) ξ_1 weighting on α_1 and (e) weighting on α_1 and α_3. (f), (g) show similar behaviour for the ξ_2 direction and (h) shows weighting on all faces including the area term α_5.

associated with the cross-derivative term is set to a high value, one might expect to end up with a mesh with a smaller face area.

The effect of these parameters is illustrated for a simple two-dimensional surface in Figure 2.7(c)–(h). Figure 2.7(c) shows the effect with no smoothing. Increasing α_1 reduces the arclength in the ξ_1 direction (Figure 2.7(d)) and including α_3 reduces the ξ_1 curvature which is seen as a flattening in this direction (Figure 2.7(e)). Similar behaviour is observed for the ξ_2 direction (Figure 2.7(f), (g)). Placing weights on all coefficients including the area term (α_5) causes a smooth reduction in the face area (Figure 2.7(h)).

3. Continuum modelling

The development of whole organ models using the above anatomically based finite element descriptions of geometry and tissue microstructure requires two further steps. Firstly, the cellular-based processes which produce a change in the state of the organ system, independent of the variations in boundary conditions, must be characterized. These cellular-based models are introduced in Section 4. Secondly, governing equations must be proposed which relate material properties to continuum tissue behaviour. Specifically, the numerical frameworks for applying equations determining deformation and activation are outlined below in Sections 3.1 and 3.2.

3.1. Finite deformation mechanics

Understanding and predicting the deformation of tissue under varying distributions of applied stresses and strains is fundamental to understanding its function. To deal with the nonlinear constitutive mechanical behaviour requires a framework based on finite deformation mechanics.

Kinematics
We begin by introducing the kinematic relations required to track material based properties through large deformations. Let $\mathbf{x} = (x_1, x_2, x_3)$ give the present position in rectangular Cartesian coordinates of a material particle that occupied the position $\mathbf{X} = (X_1, X_2, X_3)$ in the reference state.

In standard finite deformation theory (x_1, x_2, x_3) are considered as material coordinates and a *deformation gradient tensor* $\mathbf{F}$ is defined, which carries the undeformed line segment, $\mathrm{d}\mathbf{X}$, to the corresponding deformed line segment $\mathrm{d}\mathbf{x} = \mathbf{F}\,\mathrm{d}\mathbf{X}$, or in component form:

$$\mathrm{d}x^i = F^i_M \, \mathrm{d}X^M, \tag{3.1}$$

where

$$F^i_M = \frac{\partial x_i}{\partial X_M}. \tag{3.2}$$

Polar decomposition, $\mathbf{F} = \mathbf{RU}$, splits $\mathbf{F}$ into the product of an orthogonal rotation tensor, $\mathbf{R}$, and a symmetric positive definite stretch tensor, $\mathbf{U}$, which contains a complete description of the material strain, independent of any rigid body motion (Atkin and Fox 1980).

For inhomogeneous, anisotropic materials the orientation of the material axes may vary with location, for example fibre direction changes spatially throughout the myocardium. Thus it is no longer practical to identify the material axes in the undeformed body with the reference coordinates (X_1, X_2, X_3). Instead, a new material coordinate system (ν_1, ν_2, ν_3) is introduced which is aligned with the microstructural features of the material, as described above in Section 2.1. For myocardium, a natural set of material axes are formed by identifying ν_1 with the muscle fibre direction, ν_2 with the sheet direction and ν_3 with the sheet-normal direction.

It is useful to choose the base vectors for the ν_α-coordinate system to be orthogonal in the reference state. This is convenient in myocardium, for example, where the ν_α-coordinates are chosen to line up with the fibre, sheet and sheet-normal directions, which are defined to be orthogonal in the undeformed state. However, the ensuing deformation means that they are not, in general, orthogonal in the deformed configuration.

$\mathbf{A}_\alpha^{(\nu)}$, $\mathbf{A}_{(\nu)}^\alpha$ and $\mathbf{a}_\alpha^{(\nu)}$, $\mathbf{a}_{(\nu)}^\alpha$ denote the covariant and contravariant base vectors in the undeformed and deformed configurations, respectively. The corresponding metric tensors are denoted by $A_{\alpha\beta}^{(\nu)}$, $A_{(\nu)}^{\alpha\beta}$ and $a_{\alpha\beta}^{(\nu)}$, $a_{(\nu)}^{\alpha\beta}$. The undeformed covariant base vectors, $\mathbf{A}_\alpha^{(\nu)}$, are defined to be unit vectors by choosing the ν_α-coordinates to be a measure of physical arclength in the undeformed state. The base vectors and metric tensors for the ν_α-coordinate system are

$$\mathbf{A}_\alpha^{(\nu)} = \frac{\partial X_k}{\partial \nu_\alpha}\mathbf{g}_k^{(x)}, \qquad \mathbf{a}_\alpha^{(\nu)} = \frac{\partial x_k}{\partial \nu_\alpha}\mathbf{g}_k^{(x)}, \qquad (3.3)$$
$$A_{\alpha\beta}^{(\nu)} = \mathbf{A}_\alpha^{(\nu)} \cdot \mathbf{A}_\beta^{(\nu)}, \qquad a_{\alpha\beta}^{(\nu)} = \mathbf{a}_\alpha^{(\nu)} \cdot \mathbf{a}_\beta^{(\nu)},$$

where $\mathbf{g}_k^{(x)}$ are the base vectors of the rectangular Cartesian reference axes. The Green strain tensor, defining the kinematics of large deformation for an inhomogeneous anisotropic material, is then

$$E_{\alpha\beta} = \frac{1}{2}\left(a_{\alpha\beta}^{(\nu)} - A_{\alpha\beta}^{(\nu)}\right). \qquad (3.4)$$

Stress equilibrium and the principle of virtual work

The governing equations for elastostatics can be derived from a physically appealing argument. For equilibrium, the work done by the external surface forces in moving through a virtual displacement is equal to the work done by the stress vector in moving through a compatible set of virtual displacements. Using this principle (known as the principle of virtual work) the stress equilibrium can be expressed via the following equation (Nash and Hunter 2000):

$$\int_{V_0} T^{\alpha\beta} F^j_\beta \delta v_j \big|_\alpha \, dV_0 = \int_{V_0} \rho_0 \left(b^j - f^j\right) \delta v_j \, dV_0 + \int_{S_2} p_{(\text{appl})} \frac{g_{(\xi)}^{3M}}{\sqrt{g_{(\xi)}^{33}}} \frac{\partial x_j}{\partial \xi_M} \delta v_j \, dS \quad (3.5)$$

where $T^{\alpha\beta}$ are second Piola–Kirchhoff stresses expressed relative to the fibre-sheet material coordinates; $|_\alpha$ is a covariant derivative; $\delta \mathbf{v} = \delta v_j \mathbf{i}_j$ are virtual displacements expressed relative to the reference coordinate system (Malvern 1969); b^j and f^j are the components of the body force and acceleration vectors, respectively; $p_{(\text{appl})}$ is the pressure applied to the surface S_2 with normal direction ξ_3; ρ_0 is the tissue density; and $g_{(\xi)}^{MN}$ are contravariant metric tensors for the ξ_i-coordinate system. Covariant base vectors and metric tensors for the ξ_M-coordinate system are defined for the undeformed and deformed states as follows:

$$\mathbf{G}_M^{(\xi)} = \frac{\partial X_k}{\partial \xi_M} \mathbf{g}_k^{(x)}, \qquad \mathbf{g}_M^{(\xi)} = \frac{\partial x_k}{\partial \xi_M} \mathbf{g}_k^{(x)},$$

$$G_{MN}^{(\xi)} = \mathbf{G}_M^{(\xi)} \cdot \mathbf{G}_N^{(\xi)} = \frac{\partial X_k}{\partial \xi_M} \frac{\partial X_k}{\partial \xi_N}, \qquad g_{MN}^{(\xi)} = \mathbf{g}_M^{(\xi)} \cdot \mathbf{g}_N^{(\xi)} = \frac{\partial x_k}{\partial \xi_M} \frac{\partial x_k}{\partial \xi_N}. \quad (3.6)$$

Equation (3.5) is the starting point for the analysis of a body undergoing large elastic deformations. For further detail see Costa *et al.* (1996*a*, 1996*b*).

Finite element solution techniques
Using the interpolation functions ψ_n from Section 2.1 we can define virtual displacement fields δv_j as

$$\delta v_j = \psi_n \left(\xi_1, \xi_2, \xi_3\right) \delta v_j^n, \quad (3.7)$$

where δv_j^n are arbitrary virtual nodal displacements. Substituting equation (3.7) into the equilibrium equations (equation (3.5)) and setting the coefficient of each component δv_j^n to zero, gives

$$\int_{V_0} T^{\alpha\beta} F^j_\beta \psi_n \big|_\alpha \, dV_0 = \int_{V_0} \rho_0 \left(b^j - f^j\right) \psi_n \, dV_0 + \int_{S_2} p_{(\text{appl})} \frac{g_{(\xi)}^{3M}}{\sqrt{g_{(\xi)}^{33}}} \frac{\partial x_j}{\partial \xi_M} \psi_n \, dS. \quad (3.8)$$

To evaluate the integrals in equation (3.8), they must first be transformed from the reference coordinate space to the ξ_M-coordinate space using the appropriate Jacobian. The transformed integrals are written in (3.9):

$$\iiint_{V_0} T^{\alpha\beta} F^j_\beta \psi_n\big|_\alpha \sqrt{G^{(\xi)}} \, d\xi_3 \, d\xi_2 \, d\xi_1 \qquad (3.9)$$

$$= \iiint_{V_0} \rho_0 (b^j - f^j) \, \psi_n \sqrt{G^{(\xi)}} \, d\xi_3 \, d\xi_2 \, d\xi_1$$

$$+ \iint_{S_2} P_{(\text{appl})} \, g^{3M}_{(\xi)} \frac{\partial \theta_j}{\partial \xi_M} \, \psi_n \sqrt{g^{(\xi)}} \, d\xi_2 \, d\xi_1,$$

where $\sqrt{G^{(\xi)}} = \sqrt{\det\{G^{(\xi)}_{ij}\}}$ and $\sqrt{g^{(\xi)}} = \sqrt{\det\{g^{(\xi)}_{ij}\}}$ are the Jacobians of the three-dimensional coordinate transformation with respect to the undeformed and deformed configurations, respectively. Note that the surface integral is transformed by substituting $J_{2D} \, d\xi_2 \, d\xi_1$ for dS, where the two-dimensional Jacobian with respect to deformed coordinates is given by $J_{2D} = \sqrt{g^{(\xi)} g^{33}_{(\xi)}}$ (Oden 1972).

The three-dimensional integrals in equation (3.9) are evaluated over the undeformed volume and the two-dimensional integral is computed over the portion of the deformed surface (denoted S_2) for which external pressure loads are applied. These integrals are replaced by a sum of integrals over the collection of element domains which constitute the finite element model. Element integrals are evaluated numerically using Gaussian quadrature. Components of the second Piola–Kirchhoff stress tensor, $T^{\alpha\beta}$, are evaluated at each Gauss point using (Malvern 1969)

$$T^{\alpha\beta} = \frac{1}{2}\left(\frac{\partial W}{\partial E_{\alpha\beta}} + \frac{\partial W}{\partial E_{\beta\alpha}}\right) - p a^{\alpha\beta}_{(\nu)}, \qquad (3.10)$$

where p is hydrostatic pressure and the derivatives of the strain energy function W with respect to the components of $\boldsymbol{E}$ are determined using a constitutive relation.

The strain energy functions of cardiac tissue have been characterized and applied by a number of authors (Guccione, McCulloch and Waldman 1991, Emery, Omens and McCulloch 1997, Usyk, Le Grice and McCulloch 2002) using a generic exponential relation of the form

$$W = \frac{1}{2} C(e^Q - 1), \qquad (3.11)$$

where $\boldsymbol{C}$ is the right Cauchy–Green strain tensor and Q is a function in which the strain components of $\boldsymbol{E}$ are referred to the local structure-based coordinates. Guccione et al. (1991) defined the form of Q such that myocardium was assumed to be transversely isotropic and incompressible. More recently Usyk, Mazhari and McCulloch (2000) have developed and applied (Usyk et al. 2002) a fully orthotropic model in the form of equation (3.11) within a three-dimensional model of cardiac mechanics. Difficulties lie in assigning unique parameter values in the complex forms of Q required to fully represent the orthotropic behaviour of myocardium via equation (3.11).

The passive myocardial characteristics have also been encapsulated via an alternative *pole-zero* strain energy function for the myocardium (Hunter, Smaill and Hunter 1995) given by

$$W = k_{11}\frac{E_{11}^2}{|a_{11} - E_{11}|^{b_{11}}} + k_{22}\frac{E_{22}^2}{|a_{22} - E_{22}|^{b_{22}}} + k_{33}\frac{E_{33}^2}{|a_{33} - E_{33}|^{b_{33}}} \quad (3.12)$$
$$+ k_{12}\frac{E_{12}^2}{|a_{12} - E_{12}|^{b_{12}}} + k_{13}\frac{E_{13}^2}{|a_{13} - E_{13}|^{b_{13}}} + k_{23}\frac{E_{23}^2}{|a_{23} - E_{23}|^{b_{23}}}$$

where the constitutive parameters (as, bs and ks) are fitted from biaxial testing of tissue slices cut parallel with the fibre axis at several transmural sites throughout the myocardium (Novak, Yin and Humphrey 1994). Within equation (3.12) $a_{\alpha\beta}$ denote limiting strain or poles, $b_{\alpha\beta}$ relate the curvature of the uni-axial stress–strain relationships and $k_{\alpha\beta}$ weight the contribution of the corresponding mode of deformation to the total strain energy of the material.

For incompressible materials, an additional scalar hydrostatic pressure field is introduced into the constitutive equations. The extra constraint necessary to determine the parameters of the hydrostatic pressure field arise from the requirement that the third strain invariant (I_3) equals one for incompressible materials.

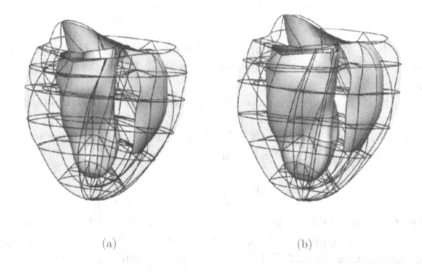

Figure 3.1. (a) Canine finite element model in its reference state. (b) Geometry calculated from application of cavity pressures to simulate inflation.

For a Galerkin formulation, the form of the incompressibility constraints is given in equation (3.13):

$$\iiint_{V_e} (\sqrt{I_3} - 1)\psi^p \sqrt{G^{(\xi)}}\, d\xi_3\, d\xi_2\, d\xi_1 = 0, \qquad (3.13)$$

where V_e denotes the domain of the element and ψ^p are the basis functions used to approximate the three-dimensional hydrostatic pressure field. Note that the undeformed three-dimensional Jacobian, $\sqrt{G^{(\xi)}}$, is introduced since the integrals are evaluated with respect to the undeformed configuration.

The methods outlined above can be applied to calculate the deformations associated with inflation of the finite element mesh of the canine ventricular model shown in Figure 3.1. The reference geometry is shown in Figure 3.1(a), and the deformed geometry produced by the application of left and right cavity ventricular pressures of 0.2 kPa and 1 kPa, respectively, is shown in Figure 3.1(b).

Numerical and computational issues

There are a number of numerical and computational challenges which must be overcome for the effective and efficient application of the finite element method to the finite deformation mechanics equations described above. The first of these is in the implementation of numerical techniques to find the roots of the system of nonlinear equations obtained by evaluating the integrals in equations (3.9) and (3.13) for each element in the mesh. The Newton–Raphson method can be used for this root-finding process. The calculation of the Newton step vector at each iteration requires the solution of a sparse set of linear equations, for which the *SuperLU* method (Demmel et al. 1999a, Demmel, Gilbert and Li 1999b) has been found to perform well.

Oden (1972) suggests that the interpolation scheme chosen to describe the deformed geometric coordinates (ψ_n in equation (3.9)) should be of higher order than those chosen to approximate the hydrostatic pressure field (ψ^p in equation (3.13)). The strain energy contribution to stress components is related to the first derivatives of the geometric displacement fields, whereas the hydrostatic pressure directly contributes to the stress components. For consistency, and to avoid numerical ill-conditioning when calculating components of the stress tensor, the two contributions should vary in a similar manner.

The computational issues associated with modelling finite deformation mechanics centre around the exploitation of parallel architectures. In particular, the determination of groups of element stiffness matrices can be allocated across multiple processors while incurring only a small computational overhead. Furthermore, the *SuperLU* algorithm provides close-to-linear scalability for the Newton step calculations. The scalability of two such major components of the method means that, with increased processor count, close-to-linear speed-up is exhibited for the whole algorithm. The ease of implementation and code maintenance for shared memory architectures is well suited to finite deformation mechanics simulations, and indeed this has been the preferred platform. However, recent improvements in availability and the ongoing optimization of specific algorithms (Li and Demmel 2003) mean that distributed memory systems now provide an increasingly attractive alternative computational platform.

3.2. Myocardial activation

Cardiac tissue consists of discrete cells but we can model the electrical behaviour of the tissue using a continuum approach that averages the electrical properties over a length scale greater than that of single cells. In the continuum approach we assume that cardiac tissue has three orthogonal conductivity directions following the microstructural organization of the tissue. The fastest conductivity is along the fibre axis; there is slower conductivity in the plane of the sheet, and finally the slowest propagation is along the sheet-normal axis.

The continuum approach uses the bidomain model of multicellular volume conductors, which has been used extensively in models of the spread of electrical activity in excitable tissues (Fischer *et al.* 2000, Henriquez 1993, Muzikant and Henriquez 1998, Plonsey and Barr 1984, Roth and Wikswo 1986, Roth 1997, Skouibine, Trayanova and Moore 2000, Trayanova 1994). The bidomain model of cardiac tissue consists of two interpenetrating domains representing cells and the space surrounding them. The intracellular domain represents the region inside the cells and is given the subscript i, and the extracellular domain represents the space between cells and is given the subscript e. The key to the model is that these two domains are assumed to exist at all points in the physical solution domain. Detailed derivation of the bidomain model can be found elsewhere (Schmitt 1969, Tung 1978, Krassowska and Neu 1994). Here we state the equations and show how they are transformed into finite element coordinates in order to study electrical propagation on a deforming finite element mesh.

Let ϕ_i and ϕ_e be the electric potentials in the intracellular and extracellular domains, respectively, and $V_m = \phi_i - \phi_e$ is the transmembrane potential. Associated with these domains are the macroscopic tensor quantities σ_i and σ_e, representing the local volume averaged conductivities in the intra- and extracellular spaces, respectively. Tensors σ_i and σ_e are each separately anisotropic and are assumed to be diagonal in the material coordinates based on the fibrous structure of muscle tissue.

The bidomain model consists of two equations. The first describes the conservation of current,

$$\nabla \cdot ((\sigma_i + \sigma_e)\nabla \phi_e) = -\nabla \cdot (\sigma_i \nabla V_m) + I_{s1}, \tag{3.14}$$

that given a transmembrane potential distribution is used to solve for the extracellular potential. The second equation describes the current flow across the cellular membrane composed of ionic and capacitive currents (the cell membrane acts as a parallel capacitance), that is,

$$\nabla \cdot (\sigma_i \nabla V_m) + \nabla \cdot (\sigma_i \nabla \phi_e) = A_m \left(C_m \frac{\partial V_m}{\partial t} + I_{\text{ion}} \right) - I_{s2}, \quad (3.15)$$

and is used to calculate the transmembrane potential distribution. A_m is the surface to volume ratio of the cell membrane, C_m is the membrane capacitance per unit area, and I_{ion} is a nonlinear function representing the sum of all the transmembrane ionic currents. Externally applied volume stimulus currents can be imposed in both the extracellular (I_{s1}) and intracellular (I_{s2}) domains.

It is assumed that there is no current flow between the intracellular domain and the external region so the boundary condition applied to V_m on the solution domain boundary is

$$(\sigma_i \nabla V_m) \cdot \boldsymbol{n} = -(\sigma_i \nabla \phi_e) \cdot \boldsymbol{n}, \quad (3.16)$$

where $\boldsymbol{n}$ is a unit vector outwardly normal to the domain boundary. For the extracellular domain the current must balance between the extracellular domain and the surrounding external regions, that is,

$$(\sigma_e \nabla \phi_e) \cdot \boldsymbol{n} = -(\sigma_o \nabla \phi_o) \cdot \boldsymbol{n}, \quad (3.17)$$

where σ_o signifies the conductivity of the surrounding region. The negative sign accounts for the direction of current flow as both sides of the equations use outward normal vectors. The boundary extracellular potential must also match the potential of the boundary of the external regions,

$$\phi_e = \phi_o. \quad (3.18)$$

In the absence of an external region any combination of current and potential boundary conditions can be used to specify the required physical problem, with the restriction that at least one extracellular boundary point has a potential boundary condition to provide a reference potential (and hence a unique solution to the bidomain equations).

The monodomain model of activation

In an effort to further reduce the computational cost of the activation modelling, the extracellular domain is sometimes assumed to be highly conducting or, alternatively, both domains are assumed equally anisotropic. With either of these assumptions the transmembrane potential is equal to the intracellular potential, and the bidomain equations simplify to

$$\nabla \cdot (\sigma \nabla V_m) = A_m \left(C_m \frac{\partial V_m}{\partial t} + I_{\text{ion}} \right) - I_s, \quad (3.19)$$

as the gradient of the extracellular potential field is effectively zero in either of these approximations. This monodomain model is suitable for situations such as computations on an isolated heart. When the extracellular electrical state is important the full bidomain model needs to be used, for example electrical current propagating from the heart to the torso in body surface potential forward simulations or the application of defibrillation-type extracellular stimuli.

For the monodomain model, there is no connection between the intracellular domain and any surrounding media. Therefore, no current can flow out of the solution domain, giving rise to the boundary condition

$$(\sigma \nabla V_m) \cdot \boldsymbol{n} = 0. \tag{3.20}$$

Finite element-derived finite difference method

In order to solve the bidomain or monodomain models on realistic ventricular geometry domains, the equations need to be solved numerically, dividing the solution domain into smaller subdomains over which the equations can be integrated. The method reviewed here for the numerical integration of the bidomain or monodomain models is a collocation method known as the finite element-derived finite difference method (Buist, Sands, Hunter and Pullan 2003). In this method, finite elements (FEs) are used to describe the tissue geometry and fibrous microstructure, and the activation equations are solved on a high-resolution nonuniform finite difference (FD) grid which is defined from and embedded in the material space of the FEs. This allows for the much smaller space constant required for the resolution of local behaviour and steep spatial gradients of the activation model while having the solution mesh defined by the geometry of the problem, including any geometric deformation applied to the host FE mesh as the points remain invariant in material space.

A significant advantage of this method is that the FD collocation points are embedded in the FE geometric mesh, so when the geometric mesh deforms (*i.e.*, due to contraction of muscle) the FD points move with the deformation. This is an essential feature when modelling coupled electromechanics, as seen in the following section (Section 3.3). Figure 3.2 illustrates the definition of such FD meshes and their mapping to the local quadratic solution space (Buist *et al.* 2003).

To solve the bidomain equations (3.14) and (3.15) or monodomain equation (3.19) we need to express the diffusion terms $\nabla \cdot (\sigma \nabla \phi)$ in the curvilinear material coordinates (ξ_1, ξ_2, ξ_3). The expansion of this term using domain metrics and standard tensor notation, with a comma denoting partial differentiation, gives

$$\nabla \cdot (\sigma \nabla \phi) = \left(\sigma^k_{\alpha,\beta} \phi_{,k} + \sigma^k_\alpha \phi_{,k\beta} - \sigma^k_l \phi_{,k} \Gamma^l_{\alpha\beta} \right) a^{\beta\alpha}_{(\nu)}, \tag{3.21}$$

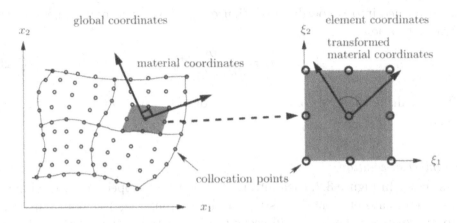

Figure 3.2. An illustration of the collocation points defined on a two-dimensional finite element mesh. Each interior point together with its 8 surrounding points (in 2D (x_1, x_2)-space) is mapped on to a unit square in (ξ_1, ξ_2)-space. Note that the material coordinates which are orthogonal in (x_1, x_2)-space become non-orthogonal in (ξ_1, ξ_2)-space.

where $a^{\beta\alpha}_{(\nu)}$ are the components of the contravariant metric tensor in the deformed configuration which is defined in terms of the contravariant base vectors

$$a^{\beta\alpha}_{(\nu)} = \boldsymbol{a}^{\beta}_{(\nu)} \cdot \boldsymbol{a}^{\alpha}_{(\nu)}. \tag{3.22}$$

See equation (3.3) for more detail. The Christoffel symbol ($\Gamma^l_{\alpha\beta}$) is used to represent the base vector derivatives and is defined as

$$\Gamma^l_{\alpha\beta} = \boldsymbol{a}^{(\xi)}_{\alpha,\beta} \cdot \boldsymbol{a}^l_{(\xi)}. \tag{3.23}$$

Tissue microstructure

The tissue microstructure is included in the activation model through the extra- and intracellular conductivity tensors (σ_e, σ_i). Using the material coordinate system (ν_1, ν_2, ν_3) defined previously, the original conductivity tensor (σ^{*a}_b) is a diagonal tensor given by

$$\sigma^{*a}_b = \begin{pmatrix} \sigma_f & 0 & 0 \\ 0 & \sigma_s & 0 \\ 0 & 0 & \sigma_c \end{pmatrix}, \tag{3.24}$$

where σ_f is the fibre direction conductivity, σ_s the sheet, and σ_c the cross-sheet conductivity. For direct inclusion into equation (3.21) σ^{*a}_b must be transformed into ξ_j coordinates (Figure 3.2). This is accomplished with the transformation

$$\sigma_j^k = \frac{\partial \xi_k}{\partial \nu_a} \frac{\partial \nu^b}{\partial \xi_j} \sigma^{*a}_{b}. \qquad (3.25)$$

The resulting effective conductivity tensor σ_j^k is, in general, no longer diagonal.

Numerical solution
As shown in Figure 3.2, each interior FD point is mapped into a local quadratic template element consisting of the surrounding FD points in order to approximate the first- and second-order partial derivatives of potential and the first-order partial derivative of the conductivity tensors. The quadratic basis is chosen simply to ensure second-order accuracy in the spatial derivatives used for the reaction–diffusion equations. The three quadratic Lagrange basis functions in one dimension are

$$\psi_1(\xi) = 2\left(\xi - \frac{1}{2}\right)(\xi - 1),$$
$$\psi_2(\xi) = -4\xi(\xi - 1), \qquad (3.26)$$
$$\psi_3(\xi) = 2\xi\left(\xi - \frac{1}{2}\right).$$

Basis functions for higher dimensions are created through the tensor product of these one-dimensional basis functions. Quadratic basis functions of the appropriate dimension are then used to interpolate over the local quadratic element, in a manner similar to that already discussed in Section 2.1.

From the central node in the local quadratic element, the indices i, j, and k are used to denote steps in the ξ_1, ξ_2, and ξ_3 directions, respectively, where a single step can be made in each of the positive and negative directions. With this template, each of the partial derivatives in equation (3.21) can be approximated by quadratic basis function derivatives where the central node has a location of $\xi = \frac{1}{2}$ in each of the local element directions. The three first-order derivatives of potential are

$$\frac{\partial \phi}{\partial \xi_1} = \phi_{(i+1)(j)(k)} - \phi_{(i-1)(j)(k)},$$
$$\frac{\partial \phi}{\partial \xi_2} = \phi_{(i)(j+1)(k)} - \phi_{(i)(j-1)(k)}, \qquad (3.27)$$
$$\frac{\partial \phi}{\partial \xi_3} = \phi_{(i)(j)(k+1)} - \phi_{(i)(j)(k-1)}.$$

The first-order derivatives of the conductivity tensor components are found in a similar way. The second-order derivatives are defined as

$$\frac{\partial^2 \phi}{\partial \xi_1^2} = 4\big(\phi_{(i+1)(j)(k)} - 2\phi_{(i)(j)(k)} + \phi_{(i-1)(j)(k)}\big),$$

$$\frac{\partial^2 \phi}{\partial \xi_2^2} = 4\big(\phi_{(i)(j+1)(k)} - 2\phi_{(i)(j)(k)} + \phi_{(i)(j-1)(k)}\big), \quad (3.28)$$

$$\frac{\partial^2 \phi}{\partial \xi_3^2} = 4\big(\phi_{(i)(j)(k+1)} - 2\phi_{(i)(j)(k)} + \phi_{(i)(j)(k-1)}\big),$$

with three cross-derivative terms:

$$\frac{\partial^2 \phi}{\partial \xi_1 \partial \xi_2} = \phi_{(i+1)(j+1)(k)} - \phi_{(i+1)(j-1)(k)} - \phi_{(i-1)(j+1)(k)} + \phi_{(i-1)(j-1)(k)},$$

$$\frac{\partial^2 \phi}{\partial \xi_1 \partial \xi_3} = \phi_{(i+1)(j)(k+1)} - \phi_{(i+1)(j)(k-1)} - \phi_{(i-1)(j)(k+1)} + \phi_{(i-1)(j)(k-1)},$$

$$\frac{\partial^2 \phi}{\partial \xi_2 \partial \xi_3} = \phi_{(i)(j+1)(k+1)} - \phi_{(i)(j-1)(k+1)} - \phi_{(i)(j+1)(k-1)} + \phi_{(i)(j-1)(k-1)}.$$
(3.29)

To calculate the metric tensors, the base vectors must also be approximated numerically. The base vectors are the first-order spatial partial derivatives and the quadratic template is used for the approximation in the same manner as that used for the solution variable, ϕ. The second derivatives of the basis functions are the first derivatives of the base vectors and they are used to generate the base vector derivatives, Γ (equation (3.23)).

Implicit and explicit formulations
The transmembrane potential (V_m) equation contains a time derivative that can be approximated by a first-order forward time approximation

$$\frac{\partial V_m}{\partial t} = \frac{V_m^{t+\Delta t} - V_m^t}{\Delta t}. \quad (3.30)$$

We present two methods to add the discrete form of the time derivative to the bidomain or monodomain models, giving explicit and implicit formulations. If the two bidomain diffusion terms are represented by I_{V_m} and I_{ϕ_e}, the transmembrane equation can be written

$$I_{V_m} + I_{\phi_e} = A_m\left(C_m \frac{V_m^{t+\Delta t} - V_m^t}{\Delta t} + I_{\text{ion}}^t\right) - I_{s2}^t. \quad (3.31)$$

The explicit formulation is created by setting the time of I_{V_m} to be t and the update for the transmembrane potential is then

$$V_m^{t+\Delta t} = V_m^t + \frac{\Delta t}{A_m C_m}(I_{V_m}^t + I_{\phi_e}^t + I_{s2}^t) - \frac{\Delta t}{C_m} I_{\text{ion}}^t. \quad (3.32)$$

For each FD node, the group of points used to create the local quadratic element are stored along with the corresponding diffusion coefficients. This allows the evaluation of the two diffusion terms to be reduced to two inner products of the stored coefficients with the appropriate potential values – one inner product for the I_{V_m} term and one for the I_{ϕ_e} term.

The implicit formulation uses I_{V_m} based at the new $t + \Delta t$ time-step, and is given by

$$V_m^{t+\Delta t} - \frac{\Delta t}{A_m C_m} I_{V_m}^{t+\Delta t} = V_m^t + \frac{\Delta t}{A_m C_m}(I_{\phi_e}^t + I_{s2}^t) - \frac{\Delta t}{C_m} I_{\text{ion}}^t. \quad (3.33)$$

The left-hand side of this equation is written into a matrix system where one row is generated for each FD point, and the column positions correspond to the difference point numbers that define the local quadratic element.

Numerical and computational issues

The wave of electrical activation travels through cardiac tissue with a very steep wavefront owing to the fast response of ventricular cells to electrical stimulation (the rapid upstroke of the cellular action potential, as will be described in Section 4 – see Figure 4.3). This leads to high spatial gradients of electrical potential in the region around the wavefront, requiring the use of a very high-density finite difference grid to resolve the wavefront accurately. Ahead of the wavefront and behind the activation wave, however, tissue is inactive and recovering, respectively, and a lower-density grid is sufficient to resolve the slower electrical activity. Therefore the use of a uniformly high-resolution finite difference grid results in a large computational overhead for the inactive and recovering regions of tissue. Adaptive grid techniques are well suited to this type of problem, where only the region of tissue about the activation wavefront is solved using the high-resolution grid, while inactive regions use lower-resolution meshes and recovering regions use a medium-density mesh. Currently we are testing an implementation of a multigrid technique (McCormick 1989) which allows for the specification of multiple grid *levels*, with stepping between these levels during model simulation.

Numerical solution of the electrical activation model requires two steps: integration of a system of ordinary differential equations for the cellular processes at each point to calculate I_{ion} and solution of the bidomain equations (3.14)–(3.15) or monodomain model (3.19) for V_m. Using anything other than the simplest cellular models, this first step will have significantly greater computational demands than the solution of the advection–diffusion equation. This is also a target for parallelization. The cellular ODEs at each grid point can be integrated independently for a given time-step, suggesting that this integration is an ideal candidate for multiprocessing. Given no communication between grid points during a time-step the integration of the cell model should scale linearly with the number of processors in a shared memory system, while the communication between the cellular and continuum models represents a fixed cost which restricts the scalability on distributed memory systems.

3.3. Coupled electro-mechanics

In the previous sections, we have seen how a finite element model of finite deformation elasticity can be formulated, and the same finite element description can be used to model the spread of electrical excitation. Previous work (Hunter, Nash and Sands 1997) has modelled the electrical excitation and mechanical contraction as two separate processes weakly coupled together through the use of the excitation wavefront to initiate active contraction of the tissue. However, in cardiac tissue these two processes are tightly coupled at both the cellular and tissue levels and essential physiology is left out of such weak-coupling models. Not only does the electrical excitation initiate mechanical contraction but mechanical deformation alters the electrophysiology at both the tissue and cellular levels. The cellular level models described in Section 4 describe these effects in cell models; here we concentrate on the continuum tissue level models.

The collocation technique introduced in the previous section allows macroscopic changes in electrical activation processes due to the mechanical deformation of the tissue. Active contraction of muscle fibres generates force only in the direction of the fibre axis (aligned with the ν_1-coordinate). Therefore only one term from the stress tensor needs to be modified, and so if the stress tensor is expressed with respect to the microstructural material axes, only the T^{11} component is modified:

$$T^{\alpha\beta} = \frac{1}{2}\left(\frac{\partial W}{\partial E_{\alpha\beta}} + \frac{\partial W}{\partial E_{\beta\alpha}}\right) - pa_{(\nu)}^{\alpha\beta} + Ta^{11}\delta_1^\alpha \delta_1^\beta, \quad (3.34)$$

where $T = T(t, \lambda_{11}, \ldots)$ is the active tension generated by a fibre at time t. The fibre extension ratio, $\lambda_{11} = \sqrt{2E_{11} + 1}$, defines the current stretch or compression at a given collocation grid point and is determined from a cellular model discussed in Section 4. The value of the active tension at a given Gauss point in the finite element scheme will be defined by the cellular model parameters interpolated from the nearest neighbour collocation grid points, with the collocation points closest to the Gauss point having greater influence than those distant to it. Given the finite element description of strain in the tissue model, the fibre axis strain is readily estimated, and hence the extension ratio can be calculated at any grid point.

Numerical simulation

Although the electrical excitation and mechanical contraction are physiologically interdependent, the quasi-static techniques used for the solution of

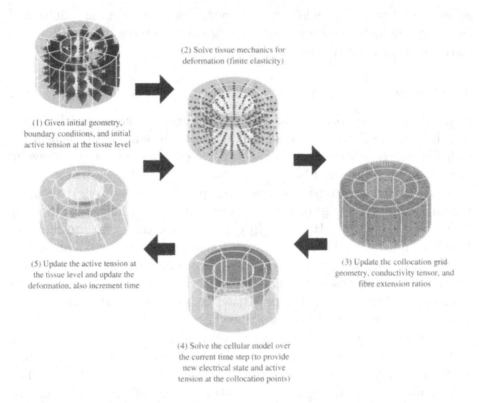

Figure 3.3. Schematic illustration of the iterative solution process for coupled cardiac electromechanics simulations of continuum tissue models. (See text for more detail.)

the finite elasticity tissue models and the fact that the electrical and cellular processes take place on much faster time-scales allows for some uncoupling in the continuum model simulation. Figure 3.3 highlights the algorithm used for the solution of coupled electromechanics in continuum tissue models. This algorithm allows for the independent solution of the finite elasticity and electrical propagation models at a given instant in time, resulting in two solution steps separated by an update step in which information is either transferred from the FE nodes to the FD grid points or *vice versa*.

Any simulation starts with a description of the model and some initial conditions. In this case the model definition includes a finite element geometry (including the fibrous microstructure continuum representation), the constitutive material law for the mechanical model, the continuum material parameters, and specification of the cellular model. The simulation framework developed allows for spatial variation of cellular models and parameters within models. Initial conditions required for solution of the finite deformation elasticity and electrical activation are covered in the previous sections.

Given an initial state for the tissue, an iteration of the finite element algorithm can be performed to give the deformed geometry at time t.

After mechanical deformation of the finite element geometry the location of the collocation points remains unchanged in solution space, but their global position and microstructural axes need to be updated. The metric tensors (equation (3.22)) and the effective conductivity tensor (equation (3.25)) are also updated before the diffusion coefficients (equation (3.21)) are recomputed. The extension ratio values at the collocation grid points are updated by interpolation of the finite element strain field.

With the collocation grid points updated, the cellular models can be integrated forward by one time-step from time t to $t + \Delta t$ and the bidomain equations solved to determine the spread of activation. The size of the time-step (Δt) can be significantly larger than the actual integration time-step required for a converged integration of the cellular differential equations and the bidomain equations. This is due to the much shorter time-scales for kinetics of the cellular level models and the propagation of electrical activation compared to the mechanical contraction. As an indication, the time-step typically required for integration of the the bidomain equations is on the order of 0.01 ms while the mechanics can typically be updated and recomputed on the order of 1.0 ms. The cellular models may require even finer time resolution, which can be accomplished with the use of adaptive time-stepping ODE integrators within a bidomain model time-step (*e.g.*, LSODA (Hindmarsh 1983)).

The solution of a time-step in the cellular model produces a new value for the active tension T at each collocation point. This active tension is then interpolated from the collocation grid points back to the finite element Gauss points as described. With the new active tension values the finite deformation elasticity model can be solved anew, resulting in the deformed geometry at time $t + \Delta t$.

4. Cellular modelling

The mechanisms by which tissue contracts in response to electrical stimulation, consuming energy provided by metabolism in the process, reflect events taking place at the sub-cellular level, and thus necessitate modelling at much smaller spatial scales from those outlined above. For continuum scale investigations of tissue and whole organ function, cellular processes may be included in simulations using simple empirical representations, in some instances using only a single variable, for example, to represent active tension, or the state of electrical activation (via the I_{ion} current term in equation (3.15)). However, a detailed mathematical quantification of the subcellular biochemistry is necessary to couple the activity of processes operating across the range of spatial scales of organization that are characteristic of complex organs such as the heart, for example, to study the effects of pharmaceutical interventions on pump function. Heart cells, like any other cell, are hugely complicated, highly organized and regulated entities. Therefore, cellular models are typically developed with the competing demands of biophysical accuracy and computational simplicity. This challenge has not entirely been met. While progress in the key areas of electrical activation and cellular mechanics has led to highly sophisticated models, which we outline below, the energetics component of cellular physiology has largely been neglected until very recently, and a satisfactory framework for coarse graining and averaging cell properties for incorporation into tissue and organ simulations is yet to emerge. It is clear that such approximation techniques must be developed in order to make fully coupled simulation of the heart, with some 10^9 cells, computationally tractable.

4.1. Active tension development

Heart tissue is made up of sheets of muscle fibres, myofibres, bound together with connective tissue, as discussed in Section 1.1. Myofibres comprise muscle cells, myocytes, which contract in response to electrical stimulation. Inside these striated muscle cells are myofibrils, bundles of contractile proteins, which can be resolved into sarcomeres, the repeating structural elements which are the functional contractile units of the cell. Sarcomeres consist of parallel and overlapping filaments of actin and myosin protein, organized into a regular cross-sectional lattice, in which a hexagonal array of thin actin filaments surrounds each thick myosin filament (see Figure 4.1). The molecular mechanism which generates active tension is due to the interaction of these proteins. Projections from the thick myosin filaments attach to thin actin filaments to form tension-bearing cross-bridges. These cross-bridges form, undergo conformational change to a load-bearing state, dissociate and reattach in a cycle to propel the thick filaments past the thin filaments, shortening the sarcomere and contracting the cell.

This process takes place in response to electrical activation via a sequence of events in the cell. Calcium ions are required to trigger active contraction. At rest, the intracellular concentration of calcium is maintained at very low levels. One of the consequences of activation of the cell is that calcium ions enter through the cell membrane. This small influx of calcium however triggers much larger release from storage sites within the cell, in a high-gain process known as calcium-induced calcium release. Calcium ions must bind to the muscle fibres in order to allow cross-bridges between the thick and thin filaments to be formed. Therefore the contraction is dependent on the calcium level in the cell. However, calcium ions are actively removed from inside the cell, both via re-uptake into the intracellular calcium stores,

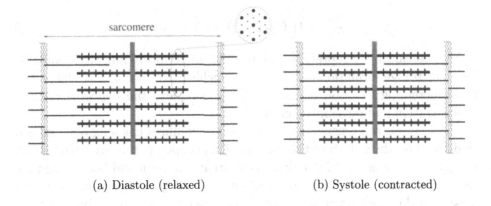

Figure 4.1. Schematic illustration of the overlapping thick (myosin) and thin (actin) filaments of a sarcomere, viewed along its length. (a) shows the relaxed sarcomere, and (b) the sarcomere in a state of contraction, where the thick and thin filaments have moved past each other due to the action of cross-bridges (here the myosin heads are represented as spurs protruding from the thick filaments). In (a) the cross-sectional organization of the filaments is represented (inset).

where it is unavailable to the muscle fibres, and by extrusion from the cell itself. As a result the level of calcium in the cell is only transiently raised following electrical activation, and so the tension development follows the same periodicity as the electrical stimuli received by the cell.

Models of active force generation in cardiac muscle can therefore be developed in three parts: (i) the calcium transient and calcium binding to the thin filaments, (ii) the availability of thin filament binding sites for cross-bridges to form, and (iii) the kinetics of actin-myosin cross-bridges themselves. Calcium binding to accessory proteins (Troponin-C) is a saturating function of the calcium concentration [Ca],

$$[\text{Ca-TnC}] = \frac{\alpha\,[\text{Ca}]}{[\text{Ca}] + \beta(T)}, \tag{4.1}$$

where [Ca-TnC] is the concentration of Ca-bound Troponin-C (TnC) and $\beta(\cdot)$, the rate of unbinding from TnC, is a decreasing function of the developed tension T. As a result of calcium binding to TnC, a sequence of events is initiated which results in the conformational change of a protein complex, removing a physical obstruction to formation of the actin-myosin cross-bridges. The kinetics of this process are essentially governed by a first-order equation of the form

$$\frac{dz}{dt} = f([\text{Ca-TnC}])(1-z) - bz, \tag{4.2}$$

where the variable z represents the fraction of actin binding sites available for cross-bridge formation and the rate at which sites become available $f(\cdot)$ is an increasing function of the [Ca-TnC] complex, and hence of calcium concentration and developed tension.

The steady state tension developed in muscle fibres has long been known to depend on the length of the muscle. This property in heart tissue (Guz, Bergel and Brutsaert 1974) reflects that greater tension and hence pumping capacity is generated in the heart when distended, *i.e.*, with increased extension of the myofilaments. The steady-state tension T_o at a given calcium concentration is linearly dependent on both sarcomere length, λ, and the available fraction of actin binding sites, z:

$$T_o \propto \left(1 + \beta_0(\lambda - 1)\right)z, \tag{4.3}$$

where β_0 is the slope of the linear tension-length relation.

Actin–myosin cross-bridge kinetics

Cross-bridge kinetics, first modelled by Huxley (1957), are the result of transitions between attached force-bearing configurations, and non-force-bearing unattached states of the actin-myosin cross-bridges within each myocyte. The rates of transition between each state are dependent on the strain carried by a given cross-bridge head, and the concentrations of metabolite molecules such as ATP (Adenosine Tri-Phosphate) whose chemical free energy of hydrolysis is converted to work via the cross-bridge cycle. At the level of a single cross-bridge, the time-dependence of transition rates on cross-bridge strain x, the extension of the myosin head when attached to an actin binding site, requires the system be modelled as a system of partial differential equations (Huxley 1957, Hill 1975). The conservation laws for cross-bridges which can be in one of n biochemically or mechanically distinguished states with probability $p_i(x,t)$, with muscle shortening velocity $v(t)$, are given by

$$\frac{\partial p_i}{\partial t} - v(t)\frac{\partial p_i}{\partial x} = \sum_{\substack{j=1 \\ j \neq i}}^{n} p_j(x,t)\alpha_{ji}(x) - p_i \sum_{\substack{j=1 \\ j \neq i}}^{n} \alpha_{ij}(x), \tag{4.4}$$

where α_{ij} is the transition rate between the ith and jth states, which may depend on the cross-bridge strain x (for transitions between unattached and attached states, or for a change of conformation between two attached states, where chemical energy from ATP is used to change the protein conformation and increase the strain in the cross-bridge). In practice it is usually assumed that kinetics are restricted to a cycle, the cross-bridge cycle, and so the rates of transition between non-adjacent states in the cycle are zero. Conservation of the total number of cross-bridges leads to a system of $n-1$ coupled first-order hyperbolic PDEs, along with the relation

$$p_n(x,t) = 1 - \sum_{i=1}^{n-1} p_i(x,t). \qquad (4.5)$$

Typically in this model cross-bridges are assumed to be independent force-generating elements, and also it is assumed that strain x is a continuous variable. In fact, myosin heads and actin binding sites are discretely spaced, but in the limit of large numbers for averages taken over large populations of cross-bridges this is a reasonable approximation. Furthermore, cooperative interactions between cross-bridges have been suggested to explain some muscle data, in particular the steady state force-calcium curve, which shows a sigmoidal-type relationship between the tension generated and the concentration of calcium ions (Rice and de Tombe 2004).

Huxley considered a two-state model, with one attached and one unattached state, which reduces to a single first-order hyperbolic PDE. For suitably chosen piecewise linear functions for attachment and detachment rates this model can be solved analytically, for example, to produce steady state distributions of attached cross-bridges for different contraction velocities v. More complicated models have been proposed which include numerous states, reflecting changing views about the detailed biochemistry underlying the cross-bridge cycle (Piazzesi and Lombardi 1995, Smith 1998). Hill has shown from principles of chemical free energy transduction that the strain-dependent transition rates for these models are not in fact independent of one another, as their ratios are related to the free energy change associated with the transition, which must also reflect the change in energy associated with the change in cross-bridge strain (Hill 1975, 1989). Typically this is calculated by assuming that each cross-bridge functions as an elastic element which develops tension $k(x)$ which is a function of its displacement from the actin binding site. The tension generated in a population of cross-bridges is calculated by integrating over the distributions of tension-bearing states (assuming uniform distribution of cross-bridges along filaments and hence uniform probability of strain x)

$$T(t) = \int_{-\infty}^{\infty} k(x) \left[\sum_{j_{\mathrm{att}}} p_j(x,t) \right] \mathrm{d}x, \qquad (4.6)$$

where the sum is over attached cross-bridge states and $k(x)$ is the force developed by a cross-bridge at strain x. Often it is assumed that the myosin heads are linear elastic elements, in which case the tension developed per sarcomere is proportional to the first moment of the distributions, summed over attached states.

Averaging and empirical modelling

For solving continuum-scale tissue mechanics problems, the active tension generated by cellular contraction is required at a large number of spatially distributed points in the model. Solving systems of partial differential equations at each point is computationally prohibitive in this context. However, to accurately model contraction, and in particular the energetics of this process in tissue we need to retain many details from this molecular description of the contractile apparatus. Two suggestions have been made for tissue level simulations. Zahalak has proposed that ODE approximations for the zeroth, first and second moments of the PDEs can be made if it is assumed that the distributions of attached and unattached states with strain are Gaussian in form (Zahalak 1981). It is a well-known property of the normal distribution that the first three moments can be expressed in closed form in terms of each other. This approach has been proposed for both skeletal (Zahalak 1981, Zahalak and Ma 1990) and cardiac cells (Guccione, Motabarzadeh and Zahalak 1998). The validity of the assumption of normal distributions is questionable, however, particularly for perturbations away from isometric (*i.e.*, zero velocity) equilibrium. An alternative view of this problem is to produce an empirical model of tension development in muscle cells which can be related back to molecular processes.

The model of Hunter *et al.* (Bergel and Hunter 1979, Hunter 1995, Hunter, McCulloch and ter Keurs 1998) couples a linear time-dependent component for contraction with a static nonlinearity to produce a phenomenological description of tension development associated with cross-bridge kinetics:

$$\frac{T}{T_o} = \frac{1 + aQ(v)}{1 - Q(v)}, \qquad (4.7)$$

where a is a parameter fitted from the steady state tension-velocity relationship and T_o is the steady state isometric (zero velocity) tension (see equation (4.3)). $Q(v)$ is a linear response function of the shortening velocity $v(t)$ with m components, which can be represented as an hereditary time integral or *fading memory* model

$$Q = \sum_{i=1}^{m} Q_i = \sum_{i=1}^{m} A_i \int_{-\infty}^{t} e^{-\alpha_i(t-\tau)} v(\tau) \, d\tau, \qquad (4.8)$$

and the parameters A_i and α_i can be recovered from data on studies of muscle stiffness measured for transient length and force-step experiments,

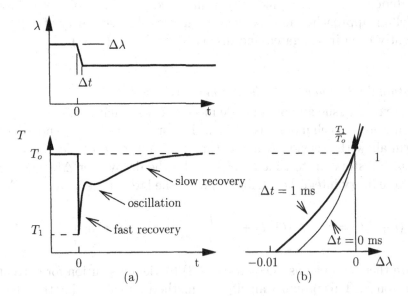

Figure 4.2. (a) Tension recovery (lower figure) following a length step of $\Delta\lambda$ of duration Δt (upper figure). Notice the different phases of the tension recovery. (b) Tension T_1 reached at the end of the length step, divided by isometric tension T_o, plotted against the magnitude of the length step $\Delta\lambda$. One curve is for a length step of 1 ms duration and the other for an idealized instantaneous step.

and for sinusoidal length perturbations about equilibrium over a range of frequencies, for both skeletal (Kawai and Brandt 1980, Kawai, Zhao and Halvorson 1993b) and cardiac (Saeki, Kawai and Zhao 1991) preparations. The form of this fading memory component is equivalent to the transfer function fitted from these sinusoidal perturbation data, and typically for cardiac muscle it is determined that three components are required to describe the response of tissue.

The model captures many features of real muscle experiments, for example the tension developed following a step change in muscle length, shown schematically in Figure 4.2. Furthermore, the dependence of the parameters of the fading memory model on metabolites such as ATP can be characterized from experimental data (Saeki et al. 1991, Kawai, Saeki and Zhao 1993a), and this is necessary in order to couple the active tension generation to models of energy metabolism in the cell. However, with this type of model a calculation of the energetic cost of contraction is not readily accessible, as the link to the molecular mechanisms of contraction is not included in the fading memory framework. The links between these two modelling approaches and the underlying cross-bridge theory of Huxley have recently been investigated and discussed by Smith (2003).

Computational efficiency of the fading memory model

In the context of tissue and organ scale mechanical simulations, the computational efficiency of cell models is critical. The properties of the exponential decay term allow the tension at a new time $(t + \Delta t)$ to be written in terms of the length history up to time t weighted by the decay over Δt, together with the hereditary integral contribution from the latest time increment Δt:

$$Q_i(t + \Delta t) = e^{-\alpha_i \Delta t} \left[Q_i(t) + A_i \int_t^{t+\Delta t} e^{-\alpha_i(t-\tau)} v(\tau) \, d\tau \right]. \quad (4.9)$$

The Q_i are then summed as in equation (4.8) to yield a solution for current active tension T. This provides an efficient method for calculating the time-dependent generation of active tension at the large number of Gauss points in a finite element mesh required to geometrically represent cardiac tissue and organ models.

4.2. Cellular electrophysiology

Contraction of the heart is triggered by a wave of electrical excitation moving through the tissue. This travelling wave is generated by the rapid depolarization of the cell membranes of the muscle cells (myocytes) from a negative resting potential, called an action potential. Muscle cells are for the most part electrically excitable. The resting potential is a stable steady state, and small displacements quickly decay back to rest. However, a super-threshold stimulus generates a large-amplitude trajectory, the action potential, before the membrane potential returns to its resting value. This response to a sufficiently large stimulation is at the basis of the electrical properties of heart muscle.

The electrophysiological properties underlying excitability were first understood in a mathematical model of a nerve cell, the squid giant axon, due to Hodgkin and Huxley (1952). As for other electrically excitable tissues, the active constituent of nerve cells and heart cells is the membrane (sarcolemma) which isolates the contents of a cell from its external environment. Electrical excitability in transmembrane potential is a consequence of the processes controlling the transport of electrically charged species across the cell membrane. This membrane is selectively permeable to charged ionic species, such as sodium, potassium and calcium ions. The passage of ions through the cell membrane is regulated by ion-specific pores, or ion channels, whose permeability may be controlled by the cell membrane electrical potential (voltage-gating) or by the binding of other ions or metabolites. The different properties of two such voltage-gated ion channels, controlling the flux of sodium and potassium ions, were sufficient to explain the excitability in the squid giant axon.

Ion channels and voltage gating

The opening and closing of ion channels in response to changes in potential underlies the excitability of the membrane. When an ion channel is open, the direction in which ions move through the channel is dictated by the electrical and chemical gradients which they experience. Ions move down a concentration gradient until the motive force is balanced by the opposing electrostatic force, produced by the potential difference across the membrane due to net movement of charge. The distribution of each ionic species on either side of the membrane determines a membrane potential, the Nernst potential, at which there is zero net motive force, for the ith ion, V_i. In many cell types including cardiac myocytes the resting potential, at which the voltage-dependent sodium channels are predominantly closed, is close to the Nernst potential for potassium.

Current balance for the capacitive membrane gives a differential equation for the membrane potential of the form (Hodgkin and Huxley 1952, Keener and Sneyd 1998)

$$C_m \frac{dV_m}{dt} = -\sum_i g_i x_i^n y_i^m (V_m - V_i) + I_s, \qquad (4.10)$$

where the sum over subscript i is over the range of different types of ion channels in the membrane. The gating variables x_i and y_i describe, respectively, the activation (opening) and inactivation (closing) of channels, taking values between 0 and 1, according to the differential equations

$$\tau_{x_i}(V_m) \frac{dx_i}{dt} = x_i^\infty(V_m) - x_i \qquad (4.11)$$

and a similar equation holds for y_i. The relaxation time-scale $\tau_{x_i}(V_m)$ and steady state value $x_i^\infty(V_m)$ are functions of the membrane potential (for voltage-gated channels). Parameters n and m reflect the number of independent gating subunits of each type making up the ion channel, which must all be in an open state for the passage of ions. The conductance g_i for the ith channel is for the open channel, and C_m is the capacitance of the membrane. I_s represents a current source applied to the cell, for example the initial stimulus from neighbouring cells. This framework is at the root of all contemporary models of cellular electrophysiology (Miura 2002).

Hodgkin and Huxley also studied the propagation of the action potential in a simple spatially extended version of their model using cable theory.

They were able to calculate the speed of the travelling wave along the axon of the nerve cell using the shooting method, in 1952, on a manually cranked calculator (a Brunsviga 20), due to the labours of Andrew Huxley (Hodgkin 1976).

Electrical excitability in the heart

The physiological principles which underlie excitability in the heart are essentially the same as for any other excitable cell type. In some cases simplified models can be produced and separation of time-scales used to reduce the complexity of a model (see, for example, Smith and Crampin (2004)). On the other hand, there are many processes involved in generating and regulating the electrical properties of cells. For example, more than 20 different types of ion channel have been identified in the myocyte. The same modelling framework can be expanded by the addition of equations representing the active (energy-requiring) transport of ions against their electrochemical gradients by ion pumps and exchangers in the membrane, which are required to maintain the resting state of the cell and to regulate cell calcium. One important feature of the cardiac action potential is that it has a so-called plateau phase during repolarization back to the resting potential, during which calcium ions enter the cell triggering a larger release from intracellular calcium stores (see Figure 4.3). This calcium-induced calcium release initiates the contraction of the muscle filaments, as discussed above. Equations describing these processes can be incorporated to produce comprehensive models of cellular electrophysiology.

The most extensive of these cardiac models are the ion channel models pioneered by Noble and co-workers (Noble 1960, DiFrancesco and Noble 1985, Noble *et al.* 1998), and the Luo and Rudy models (Luo and Rudy 1991, 1994), focusing on the physiological behaviours of premature stimulation and arrhythmogenic activity of the single myocyte. Their model has since been extended by Jafri, Rice and Winslow (1998), among others, to accommodate more complex calcium kinetics which are important for contraction coupling, as discussed above. One important consideration that is introduced by more realistic models of calcium handling is that the cellular compartment into which calcium channels empty (the diadic space) is very small, and furthermore the release of calcium from the internal stores has been shown to be in discrete quantities, or sparks, introducing very short time-scales and a stochastic element to cell models, which may be important in some circumstances (Greenstein and Winslow 2002).

4.3. Metabolic models

Muscle cells convert metabolic (chemical) energy to work via a sequence of biochemical processes involving the breakdown (or hydrolysis) of the

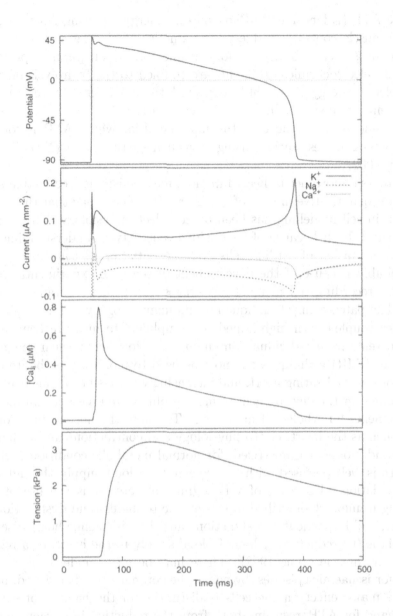

Figure 4.3. Figure showing, from top to bottom, ventricular action potential, ionic currents, intracellular calcium transient and active tension development from a cardiac myocyte model. Note that the rapid inward (negative) component of the sodium current (Na$^+$; dashed line), which initiates the action potential upstroke, reaches a much greater maximal amplitude than is shown in the figure (where the scale is truncated at -0.1μAmm^{-2}). These traces were generated using the ventricular electrophysiology model of Noble et al. (1998) coupled to the HMT mechanics model of Hunter et al., as described in Nickerson, Smith and Hunter (2001).

molecule ATP (Adenosine Tri-Phosphate). Energy-requiring reactions, for example muscle contraction due to the kinetics of binding and unbinding of actin–myosin cross-bridges, and transmembrane ion pumps operating against an electrochemical gradient, are coupled to the hydrolysis of ATP which releases energy. As might be expected, there is an intricate feedback of energy demand to supply in the cell, particularly acting at the major sites of ATP generation in the cell, the mitochondria, where ATP is formed by the process of respiration, using oxygen from the blood (Nicholls and Ferguson 2002).

Despite its importance to heart function, metabolism has received less attention for quantitative models of the heart. It is fair to say that the major attention in cell modelling has been on cell electrophysiology. This is perhaps natural, both because of its obvious centrality in understanding the electrical activation of cardiac cells and the initiation of contraction, but also probably because of the development of sensitive experimental techniques for recording cell membrane voltages and transmembrane ion currents. (The patch-clamp technique and its many progeny use micropipette electrodes coupled with high-impedance amplifiers to measure transmembrane currents from individual ion channels.) However, nuclear magnetic resonance (NMR) techniques can now be used to monitor metabolite concentrations in the beating heart, and increasingly data is becoming available which allows for the development of more sophisticated metabolism models as components for whole-cell modelling. To study the normal function of organs such as the heart, electrophysiology and contraction can be characterized, and models parametrized, for normal metabolic conditions, where the heart is well perfused with an oxygenated blood supply through the microcirculation. However, of very significant interest is the use of the modelling framework described in this article to understand dysfunction in the heart, and in particular dysfunction caused by ischaemic heart disease. Ischaemia is the reduction or loss of blood supply to the heart, or a region of the heart muscle (Opie 1998). There may be many underlying clinical reasons for ischaemic episodes, by which the coronary flow is reduced, however, the major effect on the cells is disruption of the balance of supply and demand for ATP, resulting both from the reduction in oxygen supply and from the reduction in blood flow to transport energy substrates to, and waste products from the cells.

The result of ischaemia on the muscle is an increase in the acidity of the cells (acidosis), changes to the distribution of ionic species across the cell membrane, which affect the electrophysiological properties of the cells, and eventually the loss of ATP to drive energy-requiring reactions. The final consequence of these disturbances is the loss of the contraction cycle, pump failure and cell death (infarction). However, before this occurs the changes to the ionic milieu can also give rise to pathological electrical behaviours, or arrhythmias. The origins of these arrhythmogenic processes are not well understood, and detailed cell and tissue modelling provides a testbed for different theories and potential therapies. ATP and its hydrolysis products, and in particular the cell acidity levels, regulate many of the electrophysiological and calcium handling processes in the cell. This network of regulatory interactions is now being incorporated into detailed cell models for the heart. The additional complexity of accounting for metabolites and acidity notwithstanding; the major difficulty again with numerical simulations involving metabolism is the inclusion of processes which take place over time-scales of minutes to hours, *i.e.*, many heart beats, rather than the millisecond time-scales of most electrophysiological processes. Furthermore, the loss of blood supply is usually localized to one region of the muscle, with a gradation of effects moving laterally away from the fully ischaemic zone. The resulting spatial heterogeneity of cell properties is thought to be strongly influential in the developing pathophysiology, and is a significant further challenge to modelling studies. A description of continuum modelling approaches for the coronary vasculature and microcirculation blood supply to the heart is presented in Section 5.

Computational and modelling issues
These models of cellular electrophysiology and metabolism are described by coupled systems of nonlinear ordinary differential equations, possibly with the addition of stochastic components for certain ion channel gating and calcium handling processes. An essential characteristic of these components for integrated modelling of physiological systems is the wide range of time-scales that are introduced. For example, the upstroke of the action potential takes place on the sub-millisecond time-scale, whereas the duration of the cardiac action potential itself is on the order of several tenths of a second. The introduction of metabolic processes into cell models introduces components with time constants of seconds and longer – several orders of magnitude larger than the (sub-) millisecond electrophysiological time-scales. This gives rise to large stiff systems of coupled ODEs to describe the kinetics of the various processes which underlie the action potential and its regulation by metabolism. While numerical simulation over this range of time-scales for a single cell presents few difficulties for modern computers, the coupling of large numbers of cells in a continuum tissue framework represents a significant hurdle to simulation studies (there are 10^9 or so cells in the human heart). Faster algorithms for very stiff ODE systems, along with refinements in modelling which allow course graining or adiabatic approximations for slowly varying properties are therefore needed.

5. Vascular blood flow

A third important continuum element of tissue models, following mechanical and electrical activation, is representation of the network of coronary blood vessels (vasculature) which supply cardiac tissue with oxygenated blood. The spatio-temporal distribution of the supply of oxygen and metabolic substrates (perfusion) is an important determinant of heart function, typically via its influence on the state of cellular metabolism (models of which were introduced in Section 4.3). Interactions between vascular perfusion and organ function are also relevant at the continuum modelling scale.

The arterial and venous vessels are elastic tubes, and so their volume is a function of the difference between the blood pressure and the mechanical state of the tissue in which the vessels are embedded. Thus the vascular fluid dynamics are integrally coupled to tension generation and deformation of the tissue.

Several recent studies have used statistical morphometric data to reconstruct vascular network geometries using measurements from casts (Kassab et al. 1993, Kassab and Fung 1994), image segmentation (Kantor et al. 1999) and spatial distributing algorithms (Smith, Pullan and Hunter 2000). These geometries provide the foundation for an anatomically based model of vascular blood flow. Computational limitations motivate the reduction of the full Navier–Stokes equations to produce a one-dimensional model of flow in a single vessel segment, which we describe below. The use of constraints to maintain conservation relations across bifurcations in the branching network is introduced (in Section 5.2). A lumped parameter model (Section 5.3) is used to represent small vessel microcirculation networks distributed at fine spatial resolution. Finally, modelling the effect of tension generation and deformation on blood flow in the coronary network is described in Section 5.4.

5.1. Single vessel model

There are a number of fundamental assumptions about vascular blood flow used in deriving the governing equations for the model presented here. The studies of Perktold, Resch and Peter (1991) and Cho and Kensey (1991) indicate that the shear thinning properties of blood do not play a significant role in large diameter vessels. This is because the relative size of red blood cells to vessel diameter is small and therefore blood can be modelled as a continuum. The distensibility of a coronary vessel wall is assumed to dominate any effects due to the compressibility of blood. Thus, in the equations below, blood is modelled as an incompressible, homogeneous, Newtonian fluid. The low Reynolds number applicable to the majority of the circulation means that all flows are assumed to be laminar.

With these assumptions the Navier–Stokes equations can be expressed using a cylindrical coordinate system (r, θ, x), where the x axis is aligned with the local vessel axial direction. Assuming velocity in the circumferential direction is zero and that the flow is axi-symmetric, the equations governing fluid flow reduce to

$$\frac{\partial v_x}{\partial t} + v_r \frac{\partial v_x}{\partial r} + v_x \frac{\partial v_x}{\partial x} + \frac{1}{\rho}\frac{\partial p}{\partial x} = \nu \left(\frac{\partial^2 v_x}{\partial r^2} + \frac{1}{r}\frac{\partial v_x}{\partial r} + \frac{\partial^2 v_x}{\partial x^2} \right), \quad (5.1)$$

$$\frac{\partial v_r}{\partial t} + v_r \frac{\partial v_r}{\partial r} + v_x \frac{\partial v_r}{\partial x} + \frac{1}{\rho}\frac{\partial p}{\partial r} = \nu \left(\frac{\partial^2 v_r}{\partial r^2} + \frac{1}{r}\frac{\partial v_r}{\partial r} - \frac{v_r}{r^2} + \frac{\partial^2 v_r}{\partial x^2} \right). \quad (5.2)$$

In these equations $v_x(x,r)$ and $v_r(x,r)$ are the axial and radial velocities. Pressure, viscosity and density are denoted by $p(x)$, ν and ρ respectively. Conservation of mass is governed by

$$\frac{\partial v_x}{\partial x} + \frac{1}{r}\frac{\partial (r v_r)}{\partial r} = 0. \quad (5.3)$$

We make the further assumption that radial velocity is small compared to axial velocity and, consistent with experimental observation, that the radial velocity profile can be represented in the form

$$v_x = \frac{\gamma + 2}{\gamma} V \left[1 - \left(\frac{r}{R}\right)^\gamma \right], \quad (5.4)$$

where $V(x)$ is average flow velocity, $R(x)$ is the internal vessel radius and γ is an empirical fitting parameter (Smith, Pullan and Hunter 2002).

Equations (5.1)–(5.3) can be reduced such that conservation of mass is governed by

$$\frac{\partial R}{\partial t} + V \frac{\partial R}{\partial x} + \frac{R}{2}\frac{\partial V}{\partial x} = 0, \quad (5.5)$$

and conservation of momentum equals

$$\frac{\partial V}{\partial t} + (2\alpha - 1) V \frac{\partial V}{\partial x} + 2(\alpha - 1)\frac{V^2}{R}\frac{\partial R}{\partial x} + \frac{1}{\rho}\frac{\partial p}{\partial x} = -2\frac{\nu \alpha}{\alpha - 1}\frac{V}{R^2} \quad (5.6)$$

(see Smith et al. (2002) for details). Equations (5.5) and (5.6) provide two equations in the three unknowns, (V, p, R). A third constitutive equation which describes the relationship between pressure and radius, R, must be established. Ignoring any transient or visco-elastic properties of the vessel wall, an empirical relationship between transmural pressure and the radius can be established, which is of the form

$$p(R) = G_o \left[\left(\frac{R}{R_o}\right)^\beta - 1 \right]. \quad (5.7)$$

The form of equation (5.7) was chosen to provide a good fit to experimental pressure–radius data. The solution to equations (5.5)–(5.7) characterizes transient blood flow in the coronary network.

5.2. Branching in vessel networks

Equations governing fluid flow at bifurcations within a network must now be introduced in order to model blood flow through branching structures such as the coronary network. The bifurcation model is based around an approximation of the junction between vessels as three short elastic tubes of radius R_a, R_b and R_c, respectively. If the tubes are assumed to be sufficiently short then the velocity along them is constant, i.e., they are parallel-sided and losses due to fluid viscosity are negligible. Furthermore, no fluid is assumed to be stored within the junction. The location of the parent vessel entering the junction is denoted by a and the points at the beginning of the daughter vessels are labelled as b and c. If F_a, F_b and F_c are the rates of flow through each junction segment and P_o is the pressure at the junction centre, then conservation of mass through the junction is governed by

$$F_a - F_b - F_c = 0. \tag{5.8}$$

The conservation of momentum for tube a is governed by the resultant axial force equalling the rate of change of momentum of fluid in a segment length l_a of radius R_a, i.e.,

$$R_a^2(p_a - p_b) = \rho\left(\frac{\partial(l_a R_a^2 V_a)}{\partial t} + \frac{\partial(l_b R_b^2 V_b)}{\partial t}\right), \tag{5.9}$$

along with similar expressions for the flows between tubes a and c, and tubes b and c. Having introduced equations (5.8) and (5.9), which govern the conservation of mass and momentum across the bifurcation, the next step is to couple these to the single vessel equations (5.5)–(5.7). This is achieved by manipulating the pressures at the three segment ends that form a bifurcation such that the flows F_a, F_b and F_c simultaneously satisfy equations (5.8) and (5.9). This manipulation can be cast as a root-finding problem. The boundary conditions from the single vessel elements are used to calculate flows in each segment for a given pressure and partial derivatives of flow with respect to pressure $\frac{\partial F}{\partial p}$. From these equations a Newton–Raphson step can be calculated to determine the change in the pressure increments in tubes a, b and c required to satisfy equations (5.8) and (5.9) to linear order. To account for the nonlinearities this calculation is iterated using the new pressure values until the solution converges.

5.3. Microcirculation

The microcirculation network is formed by the terminal branching arterial and venous networks, called arterioles and venules, and the capillaries that connect them. This network is both topologically and functionally different from the network of large-conduit vessels. These vessels have diameters

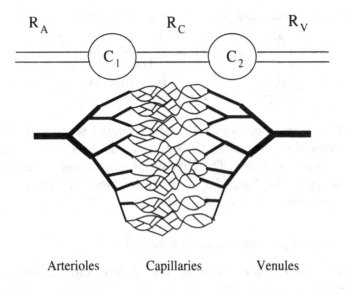

Figure 5.1. Schematic of the lumped parameter microcirculation model with the lumped comments for volume capacitance and resistance to flow shown.

comparable to the size of a red blood cell. At this spatial scale blood can no longer be considered as a homogeneous Newtonian fluid and the flow properties are strongly influenced by the individual red blood cells (Pries, Neuhaus and Gaehtgens 1992). This affects fluid viscosity (known as the Fahraeus effect), flow profiles and distribution of flow at bifurcations (Pries, Ley, Claasen and Gaehtgens 1989). Thus the equations used to model flow through the larger vessels in this study are no longer valid. The large number of microcirculation networks connecting each arteriole to a venule also makes the method of discretely modelling individual vessel segments for each microcirculation network computationally prohibitive. To overcome these problems a lumped parameter model of microcirculation has been developed based on the intramyocardial pump of Spaan, Breuls and Laird (1981). This model is used to reproduce the observed flow responses to arteriole and venule pressure of an anatomically based model combining nonlinear resistive and capacitive elements, reproducing experimentally observed behaviour in a computationally efficient way, while retaining some of the fundamental physics of the problem. A five-element lumped parameter model is shown schematically in Figure 5.1. R_A, R_C and R_V are arterial, capillary and venule resistances, respectively, and C_1 and C_2 represent the proximal and distal capacitances of the microcirculation model. The rate of change of pressure is related to the net flow across the capacitive components, and hence the pressure drop, according to

$$C_1 \frac{dp_1}{dt} = F_A - F_C = \frac{P_A - P_1}{R_A} - \frac{P_1 - P_2}{R_C}, \tag{5.10}$$

$$C_2 \frac{dp_2}{dt} = F_C - F_V = \frac{P_1 - P_2}{R_C} - \frac{P_2 - P_V}{R_V}. \tag{5.11}$$

These equations are then coupled to the terminal segment end of the arteriole and venule models using the same root-find techniques employed to determine bifurcation flows. The Newton–Raphson method is used to iteratively converge to the P_A and P_V pressure values that, via single segment boundary condition, give flows F_A and F_V.

5.4. Coupled blood flow and tissue mechanics

The effect of contraction of the heart and skeletal muscles around the embedded vessels is an important determinant of blood flow (Downey and Kirk 1975, Spaan et al. 1981, Krams, Sipkema and Westerhof 1989). There are two distinct steps to coupling coronary blood flow to the pressure exerted by the other host organ on the embedded vessels, firstly calculation of the pressure exerted on the vessel wall and secondly to include this pressure in the blood flow models presented above.

For each vessel, the pressure exerted on the vessel wall as a function of time and distance along the vessel must be calculated. The pressure acting on a vessel wall at a given point along the length of a vessel is assumed to be the average of the radial forces acting normal to the wall. The starting point in determining this pressure is the second Piola–Kirchoff stress tensor $T^{\alpha\beta}$, calculated by solving the finite deformation equations which govern the deformation of the host medium (Section 3.1). This stress tensor is defined in the local material coordinates (ν_1, ν_2, ν_3) (see Figure 5.2).

$T^{\alpha\beta}$ does not directly provide information about the physical stresses. The objective is to use the second Piola–Kirchoff stress tensor to determine the Cauchy stress tensor σ^{ij} in a deformed rectangular Cartesian coordinate system. Using σ^{ij} the physical stresses can then be calculated. The Cauchy stress tensor is related to the 2nd Piola–Kirchoff stress tensor in this Cartesian coordinate system (Malvern 1969) by the following relation:

$$\sigma^{(x)} = \frac{1}{J} F T^{(\nu)} F^T. \tag{5.12}$$

$\mathbf{F}$ is evaluated using the material coordinates of the host finite element mesh ξ and the reference coordinate system of that host mesh θ:

$$\frac{\partial x_i}{\partial \nu_M} = \frac{\partial x_i}{\partial \theta_j} \frac{\partial \theta_j}{\partial \xi_k} \frac{\partial \xi_k}{\partial \nu_M}. \tag{5.13}$$

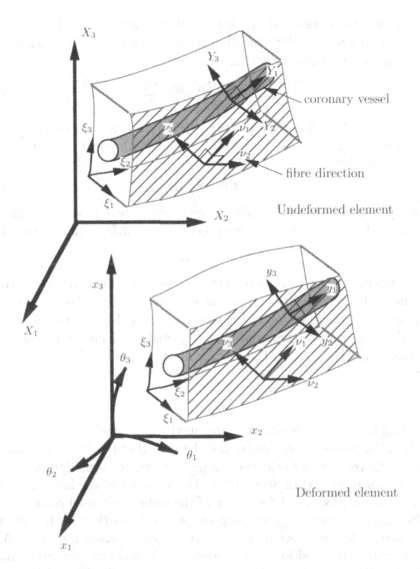

Figure 5.2. Coordinate systems used in the kinematic analysis of large-deformation finite element host mesh with embedded coronary elements. The axes (X_1, X_2, X_3) are the material Cartesian coordinates aligned with the reference coordinates (x_1, x_2, x_3) in the undeformed state. (ν_1, ν_2, ν_3) are the material coordinates aligned with the tissue microstructure; these are defined to be orthogonal in the undeformed state. (ξ_1, ξ_2, ξ_3) are the local finite element material coordinates. (Y_1, Y_2, Y_3) and (y_1, y_2, y_3) are the Cartesian set of coordinates which are orthogonal and aligned with the vessel direction in the undeformed and deformed states respectively. $(\theta_1, \theta_2, \theta_3)$ are the curvilinear reference coordinate system.

To evaluate the stresses on the vessel wall the Cauchy stress tensor is transformed into a rectangular Cartesian coordinate system y in which the local deformed vessel geometry is referenced:

$$\sigma^{ij}(y) = \frac{\partial y_i}{\partial x_k} \sigma^{kl}(x) \frac{\partial y_j}{\partial x_l}. \tag{5.14}$$

According to Cauchy's formula, the stress vector in Cartesian coordinates t (the force per unit area acting on a surface with unit normal n) is given by (Spencer 1980)

$$t = \sigma^{ij} n_i g_j, \tag{5.15}$$

where g_j are the base vectors of the coordinate system. From this expression, the normal stress, t_n, acting perpendicular to a surface with a normal n_i is

$$t_n = \sigma^{ij} n_i n_j. \tag{5.16}$$

By determining a vector normal to the vessel wall the normal stress acting on the wall can therefore be calculated. This can be further simplified by aligning y_1 in the direction of the vessel, y_2 normal to y_1 and to the x axis and choosing y_3 as the cross-product of y_1 and y_2. Thus both y_2 and y_3 are normal to the vessel axis. The average normal stress in the radial direction is then

$$t_{\text{wall}} = \frac{1}{2}(\sigma^{22}(y) + \sigma^{33}(y)) \tag{5.17}$$

and t_{wall} is taken as the stress acting on the vessel wall.

Once the pressures are determined the second step is to include the reaction of the vessel wall to the varying pressure and deformation in the blood flow model presented in Section 5.1. Deformation of the host organ finite element mesh produces deformation of the embedded vessel geometry. This deformation is calculated by using one set of ξ positions within the mesh geometry to fix vessel segments at material points inside the mesh. As the mesh deforms the position of the vessels is calculated by reinterpolating the basis functions using the deformed geometry at the fixed ξ positions. Using this method axial vessel strain (denoted by λ) can be determined from the solution of the finite deformation equations (an example of the calculated deformation of embedded vessels is shown in Figure 5.3).

The elastic pressure–radius relationship must be modified such that pressure within a vessel is the sum of two terms: the reaction of the fluid on the elastic vessel wall and the pressure

$$P(R) = G_o \left[\left(\frac{R\sqrt{\frac{\lambda}{\lambda_o^*}}}{R_o^*} \right)^{\beta_1 \lambda + \beta_2} - 1 \right] - t_{\text{wall}}, \tag{5.18}$$

exerted on the wall by the deforming host medium, where β_1, β_2 and G_o are empirical parameters fitted from experimental data.

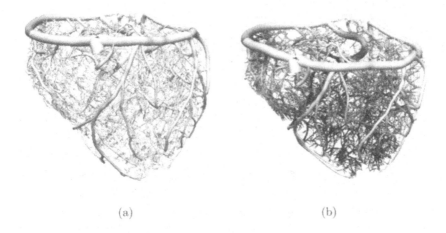

(a)   (b)

Figure 5.3. A coupled model of coronary arterial blood flow and myocardial mechanics showing calculated intramyocardial pressure mapped on to the deformed coronary arterial geometry at two stages of the heart cycle: (a) start of contraction and (b) end of contraction. Shading indicates the degree of pressure exerted on the vessels by contraction.

To account for deformation, *i.e.*, stretch or compression of vessels, the conservation equations for mass and momentum are also modified:

$$\frac{\partial(R\sqrt{\lambda})}{\partial t} + \frac{R\sqrt{\lambda}}{2}\frac{\partial V}{\partial x} + V\sqrt{\lambda}\frac{\partial R}{\partial x} = 0, \qquad (5.19)$$

$$\frac{\partial V}{\partial t} + (2\alpha - 1)V\frac{\partial V}{\partial x} + 2(\alpha - 1)\frac{V^2}{R}\frac{\partial R}{\partial x} + \frac{1}{\rho}\frac{\partial p}{\partial x} = \frac{-\alpha\nu SV(a^2 + b^2)}{(\alpha - 1)\pi a^3 b^3}. \qquad (5.20)$$

The solution techniques for these modified blood flow equations are unchanged. However, the λ and t_{wall} values vary with time and thus require updating at each time-step. This can be achieved via linear interpolation of values calculated from mechanics solutions which are usually incremented with a significantly greater time-step than that used in the flow solution.

Numerical and computational issues
Calculations of coronary blood flow for the whole heart are computationally intensive owing to the large number of finite difference grid points and the number of network bifurcations and lumped parameter microcirculation models to be solved at terminal branching of the network. The computational storage required for the implementation of an implicit method is prohibitive. The explicit two-step Lax–Wendroff finite difference scheme is a suitable solution method for the set of equations which is second-order

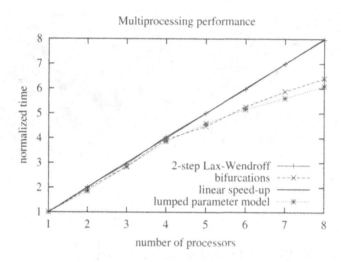

Figure 5.4. The normalized time $\frac{T_1}{T_n}$, where T_1 is the time taken using one processor and T_n is the time taken using n processors, as a function of number of processors n for 10000 time-steps solving for flow variables on the first three generations of the full coronary mesh.

accurate in both time and space, and also eliminates large numerical dissipation (Press, Teukolsky, Vetterling and Flannery 1992). For a detailed analysis of the stability properties of this approach see Smith et al. (2002).

A major advantage of the explicit Lax–Wendroff scheme, applied to the blood flow equations, is the ability to divide the computational effort at a given time-step evenly between processors with little additional work. The scalability is demonstrated in Figure 5.4 for a typical network simulation on a shared memory Silicon Graphics Power Challenge. The computation times for the finite difference calculations are very close to the theoretical ideal of linear speed-up, indicating an efficient parallelization, in this case using simple scheduling to divide grid points evenly between processors. The computation times for the network elements, both for bifurcations and microcirculation calculations, do not show linear speed-up. More iterations are needed to reach convergence in regions of the coronary network where large pressure gradients occur. Thus, to avoid computational bottlenecks, where a large subset of the processors may be idle, dynamic scheduling can be used to divide computational work between processors. The computational overhead, however, increases and efficiency drops as the number of processors is increased, as shown in Figure 5.4 by the fall away from the ideal linear speed-up.

6. Future directions

Having presented a framework for integrative modelling linking across multiple physical processes at one spatial scale (mechanics, electrical wave propagation and coronary blood flow) and across multiple spatial scales (subcellular, ion channels to the intact organ), we now explore future directions for this work and present models from two other organ systems developed using this framework.

6.1. Multiscale modelling in the heart

The framework for computational physiology presented here, and illustrated in relation to the heart, requires anatomically and biophysically based models to be developed at multiple spatial scales from genes to the intact organism.

At the organ level the models are based on physical conservation laws and are solved with numerical techniques that are typically finite element or boundary element methods, since these are readily able to cope with the complex three-dimensional geometries. A feature of the models at this level is that they must deal with more than one type of physical problem. Large deformation mechanics in the heart, for example, is solved with Galerkin finite element techniques coupled closely with the solution of the reaction–diffusion equations governing electric current flow around the heart. This latter technique uses a computational grid which is tied to material points of the moving finite element mesh, but the spatial resolution requirements for these two processes are quite different. Spatial convergence for the mechanics is achieved with a finite element mesh of around 100 tricubic-Hermite elements but convergence for the reaction–diffusion equations requires a mesh resolution of 0.1–0.2 mm, corresponding to 30 million grid points if implemented for the whole heart. The challenge now is to develop adaptive meshing techniques (multigrid, for example) which use the finely resolved meshes only in the vicinity of the moving wavefronts.

Another priority for model development at the organ level is to incorporate the measured spatial distribution of material parameters, such as the density variation of various ion channels in the heart (Akar and Rosenbaum 2003).

The organ systems for which models are currently at an advanced stage of development are the heart and circulation, the lungs, the musculo-skeletal system and the digestive system. The next-highest priorities are probably the kidney, endocrine pancreas and liver.

Tissue properties are included in the conservation laws of continuum mechanics (equations (3.10) and (3.5)) or electrophysiology (equations (3.14) and (3.15)) via material constitutive laws that express stress–strain relations (equation (3.12)) or current–voltage relations (equation (4.10)). The next challenge at the tissue level is to relate these macroscopic continuum descriptions to detailed structural models of the tissue. Linking the parameters of these constitutive laws to the underlying tissue structure requires tissue models at the millimetre scale which include the distributions of type, orientation, density and cross-linking for collagen, proteoglycans and other extracellular matrix components. The properties of these tissue types must be linked to cellular properties within appropriate cell types. Similarly, the cellular processes such as ion transport, signal transduction, metabolic pathways, *etc.*, must be linked to the spatial distribution of proteins within cells and their enzyme and substrate-dependent binding reactions. In some cases these must also account for the reaction–diffusion behaviour of intracellular messengers.

Another important goal in multiscale heart modelling is the development of a three-dimensional (3D) myocyte cell model which can link the function of individual proteins to the integrated function of cells operating within the extracellular matrix (ECM). The forces developed by myofilaments during active contraction are conveyed to adjacent cells via intercellular gap junctions (formed from the protein connexin) and the surrounding ECM of collagen (the primary structural protein) and proteoglycans (with their electrically charged water binding groups). The internal structure of the myocytes is maintained by a network of intermediate filaments (primarily the protein desmin) (Balogh *et al.* 2002). Three-dimensional models that incorporate the spatial distribution and material properties of these ECM and intracellular structural proteins are needed. The 3D models also need to include the reaction–diffusion kinetics of mobile ions like H^+ and Ca^{2+} and the spatially localized action of signal transduction pathways.

6.2. *Markup languages and ontologies*

Mathematical modelling of physiological function at the level of tissues and cells often requires that independently developed models be combined. Signalling pathways within cells, for example, are highly interdependent and cannot be treated in isolation. This raises two very important issues for biological modellers. The first is the need for a standard format (a 'markup language') for encoding models in a robust, parsible, electronic form. Fortunately the recent development of XML[1] standards provides a platform for the development of modelling standards such as CellML (www.cellml.org) and SBML (www.sbml.org). The second is the need for consistent names for all the biological components (an 'ontology') and a strong typing con-

---

[1] eXtensible Markup Languages developed by the W3C (Worldwide web consortium www.w3c.org)

vention so that a particular protein such as PKA, for example, is labelled as a kinase (a type of enzyme) that can participate in a particular type of reaction (phosphorylation of a binding site on the target protein). This second requirement is being addressed by a number of international consortiums such as TAMBIS (imgproj.cs.man.ac.uk/tambis) and BioPAX (www.biopax.org).

6.3. Other organ systems

There are twelve organ systems in the body[2] and each component of these organ systems involves one or more of the four basic tissue types[3] which contain, depending on their location, a fraction of the approximately 200 different cell types. Here we briefly describe progress on applying the computational framework described above to two other organ systems.

The lungs

The dominant physical processes occurring at the organ level in the lungs are gas transport through the airways, blood flow in the pulmonary circulation and large-deformation mechanics of the soft parenchymal tissue. The airways are divided, on average, into 16 generations of conducting airways, which occupy about 150 ml of the 4 litre total air capacity, and 9 (on average) generations of respiratory airways in which gas exchange with the blood flow takes place. A finite element model of the conducting airways, based on CT data is shown in Figure 6.1 (Tawhai, Pullan and Hunter 2000). This model has been used to study gas transport (Tawhai and Hunter 2001a, 2001b) and the transport of heat and humidity in the conducting airways (Tawhai, Rankin, Ryan and Hunter 2002, Tawhai and Hunter 2004).

The challenge now is to develop tissue models that incorporate the structure and functional properties of the epithelial and connective tissue surrounding the airways and pulmonary blood vessels. Once the three major physical processes dominating lung function (air flow, blood flow and tissue mechanics) have been modelled, the spatial distribution of cell types can be included and the link made to diseases of the lungs such as COPD (chronic obstructive pulmonary disease) and asthma.

The musculo-skeletal system

Anatomically based models from the musculo-skeletal system are shown in Figure 6.2. This includes tricubic-Hermite models of bones and muscle

---

[2] Cardiovascular system, respiratory system, muscular-skeletal system, skin (integument), digestive system, urinary system, nervous system, endocrine system, lymphoid system, male reproductive system, female reproductive system and the special sense organs.

[3] Connective tissue, epithelial tissue, muscle tissue and nervous tissue.

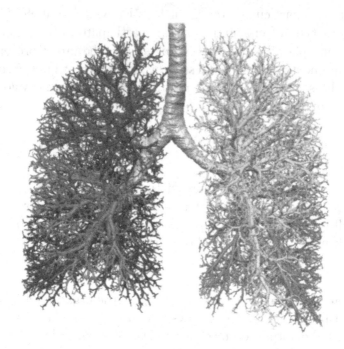

Figure 6.1. A model of the human airways for studying gas transport and exchange in the lungs. From Crampin *et al.* (2004), with permission

(a) Three-dimensional bone structure.

(b) The fitted femur bone model.

(c) Finite element models of the muscles and tendons of the human forearm.

Figure 6.2. Examples of images and finite element models from the musculo-skeletal system.

which incorporate the underlying microstructure using the method previously applied to the heart. For certain regions of the body, such as the legs, the fibrous structure of the muscles has also been fitted to experimental data and the tendons, ligaments and cartilage are also included. One current application is examining the distribution of stress over the head of the femur as the patella (knee cap) rolls over it. Note that the geometric fitting processes used here are very similar to those described in Section 2.2 for the heart and the equations governing mechanics are also similar to those described in Section 3.1. Bone, which is a linearly elastic compressible material, can be treated within the same nonlinear elasticity framework used for muscle, tendons and cartilage provided that the strain energy function equation (3.12) is spatially modified.

Again, the goal here is to link models of detailed structure (bone trabeculae and the soft connective tissues of skeletal muscle and cartilage) to continuum constitutive law parameters and then to incorporate the distribution of cell types and cell level processes to link the tissue and organ models to diseases such as osteoporosis and rheumatoid arthritis.

7. Conclusion

In this paper we have discussed computational modelling of organ systems (the heart in particular) at several spatial scales. At the level of the intact organ this requires the solution of partial differential equations expressing physical conservation laws, such as conservation of mass and momentum for mechanics and conservation of current for electrical activation.

The challenges from a numerical perspective are finding efficient ways of coupling the various physical systems (large-deformation mechanics, electrical wave propagation, fluid flow, *etc.*), each of which has its own characteristic length and time scales. Another challenge is linking across spatial scales from subcellular protein pathways to integrated cell function and then to tissue and organ behaviour. For electrical activation in the heart we have shown in this review how the solution of equations governing the electrophysiology of subcellular ion channels can be linked to the reaction–diffusion equations at the tissue and whole organ level.

The major challenges for the heart modelling work are: (i) to extend the organ level models to include other anatomical structures such as the atria and heart valves, (ii) to extend the tissue level models to incorporate more detail on extracellular matrix structure and how this changes in disease, and (iii) to extend cellular level models to incorporate signal transduction and metabolic pathways and how these are altered by disease.

The physiological accuracy of these continuum models depends critically on the constitutive laws (both the functional form and spatially varying parameter values) that define the relationships between stress and strain, current and voltage gradient, *etc.* The fact that the constitutive law parameters do vary spatially is a reflection of the inhomogeneous structure of the underlying tissue, which grows and remodels to suit its local functional requirements. Since it is generally not possible to measure experimentally the constitutive law parameters throughout an organ, an alternative approach is to establish a relationship between the constitutive law and the underlying microstructure of the tissue. Constitutive parameters can then be inferred from observations of structure throughout an organ. Establishing the relationship between microstructure and the continuum constitutive law requires micromechanical models which incorporate extracellular matrix structure as well as subcellular cytoskeletal structure. Such models are now under development.

Finally, the multi-physics and multiscale continuum modelling framework presented here is applicable across all organ systems and tissue types. Models of the lungs, musculo-skeletal system and digestive system are well under way.

Acknowledgements

The authors gratefully acknowledge many contributions by colleagues in the Auckland Bioengineering Institute. We are also grateful for the support of grants from the NZ Royal Society (to the Centre for Molecular Biodiscovery and the New Zealand Institute of Mathematics and its Applications) and the NZ Foundation for Research Science and Technology.

REFERENCES

M. J. Ackerman (1998), 'The Visible Human project', *Proc. IEEE* **86**, 504–511.

F. G. Akar and D. S. Rosenbaum (2003), 'Transmural electrophysiological heterogeneities underlying arrhythmogenesis in heart failure', *Circ. Res.* **93**, 638–645.

R. J. Atkin and N. Fox (1980), *An Introduction to the Theory of Elasticity*, Longman, London.

J. Balogh, M. Merisckay, Z. Li, D. Paulin and A. Arner (2002), 'Hearts from mice lacking desmin have a myopathy with impaired active force generation and unaltered wall compliance', *Cardiovasc. Res.* **53**, 439–450.

D. A. Bergel and P. J. Hunter (1979), The mechanics of the heart, in *Quantitative Cardiovascular Studies, Clinical and Research Applications of Engineering Principles* (N. H. C. Hwang, D. R. Gross and D. J. Patel, eds), University Park Press, Baltimore, chapter 4, pp. 151–213.

C. P. Bradley, A. J. Pullan and P. J. Hunter (1997), 'Geometric modelling of the human torso using cubic Hermite elements', *Ann. Biomed. Eng.* **76**, 96–111.

M. Buist, G. B. Sands, P. J. Hunter and A. J. Pullan (2003), 'A deformable finite element derived finite difference method for cardiac activation problems', *Ann. Biomed. Eng.* **31**, 577–588.

Y. I. Cho and R. Kensey (1991), 'Effects of the non-Newtonian viscosity of blood flows in a diseased arterial vessel', *Biorheology* **28**, 241–262.

K. D. Costa, P. J. Hunter, J. M. Rogers, J. M. Guccione, L. K. Waldman and A. D. McCulloch (1996*a*), 'A three-dimensional finite element method for large elastic deformations of ventricular myocardium, Part I: Cylindrical and spherical polar coordinates', *ASME J. Biomech. Eng.* **118**, 452–463.

K. D. Costa, P. J. Hunter, J. S. Wayne, L. K. Waldman, J. M. Guccione and A. D. McCulloch (1996*b*), 'A three-dimensional finite element method for large elastic deformations of ventricular myocardium, Part II, Prolate spheroidal coordinates', *ASME J. Biomech. Eng.* **118**, 464–472.

E. J. Crampin, M. Halstead, P. J. Hunter, P. Nielsen, D. Noble, N. Smith and M. Tawhai (2004), 'Computational physiology and the Physiome project', *Exp. Physiol.* **89**, 1–26.

J. Demmel, S. Eisenstat, J. Gilbert, X. Li and J. Liu (1999*a*), 'A supernodal approach to sparse partial pivoting', *SIAM J. Mat. Anal. Appl.* **20**, 720–755.

J. Demmel, J. Gilbert and X. Li (1999*b*), 'An asynchronous parallel supernodal algorithm for sparse Gaussian elimination', *SIAM J. Mat. Anal. Appl.* **20**, 915–952.

D. DiFrancesco and D. Noble (1985), 'A model of cardiac electrical activity incorporating ionic pumps and concentration changes', *Phil. Trans. R. Soc. Lond. B* **307**, 353–398.

J. M. Downey and E. S. Kirk (1975), 'Inhibition of coronary blood flow by a vascular waterfall mechanism', *Circ. Res.* **36**, 753–760.

J. L. Emery, J. H. Omens and A. D. McCulloch (1997), 'Biaxial mechanics of the passively overstretched left ventricle', *Amer. J. Physiol.* **272**, H2299–H2305.

G. Fischer, B. Tilg, R. Modre, G. J. M. Huiskamp, J. Fetzer, W. Rucker and P. Wach (2000), 'A bidomain model based BEM-FEM coupling formulation for anisotropic cardiac tissue', *Ann. Biomed. Eng.* **28**, 1229–1243.

J. L. Greenstein and R. L. Winslow (2002), 'An integrative model of the cardiac ventricular myocyte incorporating local control of ca2+ release', *Biophys. J.* **83**, 2918–2945.

J. M. Guccione, A. D. McCulloch and L. K. Waldman (1991), 'Passive material properties of intact ventricular myocardium determined from a cylindrical model', *ASME J. Biomech. Eng.* **113**, 42–55.

J. M. Guccione, I. Motabarzadeh and G. I. Zahalak (1998), 'A distribution-moment model of deactivation in cardiac muscle', *J. Biomech.* **31**, 1069–73.

A. Guz, D. H. Bergel and D. L. Brutsaert (1974), *The Physiological Basis of Starling's Law of the Heart*, Elsevier.

C. S. Henriquez (1993), 'Simulating the electrical behaviour of cardiac tissue using the bidomain model', *Crit. Rev. Biomed. Eng.* **21**, 1–77.

T. L. Hill (1975), 'Theoretical formalism for the sliding filament model of contraction of striated muscle, part II', *Prog. Biophys. Molec. Biol.* **29**, 105–159.

T. L. Hill (1989), *Free Energy Transduction and Biochemical Cycle Kinetics*, Springer, New York.

A. C. Hindmarsh (1983), ODEPACK, a systematized collection of ode solvers, in *Scientific Computing* (R. S. Stepleman, ed.), North-Holland, Amsterdam, pp. 55–64.

A. L. Hodgkin (1976), 'Chance and design in electrophysiology: An informal account of certain experiments on nerve carried out between 1934 and 1952', *J. Physiol. (Lond.)* **263**, 1–21.

A. L. Hodgkin and A. F. Huxley (1952), 'A quantitative description of membrane current and its application to conduction and excitation in nerve', *J. Physiol. (Lond.)* **117**, 500–544.

P. J. Hunter (1975), Finite element analysis of cardiac muscle mechanics, DPhil thesis, University of Oxford.

P. J. Hunter (1995), Myocardial constitutive laws for continuum mechanics models of the heart, in *Molecular and Subcellular Cardiology: Effects of Structure and Function* (S. Sideman and R. Beyar, eds), Plenum Press, chapter 30, pp. 303–318.

P. J. Hunter, A. D. McCulloch and H. E. D. J. ter Keurs (1998), 'Modelling the mechanical properties of cardiac muscle', *Prog. Biophys. Molec. Biol.* **69**, 289–331.

P. J. Hunter, M. P. Nash and G. B. Sands (1997), Computational electromechanics of the heart, in *Computational Biology of the Heart* (A. V. Panfilov and A. V. Holden, eds), Wiley, Chichester, chapter 12, pp. 345–407.

P. J. Hunter, B. H. Smaill and I. W. Hunter (1995), 'A 'pole-zero' constitutive law for myocardium', *ASME J. Biomech. Eng.* **382**, 303–18.

A. F. Huxley (1957), 'Muscle structure and theories of contraction', *Prog. Biophys. Chem.* **7**, 255–318.

S. Jafri, J. Rice and R. Winslow (1998), 'Cardiac Ca^{2+} dynamics: The role of ryanodine receptor adaptation and sarcoplasmic reticulum load', *Biophys. J.* **74**, 1149–1168.

B. Kantor, H. Kwon, E. Ritman, D. Holmes and R. Schwartz (1999), 'Images in cardiology imaging the coronary microcirculation: 3d Micro-CT of coronary vasa vasorum', *Internat. J. Cardiovasc. Intervent.* **2**, 79.

G. S. Kassab and Y. C. Fung (1994), 'Topology and dimensions of pig coronary capillary network', *Amer. J. Physiol.* **267**, H319–H325.

G. S. Kassab, C. A. Rider, N. J. Tang and Y. C. Fung (1993), 'Morphometry of pig coronary arterial trees', *Amer. J. Physiol.* **265**, H350–H365.

M. Kawai and W. P. Brandt (1980), 'Sinusoidal analysis: A high resolution method for correlating biochemical reactions with physiological processes in activated skeletal muscles of rabbit, frog and crayfish', *J. Mus. Res. Cell Motil.* **3**, 279–303.

M. Kawai, Y. Saeki and Y. Zhao (1993*a*), 'Crossbridge scheme and the kinetic constants of elementary steps deduced from chemically skinned papillary and trabecular muscles of the ferret', *Circ. Res.* **73**, 35–50.

M. Kawai, Y. Zhao and H. Halvorson (1993*b*), 'Elementary steps of contraction probed by sinusoidal analysis technique in rabbit psoas fibers', *Circ. Res.* **332**, 567–580.

J. Keener and J. Sneyd (1998), *Mathematical Physiology*, Springer, New York.

R. Krams, P. Sipkema and N. Westerhof (1989), 'Varying elastance concept may explain coronary systolic flow impediment', *Amer. J. Physiol.* **257**, H1471–H1479.

W. Krassowska and J. C. Neu (1994), 'Effective boundary conditions for syncytial tissues', *IEEE Trans. Biomed. Eng.* **41**, 143–150.

I. J. Le Grice, B. H. Smaill, L. Z. Chai, S. G. Edgar, J. B. Gavin and P. J. Hunter (1995), 'Laminar structure of the heart: Ventricular myocyte arrangement and connective tissue architecture in the dog', *Amer. J. Physiol.* **269**, H571–H582.

X. S. Li and J. W. Demmel (2003), 'SuperLU_DIST: A scalable distributed-memory sparse direct solver for unsymmetric linear systems', *ACM Trans. Math. Software* **29**, 110–140.

C.-H. Luo and Y. Rudy (1991), 'A model of the ventricular cardiac action potential: Depolarisation, repolarisation, and their interaction', *Circ. Res.* **68**, 1501–1526.

C.-H. Luo and Y. Rudy (1994), 'A dynamic model of the cardiac ventricular action potential, I: Simulations of ionic currents and concentration changes', *Circ. Res.* **74**, 1071–1096.

L. E. Malvern (1969), *Introduction to the Mechanics of a Continuous Medium*, Prentice-Hall, New Jersey.

S. F. McCormick (1989), *Multilevel Adaptive Methods for Partial Differential Equations*, SIAM.

R. M. Miura (2002), 'Analysis of excitable cell models', *J. Comput. Appl. Math.* **144**, 29–47.

A. L. Muzikant and C. S. Henriquez (1998), 'Bipolar stimulation of a three-dimensional bidomain incorporating rotational anisotropy', *IEEE Trans. Biomed. Eng.* **45**, 449–462.

M. P. Nash and P. J. Hunter (2000), 'Computational mechanics of the heart', *J. Elasticity* **61**, 112–141.

D. G. Nicholls and S. J. Ferguson (2002), *Bioenergetics 3*, Academic Press, London.

D. P. Nickerson, N. P. Smith and P. J. Hunter (2001), 'A model of cardiac cellular electromechanics', *Phil. Trans. R. Soc. Lond. A* **359**, 1159–1172.

P. M. F. Nielsen, I. J. Le Grice, B. H. Smaill and P. J. Hunter (1991), 'Mathematical model of geometry and fibrous structure of the heart', *Amer. J. Physiol.* **260**, H1365–H1378.

D. Noble (1960), 'Cardiac action and pacemaker potentials based on the Hodgkin–Huxley equations', *Nature* **188**, 495–497.

D. Noble, A. Varghese, P. Kohl and P. Noble (1998), 'Improved guinea-pig ventricular model incorporating diadic space, i_{kr} and i_{ks}, length and tension-dependent processes', *Canad. J. Cardiol.* **14**, 123–134.

V. P. Novak, F. C. P. Yin and J. D. Humphrey (1994), 'Regional mechanical properties of passive myocardium', *J. Biomech.* **27**, 403–412.

J. T. Oden (1972), *Finite Elements of Nonlinear Continua*, McGraw-Hill, New York.

L. H. Opie (1998), *The Heart: Physiology, from Cell to Circulation*, 3rd edn, Lippincott-Raven.

K. Perktold, M. Resch and R. O. Peter (1991), 'Three-dimensional numerical analysis of the pulsatile flow and wall shear stress in the carotid artery bifurcation', *J. Biomech.* **24**, 409–420.

G. Piazzesi and V. Lombardi (1995), 'A cross-bridge model that is able to explain mechanical and energetic properties of shortening muscle', *Biophys. J.* **68**, 1966–1979.

R. Plonsey and R. C. Barr (1984), 'Current flow patterns in two-dimensional anisotropic bisyncytia with normal and extreme conductivities', *Biophys. J.* **45**, 557–571.

W. H. Press, S. A. Teukolsky, W. T. Vetterling and B. P. Flannery (1992), *Numerical Recipes in FORTRAN: The Art of Scientific Computing*, 2nd edn, Cambridge University Press, Cambridge.

A. R. Pries, K. Ley, M. Claasen and P. Gaehtgens (1989), 'Red cell distribution at microvascular bifurcations', *Microvasc. Res.* **38**, 81–101.

A. R. Pries, D. Neuhaus and P. Gaehtgens (1992), 'Blood viscosity in tube flow: Dependence on diameter and hematocrit', *Amer. J. Physiol.* **263**, H1770–H1778.

J. J. Rice and P. P. de Tombe (2004), 'Approaches to modeling crossbridges and calcium-dependent activation in cardiac muscle', *Prog. Biophys. Mol. Biol.*, DOI 10.1016/j.pbiomolbio.2004.01.011. In press.

B. J. Roth (1997), 'Electrical conductivity values used with the bidomain model of cardiac tissue', *IEEE Trans. Biomed. Eng.* **44**, 326–328.

B. J. Roth and J. P. Wikswo, Jr. (1986), 'A bidomain model for the extracellular potential and magnetic field of cardiac tissue', *IEEE Trans. Biomed. Eng.* **33**, 467–469.

S. Rusinkiewicz and M. Levoy (2001), 'Efficient variants of the icp algorithm', *Third International Conference on 3D Digital Imaging and Modeling (3DIM 2001 Proceedings)*, pp. 145–152.

Y. Saeki, M. Kawai and Y. Zhao (1991), 'Comparison of crossbridge dynamics between intact and skinned myocardium from ferret right ventricles', *Circ. Res.* **68**, 772–81.

O. H. Schmitt (1969), Biological information processing using the concept of interpenetrating domains, in *Information Processing in the Nervous System* (K. N. Leibovic, ed.), Springer, New York, pp. 325–331.

K. B. Skouibine, N. A. Trayanova and P. K. Moore (2000), 'A numerically efficient model for simulation of defibrillation in an active bidomain sheet of myocardium', *Math. Biosci.* **166**, 85–100.

D. A. Smith (1998), 'A strain-dependent ratchet model of [phosphate]- and [atp]-dependent muscle contraction', *J. Mus. Res. Cell Motil.* **19**, 189–211.

N. P. Smith (2003), 'From sarcomere to cell: An efficient algorithm for linking mathematical models of muscle contraction', *Bull. Math. Biol.* **65**, 1141–1162.

N. P. Smith and E. J. Crampin (2004), 'Development of models of active ion transport for whole-cell modelling: Cardiac sodium–potassium pump as a case study', *Prog. Biophys. Molec. Biol.*, DOI 10.1016/j.pbiomolbio.2004.01.010. In press.

N. P. Smith, A. J. Pullan and P. J. Hunter (2000), 'The generation of an anatomically accurate geometric coronary model', *Ann. Biomed. Eng.* **28**, 14–25.

N. P. Smith, A. J. Pullan and P. J. Hunter (2002), 'An efficient finite difference model of transient coronary blood flow in the heart', *SIAM J. Appl. Math.* **62**, 990–1018.

J. A. E. Spaan, N. P. W. Breuls and J. D. Laird (1981), 'Diastolic -systolic coronary flow differences are caused by intramyocardial pump action in the anesthetised dog', *Circ. Res.* **49**, 584–593.

A. J. M. Spencer (1980), *Continuum Mechanics*, Longman, London.

M. Tawhai and P. J. Hunter (2001a), 'Characterising respiratory airway gas mixing using a lumped parameter model of the pulmonary acinus', *Respir. Physiol.* **127**, 241–248.

M. Tawhai and P. J. Hunter (2001b), 'Multibreath washout analysis: Modelling the influence of conducting airway asymmetry', *Respir. Physiol.* **127**, 249–258.

M. Tawhai and P. J. Hunter (2004), 'Modeling water vapour and heat transfer in the normal and the intubated airway', *Ann. Biomed. Eng.* In press.

M. H. Tawhai, A. J. Pullan and P. J. Hunter (2000), 'Generation of an anatomically based three-dimensional model of the conducting airways', *Ann. Biomed. Eng.* **28**, 793–802.

M. Tawhai, N. Rankin, S. Ryan and P. J. Hunter (2002), 'Measurement and mathematical modelling of thermodynamics within the intubated airway', *Eur. Respir. J.* **20**, 602s.

N. A. Trayanova (1994), 'A bidomain model for ring stimulation of a cardiac strand', *IEEE Trans. Biomed. Eng.* **41**, 393–397.

L. Tung (1978), A bidomain model for describing ischemic myocardial D-C potentials, PhD thesis, MIT, Boston, MA.

T. P. Usyk, I. J. Le Grice and A. D. McCulloch (2002), 'Computational model of three-dimensional cardiac electromechanics', *Comput. Visual Sci.* **4**, 249–257.

T. P. Usyk, R. Mazhari and A. D. McCulloch (2000), 'Effect of laminar orthotropic myofiber architecture on regional stress and strain in the canine left ventricle', *J. Elasticity* **61**, 143–164.

F. J. Vetter and A. D. McCulloch (2000), 'Three-dimensional stress and strain in passive rabbit left ventricle: A model study', *Ann. Biomed. Eng.* **28**, 781–92.

A. A. Young, Z. A. Fayad and L. Axel (1996), 'Right ventricular midwall surface motion and deformation using magnetic resonance tagging', *Amer. J. Physiol.* **271**, H2677–H2688.

A. A. Young, P. J. Hunter and B. H. Smaill (1989), 'Epicardial surface estimation from coronary cinéangiograms', *Comput. Vis. Graph. Image Proc.* **47**, 111–127.

G. I. Zahalak (1981), 'A distribution-moment approximation for kinetic theories of muscular contraction', *Mathematical Biosciences* **55**, 89–114.

G. I. Zahalak and S.-P. Ma (1990), 'Muscle activation and contraction: Constitutive relations based directly on cross-bridge kinetics', *ASME J. Biomech. Eng.* **112**, 52–62.

O. C. Zienkiewicz and R. L. Taylor (1994), *The Finite Element Method, I: Basic Formulation and Linear Problems*, 4th edn, McGraw-Hill, Berkshire.

Printed in the United States
By Bookmasters